Lecture Notes in Computer Science　9647

Commenced Publication in 1973
Founding and Former Series Editors:
Gerhard Goos, Juris Hartmanis, and Jan van Leeuwen

More information about this series at http://www.springer.com/series/7412

Nicolas Normand · Jeanpierre Guédon
Florent Autrusseau (Eds.)

Discrete Geometry for Computer Imagery

19th IAPR International Conference, DGCI 2016
Nantes, France, April 18–20, 2016
Proceedings

 Springer

Editors
Nicolas Normand
IRCCyN
University of Nantes
Nantes
France

Jeanpierre Guédon
IRCCyN
University of Nantes
Nantes
France

Florent Autrusseau
IRCCyN
University of Nantes
Nantes
France

ISSN 0302-9743   ISSN 1611-3349   (electronic)
Lecture Notes in Computer Science
ISBN 978-3-319-32359-6   ISBN 978-3-319-32360-2   (eBook)
DOI 10.1007/978-3-319-32360-2

Library of Congress Control Number: 2016935975

LNCS Sublibrary: SL6 – Image Processing, Computer Vision, Pattern Recognition, and Graphics

Printed on acid-free paper

This Springer imprint is published by Springer Nature
The registered company is Springer International Publishing AG Switzerland

Preface

Following DGCI 2014 held in Siena, Italy, DGCI 2016, the 19th in a series of international conferences on Discrete Geometry for Computer Imagery, was held in Nantes, France, April 18–20, 2016. DGCI 2016 attracted a good number of research contributions from academic and research institutions in our field. A total of 51 papers were submitted from all over the word. After the reviewing process, 32 papers were accepted with 22 for oral presentation and 10 for poster presentation, all in a single-track session. Three international well-known speakers were invited for specific lectures: Matthias Beck (San Francisco State University), Hugues Talbot (ESIEE Paris), and Jacques-Olivier Lachaud (Université Savoie Mont Blanc).

We are pleased that DGCI was held under the sponsorship of the International Association of Pattern Recognition (IAPR). For the first time, DGCI also had the honor of producing a special issue of the *Journal of Mathematical Imaging and Vision*.

We would like to thank all contributors, the invited speakers, all reviewers, and members of the Steering Committees, and all the people who made this conference happen. We are grateful to our financial support institutions—the Region Pays-de-la-Loire, Nantes Métropole, the University of Nantes, the CNRS, and Polytech Nantes. We also thank our industry support: Rozo Systems and Keosys.

Last, but not least, we thank all the participants and hope that everyone found great interest in the DGCI 2016 program and also had a very good time in our city of Nantes.

April 2016

Nicolas Normand
Jeanpierre Guédon
Florent Autrusseau

Organization

Nicolas Normand   IRCCyN, University of Nantes, France
Jeanpierre Guédon   IRCCyN, University of Nantes, France
Florent Autrusseau   IRCCyN, University of Nantes, France

Program Chairs

Nicolas Normand   IRCCyN, University of Nantes, France
Jeanpierre Guédon   IRCCyN, University of Nantes, France
Florent Autrusseau   IRCCyN, University of Nantes, France
David Coeurjolly   LIRIS, CNRS, Lyon, France

Program Committee

Joost Batenburg   CWI, Amsterdam, The Netherlands
Valerie Berthé   LIAFA, CNRS, Paris, France
Gilles Bertrand   LIGM, ESIEE Paris, Noisy-le-Grand, France
Isabelle Bloch   LTCI, Telecom ParisTech, Paris, France
Sara Brunetti   University of Siena, Italy
Michel Couprie   LIGM, ESIEE Paris, Noisy-le-Grand, France
Guillaume Damiand   LIRIS, CNRS, Lyon, France
Laurent Fuchs   XLIM, University of Poitiers, France
Yan Gerard   ISIT, Auvergne University, France
Rocio Gonzalez-Diaz   University of Seville, Spain
Yukiko Kenmochi   LIGM, University of Paris-Est, France
Bertrand Kerautret   Loria, University of Lorraine, France
Christer Kiselman   Uppsala University, Sweden
Walter Kropatsch   Vienna University of Technology, Austria
Xavier Provençal   LAMA, University Savoie Mont Blanc, France
Robin Strand   Centre for Image Analysis, Uppsala, Sweden
Imants Svalbe   Monash University, Melbourne, Australia
Edouard Thiel   LIF, University of Aix-Marseille, France

Contents

Discrete and Combinatorial Topology

Shape Descriptors

Models for Discrete Geometry

Invited Talks

Convergent Geometric Estimators with Digital Volume and Surface Integrals

Jacques-Olivier Lachaud[(✉)]

Laboratoire de Mathématiques (LAMA),
CNRS, UMR 5127, Université Savoie Mont Blanc, Chambéry, France
jacques-olivier.lachaud@univ-savoie.fr

Abstract. This paper presents several methods to estimate geometric quantities on subsets of the digital space $\mathbb{Z}^d$. We take an interest both on global geometric quantities like volume and area, and on local geometric quantities like normal and curvatures. All presented methods have the common property to be *multigrid convergent*, i.e. the estimated quantities tend to their Euclidean counterpart on finer and finer digitizations of (smooth enough) Euclidean shapes. Furthermore, all methods rely on digital integrals, which approach either volume integrals or surface integrals along shape boundary. With such tools, we achieve multigrid convergent estimators of volume, moments and area in $\mathbb{Z}^d$, of normals, curvature and curvature tensor in $\mathbb{Z}^2$ and $\mathbb{Z}^3$, and of covariance measure and normals in $\mathbb{Z}^d$ even with Hausdorff noise.

Keywords: Digital geometry · Volume estimation · Moments estimation · Normal estimation · Curvatures estimation · Area estimation · Multigrid convergence · Digital integration · Integral invariants · Digital moments · Voronoi covariance measure · Stability

1 Introduction

Objectives. We are interested in the geometry of subsets of the digital space $\mathbb{Z}^d$, where $\mathbb{Z}$ is the set of integer numbers. More precisely, when seeing these subsets as a sampling of a Euclidean shape, say X, we would like to recover an approximation of the geometry of X with solely the information of its sampling. It is clear that this task cannot be done without further hypothesis on X and on the sampling method. First, at a fixed sampling resolution, there are infinitely many shapes having the same sampling. Second, subsets of $\mathbb{Z}^d$ have no canonic tangent plane or differential geometry. To address the first issue, we will take a look at specific families of Euclidean shapes, generally by requiring smoothness properties. We will then show that we can achieve *multigrid convergence* properties for some estimators on such shapes, i.e. when the sampling gets finer and finer, the estimation gets better. The second issue is addressed by using digital integrals, i.e. well-chosen sums.

This work has been mainly funded by DigitalSnow ANR-11-BS02-009, KIDICO ANR-2010-BLAN-0205 research grants.

© Springer International Publishing Switzerland 2016
N. Normand et al. (Eds.): DGCI 2016, LNCS 9647, pp. 3–17, 2016.
DOI: 10.1007/978-3-319-32360-2_1

This paper presents the main ingredients and results of three methods that provide multigrid convergent estimators of the most common geometric quantities: volume, area, tangent, normal, curvatures, etc. Their common denominator is to use digital integrals, i.e. sums that approach integrals defined on the Euclidean shape. The stability of these integrals in turns induces the multigrid convergence of these estimators.

This topic is in fact rather old, since Gauss and Dirichlet already knew that the volume of a convex set can be approximated by counting digital points within (reported in [KR04]). Furthermore, it is related to numerical integration. The purpose of this paper is not to provide an exhaustive lists of multigrid convergent digital estimators. We may point several sources and surveys in the literature that provides many references or comparisons: [KŽ00, CK04, dVL09, CLR12]. This work compiles methods and results developed in several papers [CLL13, CLL14, LCLng, LT15, CLT14, CLMT15]. Note that the topic of digital geometric estimation through integrals is very active at the present time. Among the very recent works, we may quote the varifold approach of [Bue14, BLM15] for normal and mean curvature estimation, the estimation of intrinsic volumes of [EP16] with persistent homology, or the estimation of Minkowski tensors with an extension of the Voronoi covariance measure [HKS15].

Main Definitions and Notations. A *digitization process* is a family of maps mapping subsets of $\mathbb{R}^d$ towards subsets of $\mathbb{Z}^d$, parameterized by a positive real number h. The parameter h defines the *gridstep* of the digitization, a kind of sampling distance. The digitization process D_h is *local* whenever $\mathbf{z} \in \mathsf{D}_h(X)$ depends solely on $X \cap N(h\mathbf{z})$, where $N(h\mathbf{z})$ is a neighborhood of radius $O(h)$ around point $h\mathbf{z} \in \mathbb{R}^d$.

In this work, we consider three simple local digitization processes, defined below after a few notations. If $\mathbf{z} \in \mathbb{Z}^d$, then $Q(\mathbf{z})$ denotes the (closed) unit d-dimensional cube of $\mathbb{R}^d$ centered on $\mathbf{z}$ and aligned with the axes of $\mathbb{Z}^d$. We further define $Q_h(\mathbf{z}) := h \cdot Q(\mathbf{z})$, so-called $h-cube$, as the scaled version of $Q(\mathbf{z})$ by factor h, *i.e.* $Q_h(\mathbf{z})$ is a d-dimensional cube centered at $h\mathbf{z}$ with edge length h. Then we define the *Gauss digitization* $\mathsf{G}_h(X)$, the *inner Jordan digitization* $\mathsf{J}_h^-(X)$ and *outer Jordan digitization* $\mathsf{J}_h^+(X)$ at step h of a shape $X \in \mathbb{X}$ as (see Fig. 1):

$$\mathsf{G}_h(X) := \left\{ \mathbf{z} \in \mathbb{Z}^d, h\mathbf{z} \in X \right\}, \tag{1}$$

$$\mathsf{J}_h^-(X) := \left\{ \mathbf{z} \in \mathbb{Z}^d, Q_h(\mathbf{z}) \subset X \right\}, \tag{2}$$

$$\mathsf{J}_h^+(X) := \left\{ \mathbf{z} \in \mathbb{Z}^d, Q_h(\mathbf{z}) \cap X \neq \emptyset \right\}. \tag{3}$$

First Relations Between Shape and Its Digitization. We will need to compare the geometry of the Euclidean shape X and its topological boundary ∂X with the "geometry" of their digitizations. So, for $Z \subset \mathbb{Z}^d$, we define the *body* of Z at step h as $[Z]_h := \bigcup_{\mathbf{z} \in Z} Q_h(\mathbf{z})$. We call *Jordan strip* the digitization $\mathsf{J}_h^0(X) := \mathsf{J}_h^+(X) \backslash \mathsf{J}_h^-(X)$, which is a kind of digitization of ∂X. The following properties clarify the relations between $X, \partial X$ and their digitizations and are easy to derive [LCLng].

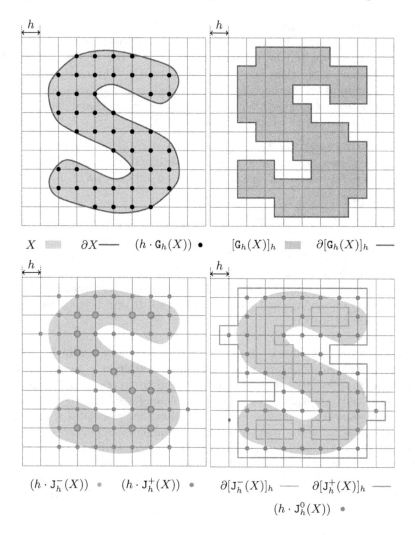

Fig. 1. Illustration of digitization processes and notations (Color figure online).

Lemma 1. $J_h^-(X) \subset G_h(X) \subset J_h^+(X)$ *and* $[J_h^-(X)]_h \subset X \subset [J_h^+(X)]_h$.

Lemma 2. $[J_h^0(X)]_h = [J_h^+(X)]_h \backslash \text{Int}([J_h^-(X)]_h)$ *and* $\partial X \subset [J_h^0(X)]_h$.

In fact, we can be more precise and relate these sets with the Hausdorff distance. Recall that the ϵ-*offset* of a shape X, denoted by X^ϵ, is the set of points of $\mathbb{R}^d$ at distance lower or equal to ϵ from X. We can state with some elementary arguments that boundaries of Jordan digitizations are close to the boundary of the shape in the Hausdorff sense:

Lemma 3 ([LCLng], Lemma 3). *Let* X *be a compact domain of* $\mathbb{R}^d$. *Then* $[J_h^0(X)]_h \subset (\partial X)^{\sqrt{d}h}$, $\partial[J_h^-(X)]_h \subset (\partial X)^{\sqrt{d}h}$ *and* $\partial[J_h^+(X)]_h \subset (\partial X)^{\sqrt{d}h}$.

The remarkable point here is that the sole requirement on X is compactness ! If the shape X has a smoother boundary, we can get tighter bounds for the Gauss digitization. Therefore, let the *medial axis* $\mathrm{MA}(\partial X)$ of ∂X be the subset of $\mathbb{R}^d$ whose points have more than one closest point to ∂X. The *reach* $\mathrm{reach}(X)$ of X is the infimum of the distance between ∂X and its medial axis. Shapes with positive reach have a C^2 smooth boundary almost everywhere and have principal curvatures bounded by $\pm 1/\mathrm{reach}(X)$. We then have:

Theorem 1 ([LT15]). *Let X be a compact domain of $\mathbb{R}^d$ such that the reach of ∂X is greater than ρ. Then, for any digitization step $0 < h < 2\rho/\sqrt{d}$, the Hausdorff distance between sets ∂X and $\partial[\mathsf{G}_h(X)]_h$ is less than $\sqrt{d}h/2$. In particular,* $\partial[\mathsf{G}_h(X)]_h \subset (\partial X)^{\frac{\sqrt{d}}{2}h}$.

The *projection* π^X of $\mathbb{R}^d \backslash \mathrm{MA}(\partial X)$ onto ∂X is the map which associates to any point its closest point on ∂X. From the properties of the medial axis, the projection is defined almost everywhere in $\mathbb{R}^d$. We may thus associate to any point $\hat{\mathbf{x}} \in \partial[\mathsf{G}_h(X)]_h$ the point $\mathbf{x} := \pi^X(\hat{\mathbf{x}}) \in \partial X$, such that the distance between $\hat{\mathbf{x}}$ and its projection $\mathbf{x}$ is smaller than $\sqrt{d}h/2$. We have just constructed a mapping between a shape boundary and its digitization, which will help us for defining local geometric estimators.

Multigrid Convergence. Let V be any vector space (generally $\mathbb{R}$ or $\mathbb{R}^d$). A *geometric quantity* is an application that associates a value in V to any subset of $\mathbb{R}^d$, with the property that it is invariant to some group operations, most often the group of rigid transformations. Notable examples are the volume and the area. A *local geometric quantity* is an application that associates a value in V to a subset X of $\mathbb{R}^d$ and a point $\mathbf{x}$ on ∂X. Common examples are the normal vector, the mean curvature or principal curvatures and directions. A *discrete geometric estimator* is an application that associates a value in V to a subset of $\mathbb{Z}^d$ and a gridstep $h \in \mathbb{R}^+$. A *local discrete geometric estimator* is an application that associates a value in V to a subset Z of $\mathbb{Z}^d$, a point in $\partial[Z]_h$ and a gridstep $h \in \mathbb{R}^+$.

Definition 1 (Multigrid convergence). *A discrete geometric estimator $\hat{E}$ (resp. a local discrete geometric estimator $\hat{F}$) is said to be* multigrid convergent *to some geometric quantity E (resp. to some local geometric quantity F) for the family of shapes $\mathbb{X}$ and digitization process D, if and only if, for any $X \in \mathbb{X}$, there exists a gridstep $h_X > 0$, such that $\forall h, 0 < h < h_X$, we have:*

$$|E(X) - \hat{E}(\mathsf{D}_h(X), h)| \leq \tau_X(h), \qquad (4)$$

respectively $\forall \mathbf{x} \in \partial X, \forall \hat{\mathbf{x}} \in \partial[\mathsf{D}_h(X)]_h, \|\hat{\mathbf{x}} - \mathbf{x}\|_\infty \leq h$,

$$|F(X, \mathbf{x}) - \hat{F}(\mathsf{D}_h(X), \hat{\mathbf{x}}, h)| \leq \tau_X(h), \qquad (5)$$

where the function $\tau_X : \mathbb{R}^+ \backslash \{0\} \to \mathbb{R}^+$ has null limit at 0 and defines the speed of convergence *of the estimator.*

In both definitions, the multigrid convergence property characterizes estimators that give better and better geometric estimates as the grid sampling gets finer and finer. We have now all the notions to study the multigrid convergence of several discrete geometric estimators.

2 Volume and Moments Estimators

In this section, X is some compact domain of $\mathbb{R}^d$ and Z is a subset of $\mathbb{Z}^d$. We take here an interest in estimating volume and moments from digital sets. These results will be used to define digital integral invariant estimators of curvatures in the following section. In the whole section, let $(p_i)_{i=1...d}$ be the integers defining the moment exponents, with $0 \leq p_i \leq 2$, and let $\sigma := p_1 + \cdots + p_d$, with $\sigma \leq 2$.

Moments and Digital Moments. The $p_1 \cdots p_d$-*moment of* X is defined as

$$m^{p_1 \cdots p_d}(X) := \int \cdots \int_X x_1^{p_1} \cdots x_d^{p_d} dx_1 \ldots dx_d.$$

The $0 \cdots 0$-moment of X is the volume of X (denoted $\mathrm{Vol}^d(X)$). The $p_1 \cdots p_d$-*digital moment of* Z *at step* h is defined as

$$\hat{m}_h^{p_1 \cdots p_d}(Z) := h^{d+p_1+\cdots+p_d} \sum_{(z_1,\ldots,z_d) \in Z} z_1^{p_1} \cdots z_d^{p_d}.$$

The $0 \cdots 0$-digital moment of Z is the digital volume of Z (denoted by $\widehat{\mathrm{Vol}}^d(Z,h)$).

It is well known that $\widehat{\mathrm{Vol}}^d$ is multigrid convergent toward Vol^d for the family of convex shapes and the Gauss digitization, with a convergence speed of $O(h)$, and even faster for smoother shapes [Hux90, KN91, M99, Guo10]. We wish to go further on multigrid convergence of moments, so we take a special interest in (digital) moments of h-cubes. The following equalities, obtained by simple integration, show that discrepancies between digital and continuous moments begin with order two, and only when one $p_i = 2$.

Lemma 4. *Let* $\mathbf{z} \in \mathbb{Z}^d$. *Point* $\mathbf{z}$ *is the Gauss digitization of* h-*cube* $Q_h(\mathbf{z})$, *but also its inner or outer Jordan digitization. Moments and digital moments of* h-*cubes satisfy* $\hat{m}_h^{p_1 \cdots p_d}(\{\mathbf{z}\}) = m^{p_1 \cdots p_d}(Q_h(\mathbf{z})) + E(p_1, \ldots, p_d)$, *where* $E = \frac{h^{d+4}}{12}$ *when one* p_i *equals 2 and otherwise* $E = 0$.

Errors in Volume Estimation. The following volume "convergence" theorem is remarkable since it requires only the compactness of X. Its proof requires Lemma 4, a volume relation on symmetric difference of sets, the definition of Jordan strip, and Lemma 3.

Theorem 2 ([LCLng]). *Let* X *be a compact domain of* $\mathbb{R}^d$. *Let* D *be any digitization process such that* $\mathsf{J}_h^-(X) \subset \mathsf{D}_h(X) \subset \mathsf{J}_h^+(X)$. *Digital and continuous volumes are related as follows:*

$$\left| \mathrm{Vol}^d(X) - \widehat{\mathrm{Vol}}^d(\mathsf{D}_h(X), h) \right| \leq \mathrm{Vol}^d(\partial X^{\sqrt{d}h}). \tag{6}$$

Another proof, written independently, is in [HKS15]. This theorem states a multigrid convergence property whenever ∂X is $d-1$-rectifiable, but not in the general case: consider for instance the set X of rational numbers in the unit cube. A more useful — but more restricted — convergence theorem relates this error bound to the area of ∂X. It uses Theorem 2 but also the fact that, for sets X with positive reach, the volume of some ϵ-offset of ∂X is upper-bounded by a constant times the area of ∂X (Proof of Lemma 10 [LT15]).

Theorem 3. *With the same hypotheses as Theorem 2 with the further requirement that the reach of ∂X is greater than some value ρ. For $h < \rho/\sqrt{d}$, the volume estimator $\widehat{\mathrm{Vol}}^d$ is multigrid convergent toward the volume Vol^d with speed $2^{d+1}\sqrt{d}\mathrm{Area}(\partial X)h$.*

Volume and Moment Estimation in a Local Neighborhood. For the digital integral invariant method we will require convergence results on sets that are the intersection of two sets with positive reach, more precisely on sets of the form $X \cap B_R(\mathbf{x})$, where $B_R(\mathbf{x})$ denotes the ball of center $\mathbf{x} \in \mathbb{R}^d$ and radius R. The previous theorem cannot be applied as is. We must first bound the volume of offsets of the boundary of $X \cap B_R(\mathbf{x})$:

Theorem 4 ([LCLng], Theorem 3). *Let X be a compact domain of $\mathbb{R}^d$ such that the reach of ∂X is greater than ρ. Let $\mathbf{x} \in \mathbb{R}^d$. Let R (the radius of the ball) and h (the gridstep) be some positive numbers such that $h \leq \frac{R}{\sqrt{2d}}$ and $2R \leq \rho$.*

$$\mathrm{Vol}^d\left((\partial(X \cap B_R(\mathbf{x})))^{\sqrt{d}h}\right) \leq K_1(d)R^{d-1}h, \tag{7}$$

where $K_1(d)$ is a constant that depends only on the dimension d.

The proof first decomposes the set with the equality $(\partial(A \cap B))^\epsilon = ((\partial A) \cap B)^\epsilon \cup (A \cap (\partial B))^\epsilon$. Then it uses differential geometry and the fact that curvatures are bounded by the reach. The good point is that the geometry of X does not intervene in the constants.

We have now all the keys to upperbound the error in volume estimation, and more generally moments estimation, within a ball around the boundary of a compact domain X.

Theorem 5 ([LCLng], Theorem 4). *We take the same hypotheses as the ones of Theorem 4 and the digitization process D defined in Theorem 2. Then digital moments within a ball $B_R(\mathbf{x})$ are multigrid convergent toward continuous moments as follows*

$$\left| m^{p_1 \cdots p_d}(X \cap B_R(\mathbf{x})) - \hat{m}_h^{p_1 \cdots p_d}\left(\mathsf{D}_h(X \cap B_R(\mathbf{x}))\right)\right|$$

$$\leq K_1(d)R^{d-1}(\|\mathbf{x}\|_\infty + 2R)^\sigma h + \frac{h^4}{12}V_d R^d, \tag{8}$$

where V_d is the volume of the unit d-dimensional ball. Furthermore, the term in h^4 is only present when one p_i is equal to 2.

The proof decomposes the errors in each h-cube induced by digitization. Errors in the interior of the set $X \cap B_R(\mathbf{x})$ are easily solved with Lemma 4. Errors close to the boundary of $X \cap B_R(\mathbf{x})$ are bounded with Theorem 4 and some care on moments with negative value.

3   Curvatures with Digital Integral Invariants

Curvature and Mean Curvature Estimation. If $\mathbf{x} \in \partial X$ and ∂X smooth enough, one easily notices that the volume of $X \cap B_R(\mathbf{x})$ is related to the local differential geometry of ∂X around $\mathbf{x}$ for infinitesimal values of R. Several authors [BGCF95, PWY+07, PWHY09] have made explicit this relation:

Lemma 5 ([PWHY09]). *For a sufficiently smooth shape X in $\mathbb{R}^2$, $\mathbf{x} \in \partial X$ (resp. smooth shape X' in $\mathbb{R}^3$, $\mathbf{x}' \in \partial X'$), we have*

$$\mathrm{Vol}^2(X \cap B_R(\mathbf{x})) = \frac{\pi}{2}R^2 - \frac{\kappa(X,\mathbf{x})}{3}R^3 + O(R^4), \tag{9}$$

$$\mathrm{Vol}^3(X' \cap B_R(\mathbf{x}')) = \frac{2\pi}{3}R^3 - \frac{\pi H(X',\mathbf{x}')}{4}R^4 + O(R^5), \tag{10}$$

where $\kappa(X,\mathbf{x})$ is the curvature of ∂X at $\mathbf{x}$ and $H(X',\mathbf{x}')$ is the mean curvature of $\partial X'$ at $\mathbf{x}'$.

Since we have seen that we can approach volumes within a ball (see Theorem 5), it is very natural to define digital curvature estimators from the volume relations Eqs. (9) and (10).

Definition 2 ([CLL13]). *For any positive radius R, we define the 2D integral digital curvature estimator $\hat{\kappa}^R$ (resp. 3D integral mean digital curvature estimator $\hat{H}^R$) of a digital shape $Z \subset \mathbb{Z}^2$ at any point $\mathbf{x} \in \mathbb{R}^2$ (resp. of $Z' \subset \mathbb{Z}^3$ at any point $\mathbf{x}' \in \mathbb{R}^3$) and for a grid step $h > 0$ as:*

$$\forall 0 < h < R, \quad \hat{\kappa}^R(Z,\mathbf{x},h) := \frac{3\pi}{2R} - \frac{3\widehat{\mathrm{Vol}}^2(B_{R/h}(\mathbf{x}/h) \cap Z, h)}{R^3}, \tag{11}$$

$$\hat{H}^R(Z',\mathbf{x}',h) := \frac{8}{3R} - \frac{4\widehat{\mathrm{Vol}}^3(B_{R/h}(\mathbf{x}'/h) \cap Z', h)}{\pi R^4}. \tag{12}$$

We can bound the error between these estimators and their associated geometric quantities as ([CLL13], but precise constants in [LCLng]):

Theorem 6. *Let X be a compact domain of $\mathbb{R}^2$ such that its boundary ∂X is C^3-smooth and has reach greater than ρ. (Respectively let X' be a compact domain of $\mathbb{R}^3$ such that its boundary $\partial X'$ is C^3-smooth and has reach greater than ρ). The following bounds hold for $R < \rho/2$:*

$$\forall 0 < h \leq R/2, \ \forall \mathbf{x} \in \partial X, \ \forall \hat{\mathbf{x}} \in \partial[\mathsf{G}_h(X)]_h \ with \ \|\hat{\mathbf{x}} - \mathbf{x}\|_\infty \leq h,$$

$$\left| \hat{\kappa}^R(\mathsf{G}_h(X), \hat{\mathbf{x}}, h) - \kappa(X, \mathbf{x}) \right| \leq \left(27\pi\sqrt{2}/4 + 3K_1(2) \right) R^{-2}h + O(R). \tag{13}$$

$$\forall 0 < h \leq R/\sqrt{6}, \ \forall \mathbf{x}' \in \partial X', \ \forall \hat{\mathbf{x}}' \in \partial[\mathsf{G}_h(X')]_h \ with \ \|\hat{\mathbf{x}}' - \mathbf{x}'\|_\infty \leq h,$$

$$\left| \hat{H}^R(\mathsf{G}_h(X'), \hat{\mathbf{x}}', h) - H(X', \mathbf{x}') \right| \leq \left(18\sqrt{3} + 4K_1(3)/\pi \right) R^{-2}h + O(R). \tag{14}$$

Chosing $R = \Theta(h^{\frac{1}{3}})$ implies the multigrid convergence of estimator $\hat{\kappa}^R$ (resp. estimator $\hat{H}^R$) toward curvature κ (resp. mean curvature H) with speed $O(h^{\frac{1}{3}})$.

The first term in each error bound comes from the error done in volume estimation of $X \cap B_R(\mathbf{x})$ because point $\hat{\mathbf{x}}$ is not exactly on $\mathbf{x}$ but at distance $O(h)$ (Theorem 1). The second term comes from the digitization in the volume estimation (Theorem 5). The third term in the error comes from the Taylor expansion of Eqs. (9) and (10). Since error terms are either decreasing with R or increasing with R, we balance error terms to minimize the sum of errors and we get convergence of curvature estimators as immediate corollaries.

Normal, Principal Curvatures and Principal Directions. The same methodology can lead to estimators of principal curvatures or normal and principal directions. The idea is to use moments of zeroth, first, and second order instead of volumes. More precisely, the eigenvalues and eigenvectors of the covariance matrix of $X \cap B_R(\mathbf{x})$ were shown to hold curvature information ([PWY+07], Theorem 2). The *covariance matrix* of a set $A \subset \mathbb{R}^3$ is easily defined from moments as:

$$\mathcal{V}(A) := \begin{bmatrix} m^{200}(A) & m^{110}(A) & m^{101}(A) \\ m^{110}(A) & m^{020}(A) & m^{011}(A) \\ m^{101}(A) & m^{011}(A) & m^{002}(A) \end{bmatrix}$$
$$- \frac{1}{m^{000}(A)} \begin{bmatrix} m^{100}(A) \\ m^{010}(A) \\ m^{001}(A) \end{bmatrix} \otimes \begin{bmatrix} m^{100}(A) \\ m^{010}(A) \\ m^{001}(A) \end{bmatrix}^T . \tag{15}$$

The definition of *digital covariance matrix* $\mathcal{V}_h(Z)$ at step h of a set $Z \subset \mathbb{Z}^3$ is similar, just replacing $m^{p_1 \cdots p_d}(A)$ by $\hat{m}_h^{p_1 \cdots p_d}(Z)$ in Eq. (15).

Definition 3. *Let $Z \subset \mathbb{Z}^3$ be a digital shape and $h > 0$ be the gridstep. For $h < R$, we define the integral principal curvature estimators $\hat{\kappa}_1^R$ and $\hat{\kappa}_2^R$ of Z at point $\mathbf{y} \in \mathbb{R}^3$, their respective integral principal direction estimators $\hat{\mathbf{w}}_1^R$ and $\hat{\mathbf{w}}_2^R$, and the integral normal estimator $\hat{\mathbf{n}}^R$ as*

$$\hat{\kappa}_1^R(Z,\mathbf{y},h) := \frac{6}{\pi R^6}(\hat{\lambda}_2 - 3\hat{\lambda}_1) + \frac{8}{5R}, \quad \hat{\mathbf{w}}_1^R(Z,\mathbf{y},h) := \hat{\nu}_1 \quad \hat{\mathbf{n}}^R(Z,\mathbf{y},h) := \hat{\nu}_3,$$
$$\hat{\kappa}_2^R(Z,\mathbf{y},h) := \frac{6}{\pi R^6}(\hat{\lambda}_1 - 3\hat{\lambda}_2) + \frac{8}{5R}, \quad \hat{\mathbf{w}}_2^R(Z,\mathbf{y},h) := \hat{\nu}_2$$

where $\hat{\lambda}_1 \geq \hat{\lambda}_2 \geq \hat{\lambda}_3$ are the eigenvalues of $\mathcal{V}_h(B_{R/h}(\mathbf{y}/h) \cap Z)$, and $\hat{\nu}_1, \hat{\nu}_2, \hat{\nu}_3$ are their corresponding eigenvectors.

Unfortunately, there is no hope in turning Theorem 5 into a multigrid convergence theorem for arbitrary moments, because a very small perturbation in the position of the ball center can lead to an arbitrary error on polynomial x^σ. However, due to their formulations, continuous and digital covariance matrices are invariant by translation, and error terms can thus be confined in a neighborhood around zero. Using this fact and error bounds of moments within the symmetric difference of two balls, we get:

Theorem 7 ([CLL14]). *Let X be a compact domain of $\mathbb{R}^3$ such that its boundary ∂X has reach greater than ρ. Then the digital covariance matrix is multigrid convergent toward the covariance matrix for Gauss digitization for any radius $R < \frac{\rho}{2}$ and gridstep $h < \frac{R}{\sqrt{6}}$, with speed $O(R^4 h)$.*

The constant in O is independent from the shape size or geometry. According to Definition 3, it remains to show that eigenvalues and eigenvectors of the digital covariance matrix are convergent toward the eigenvalues and eigenvectors of the continuous covariance matrix, when error on matrices tends to zero. Classical results of matrix perturbation theory (especially Lidskii-Weyl inequality and Davis-Kahan $\sin\theta$ Theorem) allow to conclude on the following relations:

Theorem 8 ([CLL14, LCLng]). *Let X be a compact domain of $\mathbb{R}^3$ such that its boundary ∂X has reach greater than ρ and has C^3-continuity. Then, for these shapes and for the Gauss digitization process, integral principal curvature estimators $\hat{\kappa}_1^R$ and $\hat{\kappa}_2^R$ are multigrid convergent toward κ_1 and κ_2 for small enough gridsteps h, choosing $R = kh^{\frac{1}{3}}$ and k an arbitrary positive constant. Convergence speed is in $O(h^{\frac{1}{3}})$. Furthermore, integral principal direction estimators $\hat{\mathbf{w}}_1^R$ and $\hat{\mathbf{w}}_2^R$ are also convergent toward $\mathbf{w}_1$ and $\mathbf{w}_2$ with speed $O(h^{\frac{1}{3}})$ provided principal curvatures are distinct. Last the integral normal estimator $\hat{\mathbf{n}}^R$ is convergent toward the normal $\mathbf{n}$ with speed $O(h^{\frac{2}{3}})$.*

To our knowledge, these were the first estimators of principal curvatures shown to be multigrid convergent.

4   Digital Voronoi Covariance Measure

Integral curvatures estimators are convergent. They are rather robust to the presence of Hausdorff noise in input data (see [CLL14]). However, this robustness comes with no guarantee. We present here another approach that reaches this goal, and which can even be made robust to outliers. The first idea is to use a distance function to input data, which is stable to perturbations [CCSM11]. The second idea is to notice that Voronoi cells of input data tend to align with normals of the underlying shape. To make this idea more robust, it suffices to integrate the covariance of gradient vectors of the distance function within a local neighborhood: this is called *Voronoi covariance measure* [MOG11].

Definition 4. *A function $\delta : \mathbb{R}^d \to \mathbb{R}^+$ is called* distance-like *if (i) δ is proper, i.e. $\lim_{\|x\|\to\infty} \delta(x) = \infty$, (ii) δ^2 is 1-semiconcave, that is $\delta^2(.) - \|.\|^2$ is concave.*

It is worthy to note that the standard distance d_K to a compact K is distance-like. Clearly this distance is robust to Hausdorff noise, i.e. $d_H(K, K') < \epsilon$ implies $\|d_K - d_{K'}\|_\infty < \epsilon$. Although we will not go into more details here, the *distance to a measure* [MOG11, CLMT15] is even resilient to outliers and the error is bounded by the Wasserstein-2 distance between measures. Let $\mathbf{N}_\delta := \frac{1}{2}\nabla\delta^2$.

Definition 5 (δ-VCM). *The δ-Voronoi Covariance Measure (VCM) is a tensor-valued measure. Given any non-negative probe function χ, i.e. an integrable function on $\mathbb{R}^d$, we associate a positive semi-definite matrix defined by*

$$\mathcal{V}_\delta^R(\chi) := \int_{\delta^R} \mathbf{N}_\delta(\mathbf{x}) \otimes \mathbf{N}_\delta(\mathbf{x}) \cdot \chi\left(\mathbf{x} - \mathbf{N}_\delta(\mathbf{x})\right) dx, \tag{16}$$

where $\delta^R := \delta^{-1}((-\infty, R])$.

Note that $\mathbf{N}_\delta$ is defined almost everywhere in $\mathbb{R}^d$.

In the following, we define δ as the distance to a compact d_K in all results, but one should keep in mind that these results are extensible to arbitrary distance-like functions. Then $\mathbf{N}_{d_K}$ corresponds to the vector of the projection onto K, except for the sign. The d_K-VCM corresponds then to the covariance matrix of Voronoi cells of points of K, restricted to a maximum distance R, and weighted by the probe function.

Definition 6. *Let $Z \subset \mathbb{Z}^d$ and $h > 0$. The digital Voronoi Covariance Measure of Z at step h and radius R associates to a probe function χ the matrix:*

$$\hat{\mathcal{V}}_{Z,h}^R(\chi) := \sum_{\mathbf{z} \in \mathrm{vox}(Z_h, R)} h^d \mathbf{N}_{dZ_h}(\mathbf{z}) \otimes \mathbf{N}_{dZ_h}(\mathbf{z})\chi(\mathbf{z} - \mathbf{N}_{dZ_h}(\mathbf{z})), \tag{17}$$

where $Z_h := h \cdot Z$ and $\mathrm{vox}(Z_h, R)$ designates the points of $h \cdot \mathbb{Z}^d$ whose h-cube is completely inside the R-offset of Z_h. See Fig. 2 for an illustration.

Since $\mathbf{N}_{dZ_h}$ corresponds to the projection onto Z_h, the previous formulation is easily decomposed per Voronoi cells of Z_h and can be computed exactly by simple summations. The digital VCM is shown to be close to the VCM for digitizations of smooth enough shapes [CLT14]. Errors are related first to the

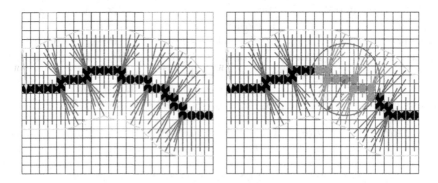

Fig. 2. Left: the limits of the R-offset of a set of digital points are drawn as a cyan contour, while vectors connecting points within the R-offset (i.e. Z_h^R) to their projection are drawn in deep blue. Right: Voronoi cells defining the VCM. The domain of integration for a kernel χ of radius r is drawn in dark orange (both germs and projection vectors). The kernel itself is drawn in red (Color figure online).

difference between ∂X and its digitization $\partial[\mathsf{G}_h(X)]_h$ (essentially bounded by Theorem 5.1 of [MOG11]). Secondly they are linked to the transformation of the integral in V_δ^R in a sum in $\hat{V}_{Z,h}^R$ (bounded by the fact that the projection is stable and that the strip $Z_h^R \backslash \text{vox}(Z_h, R)$ is negligible). Proofs ressemble the ones of Sect. 2.

Theorem 9 ([CLT14]). *For X a compact domain of $\mathbb{Z}^3$ whose boundary is C^2-smooth and has reach greater than $\rho > 0$. Then, for any $R < \rho/2$ and probe function χ with finite support diameter r and for small enough $h > 0$, letting $Z = \partial[\mathsf{G}_h(X)]_1 \cap (\mathbb{Z} + \frac{1}{2})^3$, we have:*

$$\|V_{\partial X}^R(\chi) - \hat{V}_{Z,h}^R(\chi)\|_{\text{op}} \leq O\big(\text{Lip}(\chi)(r^3 R^{\frac{5}{2}} + r^2 R^3 + rR^{\frac{9}{2}})h^{\frac{1}{2}}$$
$$+ \|\chi\|_\infty[(r^3 R^{\frac{3}{2}} + r^2 R^2 + rR^{\frac{7}{2}})h^{\frac{1}{2}} + r^2 Rh]\big).$$

As one can see, the digital VCM is an approximation of the VCM, but the quality of the approximation is related not only to the gridstep h but also to two parameters, the size r of the support of χ and the distance R to input data, which defines the computation window.

An interesting fact about the VCM is that it is in some sense complementary to integral invariants. It does not aim at fitting a tangent plane to ∂X, it aims at fitting a line to the normal vector $\mathbf{n}$ to ∂X. It carries thus information about the normal to ∂X. This is shown by the relation below ([CLT14], Lemma 2, adaptation of a relation in [MOG11]):

$$\left\| \frac{3}{2\pi R^3 r^2} V_{d\partial X}^R (\mathbf{1}_{B_r(\mathbf{x})}) - [\mathbf{n}(\mathbf{x}) \otimes \mathbf{n}(\mathbf{x})] \right\|_{\text{op}} = O(r + R^2). \qquad (18)$$

Putting together Theorem 9 and Eq. (18), as well as an error term coming from the positioning error of the kernel χ, provides a multigrid convergence theorem, that has the remarkable property to be stable to Hausdorff perturbation of digital data.

Theorem 10 ([CLT14]). *With the same hypothesis as in Theorem 9, χ bounded Lipschitz, and denoting $\hat{\mathbf{n}}_\chi^R$ the eigenvector associated to the highest eigenvalue of $\hat{V}_{Z,h}^R(\chi)$, then $\hat{\mathbf{n}}_\chi^R$ is multigrid convergent toward the normal vector to ∂X, with speed $O(h^{\frac{1}{8}})$ when both R and the support r of χ are chosen in $\Theta(h^{\frac{1}{4}})$.*

Experiments indicate a much faster convergence speed (close to $O(h)$) even in presence of noise. The discrepancy comes mainly from the fact that χ is any bounded Lipschitz function while Eq. (18) is valid for the characteristic function of $B_r(\mathbf{x})$.

5 Digital Surface Integration

Until now, convergence results where achieved by approximating well-chosen volume integrals around input data. What can we say if we wish to approach integrals defined over the shape boundary, given only the shape digitization. We focus here on Gauss digitization and we write $\partial_h X$ for $\partial[\mathsf{G}_h(X)]_h$. A natural answer is to define a mapping between the digitized boundary $\partial_h X$ and the continuous boundary ∂X. Using standard geometric integration results, a surface integral defined over ∂X can be transformed into a surface integral over $\partial_h X$ by introducing the Jacobian of this mapping. However, we have to face several difficulties in our case. The first one is that, starting from 3D, $\partial_h X$ may not even be a manifold, whatever the smoothness of ∂X and the gridstep h [SLS07]. Hence, it is not possible to define an injective mapping between the two sets. The second difficulty is that the underlying continuous surface is unknown, so we have to define a natural mapping without further hypothesis on the shape. The best candidate is the projection $\pi^{\partial X}$ onto ∂X, which is defined everywhere in the R-offset of ∂X, for R smaller than the reach of ∂X. It is nevertheless easily seen that $\pi^{\partial X}$, although surjective, is generally not injective between the digitized boundary and the continuous boundary. In the following, we bound these problematic zones in order to define convergent digital surface integrals. For simplicity, we write π for $\pi^{\partial X}$ and π' for its restriction to $\partial_h X$.

Definition 7 ([LT15]). *Let $Z \subset \mathbb{Z}^d$ be a digital set, with gridstep $h > 0$ between samples. Let $\mathrm{Bd}(Z) = \{(\mathbf{z}_1 + \mathbf{z}_2)/2, \mathbf{z}_1 \in Z, \mathbf{z}_2 \notin Z, \|\mathbf{z}_1 - \mathbf{z}_2\|_1 = 1\}$. It corresponds to centroid of $d - 1$-cells separating Z from $\mathbb{Z}^d \backslash Z$. Let $f : \mathbb{R}^d \to \mathbb{R}$ be an integrable function and $\hat{\mathbf{n}}$ be a digital normal estimator. We define the* digital surface integral *by*

$$\mathrm{DI}_{f,\hat{\mathbf{n}}}(Z, h) := \sum_{\mathbf{y}=h\mathbf{z}, \mathbf{z}\in\mathrm{Bd}(Z)} f(\mathbf{y})|\hat{\mathbf{n}}(\mathbf{y}) \cdot \mathbf{n}_h(\mathbf{y})|,$$

where $\mathbf{y}$ visits faces of $\partial[Z]_h$ and $\mathbf{n}_h(\mathbf{y})$ is the trivial normal of this face.

We show the following facts in [LT15], if X is a compact domain of $\mathbb{R}^d$ such that $\mathrm{reach}(\partial X) > \rho > 0$ (see Fig. 3):

- for $d = 3$ and $h < 0.198\rho$, $\partial_h X$ may not be a manifold only at places at distance lower than h to parts of ∂X whose normal makes an angle smaller than $1.26h/\rho$ to some axis;
- for arbitrary $d \geq 2$ and $h < \rho/\sqrt{d}$, let $\mathbf{y} \in \partial_h X$ and $\mathbf{n}_h(\mathbf{y})$ be its (trivial) normal vector; then the angle between the normal $\mathbf{n}(\mathbf{x})$ to ∂X at $\mathbf{x} = \pi(\mathbf{y})$ and $\mathbf{n}_h(\mathbf{y})$ cannot be much greater than $\pi/2$, since $\mathbf{n}(\mathbf{x}) \cdot \mathbf{n}_h(\mathbf{y}) \geq -\sqrt{3d}h/\rho$;
- let $\mathrm{Mult}(\partial X)$ be the points of ∂X that are images of several points of $\partial_h X$ by π', then it holds for at least one of the point $\mathbf{y}$ in the fiber of $\mathbf{x} \in \mathrm{Mult}(\partial X)$ under π' that $\mathbf{n}(\mathbf{x}) \cdot \mathbf{n}_h(\mathbf{y})$ is not positive;
- the Jacobian of π' is almost everywhere $|\mathbf{n}(\mathbf{x}) \cdot \mathbf{n}_h(\mathbf{y})|(1 + O(h)|$;
- areas are related with $\mathrm{Area}(\partial_h X) \leq 2^{d+2}d^{\frac{3}{2}}\mathrm{Area}(\partial X)$;
- hence, when $h \leq R/\sqrt{d}$, the area of the non-injective part of π' decreases with h: $\mathrm{Area}(\mathrm{Mult}(\partial X)) \leq K_2 \, \mathrm{Area}(\partial X) \, h$, with $K_2 \leq 2\sqrt{3} \, d^2 \, 4^d/\rho$.

The preceding properties show that problematic places on $\partial_h X$ for the integration have decreasing area. Furthermore, if $E(\hat{\mathbf{n}}, \mathbf{n}) := \sup_{\mathbf{y}\in\partial_h X} \|\mathbf{n}(\pi(\mathbf{y})) - \hat{\mathbf{n}}(\mathbf{y})\|$, errors in normal estimation induce propotional errors in surface integration. We can then prove the multigrid convergence of the digital surface integral toward the surface integral.

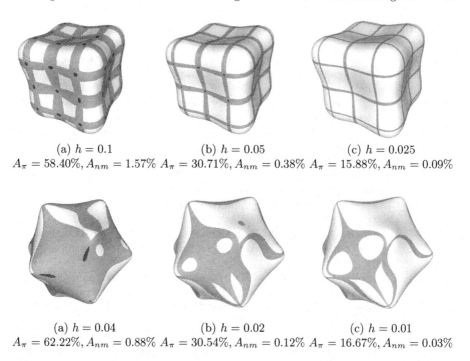

(a) $h = 0.1$ (b) $h = 0.05$ (c) $h = 0.025$
$A_\pi = 58.40\%, A_{nm} = 1.57\%$ $A_\pi = 30.71\%, A_{nm} = 0.38\%$ $A_\pi = 15.88\%, A_{nm} = 0.09\%$

(a) $h = 0.04$ (b) $h = 0.02$ (c) $h = 0.01$
$A_\pi = 62.22\%, A_{nm} = 0.88\%$ $A_\pi = 30.54\%, A_{nm} = 0.12\%$ $A_\pi = 16.67\%, A_{nm} = 0.03\%$

Fig. 3. Properties of several Gauss digitizations of two polynomial surfaces (top row displays a Goursat's smooth cube and bottom row displays Goursat's smooth icosahedron). Zones in dark grey indicates the surface parts where the Gauss digitization might be non-manifold (their relative area is denoted by A_{nm}). Zones in light grey (and dark grey) indicates the surface parts where projection π might not be a homeomorphism (their relative area is denoted by A_π). Clearly, both zones tends to area zero as the gridstep gets finer and finer, while parts where digitization might not be manifold are much smaller than parts where π might not be homeomorphic (Color figure online).

Theorem 11. *Let X be a compact domain whose boundary has positive reach ρ. For $h \leq \rho/\sqrt{d}$, the digital surface integral is multigrid convergent toward the integral over ∂X. More precisely, for any integrable function $f : \mathbb{R}^d \to \mathbb{R}$ with bounded Lipschitz norm $\|f\|_{\mathrm{BL}} := \mathrm{Lip}(f) + \|f\|_\infty$, one gets*

$$\left| \int_{\partial X} f(x)dx - \mathrm{DI}_{f,\hat{\mathbf{n}}}(\mathsf{G}_h(X), h) \right| \leq \mathrm{Area}(\partial X) \|f\|_{\mathrm{BL}} \left(O(h) + O(E(\hat{\mathbf{n}}, \mathbf{n})) \right).$$

The constant involved in the notation $O(.)$ only depends on the dimension d and the reach ρ. Note that Theorems 8 and 10 have shown that there exist normal estimators such that $E(\hat{\mathbf{n}}, \mathbf{n})$ tends to zero as h tends to zero. Experimental evaluation of the digital surface integral shows a better convergence in practice for area analysis, with convergence speed close to $O(h^2)$ both for integral normal estimator and digital VCM normal estimator.

References

[BGCF95] Bullard, J.W., Garboczi, E.J., Carter, W.C., Fullet, E.R.: Numerical methods for computing interfacial mean curvature. Comput. Mater. Sci. **4**, 103–116 (1995)

[BLM15] Buet, B., Leonardi, G.P., Masnou, S.: Discrete varifolds: a unified framework for discrete approximations of surfaces and mean curvature. In: Aujol, J.F., Nikolova, M., Papadakis, N. (eds.) SSVM 2015. LNCS, vol. 9087, pp. 513–524. Springer, Cham (2015)

[Bue14] Buet, B.: Approximation de surfaces par des varifolds discrets: représentation, courbure, rectifiabilité. Ph.D. thesis, Université Claude Bernard-Lyon I, France (2014)

[CCSM11] Chazal, F., Cohen-Steiner, D., Mérigot, Q.: Geometric inference for probability measures. Found. Comput. Math. **11**(6), 733–751 (2011)

[CK04] Coeurjolly, D., Klette, R.: A comparative evaluation of length estimators of digital curves. IEEE Trans. Pattern Anal. Mach. Intell. **26**(2), 252–258 (2004)

[CLL13] Coeurjolly, D., Lachaud, J.-O., Levallois, J.: Integral based curvature estimators in digital geometry. In: Gonzalez-Diaz, R., Jimenez, M.-J., Medrano, B. (eds.) DGCI 2013. LNCS, vol. 7749, pp. 215–227. Springer, Heidelberg (2013)

[CLL14] Coeurjolly, D., Lachaud, J.-O., Levallois, J.: Multigrid convergent principal curvature estimators in digital geometry. Comput. Vis. Image Underst. **129**, 27–41 (2014)

[CLMT15] Cuel, L., Lachaud, J.-O., Mérigot, Q., Thibert, B.: Robust geometry estimation using the generalized voronoi covariance measure. SIAM J. Imaging Sci. **8**(2), 1293–1314 (2015)

[CLR12] Coeurjolly, D., Lachaud, J.-O., Roussillon, T.: Multigrid convergence of discrete geometric estimators. In: Brimkov, V.E., Barneva, R.P. (eds.) Digital Geometry Algorithms, Theoretical Foundations and Applications of Computational Imaging. LNCVB, vol. 2, pp. 395–424. Springer, Dordrecht (2012)

[CLT14] Cuel, L., Lachaud, J.-O., Thibert, B.: Voronoi-based geometry estimator for 3D digital surfaces. In: Barcucci, E., Frosini, A., Rinaldi, S. (eds.) DGCI 2014. LNCS, vol. 8668, pp. 134–149. Springer, Heidelberg (2014)

[dVL09] de Vieilleville, F., Lachaud, J.-O.: Comparison and improvement of tangent estimators on digital curves. Pattern Recogn. **42**(8), 1693–1707 (2009)

[EP16] Edelsbrunner, H., Pausinger, F.: Approximation and convergence of the intrinsic volume. Adv. Math. **287**, 674–703 (2016)

[Guo10] Guo, J.: On lattice points in large convex bodies. arXiv e-prints (2010)

[HKS15] Hug, D., Kiderlen, M., Svane, A.M.: Voronoi-based estimation of minkowski tensors from finite point samples (2015)

[Hux90] Huxley, M.N.: Exponential sums and lattice points. Proc. Lond. Math. Soc. **60**, 471–502 (1990)

[KN91] Krätzel, E., Nowak, W.G.: Lattice points in large convex bodies. Monatshefte für Mathematik **112**, 61–72 (1991)

[KR04] Klette, R., Rosenfeld, A.: Digital Geometry: Geometric Methods for Digital Picture Analysis. Series in Computer Graphics and Geometric Modeling. Morgan Kaufmann, San Francisco (2004)

[KŽ00]   Klette, R., Žunić, J.: Multigrid convergence of calculated features in image analysis. J. Math. Imaging Vis. **13**, 173–191 (2000)

[LCLng]   Lachaud, J.-O., Coeurjolly, D., Levallois, J.: Robust and convergent curvature and normal estimators with digital integral invariants. In: Mdoern Approaches to Discrete Curvature. Lecture Notes in Mathematics. Springer International Publishing (2016, forthcoming)

[LT15]   Lachaud, J.-O., Thibert, B.: Properties of gauss digitized sets and digital surface integration. J. Math. Imaging Vis. **54**(2), 162–180 (2016)

[M99]   Müller, W.: Lattice points in large convex bodies. Monatshefte für Mathematik **128**, 315–330 (1999)

[MOG11]   Mérigot, Q., Ovsjanikov, M., Guibas, L.: Voronoi-based curvature and feature estimation from point clouds. IEEE Trans. Visual. Comput. Graph. **17**(6), 743–756 (2011)

[PWHY09]   Pottmann, H., Wallner, J., Huang, Q., Yang, Y.: Integral invariants for robust geometry processing. Comput. Aided Geom. Des. **26**(1), 37–60 (2009)

[PWY+07]   Pottmann, H., Wallner, J., Yang, Y., Lai, Y., Hu, S.: Principal curvatures from the integral invariant viewpoint. Comput. Aided Geom. Des. **24**(8–9), 428–442 (2007)

[SLS07]   Stelldinger, P., Latecki, L.J., Siqueira, M.: Topological equivalence between a 3D object and the reconstruction of its digital image. IEEE Trans. Pattern Anal. Mach. Intell. **29**(1), 126–140 (2007)

Discrete Calculus, Optimisation and Inverse Problems in Imaging

Hugues Talbot$^{(\boxtimes)}$

Université Paris-Est ESIEE, Paris, France
hugues.talbot@esiee.fr, hugues.talbot@univ-paris-est.fr

Abstract. Inverse problems in science are the process of estimating the causal factors that produced a set of observations. Many image processing tasks can be cast as inverse problems: image restoration: noise reduction, deconvolution; segmentation, tomography, demosaicing, inpaiting, and many others, are examples of such tasks. Typically, inverse problems are ill-posed, and solving these problems efficiently and effectively is a major, ongoing topic of research. While imaging is often thought of as occurring on regular grids, it is also useful to be able to solve these problems on arbitrary graphs. The combined frameworks of discrete calculus and modern optimisation allow us to formulate and provide solutions to many of these problems in an elegant way. This tutorial article summarizes and illustrates some of the research results of the last decade from this point of view. We provide illustrations and major references.

1 Introduction

Inverse problems are prevalent in science, because science is rooted in observations, and observations are nearly always indirect and certainly never perfect: noise is inevitable, instruments suffer from limited bandwidth, various artifacts and far from faultless acquisition components. Yet there is widespread interest in getting the most out of any observations one can make. Indeed, observations at the frontiers of science are typically those that are the faintest, most blurred, noisy and imperfect.

Imaging science is no exception, and this is why, for many decades [37], imaging communities have worked tirelessly to develop effective methods of noise removal, deconvolution, tomography, segmentation, inpainting, and more, in order to extract useful measurements from imperfect data.

Typically, inverse problems are ill-posed [26], meaning that their solution varies widely depending on the input data, in particular with noise. A useful approach for solving inverse problems is to use some sort of prior information on the data, called regularization. To formalize this, we turn to statistics.

2 Statistical Interpretation of Inverse Problems

We want to estimate some statistical parameter θ on the basis of some observation vector x.

© Springer International Publishing Switzerland 2016
N. Normand et al. (Eds.): DGCI 2016, LNCS 9647, pp. 18–27, 2016.
DOI: 10.1007/978-3-319-32360-2_2

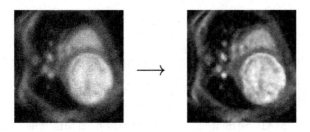

Fig. 1. An inverse problem: we observe the blurry, noisy data on the left. We would like to measure data from the image on the right.

2.1 Maximum Likelihood

If f is the sampling distribution, $f(\mathsf{x}|\mathsf{y})$ is the probability of x when the population parameter is y. The function

$$\mathsf{y} \mapsto f(\mathsf{x}|\mathsf{y}) \tag{1}$$

is called the *likelihood*. The Maximum Likelihood (ML) estimate is

$$\widehat{\mathsf{y}}_{\mathrm{ML}}(\mathsf{x}) = \underset{\mathsf{y}}{\operatorname{argmax}}\ f(\mathsf{x}|\mathsf{y}) \tag{2}$$

A very common example, assuming we have a linear operator $\mathbf{H}$ (in matrix form) and Gaussian deviates, then

$$\underset{\mathsf{y}}{\operatorname{argmax}}\ f(\mathsf{y}) = -\|\mathbf{H}\mathsf{y} - \mathsf{x}\|_2^2 = -\mathsf{y}^{\mathsf{T}}\mathbf{H}^{\mathsf{T}}\mathbf{H}\mathsf{y} + 2\mathsf{x}^{\mathsf{T}}\mathbf{H}\mathsf{y} - \mathsf{x}^{\mathsf{T}}\mathsf{x} \tag{3}$$

is a quadratic form with a unique maximum, provided by the *normal equations*:

$$\nabla f(\mathsf{y}) = -2\mathbf{H}^{\mathsf{T}}\mathbf{H}\mathsf{y} + 2\mathbf{H}^{\mathsf{T}}\mathsf{x} = 0 \Rightarrow \widehat{y} = (\mathbf{H}^{\mathsf{T}}\mathbf{H})^{-1}\mathbf{H}^{\mathsf{T}}\mathsf{x} \tag{4}$$

This simple least-square formulation is very general, yet versions of the Maximum Likelihood solution correspond to a large class of problems in statistics and imaging, from simple linear regression to the Wiener filter [28,39] used in signal and image restoration [38], tomography with the filtered back-projection method [36], and many others. When it can be used, the ML solution is fast and effective. However the ML solution requires a descriptive model (with few degrees of freedom) and a lot of data, which is often unsuitable for images because we do not have a suitable model for natural images. When we do not have all these hypotheses, sometimes the Bayesian Maximum A Posteriori approach can be used instead.

2.2 Maximum a Posteriori

If we assume that we know a *prior* distribution g over y, i.e. some *a-priori* information. Following Bayesian statistics, we can treat y as a random variable and compute the *posterior* distribution of y, via the Bayes theorem:

$$y \mapsto f(y|x) = \frac{f(x|y)g(y)}{\int_{\vartheta \in \Theta} f(x|\vartheta)g(\vartheta)d\vartheta} \qquad (5)$$

Then the Maximum a Posteriori is the estimate

$$\hat{y}_{MAP}(x) = \underset{y}{\mathrm{argmax}}\ f(y|x) = \underset{y}{\mathrm{argmax}}\ f(x|y)g(y) \qquad (6)$$

We say that the MAP estimate is a *regularization* of ML. The only difference between ML and MAP is the $g(y)$ multiplicative terms. For easier handling, we take the log, which does not change the estimator since the log function is monotonic:

$$\hat{y}_{MAP}(x) = \underset{y}{\mathrm{argmax}}\ \log f(x|y) + \log g(y). \qquad (7)$$

The first term of the right-hand side is call the log-likelihood, and the second term is the regularisation. In optimisation theory, a minimization is usually preferred, so we simply multiply (7) by -1. In particular, the log-likelihood become the negative log-likelihood [4, chap. 7].

3   Imaging Formulations

The very brief exposition of the previous section covers the basic principle of many statistical methods, including PCA, LDA, EM, Markov Random Fields, Hidden Markov Models, up to graph-cut type methods in imaging [5]. Many details can be found in classical texts on pattern analysis [22].

In the case of imaging, the log-likelihood term is often called the *data fidelity*. If we assume an image $\bar{y} \in \mathbb{R}^N$ is corrupted by noise and blur, for instance, we observe the data $x \in \mathbb{R}^Q$, and we can write:

$$x = H\bar{y} + u, \qquad H \in \mathbb{R}^{Q \times N}, \qquad (8)$$

where u is the noise. H can typically be a camera model including blur and defocus, a tomography projection matrix, an MRI analysis matrix (i.e. a Fourier transform), etc. The noise model is here additive, but with some work it is possible to express the likelihood of more complex noises: Rician, Poisson, Poisson-Gauss [27], etc. To recover an estimation $\hat{y}$ of $\bar{y}$, the ML estimator is, in the additive Gaussian noise case, the least-square estimate:

$$\hat{y} = \underset{y}{\mathrm{argmin}}\ \|Hy - x\|_2^2. \qquad (9)$$

Often this is not robust. A number of MAP regularizations can be proposed. The simplest is the Tikhonov regularisation:

$$\hat{y} = \underset{y}{\mathrm{argmin}}\ \|\Gamma y\|_2^2 + \lambda\|Hy - x\|_2^2, \qquad (10)$$

where λ is a Lagrangian multiplier, and Γ is a linear operator, which can be the identity or a spatial gradient for instance. The corresponding quadratic prior term expresses the belief that $\mathbf{\Gamma y}$ has a zero-centered Gaussian distribution, i.e. is typically smooth. This model is very easy to optimize but not realistic for most images, although it can be related to anisotropic diffusion for denoising [3,33], and the Random Walker for segmentation [24].

A more interesting approach for imaging is to define a *sparsity* prior. If we assume y to be sparsely representable, for instance in a wavelet basis, then it might be interesting to use this in a regularization prior. Ideally, one would like to use the ℓ_0 pseudo-norm to enforce sparsity, However this pseudo-norm is both non convex and non-differentiable, which makes it difficult to use in practice. A key element of *compressive sensing* [7] is based on the observation that the ℓ_1 norm is nearly as effective at promoting sparsity.

$$\widehat{\mathbf{y}} = \underset{\mathbf{y}}{\operatorname{argmin}} \ \|\mathbf{\Gamma y}\|_1 + \lambda \|\mathbf{Hy} - \mathbf{x}\|_2^2. \tag{11}$$

We will now explore some of these priors. Before we can do that, we need to propose a way to define flexible-enough linear operators well suited to imaging.

4 Linear Operators

The classical operators in continuous-domain formulations of the problems we have seen so far are the gradient and its adjoint the divergence. These can be easily discretized using finite-difference schemes [20]. Continuous and discrete versions of wavelet operators can also be considered. In the sequel, we choose to define our operators on arbitrary graphs, in the framework of discrete calculus [25].

4.1 The Incidence Matrix

Given a directed graph of N vertices and M edges, we can define the $M \times N$ incidence matrix $\mathbf{A}$, with lines containing zeros and exactly one $+1$ and one -1 so that $a_{i,k} = -1$ and $a_{i,l} = +1$ if e_i is the (k, l) edge. An illustrative example is best at this point (see Fig. 3). The matrix $\mathbf{A}$ describes the graph but can also be thought of as an operator. If p is a vector of values associated to the vertices, then $\mathbf{A}.\mathbf{p}$ is the gradient operator, associating a value to every edge. The transpose matrix $\mathbf{A}^\mathsf{T}$ is the adjoint operator, corresponding to the negative divergence (Fig. 2).

4.2 The Dual-Constrained Total Variation Model

Among the interesting regularizations, the Total Variation (TV) [35], or ROF model after the initials of its inventors, promotes sparsity of the gradient. In other words, it corresponds to a piecewise-constant image prior. This is of interest for various reasons, one of which because it is an acceptable model for texture-free natural images. Simplified versions of the Mumford-Shah model [30] for image segmentation typically use a TV term instead of the more complex piecewise-smooth prior. In [8], authors introduce TV formulations for image restoration in a MAP framework.

(a) $\overline{y}$ (b) $\times$

Fig. 2. Ideal and observed image

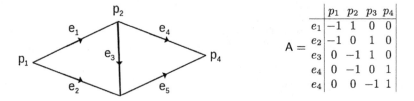

$$A = \begin{array}{c|cccc} & p_1 & p_2 & p_3 & p_4 \\ \hline e_1 & -1 & 1 & 0 & 0 \\ e_2 & -1 & 0 & 1 & 0 \\ e_3 & 0 & -1 & 1 & 0 \\ e_4 & 0 & -1 & 0 & 1 \\ e_4 & 0 & 0 & -1 & 1 \end{array}$$

Fig. 3. A small graph and its associated incidence matrix.

A weighted version of the TV model can be written in the following way [23], in the continuous framework:

$$\min_u \underbrace{\int \left(\int w_{x,y}(u_y - u_x)^2 dy \right)^{1/2} dx}_{\text{regularization } R(u)} + \underbrace{\frac{1}{2\lambda} \int (u_x - f_x)^2 dx}_{\text{data fidelity } \Phi(u)}, \qquad (12)$$

with λ a Lagrange multiplier. It is equivalent to the following min-max problem [10]

$$\min_u \max_{||p||_\infty \leq 1} \int\int w_{x,y}^{1/2}(u_y - u_x)p_{x,y}dxdy + \Phi(u), \qquad (13)$$

with p a projection vector field. Such min-max formulations are called primal-dual in optimization. The field p is introduced to achieve better speed. Constraining p can promote better results, as we will see in Sect. 6.

In discrete calculus form, we can write the same problem in this way:

$$\min_u \max_{||p||_\infty \leq 1, \, p \in \mathbb{R}^M} p^\mathsf{T}((\mathbf{A}u) \cdot \sqrt{w}) + \Phi(u). \qquad (14)$$

Introducing the projection vector $\mathbf{F} \in Rset^M = p.\sqrt{w}$, we can constrain $\mathbf{F}$ to belong to a convex set $C = \cap_{i=1}^{m-1} C_i \neq \emptyset$ where $C_1, \ldots, C_{m-1}$ closed convex sets of $\mathbb{R}^M$. Given $\mathbf{g} \in \mathbb{R}^N$, $\theta_i \in \mathbb{R}^M$, $\alpha \geq 1$, $C_i = \{\mathbf{F} \in \mathbb{R}^M \mid \|\theta_i \cdot F\|_\alpha \leq g_i\}$,

$$\min_{u \in \mathbb{R}^N} \sup_{\mathbf{F} \in C} \mathbf{F}^\mathsf{T}(\mathbf{A}u) + \Phi(u). \qquad (15)$$

The **F** constraints can be interpreted as *flow* constraints on the vertices of the connecting graph. For image denoising, we can for example propose that $g_i \in \mathbb{R}^N$ be a weight on vertex i, inversely function of the gradient of f at node i. In this case:

- Over flat areas: weak gradient implies a strong g_i, itself implyig a strong $F_{i,j} \rightarrow$ weak local variations of u.
- Near contours: strong gradient implies a weak g_i itself implying a weak $F_{i,j} \rightarrow$ large local variations of u are allowed.

This model is the dual-constrained total variation (DCTV) [17]. To optimize it, we require algorithms capable of dealing with non-differentiable convex functionals.

5   Algorithms

Optimization algorithms are numerous but research have mostly focused on differentiable methods: gradient descent, conjugate gradient, etc [4], with the exception of the simplex method for linear programming [21]. However non-differentiable optimization methods have been available at least since the 1960s. The main tool for non-differentiable optimizing convex functionals is the proximity operator [1,14,29,34]. We recall here the main points.

5.1   Proximity Operator

Let $\Gamma_0(\mathbb{R}^N)$ be the set of proper (i.e. not everywhere equal to $+\infty$), lower semi-continuous, convex functionals taking values from $\mathbb{R}^N$ to $(-\infty, +\infty]$. Such functions are necessarily quite regular. In particular, they are continuous and almost everywhere differentiable. The subdifferential of $f \in \Gamma_0(\mathbb{R}^N)$ at x is given by

$$\partial f(\mathsf{x}) = \{\mathsf{u}, \forall \mathsf{y} \in \mathbb{R}^N, (\mathsf{y} - \mathsf{x})^\mathsf{T} \mathsf{u} + f(\mathsf{x}) \le f(\mathsf{y})\} \qquad (16)$$

This definition extends the notions of tangent and thus of derivative to the non-differentiable case. Where f is differentiable, the subdifferential and the derivative are equal. We note that the subdifferential at non-differentiable points is a set, not a scalar or a vector.

The proximity operator of f in x is the operator $\mathbb{R}^N \rightarrow \mathbb{R}^N$

$$\mathrm{prox}_f(\mathsf{x}) = \underset{\mathsf{y} \in \mathrm{dom}(f)}{\mathrm{argmin}} \; f(\mathsf{y}) + \frac{1}{2}\|\mathsf{y} - \mathsf{x}\|_2^2 \qquad (17)$$

We have the following property

$$\mathsf{p} = \mathrm{prox}_f(\mathsf{x}) \Leftrightarrow \mathsf{x} - \mathsf{p} \in \partial f(\mathsf{p}), \forall (\mathsf{x}, \mathsf{p}) \in \mathbb{R}^N \times \mathbb{R}^N \qquad (18)$$

5.2   Splitting

One of the simplest cases is the situation when one wants to optimize the sum of two functions, one of which is smooth. Let $f_1 \in \Gamma_0(\mathbb{R}^N)$ and $f_2 : \mathbb{R}^N \to \mathbb{R}$ convex and differentiable with a $\beta-$ Lipschitz constant gradient ∇f_2., i.e.

$$\forall (\mathsf{x}, \mathsf{y}) \in \mathbb{R}^N \times \mathbb{R}^N, \|\nabla f_2(\mathsf{x}) - \nabla f_2(\mathsf{y})\| \le \beta \|\mathsf{x} - \mathsf{y}\|, \tag{19}$$

with $\beta > 0$. If $f_1(\mathsf{x}) + f_2(\mathsf{x}) \to +\infty$ when $\|\mathsf{x}\| \to +\infty\|$ (i.e. $f_1 + f_2$ is coercive), we wish to

$$\underset{\mathsf{x} \in \mathbb{R}^N}{\text{minimize}} \ f_1(\mathsf{x}) + f_2(\mathsf{x}). \tag{20}$$

It can be shown that this problem admits a solution and that for any $\gamma > 0$, the following fixed-point equation holds

$$\mathsf{x} = \text{prox}_{\gamma f_1}(\mathsf{x} - \gamma \nabla f_2(\mathsf{x})). \tag{21}$$

This suggests the following explicit-implicit algorithm

$$\mathsf{x}_{n+1} = \text{prox}_{\gamma f_1}(\mathsf{x}_n - \gamma \nabla f_2(\mathsf{x}_n)). \tag{22}$$

This algorithm is the forward-backward, alternating an explicit forward gradient descent step with an implicit proximity operator backward step. It can be shown [15] that this algorithm converges to a solution to (20).

This fairly simple method extends well-known ones such as gradient descent and the proximal point algorithm. It can be improved, for instance replacing the gradient descent scheme with Nesterov's method [32], which in this case yields an optimal convergence rate [31]. The corresponding method is the Beck-Teboule proximal gradient method [2].

5.3   Primal-Dual Methods

Many splitting methods exist, involving sums of two or more functions, and are detailed in [14]. In the case of (15), the presence of explicit constraints makes the analysis more difficult. Using convex analysis, and in particular the Fenchel-Rockafeller duality, and if the graph is regular, we can optimize it using the Parallel Proximal Algorithm (PPXA) [13]. In the more interesting case of an irregular graph, a primal-dual method is necessary [9]. We actually used the algorithm detailed in [6], which has since been generalized [12,16].

We now show some results obtained from solving (15) in various contexts.

6   Results

DCTV is a flexible framework. In Fig. 4(a,b,c) we restore a blurry, noisy version of an MRI scan using a local regular graph. This is the same image as in Fig. 1. In Fig. 4(d,e,f) we restore an image using an irregular, non-local graph. The fine texture of the brick wall has been restored to a high degree. In Fig. 4(g,h,i) we restore a noisy 3D mesh, with the same framework. Only the graph changes.

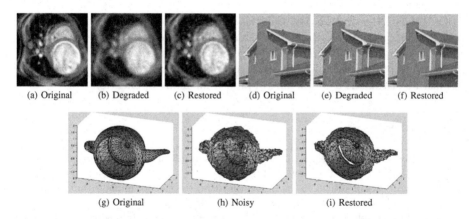

<table>
(a) Original   (b) Degraded   (c) Restored   (d) Original   (e) Degraded   (f) Restored
</table>

(g) Original   (h) Noisy   (i) Restored

Fig. 4. (a) Original MRI image, (b) Noisy, blurred image SNR = 12.3 dB, (c) DCTV result SNR = 17.2 dB. (d) Original house image, (e) Noisy image PSNR = 28.1 dB, (f) DCTV result PSNR = 35 dB. (g) Original mesh, (h) noisy mesh, (i) restored mesh

7   Discussion

Results presented here are interesting to some degree because we have kept the spatial part of the formulation fully discrete, with at its heart a graph representation for the numerical operators we use. However an important point is our assumption that the distribution of image values is continuous. In practice this is not the case and our approach is a relaxation of the reality, since images are typically discretized to 8 or 16-bit values. If we require to keep discretized values throughout the formulation, for instance to deal with labeled images, then the approach proposed here would not work. In this case, MRF formulations could be used [18,19].

We have also kept the discussion in the convex framework. Many important problems are not convex, for instance blind deblurring, where the degradation kernel must be estimated at the same time as the restoration. There exist methods for dealing with non-convex image restoraton problems, for instance [11], but dealing with non-convexity and non-differentiability together remains a challenge in the general case.

8   Conclusion

In this short overview article, we have introduced inverse problems in imaging and a statistical interpretation: the MAP principle for solving inverse problems such as image restoration using a-priori information. We have shown how we can use a graph formulation of numerical operators using discrete calculus to propose a general framework for image restoration. This DCTV framework can be optimized using non-differentiable convex optimization techniques, in particular the proximity operator. We have illustrated this approach on several examples.

DCTV is by no means the only framework available but it is one of the most flexible, fast and effective. With small changes we can tackle very different problems such as mesh or point cloud regularization. In general, the combination of powerful optimization methods, graph representations of spatial information and fast algorithms is a compelling approach for many applications.

References

1. Bauschke, H., Combettes, P.: Convex Analysis and Monotone Operator Theory in Hilbert Spaces. Springer, New York (2011)
2. Beck, A., Teboulle, M.: A fast iterative shrinkage-thresholding algorithm for linear inverse problems. SIAM J. Imaging Sci. **2**(1), 183–202 (2009)
3. Black, M.J., Sapiro, G., Marimont, D.H., Heeger, D.: Robust anisotropic diffusion. IEEE Trans. Image Process. **7**(3), 421–432 (1998)
4. Boyd, S., Vandenberghe, L.: Convex Optimization. Cambridge University Press, New York (2004)
5. Boykov, Y., Veksler, O., Zabih, R.: Fast approximate energy minimization via graph cuts. IEEE Trans. Pattern Anal. Mach. Intell. **23**(11), 1222–1239 (2001)
6. Briceno-Arias, L.M., Combettes, P.L.: A monotone+ skew splitting model for composite monotone inclusions in duality. SIAM J. Optim. **21**(4), 1230–1250 (2011)
7. Candes, E.J., Romberg, J.K., Tao, T.: Stable signal recovery from incomplete and inaccurate measurements. Commun. Pure Appl. Math. **59**(8), 1207–1223 (2006)
8. Chambolle, A., Caselles, V., Cremers, D., Novaga, M., Pock, T.: An introduction to total variation for image analysis. Theor. Found. Numer. Methods Sparse Recovery **9**(263–340), 227 (2010)
9. Chambolle, A., Pock, T.: A first-order primal-dual algorithm for convex problems with applications to imaging. J. Math. Imaging Vis. **40**(1), 120–145 (2011)
10. Chan, T.F., Golub, G.H., Mulet, P.: A nonlinear primal-dual method for total variation based image restoration. SIAM J. Sci. Comput. **20**(6), 1964–1977 (1999)
11. Chouzenoux, E., Jezierska, A., Pesquet, J.-C., Talbot, H.: A majorize-minimize subspace approach for ℓ_2-ℓ_0 image regularization. SIAM J. Imaging Sci. **6**(1), 563–591 (2013)
12. Combettes, P., Pesquet, J.: Primal-dual splitting algorithm for solving inclusions with mixtures of composite, lipschitzian, and parallel-sum type monotone operators. Set-Valued Variational Anal. **20**(2), 307–330 (2012)
13. Combettes, P.L., Pesquet, J.-C.: A proximal decomposition method for solving convex variational inverse problems. Inverse Prob. **24**(6), 065014 (2008)
14. Combettes, P.L., Pesquet, J.-C.: Proximal splitting methods in signal processing. In: Bauschke, H.H., Burachik, R., Combettes, P.L., Elser, V., Luke, D.R., Wolkowicz, H. (eds.) Fixed-Point Algorithms for Inverse Problems in Science and Engineering. Springer, New York (2010)
15. Combettes, P.L., Wajs, V.R.: Signal recovery by proximal forward-backward splitting. Multiscale Model. Simul. **4**(4), 1168–1200 (2005)
16. Condat, L.: A primal-dual splitting method for convex optimization involving lipschitzian, proximable and linear composite terms. J. Optim. Theory Appl. **158**(2), 460–479 (2013)
17. Couprie, C., Grady, L., Najman, L., Pesquet, J.-C., Talbot, H.: Dual constrained TV-based regularization on graphs. SIAM J. Imaging Sci. **6**(3), 1246–1273 (2013)

18. Couprie, C., Grady, L., Najman, L., Talbot, H.: Power watersheds: a new image segmentation framework extending graph cuts, random walker and optimal spanning forest. In: Proceedings of ICCV, Kyoto, Japan, 2009, pp. 731–738. IEEE (2009)
19. Couprie, C., Grady, L., Najman, L., Talbot, H.: Power watersheds: a unifying graph-based optimization framework. IEEE Trans. Pattern Anal. Mach. Intell. **33**(7), 1384–1399 (2011)
20. Courant, R., Friedrichs, K., Lewy, H.: On the partial difference equations of mathematical physics. IBM J. **11**(2), 215–234 (1967)
21. Dantzig, G.B.: Linear Programming and Extensions, 11th edn. Princeton University Press, Princeton (1998)
22. Duda, R.O., Hart, P.E., Stork, D.G.: Pattern Classification, 2nd edn. John Wiley & Sons, New York (2001)
23. Gilboa, G., Osher, S.: Nonlocal linear image regularization and supervised segmentation. SIAM J. Multiscale Model. Simul. **6**(2), 595–630 (2007)
24. Grady, L.: Random walks for image segmentation. IEEE Trans. Pattern Anal. Mach. Intell. **28**(11), 1768–1783 (2006)
25. Grady, L., Polimeni, J.: Discrete Calculus: Applied Analysis on Graphs for Computational Science. Springer Publishing Company, London (2010)
26. Hadamard, J.: Sur les problèmes aux dérivées partielles et leur signification physique. Princeton university bulletin **13**(49–52), 28 (1902)
27. Jezierska, A., Chouzenoux, E., Pesquet, J.-C., Talbot, H., et al.: A convex approach for image restoration with exact poisson-gaussian likelihood. IEEE Trans. Image Process. **22**(2), 828 (2013)
28. Kolmogorov, A.: Stationary sequences in hilbert space. In: Linear Least-Squares Estimation, p. 66 (1941)
29. Moreau, J.-J.: Fonctions convexes duales et points proximaux dans un espace hilbertien. CR Acad. Sci. Paris Sér. A Math. **255**, 2897–2899 (1962)
30. Mumford, D., Shah, J.: Optimal approximations by piecewise smooth functions and associated variational problems. Commun. Pure Appl. Math. **42**(5), 577–685 (1989)
31. Nemirovsky, A.-S., Yudin, D.-B., Dawson, E.-R.: Problem Complexity and Method Efficiency in Optimization. John Wiley & Sons Ltd, New York (1982)
32. Nesterov, Y.: Smooth minimization of non-smooth functions. Math. Program. **103**(1), 127–152 (2005)
33. Perona, P., Malik, J.: Scale-space and edge detection using anisotropic diffusion. IEEE Trans. Pattern Anal. Mach. Intell. **12**(7), 629–639 (1990)
34. Rockafellar, R.T.: Convex Analysis. Princeton University Press, New Jersey (1970). Reprinted 1997
35. Rudin, L.I., Osher, S., Fatemi, E.: Nonlinear total variation based noise removal algorithms. Phys. D **60**(1–4), 259–268 (1992)
36. Smith, S.W.: The Scientist and Engineer's Guide to Digital Signal Processing. California Technical Publishing, San Diego (1999)
37. Twomey, S.: Introduction to the Mathematics of Inversion in Remote Sensing and Indirect Measurements. Elsevier, New York (1977)
38. Widrow, B., Stearns, S.D.: Adaptive Signal Processing. Prentice-Hall Inc, Englewood Cliffs (1985)
39. Wiener, N.: Extrapolation, Interpolation, and Smoothing of Stationary Time Series, vol. 2. MIT press, Cambridge (1949)

Combinatorial Tools

Number of Words Characterizing Digital Balls on the Triangular Tiling

Benedek Nagy[1,2(✉)]

[1] Department of Computer Science, Faculty of Informatics,
University of Debrecen, Debrecen, Hungary
[2] Department of Mathematics, Faculty of Arts and Sciences,
Eastern Mediterranean University, Famagusta, North Cyprus, Turkey
nbenedek.inf@gmail.com

Abstract. In this paper, digital balls on two regular tessellations of the plane, on the square and on the triangular grids are analyzed. The digital balls are defined by digital, i.e., path based distance functions. The paths (built by steps to neighbor pixels) from the center to the points (pixels) of the balls are described as traces and generalized traces, respectively, on these grids. On the square grid, there are two usual types of neighborhood, and thus, the number of linearizations of these traces is easily computed by a binomial coefficient. The number of linearizations gives the number of words that describe the same digital ball. In the triangular tiling there are three types of neighborhood, moreover, this grid is not a lattice, therefore, the possible paths that define a ball form a more complicated set, a kind of generalized trace. The linearizations of these traces are described by an associative rewriting system, and, as a main combinatorial result, the number of words that define the same ball is computed.

Keywords: Digital distance · Shortest paths · Neighborhood sequences · Non-traditional grids · Digital disks · Combinatorics · Traces · Trajectories · Generalized traces

1 Introduction

The history of digital geometry has begun in the 1960's by the papers [18,19] in which two types of widely used neighborhood, namely the cityblock and chessboard neighborhood, on the square grid were described. Based on them digital distances are defined. Since both of them are very rough approximations of the Euclidean distance, it is recommended to use them alternately in a path. This octagonal distance [19] (the name comes from the shape of the digital balls defined by this digital distance) was generalized by allowing any predefined sequence of neighborhood in paths [1,12]. The theory of neighborhood sequences was also generalized to the triangular grid (i.e., the triangular tiling), see, e.g., [9,11,15]. Already in [19] combinatorial problems based on digital distances were investigated, namely 'path counting'. Our aim here is somewhat similar, a related

© Springer International Publishing Switzerland 2016
N. Normand et al. (Eds.): DGCI 2016, LNCS 9647, pp. 31–44, 2016.
DOI: 10.1007/978-3-319-32360-2_3

combinatorial property of some digital distances will be analyzed. We note that digital balls (or disks) based on digital distances are usually used to show some properties of the digital distances [5,13,17]. Digital grids and digital distances can be used in various fields of computer science and information technology, e.g., in image processing, in computer graphics, in networks. There are somewhat similar concepts, the digitized circles and disks: they are representing the Euclidean circles and disks on a digital grid. In this paper disks based on digital distances are examined and the term 'digital ball' is used to refer them.

Sequential and parallel neighborhood operations play essential roles in image processing as it is already mentioned in [18]. However, a concept for formal description of sequential and parallel executions of various operations comes from another source. In 1977, analyzing basic networks, Mazurkiewicz introduced the concept of partial commutations (see, e.g., [8]). Two parallel events are independent, they commute if their executing order can be arbitrary in a sequential simulation. By using the concept of commutations, the work of the concurrent systems can be described by traces. In these systems some (pairs of) elementary processes (i.e., atomic actions; they are represented by the letters of the alphabet) may depend on each other, and some of them can be (pairwise) independent. The order of two consecutive independent letters can be arbitrary, in this way traces are a kind of generalizations of words. Traces and trace languages play important roles in describing parallel events and processes. For simplifying the descriptions linearizations of trace languages is frequently used; they are sets of words representing traces of the trace languages. A special two-dimensional representation of some traces is given by trajectories (usually on the square grid). These trajectories were also used to describe syntactic constraints for shuffling two parallel events (described by words) in [7]. We believe that trajectories connected to other regular grids are also interesting. A kind of generalization of traces is presented in [6], where apart from the usual permutation rules, some other rewriting rules, namely serialisation equations, are also allowed. As we will see, the sets of words describing the same digital balls on the triangular tiling are generalized traces.

In this paper, two regular tessellations of the two-dimensional plane are used: the square and the triangular grids. Traces and words describing digital balls are analyzed. As we will see, the case of the square grid is very simple, it is shown only for the analogy. Descriptions of a ball on the square grid form a trace, in which the order of the two types of steps can be arbitrary. The triangular grid (tiling) is not a lattice, therefore, the sets of words describing the same balls form more complicated sets.

2   Preliminaries

We assume that the reader is familiar with some of the basic concepts of formal language theory. However, to have a self-contained paper, we recall briefly the concepts that are necessary. Let V be a finite set of symbols, called alphabet, its elements are called letters. The set of all (finite length) words over V is denoted by V^*. This set also includes the empty word λ. For any $w \in V^*$, $|w|$ denotes the length of w. Let the alphabet be an ordered set $\{a_1, \ldots, a_n\}$. For any $a_i \in V$, $|w|_{a_i}$ denotes the number of occurrences of letter a_i in w. Consequently, $\sum_{i=1}^{n} |w|_{a_i} = |w|$ for every word $w \in V^*$. The Parikh mapping assigns the *Parikh-vector*, $W(|w|_{a_1}, \ldots, |w|_{a_n})$, to every word w.

There are three regular tessellations of the plane: the square-, the triangular- and the hexagonal grids. In this paper, we are working on the first two of them. The types and number of neighbors present depend on the grid [2,11]: usually two types are used in the square grid and three types in the triangular tiling (as we detail in Sects. 3 and 4, respectively). Let V denote the set of possible neighborhood relations. The infinite sequence $B = (b(i))_{i=1}^{\infty}$ where $b(i) \in V$ represents a possible neighborhood for all $i \in \mathbb{N}$, is called a *neighborhood sequence*. In fact, neighborhood sequences are infinite words over V, thus, the set of them is denoted by V^{∞}. Let p and q be two pixels (also called points) of the grid and $B = (b(i))_{i=1}^{\infty}$ be a neighborhood sequence. A finite point sequence $\Pi(p, q; B)$ of the form $p = p_0, p_1, \ldots, p_m = q$, where p_{i-1}, p_i are $b(i)$-neighbors for $1 \leq i \leq m$, is called a *B*-path from p to q. We write $m = |\Pi(p, q; B)|$ for the length of the path. Denote by $\Pi^*(p, q; B)$ a shortest path from p to q, and set $d(p, q; B) = |\Pi^*(p, q; B)|$. We call $d(p, q; B)$ the *B*-distance from p to q. For $k \in \mathbb{N}$, let $D_c^B(k) = \{p \mid d(c, p; B) \leq k\}$ be the digital ball occupied by B in k steps starting from the pixel c: a digital ball $D_c^B(r)$ is the set of exactly those pixels (points) that have digital distance from a point c, called center, at most a given value r, called radius.

If the reader is not familiar with traces, she or he is referred to, e.g., [3,4]; here we recall only the basic concepts. Let the dependency δ be a reflexive and symmetric binary relation on V. The independency relation, $\iota = (V \times V) \backslash \delta$ is also called commutation. If $(a, b) \in \iota$, then the words $uabv$ and $ubav$ are equivalent to each other for any $u, v \in V^*$. The equivalence relation induced by ι on V^* is called partial commutation. The set of words that are equivalent to w is denoted by $[w]$ and called a *trace*. The linearizations of the traces are exactly those words that belongs to the trace. A trace is represented by the corresponding set of these words. Thus, in traces some of the neighbor letters can be permuted, e.g., the interleaving equation $ab = ba$ can be written for some letters $(a, b) \in \iota$. In generalized traces [6], apart from these equations, serialisation equations of the form $c = de$ $(c, d, e \in V)$ are also allowed, representing the serialisation of c into two consecutive steps (letters) d and e. Consequently, a *generalized trace*, denoted by $[[w]]$, for a word $w \in V^*$ is a set of words equivalent to w. Linearizations of generalized traces are understood analogously. Notice that without using any serialisation equations, i.e., with only interleaving equations (i.e., partial commutations, permutations) $[[w]] = [w]$ for every word w.

We also recall the definition of associative calculus: $\mathcal{C} = (V, P)$ is an *associative calculus*, where V is a finite alphabet and P is a finite set of productions (rewriting rules). Each rewriting rule is an element of $V^* \times V^*$. A rule is usually written in the form $u \leftrightharpoons v$, where $u, v \in V^*$. Let $w \in V^*$ be given, we say that w' is obtained from w applying the rewriting rule $u \leftrightharpoons v$, if there exist $w_1, w_2 \in V^*$

such that either $w = w_1 u w_2$ and $w' = w_1 v w_2$, or $w = w_1 v w_2$ and $w' = w_1 u w_2$. Actually, w can also be obtained from w' by the same production, thus we may use the notation $w \Leftrightarrow w'$. By the reflexive and transitive closure of $\Leftrightarrow$, the relation $\Leftrightarrow^*$ is defined. Observe that the calculus $\mathcal{C}$ defines an equivalence relation on V^*. The equivalence class represented by w is denoted by $\mathcal{C}(w) = \{w' \mid w \Leftrightarrow^* w'\}$. An example is $\mathcal{C} = (\{a, c, d, g, f, o, r, t\}, \{at \leftrightharpoons rog, c \leftrightharpoons f, d \leftrightharpoons fr\})$, in which $\mathcal{C}(cat) = \{cat, fat, crog, frog, dog\}$. One may observe that the associative calculus is very general in the sense that a rule may allow to rewrite any word by any other.

3   Traces and Trajectories on the Square Grid

Usually the term digital plane or digital grid refers to $\mathbb{Z}^2$, as the square grid is the most known and most used grid both in theory and in practice (e.g., in image processing, in computer graphics and at 2D languages). The two basic neighborhood, the cityblock and the chessboard neighborhood, on the square grid were described by Rosenfeld and Pfaltz in [19]. We use the terms, 1-neighbor and 2-neighbor for them, respectively (since in this way, they can easily be generalized to other grids). Using the Cartesian coordinate frame, one can define these neighborhood relations as follows. Let $p(p(1), p(2))$ and $q(q(1), q(2))$ be two points (pixels), they are m-neighbors ($m \in \{1, 2\}$) if $|p(i) - q(i)| \leq 1$, for $i \in \{1, 2\}$, and $|p(1) - q(1)| + |p(2) - q(2)| \leq m$.

The theory of neighborhood sequences is well developed for $\mathbb{Z}^2$ [1,12,16,21].

Neighborhood sequences are infinite, but a given ball depends only on the first r elements of the sequence [13], and therefore, here we use finite words over $V = \{1, 2\}$ to describe digital balls, hence redefining the digital balls as follows. A word $w \in V^*$ defines a digital ball:

$$D_c(w) = \{p \mid \text{there is a prefix of } w \text{ such that it describes a path from } c \text{ to } p\}.$$

Figure 1 shows an example. We note here that a greedy algorithm to compute a shortest path and formula to compute the distance for any neighborhood sequence can be found in [12,14,16].

3.1   Number of Words Describing the Same Ball on the Square Grid

Let us start this subsection with some relatively simple results about digital balls on the square grid. Let the center of the ball be $c(c(1), c(2))$. Then, (based on the formula for distances, e.g., in [14]),

$$D_c(w) = \{p(p(1), p(2)) \mid |p(1) - c(1)| \leq |w|, |p(2) - c(2)| \leq |w|, \qquad (1)$$
$$|p(1) - c(1)| + |p(2) - c(2)| \leq |w| + |w|_2\}.$$

Consequently, the number of pixels in the ball obtained from w is $(2|w| + 1)^2 - 4\frac{|w|_1(|w|_1 + 1)}{2}$.

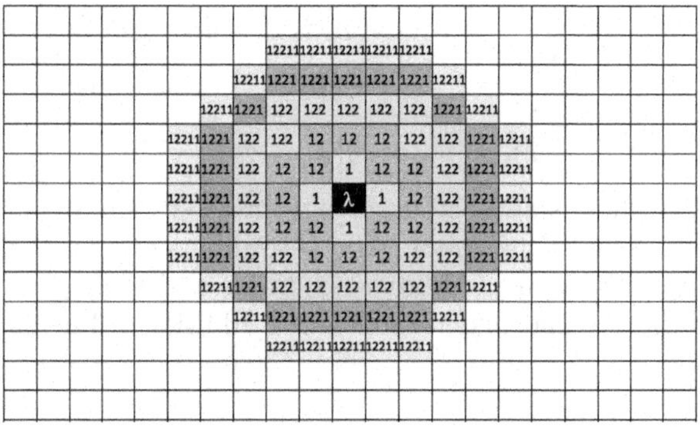

		12211	12211	12211	12211	12211				
	12211	1221	1221	1221	1221	1221	12211			
	12211	1221	122	122	122	122	122	1221	12211	
12211	1221	122	122	12	12	12	122	122	1221	12211
12211	1221	122	12	12	1	12	12	122	1221	12211
12211	1221	122	12	1	λ	1	12	122	1221	12211
12211	1221	122	12	12	1	12	12	122	1221	12211
12211	1221	122	122	12	12	12	122	122	1221	12211
	12211	1221	122	122	122	122	122	1221	12211	
	12211	1221	1221	1221	1221	1221	12211			
		12211	12211	12211	12211	12211				

Fig. 1. The digital ball defined by the word $w = 12211$. Each pixel contains a prefix of w that gives a path to there from the center marked by λ. Same color is used to highlight pixels that have the same distance from the center (Color figure online).

Now, let us turn to the main topic. By Eq. (1), it is clear that the order of the letters of w does not matter, only the numbers of their occurrences are important. Consequently, instead of a word w, its Parikh-vector, i.e., the vector $W(|w|_1, |w|_2)$ uniquely determines the digital ball $D_c(w)$. In this way, actually, the balls are represented by traces. The two letters of V are independent of each other. Therefore, the linearizations of these traces contain the commutative closure of the words w.

Now we show how the number of words describing the same ball can be computed. The radius of a ball is $r = |w|$, and, thus, the number of words describing ball $D_c(w)$ is $\binom{|w|}{|w|_1}$. That is the number of words in the equivalence class $[w]$ defined by w.

One can easily prove using Eq. (1) that the balls with the same radius form a well-ordered set:

Proposition 1. $D_c(w) \subsetneq D_c(w')$ if and only if $|w|_2 < |w'|_2$, where $|w| = |w'| = r$.

We note one important fact about the square grid: it is a lattice, and thus, the steps in a path can freely be permuted. In the next section, the triangular tiling is considered; and that is not a lattice. Therefore, as we will show, to describe the balls on the triangular tiling is a more challenging and more complex task.

4 Describing Digital Balls on the Triangular Tiling

4.1 The Triangular Tiling

The triangular grid (i.e., the triangular tiling), preserving the symmetry of the grid, can be described by three coordinates [9,11,13,20]. There are two types of pixels (by orientation): the even pixels have zero-sum triplets, while the odd pixels have one-sum triplets. Formally, the neighborhood relations are defined as follows. Let $p(p(1), p(2), p(3))$ and $q(q(1), q(2), q(3))$ be two pixels, they are m-neighbors ($m \in \{1, 2, 3\}$) if

- $|p(i) - q(i)| \leq 1$, for $i \in \{1, 2, 3\}$, and
- $|p(1) - q(1)| + |p(2) - q(2)| + |p(3) - q(3)| \leq m$.

Various neighborhoods and the used coordinate system are shown in Fig. 2. Observe that the 3 'closest' neighbors, the 1-neighbors of a pixel have the opposite parity (shape). There are 6 more 2-neighbors and their parity is the same as the parity of the original pixel, while the 3 additional 3-neighbors have the opposite parity.

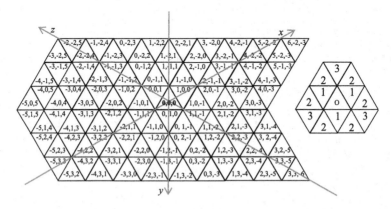

Fig. 2. The triangular tiling: pixels with coordinates (left) and various neighbors (right) that are usually considered [2].

The neighborhood sequences on the triangular tiling are infinite sequences over the alphabet $V = \{1, 2, 3\}$. Paths, distances and digital balls are defined analogously to the case of the square grid. However, there are neighborhood sequences that generate exactly the same distance functions. The next concept and result are recalled from [10].

Let $B, B' \in V^\infty$ be two neighborhood sequences. B' is called the *minimal equivalent neighborhood sequence* of B, if the following two conditions hold:

- $d(p, q; B) = d(p, q; B')$ for all pairs of points p, q, and
- for each neighborhood sequence B'', if $d(p, q; B) = d(p, q; B'')$ for all pairs of points p, q, then $b'(i) \leq b''(i)$ for all i.

Lemma 1. *The minimal equivalent neighborhood sequence B' of B is uniquely determined, and is given by*

$$
b'(i) = \begin{cases}
b(i), & \text{if } b(i) < 3; \\
3, & \text{if } b(i) = 3 \text{ and } i \text{ is the smallest index with this property;} \\
3, & \text{if } b(i) = 3 \text{ and there is an } l : l < i, b(l) = 3 \text{ and} \\
& \sum_{k=j+1}^{i-1} b'(k) \text{ is odd, with } j = \max\{l \mid l < i, b'(l) = 3\}; \\
2, & \text{otherwise.}
\end{cases}
$$

One may restrict the usage of neighborhood sequences to the minimal equivalent neighborhood sequences by representing each digital distance by only one, uniquely determined neighborhood sequence. Thus, in this paper we use only words to describe digital balls that are prefixes of some neighborhood sequences that are minimal equivalent neighborhood sequences. In this way, by counting the words describing the same ball, roughly speaking, we count the number of distance functions which could define the same ball.

Definition 1. *The subwords $u, v \in V^*$ are* region equivalent *if $D_c(w_1 u w_2) = D_c(w_1 v w_2)$ for every $w_1, w_2 \in V^*$.*

One can easily establish the fact that region equivalence is an equivalence relation among words of V^*.

Proposition 2. *The subword 2 is region equivalent to subword 11.*

Proof. By a step to a 1-neighbor at most 1 of the coordinates is changing and it is changing by ± 1 depending on the parity of the pixel. By a step to a 2-neighbor at most 2 of the coordinates are changing and if exactly 2 of them are changing then they change by $+1$ and -1. One can see that exactly those points are reached by a 2-step as the points that are reached by two consecutive 1-steps.   □

The previous proposition shows an effect that is not present on the square grid.

Proposition 3. *The subword 12 is region equivalent to subword 21.*

Proof. By Proposition 2 one can see, that 12 is region equivalent to 111, and thus, to 21.   □

Proposition 4. *The subword 32 is region equivalent to subwords 23. Moreover, they are region equivalent to 33.*

Proof. At 3-steps all the three coordinates can be changed (by ± 1). However, there are only two types of pixels, namely 0-sum (even) and 1-sum (odd) pixels. Therefore, one can see that by two consecutive 3-steps those points can be reached for which one of the coordinates is changed with at most $+2$, the other with at most -2, and the third one by at most ± 1, depending on the parity. Actually, as one may list those points, they are the same as the points reached by a 2-step and by a 3-step independently on their order.   □

One can see that by Proposition 4 not only the positions, but the letters can also be changed, and, actually, the sum of their values can also be changed. However, as we have restricted ourselves only for neighborhood sequences that are minimal equivalent neighborhood sequences, thus, the subword 33 is excluded from our analysis (together with all its related subwords having even sum between two neighbor occurrences of letter 3, e.g., 3223 or 31213).

The independence of letters 1 and 2 (Proposition 2) and of letters 2 and 3 (stated in Proposition 4) are analogous to the independence of letters 1 and 2 on the square grid. But, on the triangular tiling we have letters that are not independent of each other. To show this, one can consider the following example: $D_c(13) \neq D_c(31)$. For instance, with $c(0,0,0)$ $p(-1,2,-1) \in D_c(13)$, but $p(-1,2,-1) \notin D_c(31)$; and $q(1,-2,1) \in D_c(31)$, but $q(1,-2,1) \notin D_c(13)$. More formally, we state the following.

Proposition 5. *The subword* 13 *is not region equivalent to subword* 31. *More precisely, for any pair of words* $w, u \in V^*$, $D_c(w13u) \neq D_c(w31u)$.

Proof. The proof consists of two parts:

In the first part we prove the statement for words having 13 and 31 as suffixes. Thus, first we show that for any word w the words $w13$ and $w31$ describe different balls. Then, let us start from the ball $D_c(w)$ for any $w \in V^*$. Consider those point(s) of $D_c(w)$ that has smallest y coordinate (i.e. the second coordinate), let denote this set of points by $T_c(w)$ (top points of $D_c(w)$). Then, there are two cases, according to the parity of these point(s).

- If, there is an odd point $p(p(1), p(2), p(3))$ in $T_c(w)$, then, obviously $p(1) \geq 0, p(2) \leq 0, p(3) \geq 0, p(1) + p(2) + p(3) = 1$ because $p(2)$ can be minimal only under these conditions. Furthermore, the odd point $q(p(1)+1, p(2)-2, p(3)+1)$ is in $D_c(w13)$ (and also in $T_c(w13)$), but not in $D_c(w31)$ (since no odd point having y coordinate $p(2) - 2$ is in this ball).
- In the case, when $T_c(w)$ contains only even point(s), chose one of them, let us say, $p(p(1), p(2), p(3))$ (in this case, $p(1) \geq 0, p(2) \leq 0, p(3) \geq 0, p(1) + p(2) + p(3) = 0$). Then, the even point $q(p(1)+1, p(2)-2, p(3)+1)$ is in $D_c(w31)$ (and also in $T_c(w31)$), but not in $D_c(w13)$ (no points with y coordinate $p(2) - 2$ can be in this ball).

Thus, in both cases, the balls defined by $w13$ and $w31$ differ, for every $w \in V^*$. Moreover, the sets $T_c(w31)$ and $T_c(w13)$ are not the same, either the values of y coordinate differ or the parities of the points they contain are not the same (e.g., one of them does not contain odd points, but the other does).

The second part of the proof uses induction on the length of the word u, to show that words $w13u$ and $w31u$ are not defining the same ball, moreover the sets $T_c(w13u)$ and $T_c(w31u)$ differ either by the y coordinate of the points they contain, or by the parities of the points they contain does not match.

- The base of the induction, the case when $u = \lambda$, i.e., its length is 0 was proven in the first part.
- As induction hypothesis, let us assume that for every $w \in V^*$ and $u \in V^*$ having $|u| = n$ for some nonnegative integer n, the words $w13u$ and $w31u$ are not defining the same ball, moreover, the sets $T_c(w13u)$ and $T_c(w31u)$ have the mentioned difference.
- We state that $w31u'$ is defining another ball than $w13u'$, when $|u'| = n + 1$ and the difference of sets $T_c(w13u')$ and $T_c(w31u')$ is also preserved.

To prove this statement, observe that u' can be written exactly one of the following forms: $u' = u1$ or $u' = u2$ or $u' = u3$ for some $u \in V^*$ with $|u| = n$. Thus, for $w13u$ and $w31u$ we can apply the induction hypothesis: they are describing different balls. Now, the proof is going by cases:

Let us assume that u does not contain the letter 3.

- If $u' = u1$, then there are two cases, according to the possible differences of the sets $T_c(w13u)$ and $T_c(w31u)$:
  - If the y coordinate of the points contained in these sets differ, then it is clear that by a 1-step, if the y value is changed (decreased by 1), only even point(s) are reached. If it is not changed, then odd point(s) with the same y coordinate are also reached and become elements of the appropriate new set T_c. In this way, the sets $T_c(w13u1)$ and $T_c(w31u1)$ differ, and thus the words $w13u1$ and $w31u1$ are not describing the same ball.
  - If $T_c(w13u)$ and $T_c(w31u)$ contains points with the same value of y coordinate, then, by our assumption, they must differ by the parity of the points they contain: one of them does not contain odd points. By the additional 1-step the corresponding set T_c will be extended with odd point(s), while the other set that had already odd points will change to contain only even points with one less second coordinate value. In this way, the difference of $T_c(w13u1)$ and $T_c(w31u1)$ is proven.
- If $u' = u2$, then we can face to the same cases again. However, with a 2-step the second coordinate value is decreased by 1 in both cases when the sets $T_c(w13u)$ and $T_c(w31u)$ are changed to $T_c(w13u2)$ and $T_c(w31u2)$, but the parity of the points contained in these sets are copied from the previous sets, respectively. If $T_c(w13u)$ and $T_c(w31u)$ had points with different value of y coordinate, then so $T_c(w13u2)$ and $T_c(w31u2)$ do. If the parities of points of $T_c(w13u)$ and $T_c(w31u)$ do not match, then the parities of points of $T_c(w13u2)$ and $T_c(w31u2)$ also mismatch. Thus, this case is proven.
- Finally, the case $u' = u3$ is analyzed. Observe, that in this case both $w13u3$ and $w31u3$ contains at least twice the letter 3. Moreover, the sum of the values 1's and 2's between these two occurrences of 3 is even in one case, and odd in the other case (the sum of the values of u and one more, respectively). As such, by Lemma 1, in exactly one of the cases, this last 3 can be changed to a value 2 without changing the ball that is described by that word. Consequently, the balls $D_c(w13u3)$ and $D_c(w31u3)$ differs: some points that are in the ball having last element 3 cannot be in the ball in which this value is changed to 2, the sum of the values in these two words differ, and since both of them are prefixes of minimal equivalent neighborhood sequences, there are some points p for which the sum of the coordinate differences from point c is exactly the sum of the elements and p is contained in the respective ball (see, e.g., [15]).

This last case show also, why we can assume that u does not contain the letter 3.

Thus, the proof is finished, the subwords 13 and 31 cannot be replaced in any context without changing the described balls.   □

As we can see, on the triangular tiling we do not have a statement analogous to Proposition 1. To work with digital balls on the triangular tiling is more interesting. The order of steps becomes important, because the triangular tiling is not a lattice, i.e., some of the grid-vectors do not translate the grid to itself.

4.2 Generalized Traces Describing Digital Balls

We refer to [13,17] for description of the possible shapes of digital balls and approximation of the Euclidean disks by them. Here, we are doing related things: an associative calculus is applied to describe words representing digital balls, and it is shown how these sets of words can be seen as generalized traces.

Now we present an associative calculus that provides all the words that are region equivalent to the one the process starts with (that is, they describe the same ball). The independence relation contains pairs of letters such that one of the letters is 2. In terms of generalized traces, this fact can be concluded as:

- Letter 2 commutes with any other letters (Propositions 3 and 4).
- Moreover, on the triangular tiling the steps indicated by 2 can be broken to two steps indicated by 1's (Proposition 2). Actually, it is a serialisation.

Theorem 1. *With the associative calculus* $\mathcal{C} = (\{1, 2, 3\}, \{12 \leftrightharpoons 21, 23 \leftrightharpoons 32, 11 \leftrightharpoons 2\})$ $\mathcal{C}(w) = [[w]]$, *i.e.,* $\mathcal{C}$ *provides the equivalence classes, the generalized traces, such that* $w' \in \mathcal{C}(w)$ *if any only if* $D_c(w) = D_c(w')$.

Proof. By Propositions 2–4, one can see that the rules of the system $\mathcal{C}$ causes region-equivalent words. Moreover, if w is a prefix of a minimal equivalent neighborhood sequence, then all the words $w' \in \mathcal{C}(w)$ have this property (by Lemma 1).

On the other direction, by Proposition 5 and Lemma 1, it is clear that the parity of the sum of 1's (and 2's) before the first 3, between any two neighbor occurrences of 3 and after the last occurrence of 3 is fixed (by w) for every element w' of the generalized trace, the equivalence class $[[w]]$. However, by the calculus $\mathcal{C}$ exactly those words w' can be obtained from w for which $|w'|_3 = |w|_3$ and the parities of the sums of 1's are the same in their respective positions. Hence, $[[w]] = \mathcal{C}(w)$. □

We note here that the calculus gives the same equivalence classes without the rule $12 \leftrightharpoons 21$, since it can be simulated by two applications of $11 \leftrightharpoons 2$. However, we believe that $\mathcal{C}$ is more intuitive in the presented way, and thus, the reader understands the idea behind better in this way.

Of course, there are longer sequences of steps that are equivalent to each other, e.g., 13222 is equivalent to 112123, but their equivalence is based on the use of shorter equivalences. The equivalence of every (consecutive) sequence of steps to another sequence is based on the listed short equivalences.

The system can be understood as a generalized trace [6] as we detail below. The system has some permutative (interleaving) rules, the rules by which a step to a 2-neighbor can be interchanged in the path with the previous or the next step (if it is not the same). However, the system has another types of production (representing serialisation), namely, $11 \leftrightharpoons 2$. This rule refers the change from the unbroken 2 to two 1's. In this sense 2 represent an action in which the two 1's executed together.

In the next theorem we give a unique representation for each ball by specific regular languages.

Theorem 2. *Every digital ball $D_c(w)$ can be described by a word $w \in V^*$ such that exactly one of the following conditions holds:*

(a) $w \in 2^*$,   (b) $w \in 12^*$,

(c) $w \in 3(13)^*2^*$,   (d) $w \in 1(31)^+2^*$,

(e) $w \in (13)^+2^*$,   (f) $w \in (31)^+2^*$.

Proof. Based on Theorem 1 using only minimal equivalent neighborhood sequences, one can easily show that any possible word $w \in V^*$ has exactly one equivalent word w' in exactly one of the listed cases.   □

For all balls, their descriptions by words w fulfilling the condition of the above theorem are called *basic descriptions* or description by the *basic words*. In Fig. 3 example balls are shown for various cases. The basic words are among the shortest representations of the balls (in terms of $|w|$), the maximal (longest) representations are those words of the class that do not contain any letter 2.

4.3   The Number of Linearizations of Generalized Traces

In this subsection we count the number of words describing the same disc. We use a combinatorial approach. We distinguish six cases based on Theorem 2.

Let us start with the cases represented by words of shape 2^* and 12^*.

(a), (b): The case of balls described without 3's. In cases (a) and (b) each $w' \in [[w]]$ contains only 1's and 2's. One can also represent these balls by words of the regular language 1^*. The ball described by 1 is a triangle, and there is only one way to describe it. All other balls of these classes are hexagons (with sawtooth and straight sides, alternately, [13], see Fig. 3, first row).

Let us consider the ball described by word $w = 1^k$ (with $k \geq 1$) and let us count its possible descriptions.

- Using only 1's there is only 1 word (we may write this in the form $\binom{k}{0}$).
- Using exactly one letter 2 (in case of $k \geq 2$) we have $k - 2$ 1's and the length of these words is $k - 1$ and each permutation belongs to $[[w]]$. Their number is $\binom{k-1}{1}$

$\vdots$

- Using i letter 2 (for $i \leq \frac{k}{2}$), the length of these words is $k - i$, and their number is $\binom{k-i}{i}$.

Summing up the number of these words, one gets $\sum_{i=0}^{\frac{k}{2}} \binom{k-i}{i} = F(k)$, i.e., it is the k-th Fibonacci number.

(c): The case of balls described by basic word of the form
$w = 3(13)^{n-1}2^m$. In this case there are two parameters: $n > 0$ (gives the number of 3's) and $m \geq 0$. The shape is a hexagon with six straight sides if $n = 1$ (see Fig. 3, with $w = 322$); it is an enneagon with six hilly and three straight sides if $n > 1$, $m = 0$. In all other cases it is a dodecagon with six hilly and six straight sides [13]. By applying $\mathcal{C}$ on w, we can establish the following fact. In each word $w' \in [[w]]$, before the first occurrence of 3 there could be an even number of 1's (maybe with some 2's also), after the last occurrence of letter 3 there could also be only even number of 1's (maybe with some 2's), and between any two occurrences of 3's there could be only odd number of 1's (maybe with some 2's). Let us apply $\mathcal{C}$ on w by moving every 2 to their 'final' positions (with respect to the positions of 3's): then one may obtain a word $w'' = 2^{k_0}312^{k_1}\ldots32^{k_n}$ of the form $2^*3(12^*3)^*2^*$, where $\sum\limits_{i=0}^{n} k_i = m$. Then by applying the rule $11 \leftrightharpoons 2$ (in any directions), the number and positions of 1's and 2's can be obtained in w'. The number of region equivalent words at each maximal subword not containing letter 3 is computed analogously to the number of words in the previous case. Thus, the cardinality of the set $[[w]]$ can be computed by the formula

$$\sum_{\substack{k_0,k_1,\ldots,k_n \\ k_0+\cdots+k_n=m}} F(2k_0)F(2k_n)\prod_{i=1}^{n-1} F(2k_i + 1).$$

(d): The case of balls described by basic words $w = 1(31)^n2^m$. In this case there are two parameters again: $n > 0$ and $m \geq 0$. The shape of the ball is hexagon with six sawtooth sides if $n = 1$ (Fig. 3, $w = 131$); it is an enneagon with six hilly and three sawtooth sides if $n > 1$, $m = 0$; else it is a dodecagon with six hilly and six sawtooth sides [13].

In this case the number of 1's is odd in every maximal subword of w' not containing any letter 3. Consequently, the number of words $w' \in [[w]]$ is

$$\sum_{\substack{k_0,k_1,\ldots,k_n \\ k_0+\cdots+k_n=m}} F(2k_0 + 1)F(2k_n + 1)\prod_{i=1}^{n-1} F(2k_i + 1) = \sum_{\substack{k_0,k_1,\ldots,k_n \\ k_0+\cdots+k_n=m}} \prod_{i=0}^{n} F(2k_i + 1).$$

(e): The case of balls described by basic words $w = (13)^n2^m$. The two parameters are $n > 0$ and $m \geq 0$. The shapes of the balls are similar to the shapes at case (c) with the given values of parameters. The number of words $w' \in [[w]]$ is given by

$$\sum_{\substack{k_0,k_1,\ldots,k_n \\ k_0+\cdots+k_n=m}} F(2k_0 + 1)F(2k_n)\prod_{i=1}^{n-1} F(2k_i + 1) = \sum_{\substack{k_0,k_1,\ldots,k_n \\ k_0+\cdots+k_n=m}} F(2k_n)\prod_{i=0}^{n-1} F(2k_i + 1).$$

(f): The case of balls described by basic words $w = (31)^n2^m$, with parameters $n > 0$ and $m \geq 0$. The shapes of the balls are similar to case (d) with the same parameter values, see, e.g., Fig. 3 $w = 31312$. The number of words $w' \in [[w]]$ is computed as

$$\sum_{\substack{k_0, k_1, \ldots, k_n \\ k_0 + \cdots + k_n = m}} F(2k_0) \prod_{i=1}^{n} F(2k_i + 1).$$

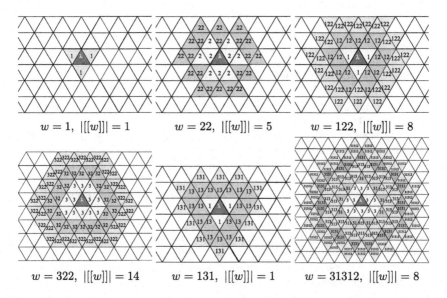

$w = 1,\ |[[w]]| = 1$ $w = 22,\ |[[w]]| = 5$ $w = 122,\ |[[w]]| = 8$

$w = 322,\ |[[w]]| = 14$ $w = 131,\ |[[w]]| = 1$ $w = 31312,\ |[[w]]| = 8$

Fig. 3. Example balls with their basic words w and the cardinality of $[[w]]$.

5   Concluding Remarks

As we have seen instead of the binomial coefficients that are obtained on the square grid (lattice), more complex formulae are required to compute the number of words defining the same ball on the triangular tiling. Thus, these numbers can be considered, as a kind of generalizations of the binomial coefficients using the triangular tiling, and actually, they are the cardinalities of the sets representing generalized traces (defined also by special associative calculus) that describe digital balls on the triangular tiling. In this way an interesting relation between some parallel processes and shortest paths in the triangular tiling is established.

References

1. Das, P.P., Chakrabarti, P.P., Chatterji, B.N.: Distance functions in digital geometry. Inf. Sci. **42**, 113–136 (1987)
2. Deutsch, E.S.: Thinning algorithms on rectangular, hexagonal and triangular arrays. Comm. ACM **15**, 827–837 (1972)

3. Diekert, V., Rozenberg, G. (eds.): The Book of Traces. World Scientific, Singapore (1995)
4. Herendi, T., Nagy, B.: Parallel Approach of Algorithms. Typotex, Budapest (2014)
5. Farkas, J., Baják, S., Nagy, B.: Approximating the Euclidean circle in the square grid using neighbourhood sequences. Pure Math. Appl. **17**, 309–322 (2006)
6. Janicki, R., Kleijn, J., Koutny, M., Mikulski, L.: Generalising traces, CS-TR-1436. Technical report, Newcastle University, Partly Presented at LATA 2015 (2014)
7. Mateescu, A., Rozenberg, G., Salomaa, A.: Shuffle on trajectories: syntactic constraints. Theoret. Comput. Sci. **197**, 1–56 (1998)
8. Mazurkiewicz, A.W.: Trace Theory: Advances in Petri Nets 1986. LNCS, vol. 255, pp. 279–324. Springer, Heidelberg (1987)
9. Nagy, B.: Finding shortest path with neighborhood sequences in triangular grids. In: 2nd IEEE R8-EURASIP International Symposium ISPA, Pula, Croatia, pp. 55–60 (2001)
10. Nagy, B.: Metrics based on neighbourhood sequences in triangular grids. Pure Math. Appl. **13**, 259–274 (2002)
11. Nagy, B.: Shortest path in triangular grids with neighbourhood sequences. J. Comput. Inf. Technol. **11**, 111–122 (2003)
12. Nagy, B.: Distance functions based on neighbourhood sequences. Publicationes Math. Debrecen **63**, 483–493 (2003)
13. Nagy, B.: Characterization of digital circles in triangular grid. Pattern Recogn. Lett. **25**, 1231–1242 (2004)
14. Nagy, B.: Metric and non-metric distances on $\mathbb{Z}^n$ by generalized neighbourhood sequences. In: 4th ISPA: International Symposium on Image and Signal Processing and Analysis, Zagreb, Croatia, pp. 215–220 (2005)
15. Nagy, B.: Distances with neighbourhood sequences in cubic and triangular grids. Pattern Recogn. Lett. **28**, 99–109 (2007)
16. Nagy, B.: Distance with generalized neighbourhood sequences in nD and ∞D. Discrete Appl. Math. **156**, 2344–2351 (2008)
17. Nagy, B., Strand, R.: Approximating Euclidean circles by neighbourhood sequences in a hexagonal grid. Theoret. Comput. Sci. **412**, 1364–1377 (2011)
18. Rosenfeld, A., Pfaltz, J.L.: Sequential operations in digital picture processing. J. ACM **13**, 471–494 (1966)
19. Rosenfeld, A., Pfaltz, J.L.: Distance functions on digital pictures. Pattern Recogn. **1**, 33–61 (1968)
20. Stojmenovic, I.: Honeycomb networks: topological properties and communication algorithms. IEEE Trans. Parallel Distrib. Syst. **8**, 1036–1042 (1997)
21. Yamashita, M., Honda, N.: Distance functions defined by variable neighborhood sequences. Pattern Recogn. **17**, 509–513 (1984)

Generation of Digital Planes Using Generalized Continued-Fractions Algorithms

Damien Jamet[1], Nadia Lafrenière[2], and Xavier Provençal[3]([✉])

[1] LORIA, Université de Lorraine, Nancy, France
[2] LaCIM, Université du Québec à Montréal, Montréal, Canada
[3] LAMA, Université Savoie Mont Blanc, Chambéry, France
xavier.provencal@univ-smb.fr

Abstract. We investigate a construction scheme for digital planes that is guided by generalized continued fractions algorithms. This process generalizes the recursive construction of digital lines to dimension three. Given a pair of numbers, Euclid's algorithm provides a natural definition of continued fractions. In dimension three and above, there is no such canonical definition. We propose a pair of hybrid continued fractions algorithms and show geometrical properties of the digital planes constructed from them.

Keywords: Digital planes generation · GCD algorithms · Generalized continued fractions algorithms · Recursive structure

1 Introduction

Over the last years, the study of digital straight lines has gathered much attention. Their properties have been exposed using different approaches and many applications were deduced from them. In particular, it is well known that a digital line has a recursive structure described by the continued fraction development of its slope. See [15] for a survey on digital straightness.

There has been much effort done in order to find analogous results and applications to those on 2D digital lines for 3D digital planes. See [8] for a survey on digital planarity. Works on dual substitutions [1] showed links between the structure of 3D digital planes and generalized continued fractions which lead, in particular, to generation and recognition techniques [3,12]. In this context, much effort has been directed to the study of the generation of digital planes with totally irrational normal vectors. See, for instance, [14] for a detailed work on this topic.

In [5,11], the authors investigate a process that, given a normal vector, computes the critical thickness and constructs the thinnest digital plane that is

This work has been partly funded by DYNA3S ANR-13-BS02-0003 research grant.
This work has been partly supported by the CNRS "Laboratoire international associé" LIRCO.

© Springer International Publishing Switzerland 2016
N. Normand et al. (Eds.): DGCI 2016, LNCS 9647, pp. 45–56, 2016.
DOI: 10.1007/978-3-319-32360-2_4

connected. This process is completely directed by the execution of the *Fully Subtractive* algorithm on the normal vector. Even if their process computes the critical thickness for any normal vector, the construction, which we call the **FS**-construction, produces a digital plane only for a measure-zero set of vectors which excludes all integer vectors [10]. In the present paper, we propose an attempt to compute a connected digital plane for all vectors $\mathbf{v} \in \mathbb{N}^3$.

In the field of combinatorics on words, the recursive structure of digital lines has been studied, in particular, via *Christoffel words*. Recent work by Labbé and Reutenauer [17] extends Christoffel words as subgraphs of the hypercubic lattice in arbitrary dimensions. The motivation for the present paper is to provide a better understanding of the self-similarities of what is called a *Christoffel parallelogram* in [17]. This is, roughly speaking, the smallest pattern with parallel sides that tiles the digital plane by translation.

We provide a recursive construction scheme for 3D digital planes that is a generic version of the **FS**-construction. It is generic in the sense that it is parametrized by a generalized continued fraction algorithm, noted GCF-algorithm for short. A GCF-algorithm is an extension of Euclid's algorithm to higher dimensions. In [5], an inclusion relation between the result of the **FS**-construction and the construction of digital planes using dual substitutions, more precisely the E_1^* formalism, was used in order to show connectedness in specific cases. The same connection between our construction and E_1^* is expected, but left for future work. We define new *hybrid* GCF-algorithms and show that they allow to build a connected digital plane for any rational normal vector.

2   Basic Notions and Notation

Let $d \geqslant 2$ be an integer. Let $\{\mathbf{e}_i \mid k \in \{1, \ldots, d\}\}$ be the canonical basis of $\mathbb{R}^d$, and let $\langle \cdot, \cdot \rangle$ stand for the usual scalar product on $\mathbb{R}^d$. We note $\mathbf{0}$ the origin and $\mathbf{1} = \sum_{i=1}^d \mathbf{e}_i$ the vector with all coordinates equal to 1. Recall that $\mathbb{N}_+ := \mathbb{N}\setminus\{\mathbf{0}\}$. We always suppose that $\mathbf{v} \in (\mathbb{N}_+)^d$ and that the coordinates of $\mathbf{v}$ are relatively prime. The 1-norm of $\mathbf{x} \in \mathbb{R}^d$, noted $\|\mathbf{x}\|_1$, is given by the sum of the absolute values of its coordinates.

Definition 2.1. *The* digital hyperplane $\mathcal{P}(\mathbf{v}, \omega)$ *with* normal vector $\mathbf{v} \in \mathbb{R}^d$ *and* thickness $\omega \in \mathbb{R}$ *is defined by:*

$$\mathcal{P}(\mathbf{v}, \omega) = \left\{ \mathbf{x} \in \mathbb{Z}^d \mid 0 \leqslant \langle \mathbf{x}, \mathbf{v} \rangle < \omega \right\}.$$

Note that the usual definition of digital hyperplanes includes a *shift* parameter μ which we decide to omit in order to lighten the notation. Moreover, we only consider digital lines ($d = 2$) and planes ($d = 3$). For any point $\mathbf{x} \in \mathbb{Z}^d$, the quantity $\langle \mathbf{x}, \mathbf{v} \rangle$ determines if $\mathbf{x}$ belongs to $\mathcal{P}(\mathbf{v}, \omega)$ or not. We call $\langle \mathbf{x}, \mathbf{v} \rangle$ the *height* of $\mathbf{x}$ (in $\mathcal{P}(\mathbf{v}, \omega)$) and note it by $\overline{\mathbf{x}}$. See Fig. 1 for an illustration.

Two points at the same height define a *period vector*, so that $\mathcal{P}(\mathbf{v}, \omega)$ is invariant under a translation by that vector. Given $(\mathbf{b}_i)_{i \in \{1, \ldots, d-1\}}$ a basis of the

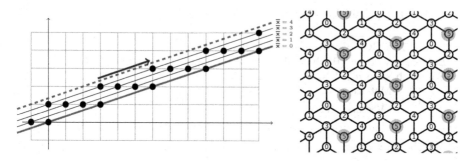

Fig. 1. The inner structure of digital lines and planes. Left: the digital line $\mathcal{P}((-1,3),4)$. The line $\bar{\mathbf{x}} = 0$ contains the lowest points, while the line $\bar{\mathbf{x}} = 3$ contains the highest points. Right: the digital plane $\mathcal{P}((1,2,3),6)$. The numbers indicate the height of each points. The highest points are highlighted showing that they form a regular lattice. In both cases, two points at same height define a period vector.

lattice $\{\mathbf{x} \in \mathbb{Z}^d \mid \bar{\mathbf{x}} = 0\}$ and a set $S \subset \mathbb{Z}^d$ such that for each integer $h \in \mathbb{Z}$, $\{\mathbf{x} \in S \mid \bar{\mathbf{x}} = h\} \neq \emptyset$ if and only if $h \in [0, \omega)$, then

$$\mathcal{P}(\mathbf{v}, \omega) = \bigcup_{\mathbf{x} \in \mathbb{Z}\mathbf{b}_1 + \cdots + \mathbb{Z}\mathbf{b}_{d-1}} S + \mathbf{x},$$

and we say that S provided with the vectors $(\mathbf{b}_i)$ *spans* $\mathcal{P}(\mathbf{v}, \omega)$.

Two points $\mathbf{x}, \mathbf{y} \in \mathbb{Z}^d$ are said to be *adjacent* if $\|\mathbf{x} - \mathbf{y}\|_1 = 1$, which means that all their coordinates are equal except for one that differs by 1. By analogy to graph theory, we say that a subset of $\mathbb{Z}^d$ is *connected* if its adjacency graph is connected. In particular, a digital plane $\mathcal{P}(\mathbf{v}, \omega)$ is always disconnected if $\omega < \min(\mathbf{v})$ and it is always connected if $\omega > \max(\mathbf{v}) + \min\{\mathbf{v}_i \mid \mathbf{v}_i \neq 0\}$ (see [6], Lemma 5.3).

3   Construction Guided by Continued Fractions

In its additive form, Euclid's algorithm can be expressed as *"given a couple of integers (a, b), subtract the smaller to the larger one, and repeat. When both numbers are equal, their value is the gcd"*. It can be expressed in a matricial expression as follows:

$$\mathbf{Euclid}\begin{pmatrix} a \\ b \end{pmatrix} = \left\{ \begin{bmatrix} 1 & 0 \\ -1 & 1 \end{bmatrix} \text{ if } b > a, \\ \begin{bmatrix} 1 & -1 \\ 0 & 1 \end{bmatrix} \text{ otherwise,} \right.$$

so that multiplying vector $\binom{a}{b}$ by the output matrix performs one step of the algorithm. Table 1 presents the general construction scheme of digital lines and planes. It requires two inputs, namely a GCF-algorithm and a normal vector. First, we focus on dimension 2. In this case, there is a canonical GCF-algorithm: **Euclid**. The construction is simply a geometrical reinterpretation of the combinatorial construction of Christoffel words by the *Christoffel tree*,

Table 1. Recursive construction of a digital lines ($d = 2$) and plane ($d = 3$).

Input: $\mathbf{X}$ a GCF-algorithm, $\mathbf{v}$ the normal vector
Let: $\mathbf{v}_0 = \mathbf{v}$, $B_0 = \{\mathbf{0}\}$, $\mathbf{h}_0 = \mathbf{0}$, $L_0 = \{\mathbf{e}_i \mid i \in \{1, \ldots, d\}\}$ and $\mathbf{a}_0 = \mathbf{1}$.
For each step $n \geqslant 1$ **let**

Matrices:	$M_n = \mathbf{X}(\mathbf{v}_{n-1})$,
Vectors:	$\mathbf{v}_n = M_n \mathbf{v}_{n-1}$,
Indexes:	$\delta_n = $ index of the coordinate of $\mathbf{v}_{n-1}$ that is subtracted,
Translation vectors:	$\mathbf{t}_n = M_1^\top M_2^\top \cdots M_n^\top \mathbf{e}_{\delta_n}$,
Bodies:	$B_n = B_{n-1} \cup (B_{n-1} + \mathbf{t}_n)$,
Highest points:	$\mathbf{h}_n = \mathbf{t}_1 + \mathbf{t}_2 + \cdots + \mathbf{t}_n$,
Legs:	$L_n = \{\mathbf{h}_n + M_1^\top M_2^\top \cdots M_n^\top \mathbf{e}_i \mid i \in \{1, \ldots, d\}\}$.
Approximations:	$\mathbf{a}_n = M_1^{-1} M_2^{-1} \cdots M_n^{-1} \mathbf{1}$,

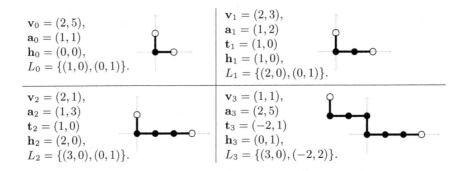

$\mathbf{v}_0 = (2, 5)$,
$\mathbf{a}_0 = (1, 1)$
$\mathbf{h}_0 = (0, 0)$,
$L_0 = \{(1, 0), (0, 1)\}$.

$\mathbf{v}_1 = (2, 3)$,
$\mathbf{a}_1 = (1, 2)$
$\mathbf{t}_1 = (1, 0)$
$\mathbf{h}_1 = (1, 0)$,
$L_1 = \{(2, 0), (0, 1)\}$.

$\mathbf{v}_2 = (2, 1)$,
$\mathbf{a}_2 = (1, 3)$
$\mathbf{t}_2 = (1, 0)$
$\mathbf{h}_2 = (2, 0)$,
$L_2 = \{(3, 0), (0, 1)\}$.

$\mathbf{v}_3 = (1, 1)$,
$\mathbf{a}_3 = (2, 5)$
$\mathbf{t}_3 = (-2, 1)$
$\mathbf{h}_3 = (0, 1)$,
$L_3 = \{(3, 0), (-2, 2)\}$.

Fig. 2. Example of the construction scheme using Euclid's algorithm and the input vector $\mathbf{v} = (2, 5)$. By convention, $\bullet \in B_n$ and $\circ \in L_n$. At step 3, $\mathbf{v}_3 = \mathbf{1}$ and thus $\mathbf{a}_3 = \mathbf{v}$ and $B_3 \cup L_3$ forms the main pattern of $\mathcal{P}((2, 5), 7)$.

see [2] (Chap. 1, Sect. 7). The result is a set of points called *pattern*. By analogy with the terminology used in [17], we say that a pattern is made of a *body*, noted B_n, and *legs*, noted

L_n. Euclid's algorithm ensures that there exists N such that $\mathbf{v}_N = (1, 1)$. The *approximations* ($\mathbf{a}_n$) correspond to the values encountered while going down in the Christoffel tree (or equivalently the Stern-Brocot tree), starting from the root $\mathbf{a}_0 = (1, 1)$ and ending after N steps at $\mathbf{a}_N = \mathbf{v}_0$. See Fig. 2 for an example.

The main difficulty in order to use this construction in dimension three is that there is no canonical extension of Euclid's algorithm. Instead, there exists a wide variety of GCF-algorithms (see for instance [18] or [16]). Among all GCF-algorithms, we only focus on the three following ones in the scope of this paper. Given three non-negative numbers,

– Selmer, noted **S** : subtract the smallest value to the largest one.
– Brun, noted **B** : subtract the middle value to the largest one.
– Fully Subtractive, noted **FS** : subtract the smallest value to the two others.

Obviously, the action of each of these GCF-algorithms may be expressed by a matrix product. Moreover, with these three algorithms, designating the index of the entry that is subtracted is unambiguous. Here are some basic properties of the construction given by Table 1 with $d = 3$.

Property 3.1. Let **X** be a GCF-algorithm such that each step is either **S**, **B** or **FS**. For each $n \geqslant 0$,

(1) $\mathbf{v}_n = M_n \cdots M_1 \mathbf{v}$.
(2) if $\mathbf{v}_n = \mathbf{1}$, then $\mathbf{a}_n = \mathbf{v}$.
(3) $B_n = \{\sum_{i \in I} \mathbf{t}_i \mid I \subseteq \{1, \ldots, n\}\}$.
(4) the coordinates of $\mathbf{v}_n$ are non-negative.
(5) for each $i \in \{1, 2, 3\}$, the height of $M_1^\top \cdots M_n^\top \mathbf{e}_i$ is equal to the i-th coordinate of $\mathbf{v}_n$.
(6) $\bar{\mathbf{t}}_n$, the height of $\mathbf{t}_n$, is the value of the coordinate of $\mathbf{v}_{n-1}$ that is subtracted to some other coordinate(s) by M_n, in order to compute $\mathbf{v}_n$.
(7) for each $i \in \{1, 2, 3\}$, let $\mathbf{x} = M_1^\top \cdots M_n^\top \mathbf{e}_i$. The vector $\mathbf{x}$ is a Bezout vector for $\mathbf{a}_n$, that is $\langle \mathbf{x}, \mathbf{a}_n \rangle = 1$.

Proof. Properties (1), (2) and (3) are straightforward. Property (4) is deduced from the definition of the continued fraction algorithm. For (5), $\langle M_1^\top \cdots M_n^\top \mathbf{e}_i, \mathbf{v} \rangle = \langle \mathbf{e}_i, M_n \cdots M_1 \mathbf{v} \rangle = \langle \mathbf{e}_i, \mathbf{v}_n \rangle$. For (6), the value of the coordinate of $\mathbf{v}_{n-1}$ that is subtracted is not modified by the action of M_n, in other words $\langle \mathbf{v}_{n-1}, \mathbf{e}_{\delta_n} \rangle = \langle \mathbf{v}_n, \mathbf{e}_{\delta_n} \rangle$. Finally, for (7), $\langle M_1^\top \cdots M_n^\top \mathbf{e}_i, \mathbf{a}_n \rangle = \langle \mathbf{e}_i, \mathbf{1} \rangle = 1$.   □

Obviously, geometric properties of the objects generated depend on the choice of the GCF-algorithm. Here are some properties that are expected from the generated patterns:

– They must be included in the digital plane $\mathcal{P}(\mathbf{v}, \|\mathbf{v}\|_1)$.
– They should form a connected set of points.
– Period vectors should be deduced from them.
– They should contain points at every height.
– They should be as small as possible, to avoid redundancy.

3.1   General Properties of the Construction

Even though the construction scheme from Table 1 produces infinite sequences, we only consider a finite number of steps. In dimension two, Euclid's algorithm always brings vector **v** to **1** (recall that **v** is assumed to have relatively prime coordinates). This is the halt condition for the algorithm. In dimension three, a GCF-algorithm may produce a sequence $(\mathbf{v}_n)_{n \geqslant 0}$ such that **1** does not appear in it. For instance, with the **FS** GCF-algorithm, if there exists n such that $\mathbf{v}_n = (1, 1, 2)$, then $\mathbf{v}_{n+1} = (1, 0, 2)$. Obviously, $\mathbf{v}_{n'} \neq \mathbf{1}$ for all $n' \geqslant n$. For now, we focus on the case where **1** does appear in $(\mathbf{v}_n)_{n \geqslant 0}$.

Definition 3.2. *Let* $\mathbf{X}$ *be a GCF-algorithm and* $\mathbf{v}$ *a vector with relatively prime coordinates. The* length, *noted* $\mathrm{length}_{\mathbf{X}}(\mathbf{v})$, *is, if it exists, the smallest integer* $N \geqslant 0$ *such that* $\mathbf{v}_N = \mathbf{1}$.

In Sect. 4.1, we provide two new GCF-algorithms and show (Lemma 4.1) that, if $\mathbf{X}$ is one of these, then $\mathrm{length}_{\mathbf{X}}(\mathbf{v})$ exists. Under such assumption, the points of the body are always included in a digital plane of normal vector $\mathbf{v}$.

Proposition 3.3. *Let* $\mathbf{X}$ *be a GCF-algorithm such that each step is either* **S**, **B** *or* **FS** *and* $N = \mathrm{length}_{\mathbf{X}}(\mathbf{v})$ *exists. For each* $n \in \{0, 1, \ldots, N\}$, $B_n \subseteq \mathcal{P}(\mathbf{v}, \|\mathbf{v}\|_1 - 2)$.

Proof. For each $n \geqslant 1$, let g_n be the value of the coordinate of $\mathbf{v}_{n-1}$ that is subtracted. From Property 3.1 (6), this value is equal to the height of $\mathbf{t}_n$. From Property 3.1 (3), given a point $\mathbf{x} \in B_n$, there exists a subset $I \subset \{1, \ldots, n\}$ such that $\mathbf{x} = \sum_{i \in I} \mathbf{t}_i$ and, by linearity, $\overline{\mathbf{x}} = \sum_{i \in I} g_i$. On one hand, each g_i being non-negative, $0 \leqslant \overline{\mathbf{x}}$. On the other hand, since each g_i is subtracted to one or two coordinates of $\overline{\mathbf{v}}_{n-1}$ while keeping its coordinates non-negative, we have that $\sum_{i \in I} g_i \leqslant \|\mathbf{v}_0\|_1 - \|\mathbf{v}_N\|_1$. Thus, $\mathbf{x} \in \mathcal{P}(\mathbf{v}, \|\mathbf{v}\|_1 - 2)$. ☐

The following proposition states that the legs define linearly independent period vectors. In Sect. 4.2, we show that these vectors may be used with the corresponding pattern in order to span a digital plane.

Proposition 3.4. *Let* $\mathbf{X}$ *be a GCF-algorithm such that each step is either* **S**, **B** *or* **FS** *and* $N = \mathrm{length}_{\mathbf{X}}(\mathbf{v})$ *exists. For each* $n \in \{0, 1, \ldots, N\}$, *let* $\{\mathbf{l}_1, \mathbf{l}_2, \mathbf{l}_3\} = L_n$ *(see Table 1). The differences* $(\mathbf{l}_2 - \mathbf{l}_1, \mathbf{l}_3 - \mathbf{l}_1)$ *form a basis of the lattice* $\{\mathbf{x} \in \mathbb{Z}^3 \mid \langle \mathbf{x}, \mathbf{a}_n \rangle = 0\}$.

Proof (Sketch). By Property 3.1 (7), we have $\langle \mathbf{l}_1, \mathbf{a}_n \rangle = \langle \mathbf{l}_2, \mathbf{a}_n \rangle = \langle \mathbf{l}_3, \mathbf{a}_n \rangle$, which implies that their differences are in the lattice $\{\mathbf{x} \in \mathbb{Z}^3 \mid \overline{\mathbf{x}} = 0\}$. The fact that $(\mathbf{l}_2 - \mathbf{l}_1, \mathbf{l}_3 - \mathbf{l}_1)$ form a basis of the lattice is deduced from the fact that the matrix $M_1^\top \cdots M_n^\top$ is unimodular, since it is a product of unimodular matrices. ☐

4   The Choice of a GCF-Algorithm

In order to build a digital plane with normal vector $\mathbf{v}$, one should use a GCF-algorithm that reduces $\mathbf{v}$ to $\mathbf{1}$. Indeed, if $\mathbf{v}_N = \mathbf{1}$, then, by Property 3.1 (2), $\mathbf{a}_N = \mathbf{v}$ and Proposition 3.4 states that the legs L_N define linearly independent period vectors of $\mathcal{P}(\mathbf{v}, \omega)$.

The **FS** GCF-algorithm appeared naturally in the study of the topological properties of digital planes [9]. Subsequent work [4,5,10,11] showed that the geometry of a digital plane is strongly linked to the execution of **FS** on its normal vector.

Given a vector $\mathbf{v}$ for which $N = \mathrm{length}_{\mathbf{FS}}(\mathbf{v})$ is defined, a point $l \in L_N$ is such that $\overline{l} = \lfloor \frac{\|\mathbf{v}\|_1}{2} \rfloor$, the set B_N is connected and contains exactly one point at each height from 0 to $\lfloor \frac{\|\mathbf{v}\|_1}{2} \rfloor - 1$. The set $B_N \cup \{l\}$ is a minimal pattern that not only spans $\mathcal{P}(\mathbf{v}, \frac{\|\mathbf{v}\|_1}{2})$, but *tiles* the digital plane, since the points of the translated copies of the pattern never overlap. That being said, for most of the normal vectors $\mathbf{v}$, **FS** *"fails"*, which means that length$_{\mathbf{FS}}(\mathbf{v})$ is not defined [13]. For instance, in dimension 3, there are two cases where this happens. The first one is when dealing with a vector $\mathbf{v}$ such that a vector $\mathbf{u}$ of the form $(a, b, a + b + c)$, with $a, b \geqslant 1$, $c \geqslant 0$, appears in the sequence $(\mathbf{v}_n)_{n \geqslant 0}$.

In other words, when one coordinate is bigger or equal to the sum of the two others at some point of the execution of **FS**. Iterating the **FS** algorithm on $\mathbf{u}$ is equivalent to running Euclid's algorithm on (a, b) since the z-coordinate always remains the biggest one. This means that the value of c has no influence on the points computed in the sequence $(B_n)_{n \geqslant 0}$, and thus the geometry of the digital plane $\mathcal{P}(\mathbf{v}, \omega)$ cannot be described by the patterns generated. Another type of vectors which **FS** does not reduce to **1** is those such that a vector of the form (a, a, b) with $a < b$ appears in the sequence $(\mathbf{v}_n)_{n \geqslant 0}$. In such case, the action of **FS** produces a vector of the form $(a, 0, b - a)$, which is a fixed point for **FS**.

4.1 Hybrid GCF-Algorithms

The idea of mixing two different GCF-algorithms appears in [7], where the authors investigate the generation of infinite words by iteration of morphisms as approximations of digital lines. We now define two new GCF-algorithms. The main idea is to emulate **FS** as much as possible, except for the cases where it "fails" and, in such case, use **S** (Selmer) or **B** (Brun) instead.

Table 2 details the **FSS** (Fully Subtractive Selmer) and **FSB** (Fully Subtractive Brun) GCF-algorithms with the matrices provided in Table 3.

The scheme of the **FSS** (resp. **FSB**) GCF-algorithm is *"if the largest coordinate is greater or equal to the sum of the two smallest ones or if the two smallest ones are equal, apply the S (resp. B) reduction; otherwise, apply the FS reduction"* (Fig. 3).

Unlike what happens with the **FS**, **S** and **B** GCF-algorithms, length$_{\mathbf{FSS}}(\mathbf{v})$ and length$_{\mathbf{FSB}}(\mathbf{v})$ always exist.

Table 2. The **FSS** and **FSB** GCF-algorithms.

Algorithm 1. Fully Subtractive Selmer	Algorithm 2. Fully Subtractive Brun
1: **function** **FSS** (**v**)	1: **function** **FSB** (**v**)
2: **Input:** $\mathbf{v} \in \mathbb{N}^3$	2: **Input:** $\mathbf{v} \in \mathbb{N}^3$
3: $(a, b, c) \leftarrow \text{SORTED}(\mathbf{v})$	3: $(a, b, c) \leftarrow \text{SORTED}(\mathbf{v})$
4: $\sigma_{\mathbf{v}} \leftarrow \min\{\sigma \in S_3 \mid \sigma(\mathbf{v}) = (a, b, c)\}$	4: $\sigma_{\mathbf{v}} \leftarrow \min\{\sigma \in S_3 \mid \sigma(\mathbf{v}) = (a, b, c)\}$
5: **if** $c \geqslant a + b$ or $a == b$ **then**	5: **if** $c \geqslant a + b$ or $a == b$ **then**
6: **return** $\mathbf{M}_\sigma^{\mathbf{S}}$	6: **return** $\mathbf{M}_\sigma^{\mathbf{B}}$
7: **else**	7: **else**
8: **return** $\mathbf{M}_\sigma^{\mathbf{FS}}$	8: **return** $\mathbf{M}_\sigma^{\mathbf{FS}}$
9: **end if**	9: **end if**
10: **end function**	10: **end function**

Table 3. The matrices for **FSS** and **FSB** GCF-algorithms. Each matrix is indexed by a permutation that indicates the relative order of the coordinates of the vector. We display only the matrices for the permutation 123, which corresponds to a vector (a, b, c) with $a \leqslant b \leqslant c$. Matrices for other permutations can be deduced from these.

$$\mathbf{M}_{123}^{\mathrm{FS}} = \begin{pmatrix} 1 & 0 & 0 \\ -1 & 1 & 0 \\ -1 & 0 & 1 \end{pmatrix}, \mathbf{M}_{123}^{\mathrm{S}} = \begin{pmatrix} 1 & 0 & 0 \\ 0 & 1 & 0 \\ -1 & 0 & 1 \end{pmatrix}, \mathbf{M}_{123}^{\mathrm{B}} = \begin{pmatrix} 1 & 0 & 0 \\ 0 & 1 & 0 \\ 0 & -1 & 1 \end{pmatrix}.$$

$\mathbf{v}_0 = (4, 5, 6),$
$\mathbf{a}_0 = (1, 1, 1),$
$\mathbf{h}_0 = (0, 0, 0),$
$L_0 = \{(1, 0, 0), (0, 1, 0), (0, 0, 1)\}.$

$M_1 = \mathbf{M}_{123}^{\mathrm{FS}},$
$\mathbf{v}_1 = (4, 1, 2),$
$\mathbf{a}_1 = (1, 2, 2)$
$\mathbf{t}_1 = (1, 0, 0),$
$\mathbf{h}_1 = (1, 0, 0),$
$L_1 = \{(2, 0, 0), (0, 1, 0), (0, 0, 1)\}.$

$M_2 = \mathbf{M}_{231}^{\mathrm{B}}$
$\mathbf{v}_2 = (2, 1, 2),$
$\mathbf{a}_2 = (2, 3, 3)$
$\mathbf{t}_2 = (-1, 1, 0),$
$\mathbf{h}_2 = (0, 0, 1),$
$L_2 = \{(2, 0, 0), (-1, 1, 1), (-1, 0, 2)\}.$

$M_3 = \mathbf{M}_{213}^{\mathrm{FS}},$
$\mathbf{v}_3 = (1, 1, 1),$
$\mathbf{a}_3 = (4, 5, 6)$
$\mathbf{t}_3 = (-1, 1, 0),$
$\mathbf{h}_3 = (-1, 1, 1),$
$L_3 = \{(2, 0, 0), (-2, 2, 1), (-1, 0, 2)\}.$

Fig. 3. Example of the construction scheme using the **FSB** GCF-algorithm and normal vector $\mathbf{v} = (4, 5, 6)$. At step 3, $\mathbf{v}_3 = \mathbf{1}$ and thus $\mathbf{a}_3 = \mathbf{v}$. For each point $\mathbf{x} \in L_3$, $\overline{\mathbf{x}} = 8$ and the body has a point at each height from 0 to 7. The spanning $\{(B_3 \cup L_3) + k(-4, 2, 1) + l(-3, 0, 2) \mid k, l \in \mathbb{Z}\}$ is the digital plane $\mathcal{P}((4, 5, 6), 9)$.

Lemma 4.1. Let $\mathbf{v} \in \mathbb{N}_+^3$, be a vector with relatively prime coordinates, then $\mathrm{length}_{\mathbf{FSS}}(\mathbf{v})$ and $\mathrm{length}_{\mathbf{FSB}}(\mathbf{v})$ both exist.

Proof. With both algorithms, $(\|\mathbf{v}_n\|_1)_{n \geqslant 0}$ forms a decreasing integer sequence. More precisely, this sequence is strictly decreasing as long as none of the coordinates is equal to zero, which forces that, inevitably, one coordinate must reach zero. Let N be the smallest integer such that $\min(\mathbf{v}_{N+1}) = 0$. For short, let $(a, b, c) = \mathbf{v}_N$ and, w.l.o.g. assume that $a \leqslant b \leqslant c$. Let $M = \mathbf{FSS}(\mathbf{v}_N)$ (resp. $\mathbf{FSB}(\mathbf{v}_N)$). There are three possibilities for M:

- $M = \mathbf{M}_1^{\mathrm{FS}}$, this case is impossible, since both **FSS** and **FSB** require that $\mathbf{v}_N$ is such that $a < b$ and $a < c$, so that $\mathbf{v}_{N+1} = (a, b - a, c - a)$ has no coordinate equal to zero.
- $M = \mathbf{M}_{123}^{\mathrm{S}}$, in such case, $\mathbf{v}_{N+1} = (a, b, c - a)$, so that $a = b = c$.
- $M = \mathbf{M}_{123}^{\mathrm{B}}$, in such case, $\mathbf{v}_{N+1} = (a, b, c - b)$, so that $b = c$. Moreover, $a = b$ since, otherwise, $a + b > c$, $a \neq b$ and $\mathbf{FSB}(\mathbf{v}_N)$ returns $\mathbf{M}_1^{\mathrm{FS}}$.

The uniqueness of N is obvious since $\mathbf{v}_N = \mathbf{1}$ implies $\min(\mathbf{v}_{N+1}) = 0$ and each coordinate is non-increasing. $\qquad\square$

4.2   Generating a Digital Plane

We now show the main geometrical properties of our construction. First, we show that the points of the patterns B_n and $B_n \cup L_n$ are connected. Since we already know (see Proposition 3.4) that, at the last step, the legs define period vectors that send points of the legs to other points of the legs, the spanning of a pattern by these period vectors is a connected set. Then, we show that B_N, for $N = \mathrm{length}_{\mathbf{FSS}}(\mathbf{v})$ (resp. **FSB**), contains at least one point at each height, which ensures that the spanning is a digital plane.

Theorem 4.2. *Let* $\mathbf{v} \in \mathbb{N}_+^3$ *be a vector with relatively prime coordinates and let* $N = \mathrm{length}_{\mathbf{FSS}}(\mathbf{v})$ *(resp.* **FSB***). For each* $n \in \{0, 1, \dots, N\}$*, the sets* B_n *and* L_n *are disjoint, and both* B_n *and* $B_n \cup L_n$ *are connected.*

Proof. Let $\mathbf{x} \in L_n$. By definition, there exists $i \in \{1, 2, 3\}$ such that $\mathbf{x} = \mathbf{h}_n + M_1^\top \cdots M_n^\top \mathbf{e}_i$. The height of $M_1^\top \cdots M_n^\top \mathbf{e}_i$ is strictly greater than zero since it is equal to the value of the i-th coordinate of $\mathbf{v}_n$. This implies that $\mathbf{x} \notin B_n$ since $\overline{\mathbf{x}} > \overline{\mathbf{h}}_n$ and $\mathbf{h}_n$ is the highest point of B_n.

In order to show that $B_n \cup L_n$ is connected, it suffices to see that, for all $n \geqslant 0$ and all $i \in \{1, 2, 3\}$,

$$\mathbf{h}_n + M_1^\top \cdots M_n^\top \mathbf{e}_i - \mathbf{e}_i \in B_n. \tag{1}$$

Indeed, Eq. (1) implies that the points of L_n are adjacent to points of B_n. It also implies that, at each step of the recursive construction, the set $B_{n-1} + \mathbf{t}_n$ contains a point, namely $\mathbf{h}_{n-1} + M_1^\top \cdots M_n^\top \mathbf{e}_{\delta_n}$, that is adjacent to a point of B_{n-1}, so that the connectedness of B_{n-1} implies the one of B_n.

By recurrence, for $n = 0$, $\mathbf{h}_0 + \mathbf{e}_i - \mathbf{e}_i = \mathbf{0} \in B_0$. Now, suppose that (1) is true for n, and we show the result for $n + 1$. There are two cases to consider.

- If the action of M_{n+1} does not modify the i-th coordinate of $\mathbf{v}_n$, that is $M_{n+1}^\top \mathbf{e}_i = \mathbf{e}_i$, then

$$\mathbf{h}_{n+1} + M_1^\top \cdots M_{n+1}^\top \mathbf{e}_i - \mathbf{e}_i = \underbrace{\mathbf{h}_n + M_1^\top \cdots M_n^\top \mathbf{e}_i - \mathbf{e}_i}_{\in B_n} + \mathbf{t}_{n+1} \in B_{n+1}.$$

- Otherwise, the action of M_{n+1} on the i-th coordinate of $\mathbf{v}_n$ is to subtract to it the δ_{n+1}-th coordinate, which implies that $M_{n+1}^\top \mathbf{e}_i = \mathbf{e}_i - \mathbf{e}_{\delta_{n+1}}$ and thus,

$$\begin{aligned}
&\mathbf{h}_{n+1} + M_1^\top \cdots M_{n+1}^\top \mathbf{e}_i - \mathbf{e}_i \\
&= \mathbf{h}_{n+1} + M_1^\top \cdots M_n^\top \mathbf{e}_i - M_1^\top \cdots M_n^\top \mathbf{e}_{\delta_{n+1}} - \mathbf{e}_i \\
&= \underbrace{\mathbf{h}_n + M_1^\top \cdots M_n^\top \mathbf{e}_i}_{\in B_n} - \underbrace{M_1^\top \cdots M_{n+1}^\top \mathbf{e}_{\delta_{n+1}}}_{= \mathbf{t}_{n+1}} + \mathbf{t}_{n+1}.
\end{aligned}$$

For the last equality, note that since M_{n+1} does not modify the δ_{n+1}-th coordinate of $\mathbf{v}_n$, we have $e_{\delta_{n+1}} = M_{n+1}^\top e_{\delta_{n+1}}$. $\qquad\square$

The second important result of the present work claims that the **FSS** and **FSB** GCF-algorithms construct an entire connected digital plane, using translations.

Theorem 4.3. *Let* $\mathbf{v} \in \mathbb{N}_+^3$, *be a vector with relatively prime coordinates, let* $N = \mathrm{length}_{\mathbf{FSS}}(\mathbf{v})$ *(resp.* $\mathrm{length}_{\mathbf{FSB}}(\mathbf{v})$*). For each* $n \in \{0, 1, \ldots, N\}$, *the set* $\{\langle \mathbf{x}, \mathbf{a}_n \rangle \mid \mathbf{x} \in B_n \cup L_n\}$ *is an integer interval.*

The proof of Theorem 4.3 relies on a technical lemma, for which we define the following notation: given a finite set $S \subset \mathbb{N}$, let $\mathrm{SUM}(S) = \sum_{s \in S} s$ be the sum of its elements and $\mathrm{PSUM}(S) = \{\sum_{x \in X} x \mid X \subseteq S\}$ be the set of its partial sums.

Lemma 4.4. *Let* $x \in \mathbb{N}$ *and* $S \subset \mathbb{N}$ *be such that* $\mathrm{PSUM}(S)$ *is an integer interval. If* $x \leqslant \mathrm{SUM}(S) + 1$, *then* $\mathrm{PSUM}(S \cup \{x\})$ *is an integer interval.*

Proof. Let $a \in \{0, 1, \ldots, \mathrm{SUM}(S) + x\}$. If $a \leqslant \mathrm{SUM}(S)$, then $a \in \mathrm{PSUM}(S) \subseteq \mathrm{PSUM}(S \cup \{x\})$. Otherwise, $0 \leqslant a - x \leqslant \mathrm{SUM}(S)$ and $a - x \in \mathrm{PSUM}(S)$. $\square$

Proof (of Theorem 4.3 – Sketch). The proof is given for **FSS** algorithm, the case **FSB** being similar. Let,

$$\mathbb{T}_{n,k} = \{\mathbf{t}_{n-k}, \ldots, \mathbf{t}_n\}, \qquad \overline{\overline{\mathbb{T}}}_{n,k} = \{\langle \mathbf{x}, \mathbf{a}_n \rangle \mid \mathbf{x} \in \mathbb{T}_{n,k}\}.$$

Hence $\overline{\overline{\mathbb{T}}}_{n,0} = \{\mathbf{0}\}$ and its partial sums is an integer interval. Let us suppose that the partial sums of the set $\overline{\overline{\mathbb{T}}}_{n,k}$, for some $k \in \{0, \ldots, n-1\}$, form an integer interval and let us show that the partial sums of the set $\overline{\overline{\mathbb{T}}}_{n,k+1}$ form an integer interval too, using

$$\mathbb{T}_{n,k+1} = \mathbb{T}_{n,k} \cup \{\mathbf{t}_{n-k-1}\}, \qquad \overline{\overline{\mathbb{T}}}_{n,k+1} = \overline{\overline{\mathbb{T}}}_{n,k} \cup \{\langle \mathbf{t}_{n-k-1}, \mathbf{a}_n \rangle\}.$$

According to Lemma 4.4, it is sufficient to show:

$$\langle \mathbf{t}_{n-k-1}, \mathbf{a}_n \rangle \leqslant 1 + \sum_{x \in \overline{\overline{\mathbb{T}}}_{n,k}} x. \tag{2}$$

Remind that $\mathbf{a}_n = M_1^{-1} M_2^{-1} \cdots M_n^{-1} \cdot \mathbf{1}$, or, equivalently, $\mathbf{1} = M_n \ldots M_1 \cdot \mathbf{a}_n$. Let $\mathbf{a}_{n,0} = \mathbf{a}_n$ and $\mathbf{a}_{n,i+1} = M_i \cdot \mathbf{a}_{n,i}$, with $i \in \{0, \ldots, n-1\}$. The *action* of M_i consists in subtracting $\langle \mathbf{t}_i, \mathbf{a}_n \rangle$ to at least one coordinate of $\mathbf{a}_{n,i-1}$ to compute $\mathbf{a}_{n,i}$, and $\langle \mathbf{t}_i, \mathbf{a}_n \rangle$ is the minimal coordinate of $\mathbf{a}_{n,i-1}$. Since $\mathbf{a}_{n,n} = \mathbf{1}$, there exists a subset $J \subseteq \{i+1, \ldots, n\}$ such that $1 = \langle \mathbf{t}_i, \mathbf{a}_n \rangle - \sum_{j \in J} \langle \mathbf{t}_i, \mathbf{a}_n \rangle$, that is:

$$\langle \mathbf{t}_i, \mathbf{a}_n \rangle = 1 + \sum_{j \in J} \langle \mathbf{t}_i, \mathbf{a}_n \rangle \leqslant 1 + \sum_{j=i+1}^{n} \langle \mathbf{t}_i, \mathbf{a}_n \rangle.$$

Setting $i = n - k - 1$, condition (2) holds. $\square$

Now, it becomes natural to investigate the thickness of the digital plane generated.

Theorem 4.5. *Let* $\mathbf{v} \in \mathbb{N}_+^3$, *be a vector with relatively prime coordinates, let* $\mathbf{X}$ *be either* **FSS** *or* **FSB** *and let* $N = \text{length}_{\mathbf{X}}(\mathbf{v})$. *The thickness is bounded:*

$$\left\lfloor \frac{\|\mathbf{v}\|_1}{2} \right\rfloor \leqslant \max\left\{ \langle \mathbf{x}, \mathbf{v} \rangle \mid \mathbf{x} \in B_N \cup L_N \right\} \leqslant \|\mathbf{v}\|_1 - 2.$$

Proof. The highest point of B_N is, by construction, $h_N = \mathbf{t}_1 + \mathbf{t}_2 + \cdots + \mathbf{t}_N$. Also,

$$\max\left\{ \langle \mathbf{x}, \mathbf{v} \rangle \mid \mathbf{x} \in B_N \right\} = \sum_{i=1}^N \overline{\mathbf{t}}_i,$$

which is, by Property 3.1 (6), the sum of the values subtracted at each step of the execution of the algorithm.

Using the definition of the algorithms, Selmer and Brun subtract each time this value to exactly one coordinate. In such case, $\|\mathbf{v}_n\|_1 = \|\mathbf{v}_{n-1}\|_1 - \overline{\mathbf{t}}_{n-1}$. For Fully Subtractive algorithm, one subtracts at each step the latter value to two different coordinates. In such case, $\|\mathbf{v}_n\|_1 = \|\mathbf{v}_{n-1}\|_1 - 2\overline{\mathbf{t}}_{n-1}$.

This means that if every step of the reduction is Selmer or Brun, then $\|\mathbf{v}\|_1 = 3 + \sum_{i=1}^N \overline{\mathbf{t}}_i$. On the other hand, if the reduction only uses the Fully Subtractive matrices, then $\|\mathbf{v}\|_1 = 3 + 2\sum_{i=1}^N \overline{\mathbf{t}}_i$.

A mix of both Fully Subtractive and Selmer (resp. Brun) reductions returns a value that lies between the bounds for iterating only **FS** or only **S** (resp. **B**). Finally, by Property 3.1 (7), and Lemma 4.1, the height of the points in L_N is exactly 1 more than the highest point in B_N, showing the result. □

Let us remind that according to Proposition 3.4, the set L_N provides linearly independent vectors allowing us to span the entire connected digital plane $\mathcal{P}(\mathbf{v}, \omega)$, with $\omega = \max\left\{ \langle \mathbf{x}, \mathbf{v} \rangle \mid \mathbf{x} \in B_N \cup L_N \right\} + 1 < \|\mathbf{v}\|_1$.

5   Conclusion and Further Work

In the present paper, we have provided a process to construct connected digital planes with integer normal vectors. This process is guided by a GCF-algorithm which mixes the Fully Subtractive and Selmer (resp. Brun) GCF-algorithms. The result is an algorithm that recursively builds a digital plane for any rational normal vector. However, although the resulting digital plane is connected, it is not, in general, the thinnest one among the connected digital planes with the same normal vector. More precisely, the more the GCF-algorithm applies the Fully Subtractive reduction, the thinner the digital plane is. The GCF-algorithms **FSS** and **FSB** construct different patterns but we are not able to state that one is better than the other. As a future work, we want to identify more geometric properties of these patterns. Moreover, we hope that new GCF-algorithms will provide patterns with even more properties like the ability to tile a digital plane or to control the anisotropy of the patterns. These properties, among others, would be of interest for practical use in the context of image analysis.

References

1. Arnoux, P., Ito, S.: Pisot substitutions and Rauzy fractals. Bull. Belg. Math. Soc. Simon Stevin **8**(2), 181–207 (2001)
2. Berstel, J., Lauve, A., Reutenauer, C., Saliola, F.V.: Combinatorics on Words: Christoffel Words and Repetitions in Words. CRM Monograph Series, vol. 27. AMS, Providence, RI (2009)
3. Berthé, V., Fernique, T.: Brun expansions of stepped surfaces. Discrete Math. **311**(7), 521–543 (2011)
4. Berthé, V., Domenjoud, E., Jamet, D., Provençal, X.: Fully subtractive algorithm, tribonacci numeration and connectedness of discrete planes. Research Institute for Mathematical Sciences, Lecture note Kokyuroku Bessatu B46, pp. 159–174 (2014)
5. Berthé, V., Jamet, D., Jolivet, T., Provençal, X.: Critical connectedness of thin arithmetical discrete planes. In: Gonzalez-Diaz, R., Jimenez, M.-J., Medrano, B. (eds.) DGCI 2013. LNCS, vol. 7749, pp. 107–118. Springer, Heidelberg (2013)
6. Berthé, V., Jamet, D., Jolivet, T., Provençal, X.: Critical connectedness of thin arithmetical discrete planes (2013). http://arxiv.org/abs/1312.7820
7. Berthé, V., Labbé, S.: Uniformly balanced words with linear complexity and prescribed letter frequencies. In: WORDS, pp. 44–52 (2011)
8. Brimkov, V., Coeurjolly, D., Klette, R.: Digital planarity–a review. Discrete Appl. Math. **155**(4), 468–495 (2007)
9. Domenjoud, E., Jamet, D., Toutant, J.-L.: On the connecting thickness of arithmetical discrete planes. In: Brlek, S., Reutenauer, C., Provençal, X. (eds.) DGCI 2009. LNCS, vol. 5810, pp. 362–372. Springer, Heidelberg (2009)
10. Domenjoud, E., Provençal, X., Vuillon, L.: Facet connectedness of discrete hyperplanes with zero intercept: the general case. In: Barcucci, E., Frosini, A., Rinaldi, S. (eds.) DGCI 2014. LNCS, vol. 8668, pp. 1–12. Springer, Heidelberg (2014)
11. Domenjoud, E., Vuillon, L.: Geometric palindromic closure. Uniform Distrib. Theory **7**(2), 109–140 (2012)
12. Fernique, T.: Generation and recognition of digital planes using multi-dimensional continued fractions. Pattern Recogn. **42**(10), 2229–2238 (2009)
13. Fokkink, R., Kraaikamp, C., Nakada, H.: On schweiger's problems on fully subtractive algorithms. Isr. J. Math. **186**(1), 285–296 (2011)
14. Jolivet, T.: Combinatorics of Pisot substitutions. Ph.D. thesis, Université Paris Diderot, University of Turku (2013)
15. Klette, R., Rosenfeld, A.: Digital straightness - a review. Discrete Appl. Math. **139**(1–3), 197–230 (2004)
16. Labbé, S.: 3-dimensional continued fraction algorithms cheat sheets (2015). http://arxiv.org/abs/1511.08399
17. Labbé, S., Reutenauer, C.: A d-dimensional extension of christoffel words. Discrete Comput. Geom. **54**(1), 152–181 (2015)
18. Schweiger, F.: Multidimensional Continued Fractions. Oxford Science Publications, Oxford University Press, Oxford (2000)

Discretization

Curve Digitization Variability

Étienne Baudrier[(✉)] and Loïc Mazo

ICube-UMR 7357, 300 Bd Sébastien Brant - CS10413, 67412 Illkirch Cedex, France
{baudrier,mazo}@unistra.fr

Abstract. This paper presents a study on the set of digitizations generated by the action of a group of transformations on a continuous curve before the digitization step. An upper bound for the cardinal of this digitization set under the translation group action is exhibited. Then this bound is tested on several functions. Finally, a representation of this digitization set is proposed and an illustration of its potential use is given on a length estimator.

1 Introduction

We propose in this paper to study all the digitizations generated from an euclidean curve by the action of a given group of transformations. This is done in order to use them for the evaluation of Euclidean geometric features digital estimators. Indeed, from the set of digitizations of a curve at a given resolution, one can get the set of values (and associated statistics) an estimator produces.

Many continuous geometric characteristics have been adapted to digital curves. This is the case of length, derivative, tangent, curvature, convexity, area. The performance of the developed estimators has been evaluated from an experimental point of view on one or more curves with one or more resolutions [1,2,5,7,8,10,12,13,16], or theoretically by the property of the multigrid convergence [3,4,7,9,12–16]. An estimator is said to be multigrid convergent on a given curve set if its estimation converges toward the true value for any curve of the set when the grid step h tends to 0. This property insures that one can obtain an estimation of the characteristic with the desired precision provided the resolution is sufficiently high. It is used for an objective comparison of the estimators but it is not made to give precise information at a given resolution.

To complete this criterion, estimator comparison criteria at a fixed resolution have been proposed to estimate the perimeter [14] and the curvature [7]. For a given digital curve S, these criteria calculate the minimum of the (continuous) estimator on a family of continuous curves whose digitization at a given resolution is S. This minimum is then used as reference. These criteria imply two arbitrary choices, (1) the family of continuous curves, which must be relatively small to enable the calculation of the minimum (2) the choice of the statistical estimate that gives the criterion. Indeed, the two proposed criteria use the minimum estimation but it could have been, e.g., the mean estimation or the maximum likelihood estimation. These choices are constrained by the feasibility of the computation.

© Springer International Publishing Switzerland 2016
N. Normand et al. (Eds.): DGCI 2016, LNCS 9647, pp. 59–70, 2016.
DOI: 10.1007/978-3-319-32360-2_5

In all experimental or theoretical computations presented above, at a given resolution, a single digitization is associated with a continuous curve. However, during the digitization from the continuous curve, a certain variability is possible (for example, the position of the shape relatively to the sensor during the acquisition) and there does not exist an absolute grid that comes with the Euclidean space. We propose to take this variability into account when considering not one but a set of digitizations generated by the action of a group of transformations on the continuous curve before the step of digitization. The study of such a set of digitizations, at least for its combinatorial aspect, can be found in [6,11] for the straight lines and the circles (actually the discs). In this article, we consider the case of function graphs and we limit ourselves to the group of translations. After a reduction of the problem, a bound for the cardinal of the set of curve digitizations up to translations is provided and an application to length estimation is presented.

2   Notations and Definitions

In this work, we restrict ourselves to the digitizations of function graphs. So, let us consider a function $f : D \to \mathbb{R}$ where D is a closed bounded interval whose width is at least 2. We write $\mathcal{C}(f)$ for the graph of f: $\mathcal{C}(f) = \{(x, f(x)) \mid x \in D\}$. The aim of this paper is to study the set of the digitizations of $\mathcal{C}(f)$ obtained using the grids generated by the action of the group of translations on the standard grid. Equivalently, we can consider a unique grid, the standard one, and let the group of translations acts on the graph $\mathcal{C}(f)$. This is the technical point of view that we have adopted in the present article. As a first consequence of this choice, we can assume without loss of generality that $\min(D) = 0$ and $f(0) = 0$.

The common methods to model the digitization of the graph $\mathcal{C}(f)$ are closely related to each other. In this paper, we assume an *object boundary quantization* (OBQ). This method associates to the graph $\mathcal{C}(f)$ the digitization set $\mathcal{D}ig(f) = \{(k, \lfloor f(k) \rfloor) \mid k \in D \cap \mathbb{N}\}$ where $\lfloor \cdot \rfloor$ denotes the floor function. We write $\mathbf{f}$ for the discrete function defined on $\Gamma = D \cap \mathbb{N}$ whose graph is $\mathcal{D}ig(f)$. The set $\mathcal{D}ig(f)$ contains the uppermost grid points which lie in the hypograph of f, hence it can be understood as a part of the boundary of a solid object. Provided the slope of f is limited by 1 in modulus, $\mathcal{D}ig(f)$ is an 8-connected digital curve. Otherwise, one generally uses symmetries on the graph $\mathcal{C}(f)$ in order to come down to the previous case. Nevertheless, in this paper, we make no assumption on the slope of f since we are not concerned with connectivity.

Let $u = (x, y) \in \mathbb{R}^2$. The translate by $-u$ of the graph $\mathcal{C}(f)$ is the graph of the function f_u defined by

$$f_u : t \in (D - x) \mapsto f(t + x) - y.$$

The digitization set $\mathcal{D}ig(f_u)$ is a finite subset of $\mathbb{Z}^2$ and we are only interested in the relative positions of its elements (in other words, $\mathbb{Z}^2$ is viewed as a geometrical subset of the Euclidean plane without any preferential origin). Thus, rather than the set $\mathcal{D}ig(f_u)$, we will consider its translate $\mathcal{D}ig_0(f_u)$ whose leftmost point is the origin $(0,0)$. We write $\mathbf{f}_u$ for the function whose graph is $\mathcal{D}ig_0(f_u)$ and we write Γ_x for the domain of $\mathbf{f}_u$ ($\Gamma_x = (D - x) \cap \mathbb{N}$).

The next proposition will allow us to reduce the space of the translation vectors that have to be considered in our study and, incidentally, we will be able to give an expression of the function $\mathbf{f}_u$.

Proposition 1. *Let $u, v \in \mathbb{R}^2$ such that $u - v \in \mathbb{Z}^2$. Then, $\mathcal{D}ig(f_v) = \mathcal{D}ig(f_u) + u - v$.*

Proof. We set $u = (u_x, u_y)$, $v = (v_x, v_y)$ and $w = u - v = (w_x, w_y)$. Then $(x, y) \in \mathcal{D}ig(f_v)$ iff $x + v_x \in \mathbb{Z}$ and $y = \lfloor f(x + v_x) - v_y \rfloor$. On the one hand $x + v_x \in \mathbb{Z}$ iff $x + u_x \in \mathbb{Z}$ for $w_x = u_x - v_x \in \mathbb{Z}$. On the other hand, since $w_y \in \mathbb{Z}$, we have

$$\lfloor f(x + v_x) - v_y \rfloor = \lfloor f(x + v_x) - u_y \rfloor + w_y = \lfloor f((x - w_x) + u_x) - u_y \rfloor + w_y.$$

Thus $(x, y) \in \mathcal{D}ig(f_v)$ iff $(x, y) \in \mathcal{D}ig(f_u) + w$.   □

As a consequence of Proposition 1, we have $\mathcal{D}ig_0(f_{u+w}) = \mathcal{D}ig_0(f_u)$ for any $w \in \mathbb{Z}^2$. Hence, in the sequel we may assume without loss of generality that the translation vector u lies in $[0, 1)^2$. Then, from the hypotheses on f we derive that the smaller integer in $D - x$, the domain of f_u, is 0. Moreover, we can now define the discrete function $\mathbf{f}_u$ as follows

$$\mathbf{f}_u : k \in \Gamma_x \mapsto \lfloor f_u(k) \rfloor - \lfloor f_u(0) \rfloor.$$

Let $I \subseteq [0, 1)^2$. The family of all the digitization sets $\mathcal{D}ig_0(f_u)$, $u \in I$, is noted $\mathcal{D}^I(f)$ and we write $\mathcal{D}(f)$ for $\mathcal{D}^{[0,1)^2}(f)$. We have the following straightforward property

$$\forall I, J \subseteq [0, 1), \ \text{card} \ \mathcal{D}^{I \cup J}(f) \leq \text{card} \ \mathcal{D}^I(f) + \text{card} \ \mathcal{D}^J(f). \tag{1}$$

3   Combinatorial Properties of $\mathcal{D}(f)$

In this section, $[\![x, y]\!]$ stands for $[x, y] \cap \mathbb{Z}$, $\langle x \rangle$ denotes the fractional part, that is $\langle x \rangle = x - \lfloor x \rfloor$, and we write $\langle\!\langle x \rangle\!\rangle$ for the 'upper fractional part': $\langle\!\langle x \rangle\!\rangle = \langle -x \rangle$. Both $\langle x \rangle$ and $\langle\!\langle x \rangle\!\rangle$ lie in $[0, 1)$.

3.1   On the Cardinal of the Digitization Set $\mathcal{D}ig(f_u)$

When translating the graph $\mathcal{C}(f)$ by a vector $u(x, y) \in [0, 1)^2$, the domain of the discrete function $\mathbf{f}_u$ can change in size. In this subsection, we describe these changes.

Lemma 1. *Let $a, b \in \mathbb{R}$. Then,*

$$\text{card} \left([a, b] \cap \mathbb{Z} \right) = \lfloor b - a \rfloor + \varepsilon$$

where $\varepsilon = 0$ if $\langle\!\langle a \rangle\!\rangle + \langle b \rangle \geq 1$ and $\varepsilon = 1$ otherwise.

Proof. Firstly, we observe that

$$\mathrm{card}(\llbracket a,b \rrbracket) = \mathrm{card}(\llbracket \lceil a \rceil, \lfloor b \rfloor \rrbracket) = \lfloor b \rfloor - \lceil a \rceil + 1 = b - \langle b \rangle - (a + \langle\!\langle a \rangle\!\rangle) + 1.$$

That is,

$$\mathrm{card}(\llbracket a,b \rrbracket) = b - a - (\langle\!\langle a \rangle\!\rangle + \langle b \rangle - 1).$$

Secondly, from the very definition of $\langle \cdot \rangle$ and $\langle\!\langle \cdot \rangle\!\rangle$, we have

$$-1 \le \langle\!\langle a \rangle\!\rangle + \langle b \rangle - 1 < 1.$$

Then, setting $\varepsilon = 0$ if $\langle\!\langle a \rangle\!\rangle + \langle b \rangle - 1 \ge 0$ and $\varepsilon = 1$ otherwise, we have

$$b - a - 1 < \mathrm{card}(\llbracket a,b \rrbracket) - \varepsilon \le b - a.$$

Taking into account that $\mathrm{card}(\llbracket a,b \rrbracket) \in \mathbb{N}$, we conclude that

$$\mathrm{card}(\llbracket a,b \rrbracket) - \varepsilon = \lfloor b - a \rfloor. \qquad \square$$

Proposition 2. *Let $D = [0, \ell]$ and $x \in [0, 1)$. Then,*

$$\mathrm{card}\,(D - x) \cap \mathbb{N} = \begin{cases} \lfloor \ell \rfloor & \text{if } x > \ell \\ \lfloor \ell \rfloor + 1 & \text{otherwise.} \end{cases}$$

Proof. From Lemma 1, we derive that

$$[-x, \ell - x] \cap \mathbb{Z} = \begin{cases} \lfloor \ell \rfloor & \text{if} \langle\!\langle -x \rangle\!\rangle + \langle \ell - x \rangle \ge 1 \\ \lfloor \ell \rfloor + 1 & \text{otherwise.} \end{cases}$$

It can be seen that $\langle\!\langle -x \rangle\!\rangle = x$ and

$$\langle \ell - x \rangle = \begin{cases} \langle \ell \rangle - x & \text{if } x \le \langle \ell \rangle, \\ \langle \ell \rangle - x + 1 & \text{otherwise.} \end{cases}$$

Then, the reader can check that $\langle\!\langle -x \rangle\!\rangle + \langle \ell - x \rangle \ge 1$ iff $\langle \ell \rangle < x$. $\qquad \square$

3.2   Translation Along the Y Axis

The number of digitizations obtained by vertical translations of the graph $\mathcal{C}(f)$ is bounded.

Proposition 3. *Let $f \colon D \to \mathbb{R}$. Then,*

$$\mathrm{card}\,\left(\mathcal{D}^{\{0\} \times [0,1)}(f)\right) \le \mathrm{card}\,(\Gamma).$$

Proof. Let $k \in \Gamma \backslash \{0\}$, $0 < y < 1$ and $\boldsymbol{u} = (0, y)$. Then

$$\mathbf{f}_u(k) = \lfloor f_u(k) \rfloor - \lfloor f_u(0) \rfloor = \lfloor f(k) - y \rfloor + 1 = \begin{cases} \lfloor f(k) \rfloor + 1 & \text{if } 0 < y \le \langle f(k) \rangle, \\ \lfloor f(k) \rfloor & \text{otherwise.} \end{cases}$$

Thus, the set of the digitizations generated by the translations along the Y axis is $\{\mathcal{D}ig_0(f_{(0,y)}) \mid y \in \{\langle f(k) \rangle\}_{k \in \Gamma}\}$ and its cardinal is less than, or equal to, $\mathrm{card}\,(\Gamma)$. $\qquad \square$

3.3 Translations Along the X Axis

Unlike vertical translations, horizontal translations may yield infinitely many digitizations when the function f is unbounded. Indeed, consider the inverse function $\mathtt{inv}\colon [0,2] \to (0,+\infty)$ extended in zero by $\mathtt{inv}(0) = 0$. Then, its family of digitization sets is $\mathcal{D}(\mathtt{inv}) = (S_i)_{i\in\mathbb{N}}$ with

$$S_0 = \{O,Q,R\} \quad \text{and, for any } i > 0, \ S_i = \{O, P_i\}$$

where O is the origin and Q, R, P_i are the points with coordinates $(1,1)$, $(2,0)$, $(1,-n)$.

If the function f is bounded, say $f(D) \subseteq [m, M]$, then it is plain that card $\mathcal{D}^{[0,1)\times\{0\}}(f) \le (M-m+1)^{\text{card }\varGamma\setminus\{0\}}$. We obtain the bound with horizontal translations of functions like the following whose graph is depicted in Fig. 1:

$$g\colon [0,\ell] \to [0,M]$$

$$x \mapsto 2M \left| \langle M^{\lfloor x \rfloor} x \rangle - \tfrac{1}{2} \right|.$$

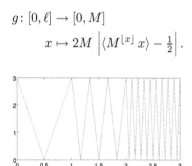

Fig. 1. Graph of the function g for $M = 3$

An amazing consequence is that any set of points which is a function graph on a finite rectangular discrete grid is the digitization of the function g for some position of the grid. Thus, any 'good' geometric feature estimator should return the value for g.

In the previous example, the total variation of the function g is $(2M(M^\ell - 1)/(M-1)$. It is about twice the cardinal of $\mathcal{D}(g)$. The factor 2 vanishes if we replace the continuous function g by the non continuous function $h\colon x \mapsto M \times \langle M^{\lfloor x \rfloor} x \rangle$. From the digitization point of view, the difference is that, with the function h, the digitizations are uniquely generated when the translation vector norm goes from 0 to 1 whereas they are produced twice with the function g. This leads us to study the links between the changes of digitizations during the translation process and the total variation of the function. Indeed, for a horizontal translation, the digitization changes depend on the function f variation. These changes are given in the sets $E_{f,i,I}$. Let $\mathcal{U}$ be the family of open sets for the usual topology of $[0,1)$. For any $i \in \varGamma$ and any $I \subseteq [0,1)$, we define the set $E_{f,i,I} \subset \{i\} \times I$ by

$$E_{f,i,I} = \{(i,x) \mid \forall U \in \mathcal{U}, x \in U \implies \exists x' \in U, \lfloor f(i+x') \rfloor \neq \lfloor f(i+x) \rfloor \}$$

and we set $E_{f,I} = \sqcup_{i\in\varGamma} E_{f,i,I}$, $E_f = E_{f,[0,1)}$.
Figure 2 illustrates this definition.

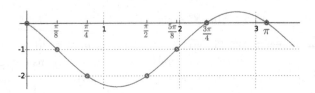

Fig. 2. The set E_f for the function $f \colon x \mapsto \sqrt{2}\sin(\pi/4-2x)-1$, $x \in [0,3.5]$. A pair (i,x) is in E_f iff $(i+x, f(i+x))$ is the centre of a green circle. We get (in the lexicographical order): $E_f = \{(0,0),(0,\frac{\pi}{8}),(0,\frac{\pi}{4}),(1,\frac{\pi}{2}-1),(1,\frac{5\pi}{8}-1),(2,\frac{3\pi}{4}-2),(3,\pi-3)\}$ (Color figure online).

As the set $E_{f,I}$ indexes all the changes of digitization for the function f when the abscissa of the translation vector $u = (x,0)$ lies in I, the number of distinct digitizations for f under horizontal translations is less than, or equal to, the cardinal of $E_{f,I}$ plus one:

$$\operatorname{card}\ \mathcal{D}^I(f) \le 1 + \sum_{i \in \Gamma} \operatorname{card}\ E_{f,i,I}. \tag{2}$$

We are now able to link the changes of digitization with the total variation of the function. Firstly, we recall that for any real function $f \colon D \to \mathbb{R}$, the total variation is defined as $\sup_\sigma \sum_i |f(x_{i+1}) - f(x_i)|$ where the supremum is over all the partitions $\sigma = (x_i)$ of the interval D. Even if f is continuous, and though D is compact, the total variation may be infinite (consider for instance the continuous completion of the function $x\sin(1/x)$). Alike, we define for any real function $f \colon D \to \mathbb{R}$ the *number of monotonicity changes* – which may also be infinite – as $\sup_\sigma \operatorname{card}\{i \mid (f(x_i) - f(x_{i-1}))(f(x_{i+1}) - f(x_i)) < 0\}$.

Lemma 2. *Let $f \colon D \to \mathbb{R}$, $i \in \Gamma$ and I a subinterval of $[0,1)$. Then,*

$$\operatorname{card}\ E_{f,i,I} \le n + V$$

where n is the number of monotonicity changes in $i+I$ and, V the total variation of f on $i+I$.

Proof. If $n = +\infty$, Lemma 2 is obvious. We now assume that n is finite. Let $i \in \mathbb{N}$ and $a_0 < a_1 < \cdots < a_{n+1}$ be a partition of the topological closure of the interval $i+I$ such that f is monotonic on each subinterval $[a_i, a_{i+1}]$ and changes its monotonicity in each a_i, $1 \le i \le n$. Then, it is plain that the cardinal of $E_{f,i,I}$, that is the number of changes of $\lfloor f(i+x)\rfloor$ when x lies in I, is equal to

$$\sum_{i \in J} 1 + \lfloor f(a_{i+1})\rfloor - \lceil f(a_i)\rceil + \sum_{i \in K} 1 + \lfloor f(a_i)\rfloor - \lceil f(a_{i+1})\rceil$$

where $J = \{i \in [\![0,n]\!] \mid f(a_i) \le f(a_{i+1})\}$ and $K = [\![0,n]\!]\setminus I$. Thus, $\operatorname{card}(E_{f,i,I}) \le n + 1 + \sum_i |f(a_{i+1}) - f(a_i)|$. We conclude straightforwardly. $\qquad\square$

Corollary 1. *Let $f\colon D \to \mathbb{R}$. Then,*

$$\text{card } \mathcal{D}^{[0,1)\times\{0\}}(f) \leq \text{card } E_f \leq 1 + n + V$$

where

- *n is the number of monotonicity changes of f and,*
- *V the total variation of f.*

In the particular case where the function is bijective and its codomain satisfies the same hypotheses as the domain – the codomain is a closed bounded interval whose width is at least 2, we can use the result of Proposition 3 by a symmetry argument. Then, we obtain card $\mathcal{D}^{[0,1)\times\{0\}}(f) \leq$ card $f(D) \cap \mathbb{Z}$ which is less than the total variation plus one. Thereby, if moreover the function is continuous ($n = 0$), Proposition 3 and Corollary 1 give the same result (but in the corollary we have only an upper bound). For a non continuous bijective function, Proposition 3 shows that we could drop the parameter n in the above corollary. Actually, the parameter n is only needed to catch small changes around an horizontal grid line that generate changes of digitization that are not captured by the total variation. Nevertheless, there can be a huge difference between the complexity of the changes of digitization and the number of digitizations. Indeed, consider a function like $x \sin(1/x)$. The translation of the function along the X axis yields infinitely many digitization changes but finitely many digitizations since almost all the oscillations around the X axis will produce only two digitizations. The following lemma and proposition is a first step to tackle this issue. The lemma is a specialization of Eq. (1) when there exists an interval $I \subseteq [0,1)$ such that no set $E_{f,i,I}$ but one contains a pair (i, x) where $x \in I$. Then, the set of digitizations produced by the translations whose vectors lie in $I \times \{0\}$ can be computed on a restriction of the domain of f.

Lemma 3. *Let $f\colon D \to \mathbb{R}$. Let I be a subinterval of $[0,1)$ and $i \in \Gamma$ such that*

$$(j, x) \in E_f \implies x \notin I \text{ or } j = i.$$

Then,
$$\forall J \subseteq [0,1), \text{ card } \mathcal{D}^{I \cup J}(f) \leq \text{card } \mathcal{D}^I(f_{|D'}) + \text{card } \mathcal{D}^J(f)$$

where $D' \subseteq D$ is equal to $[i-1, i+1]$ or $[0,2]$ if $i = 0$ (remember that we assume that the width of D is at least 2).

Proof. From the hypotheses, the values of $x \to \lfloor f(j + x) \rfloor$, $j \neq i$, is constant on I. Then,

$$\text{card } \mathcal{D}^I(f) = \text{card } \mathcal{D}^I(f_{|D'}). \tag{3}$$

From (1) we derive that

$$\text{card}(\mathcal{D}^{I \cup J}(f)) \leq \text{card } \mathcal{D}^I(f_{|D'}) + \text{card } \mathcal{D}^J(f). \qquad \square$$

Proposition 4. *Let $f : D \to \mathbb{R}$ be a function whose set of monotonicity change abscissas has exactly one limit point z. Then there exists an open interval I containing z such that the number of digitizations of f is upper bounded by*

– if there is no pair $(a, b) \in E_f \cup \{(0, 0)\}$ such that $a \neq \lfloor z \rfloor$ and $b = \langle z \rangle$:

$$\text{card } \mathcal{D}^I(f_{|D_z}) + \text{card } \Gamma + n + V + n' + V';$$

– otherwise:

$$\text{card } \mathcal{D}^{I \cap (0,z)}(f_{|D_z}) + \text{card } \mathcal{D}^{I \cap (z,1)}(f_{|D_z}) + 1 + \text{card } \Gamma + n + V + n' + V'$$

where $D_z = [\lfloor z \rfloor - 1, \lfloor z \rfloor + 1]$, or $D_z = [0, 2]$ if $z < 1$, n and V, resp. n' and V', are the number of monotonicity changes and the total variation of $f_{|[0, \inf I]}$, resp. $f_{|[\sup I, \max D]}$.

By lack of place, we only give the sketch of the proof. The reader will find further, after the proof, an example of calculus of the bound given in Proposition 4.

Proof. Let $x_0 \in [0, 1)$ and $i_0 \in \Gamma$ such that $i_0 + x_0$ is a limit point of the monotonicity change set. We assume that the total variation of f on any subinterval of D that does not contain z is finite (otherwise, the result is obvious). From the hypothesis, it can be seen that, for any $i \neq i_0$ (and, may be, $i \neq i_0 - 1$), the 'event' set $E_{f,i,[0,1)}$ is finite. This enables us to define, in the general case, an interval I around x_0 such that no event occurs, but on the grid line corresponding to i_0, when the abscissa of the translation vector lies in I. Then, we can applied Lemmas 2 and 3 to conclude. $\qquad\square$

Example 1. We consider the function f defined by $f(x) = x \sin(13/x)$ when $x \in (0, 3]$ and $f(0) = 0$ (f has an unbounded total variation). The set S of monotonicity change abscissas has one limit point, $z = 0$, and there is no other point x such that $(\lfloor x \rfloor, 0) \in E_f$. Thus, we have to compute the first formula,

$$\text{card } \mathcal{D}^I(f_{|D_z}) + \text{card } \Gamma + n + V + n' + V'.$$

Here, we have $D_z = [0, 2]$ and $I = (0, \beta)$ where β is the smallest real number in $(0, 1]$ such that, for some integer $i \in \{1, 2\}$, the curve of the function $\beta \mapsto f(i + \beta)$ passes through an horizontal grid line. We find $\beta \approx 0.13$. Then, card $\mathcal{D}^I(f_{|D_z})$ is the number of distinct functions

$$\mathbf{f}_u : x \in ([0, 2] - u) \cap \mathbb{N} \mapsto \lfloor f(x + u) \rfloor - \lfloor f(u) \rfloor$$

for $u \in I$. Since, for any $u \in I$, $([0, 2] - u) \cap \mathbb{N} = \{0, 1\}$ and $\mathbf{f}_u(0) = 0$, we see that

$$\text{card } \mathcal{D}^I(f_{|D_z}) = \text{card } \{\mathbf{f}_u(1) \mid u \in I\}.$$

From the curve of the function $u \mapsto \mathbf{f}_u(1)$, we derive that card $\mathcal{D}^I(f_{|D_z}) = 3$. As $\min I = 0$, we have $n = V = 0$. We find on a plot of the derivative of f that $n' = 30$ and a numerical calculus gives $V' \leq 26$. Thus the number of digitizations of the curve $y = f(x)$, $x \in [0, 3]$, is upper bounded by $3 + 4 + 30 + 26 = 63$.

Proposition 4 allows us to count locally the number of digitizations around the limit point while counting the digitization changes on the left and right parts of the function. Further works will have to address the problem of functions with finitely, or infinitely, many limit points.

3.4   On the Cardinal of the Set $\mathcal{D}(f)$

Proposition 5. *Let $f \colon D \to \mathbb{R}$. Let $H = \mathrm{card}\ \mathcal{D}^{[0,1) \times \{0\}}(f)$ and $\gamma = \mathrm{card}\ \Gamma$. Then,*

$$\mathrm{card}\ \mathcal{D}(f) \le H\,2^{\gamma}.$$

Proof. Let $u(x,y) \in [0,1)^2$. Let $I_x = \Gamma_x \backslash \{0\}$ and $v = \mathrm{card}\ I_x$ (from Proposition 2, $v \in \{\gamma - 2, \gamma - 1\}$).

For any $i \in I_x$, one has $\mathbf{f}_{(x,0)}(i) = \lfloor f(x+i) \rfloor - \lfloor f(x) \rfloor$ and $\mathbf{f}_u(i) = \lfloor f(x+i) - y \rfloor - \lfloor f(x) - y \rfloor$. Thus, setting $a = \mathbf{f}_{(x,0)}(i)$, we have $\mathbf{f}_u(i) \in \{a-1, a\}$ if $\lfloor f(x) - y \rfloor = \lfloor f(x) \rfloor$ and $\mathbf{f}_u(i) \in \{a, a+1\}$ when $\lfloor f(x) - y \rfloor = \lfloor f(x) \rfloor - 1$. The set $Dig_0(f_u)$, which is in bijection with $\prod_{i \in I_x} \{\mathbf{f}_u(i)\}$, is then obtained from $Dig_0(f_{(x,0)})$ by doing v "choices" in a pair than can be either $\{-1, 0\}$ or $\{0, 1\}$. Since, there are H distinct sets $Dig_0(f_{(x,0)})$ and $v \le \gamma - 1$, there are at most $H \times 2 \times 2^{\gamma - 1}$, that is $H\,2^{\gamma}$, distinct digitizations. □

Definition 1 (Dual by translation). *The dual by translation of the function f is the label image Δ defined on the torus $\mathbb{R}^2/\mathbb{Z}^2$ by*

$$\forall [u], \quad \Delta([u]) = Dig_0(f_u).$$

where $[u]$ denotes the class of $u \in [0,1)^2$ in $\mathbb{R}^2/\mathbb{Z}^2$.

The boundaries of the dual are obtained by plotting the curve $y = f(x)$ on the torus. Then, in order to get the labels, one just has to pick a point u in each region and to compute $Dig_0(f_u)$. Moreover, crossing in the positive direction of the y-axis the boundary $y = f(x+i)$, $i \in \Gamma \backslash \{0\}$ and $x \in [0,1)$, amounts to decrease the $(i+1)$-th value of the digitization and crossing the boundary $y = f(x)$, $x \in [0,1)$, amounts to increase all the values of the digitization but the first which has to stay equal to 0 according to our settings.

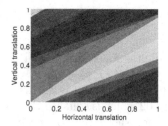
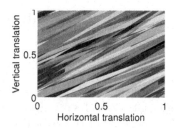

Fig. 3. Examples of dual for the function f_1 with $r = 5$ and $r = 50$. Each color stands for a digitization (Color figure online).

For geometric feature estimation, the torus is sampled so as to compute each digitization area and the corresponding estimation value.

An example of dual is shown in Fig. 3. This is the dual of the function f_1 described in Sect. 4.

4   Experiment

Cardinal of the set of digitizations up to a translation. The digitization sets have been computed on the graph of the following functions

$$f_1 : x \in [0, r] \mapsto r \ln \left(1 + \frac{x}{r}\right), \qquad f_3 : \begin{cases} x \in (0, 3] \mapsto \sin \frac{13}{x} \\ 0 \mapsto 0, \end{cases}$$

$$f_2 : x \in [0, \tfrac{11}{10}r] \mapsto r\frac{1}{2}\left(\frac{x}{r}\right)^2, \quad f_4 : x \in [0, \tfrac{11}{10}r] \mapsto \frac{r}{100}\left(\sin\frac{100x}{r} + 50\frac{x^2}{r^2}\right)$$

with $r = 50$ (r stands for the resolution).

Their cardinal and the upper bound obtained in Proposition 5 are gathered in Table 1.

Table 1. Number of digitizations of the functions f_1, f_2, f_3, f_4 under translation.

	$f1$	$f2$	$f3$	$f4$
Card$\mathcal{D}(f)$	297	593	26	2908
Upper bounds	$(1 + 0 + 35) \times 2^{51}$	$(1 + 0 + 66) \times 2^{56}$	$(3 + 4 + 30 + 26) \times 2^4$	$(1 + 32 + 43) \times 2^{56}$

Application to length estimators. In order to evaluate the quality of an estimator, it is desirable to test it on digitization sets up to translations. The produced set of estimates can be used to precise the estimate variability. Figure 4 gives an example of a dual for the function f_4 at the resolution $r = 50$. The second map gives the local estimate in function of the translation vector. Formally, we note $l(f)$ the true length of the function f and

$$\mathcal{E}^L(\mathbf{f}) = \sum_{i \in \Gamma} \left((\mathbf{f}(i+1) - \mathbf{f}(i))^2 + 1\right)^{\frac{1}{2}}$$

a local estimate of the length, using the digital function $\mathbf{f}$. Then the central map of Fig. 4 is the graph $\{x, y, \mathcal{E}^L(\mathbf{f}_{3(x,y)}) - l(f_3)), (x, y) \in [0, 1)^2\}$ of the error between the true length and the estimate. The right graphic is the distribution of the length estimate error. Its mean value is 0.02 which is equal to the digitization step $\frac{1}{50}$. The spreading of the error distribution shows that the evaluation on a single digitization is not reliable for this estimator on this graph function.

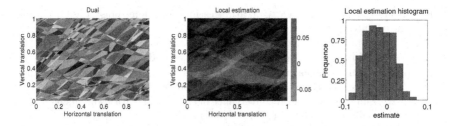

Fig. 4. (left) dual of f_4 for r=50 (on $[0, 55]$), (center) local estimation error applied to the dual, (right) distribution of the local estimation error (Color figure online).

5   Conclusion

This paper presents a first study on the set of digitizations generated by the action of the translation group over a given function graph. Bounds of the digitization set complexity are given for vertical and horizontal translations. A representation of the set of generated digitizations is defined as the dual under translation. Finally some illustrations are given and a glimpse on the potential use of this tool is presented on a local length estimator.

There are several perspectives to this work: First of all, the study of the plane curve case and the improvement of the digitization generation algorithm, then the study of its contribution in term of variability for the estimator evaluation and the extension to a multigrid study by adding dilation to the transformation group and the link to the notion of digitization scale.

References

1. Coeurjolly, D., Klette, R.: A comparative evaluation of length estimators of digital curves. IEEE Trans. Pattern Anal. Mach. Intell. **26**(2), 252–257 (2004)
2. Coeurjolly, D., Miguet, S., Tougne, L.: Discrete curvature based on osculating circle estimation. In: Arcelli, C., Cordella, L.P., Sanniti di Baja, G. (eds.) IWVF 2001. LNCS, vol. 2059, pp. 303–312. Springer, Heidelberg (2001). http://link.springer.com/10.1007/3-540-45129-3_27
3. Daurat, A., Tajine, M., Zouaoui, M.: About the frequencies of some patterns in digital planes. Application to area estimators. Comput. Graph. **33**(1), 11–20 (2009). http://linkinghub.elsevier.com/retrieve/pii/S0097849308001416
4. Daurat, A., Tajine, M., Zouaoui, M.: Patterns in discretized parabolas and length estimation. In: Brlek, S., Reutenauer, C., Provençal, X. (eds.) DGCI 2009. LNCS, vol. 5810, pp. 373–384. Springer, Heidelberg (2009)
5. Esbelin, H.A., Malgouyres, R., Cartade, C.: Convergence of binomial-based derivative estimation for 2 noisy discretized curves. Theor. Comput. Sci. **412**(36), 4805–4813 (2011). http://linkinghub.elsevier.com/retrieve/pii/S030439751000736X
6. Huxley, M.N., Zunic, J.D.: The number of n-point digital discs. IEEE Trans. Pattern Anal. Mach. Intell. **29**(1), 159–161 (2007). http://doi.ieeecomputersociety.org/10.1109/TPAMI.2007.250606

7. Kerautret, B., Lachaud, J.O.: Curvature estimation along noisy digital contours by approximate global optimization. Pattern Recogn. **42**(10), 2265–2278 (2009). http://linkinghub.elsevier.com/retrieve/pii/S0031320308004755

8. Kiryati, N., Kübler, O.: Chain code probabilities and optimal length estimators for digitized three-dimensional curves. Pattern Recogn. **28**(3), 361–372 (1995). http://www.sciencedirect.com/science/article/pii/003132039400101Q

9. Klette, R., Žunić, J.: Multigrid convergence of calculated features in image analysis. J. Math. Imaging Vis. **13**(3), 173–191 (2000)

10. Klette, R., Yip, B.: Evaluation of curve length measurements. In: Proceedings of 1st International Conference on Pattern Recognition, vol. 1, pp. 610–613 (2000)

11. Koplowitz, J., Lindenbaum, M., Bruckstein, A.: The number of digital straight lines on an n × n grid. IEEE Trans. Inf. Theor. **36**(1), 192–197 (1990)

12. Lachaud, J.-O., Vialard, A., de Vieilleville, F.: Analysis and comparative evaluation of discrete tangent estimators. In: Andrès, É., Damiand, G., Lienhardt, P. (eds.) DGCI 2005. LNCS, vol. 3429, pp. 240–251. Springer, Heidelberg (2005). http://dx.doi.org/10.1007/978-3-540-31965-8_23

13. Mazo, L., Baudrier, É.: About multigrid convergence of some length estimators. In: Barcucci, E., Frosini, A., Rinaldi, S. (eds.) DGCI 2014. LNCS, vol. 8668, pp. 214–225. Springer, Heidelberg (2014). http://dx.doi.org/10.1007/978-3-319-09955-2_18

14. Sloboda, F., Zatko, B., Stoer, J.: On approximation of planar one-dimensional continua. Advances in Digital and Computational Geometry, pp. 113–160 (1998)

15. Tajine, M., Daurat, A.: Patterns for multigrid equidistributed functions: application to general parabolas and length estimation. Theor. Comput. Sci. **412**(36), 4824–4840 (2011)

16. de Vieilleville, F., Lachaud, J.-O.: Experimental comparison of continuous and discrete tangent estimators along digital curves. In: Brimkov, V.E., Barneva, R.P., Hauptman, H.A. (eds.) IWCIA 2008. LNCS, vol. 4958, pp. 26–37. Springer, Heidelberg (2008). http://dx.doi.org/10.1007/978-3-540-78275-9_3

Discretizations of Isometries

Pierre-Antoine Guihéneuf$^{(\boxtimes)}$

Universidade Federal Fluminense, Niterói, Brazil
pierre-antoine.guiheneuf@math.u-psud.fr

Abstract. This paper deals with the dynamics of discretizations of isometries of $\mathbf{R}^n$, and more precisely the density of the successive images of $\mathbf{Z}^n$ by the discretizations of a generic sequence of isometries. We show that this density tends to 0 as the time goes to infinity. Thus, in general, all the information of a numerical image will be lost by applying many times a naive algorithm of rotation.

1 Introduction

In this paper, we consider the dynamical behaviour of the discretizations of (linear) isometries of a real and finite dimensional vector space $\mathbf{R}^n$. The goal is to understand how it is possible to rotate a numerical image (made of pixels) with the smallest loss of quality as possible. For example, in Fig. 1, we have applied 10 successive random rotations to a 220×282 pixels picture, using the software *Gimp* (linear interpolation algorithm). These discretized rotations induce a very strong blur in the resulting image.

Here, we consider the simplest algorithm that can be used to perform a discrete rotation: discretizing the rotation. More precisely, if $x \in \mathbf{Z}^2$ is a integer point (representing a pixel), then the image of x by the discretization of a rotation R will be the integer point which is the closest of $R(x)$. More precisely, in the general case of isometries we will use the following definition of a discretization.

Definition 1. *We define the projection $p : \mathbf{R} \to \mathbf{Z}$ such that for every $x \in \mathbf{R}$, $p(x)$ is the unique integer $k \in \mathbf{Z}$ such that $k - 1/2 < x \le k + 1/2$ (in other words, $p(x) = \lfloor x + 11/2 \rfloor$). This projection induces the map*

$$\pi : \quad \mathbf{R}^n \longmapsto \mathbf{Z}^n$$
$$(x_i)_{1 \le i \le n} \longmapsto \big(p(x_i)\big)_{1 \le i \le n}$$

which is an Euclidean projection on the lattice $\mathbf{Z}^n$. For $P \in O_n(\mathbf{R})$ (the set of linear isometries of $\mathbf{R}^n$), we denote by $\widehat{P}$ the discretization of P, defined by

$$\widehat{P} : \mathbf{Z}^n \longrightarrow \mathbf{Z}^n$$
$$x \longmapsto \pi(Px).$$

We will measure the loss of information induced by the action of discretizing by the *density* of the image set. More precisely, given a sequence $(P_k)_{k \ge 1}$ of linear isometries of $\mathbf{R}^n$, we will look for the density of the set $\Gamma_k = (\widehat{P_k} \circ \cdots \circ \widehat{P_1})(\mathbf{Z}^n)$.

© Springer International Publishing Switzerland 2016
N. Normand et al. (Eds.): DGCI 2016, LNCS 9647, pp. 71–92, 2016.
DOI: 10.1007/978-3-319-32360-2_6

Fig. 1. Original image (left) of size 220×282 and 10 successive random rotations of this image (right), obtained with the software *Gimp* (linear interpolation algorithm).

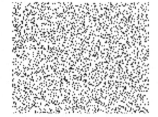

Fig. 2. Successive images of $\mathbf{Z}^2$ by discretizations of random rotations, a point is black if it belongs to $(\widehat{R_{\theta_k}} \circ \cdots \circ \widehat{R_{\theta_1}})(\mathbf{Z}^2)$, where the θ_i are chosen uniformly randomly in $[0, 2\pi]$. From left to right and top to bottom, $k = 2, 5, 50$.

Definition 2. *For $A_1, \cdots, A_k \in O_n(\mathbf{R})$, the rate of injectivity in time k of this sequence is the quantity*

$$\tau^k(P_1, \cdots, P_k) = \limsup_{R \to +\infty} \frac{\mathrm{Card}\left((\widehat{P_k} \circ \cdots \circ \widehat{P_1})(\mathbf{Z}^n) \cap [B_R]\right)}{\mathrm{Card}[B_R]} \in]0, 1],$$

where B_R denotes the infinite ball of radius R centered at 0 and $[B_R]$ the set of integral points (i.e. with integer coordinates) inside B_R. For an infinite sequence $(P_k)_{k \geq 1}$ of isometries, as the previous quantity is decreasing, we can define the asymptotic rate of injectivity

$$\tau^\infty\left((P_k)_{k \geq 1}\right) = \lim_{k \to +\infty} \tau^k(P_1, \cdots, P_k) \in [0, 1].$$

An example of the sets Γ_k for a random draw of isometries P_m is presented on Fig. 2. In particular, we observe that the density of these sets seems to decrease when k goes to infinity: the images get whiter and whiter.

This phenomenon is confirmed when we plot the density of the intersection between these image sets Γ_k and a big ball of $\mathbf{R}^n$ (see Fig. 3): this density seems to tend to 0 as the time k goes to infinity.

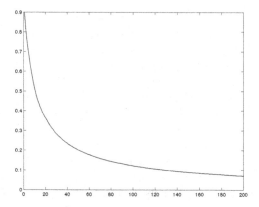

Fig. 3. Expectation of the rate of injectivity of a random sequences of rotations: the graphic represents the mean of the rate of injectivity $\tau^k(R_{\theta_k}, \cdots, R_{\theta_1})$ depending on k, $1 \le k \le 200$, for 50 random draws of sequences of angles $(\theta_i)_i$, with each θ_i chosen independently and uniformly in $[0, 2\pi]$. Note that the behaviour is not exponential.

We would like to explain theoretically this phenomenon. Of course, if we take $P_m = \mathrm{Id}$, then we will have $\Gamma_k = \mathbf{Z}^n$ and the rates of injectivity will be equal to 0. To avoid this kind of "exceptional cases", we will study the asymptotic rate of injectivity of a *generic* sequence of matrices of $O_n(\mathbf{R})$, in the following sense.

Definition 3. *We fix once for all a norm* $\|\cdot\|$ *on* $M_n(\mathbf{R})$. *For any sequence* $(P_k)_{k \ge 1}$ *of matrices of* $O_n(\mathbf{R})$, *we set*

$$\|(P_k)_k\|_\infty = \sup_{k \ge 1} \|P_k\|.$$

In other words, we consider the space $\ell^\infty(O_n(\mathbf{R}))$ *of uniformly bounded sequences of linear isometries endowed with this natural metric.*

This metric is complete, thus there is a good notion of genericity on the set of linear isometries: a set $\mathcal{U} \subset (O_n(\mathbf{R}))^{\mathbf{N}}$ is *generic* if it is a countable intersection of open and dense subsets of $\ell^\infty(O_n(\mathbf{R}))$. The main theorem of this paper studies the asymptotic rate of injectivity in this context.

Theorem 1. *Let* $(P_k)_{k \ge 1}$ *be a generic sequence of matrices of* $O_n(\mathbf{R})$. *Then* $\tau^\infty((P_k)_k) = 0$.

The proof of this theorem will even show that for every $\varepsilon > 0$, there exists an open and dense subset of $\ell^\infty(O_n(\mathbf{R}))$ on which τ^∞ is smaller than ε. This theorem expresses that for "most of" the sequences of isometries, the loss of information is total. Thus, for a generic sequence of rotations, with the naive algorithm of discretization, we will not be able to avoid the blur observed in Fig. 1.

Note that we do not know what is the rate of injectivity of a sequence of isometries made of independent identically distributed random draws (for example with respect to the Haar measure on $O_n(\mathbf{R})$).

The proof of Theorem 1 will be the occasion to study the structure of the image sets $\Gamma_k = (\widehat{P_k} \circ \cdots \circ \widehat{P_1})(\mathbf{Z}^n)$. It appears that there is a kind of "regularity at infinity" in Γ_k. More precisely, this set is an *almost periodic pattern*: roughly speaking, for R large enough, the set $\Gamma_k \cap B_R$ determines the whole set Γ_k up to an error of density smaller than ε (see Definition 6). We prove that the image of an almost periodic pattern by the discretization of a linear map is still an almost periodic pattern (Theorem 2); thus, the sets Γ_k are almost periodic patterns.

The idea of the proof of Theorem 1 is to take advantage of the fact that for a generic sequence of isometries, we have a kind of independence of the coefficients of the matrices. Thus, for a generic isometry $P \in O_n(\mathbf{R})$, the set $P(\mathbf{Z}^n)$ is uniformly distributed modulo $\mathbf{Z}^n$. We then remark that the local pattern of the image set $\widehat{P}(\mathbf{Z}^n)$ around $\widehat{P}(x)$ is only determined by P and the remainder of Px modulo $\mathbf{Z}^n$: the global behaviour of $\widehat{P}(\mathbf{Z}^n)$ is coded by the quotient $\mathbf{R}^n/\mathbf{Z}^n$. This somehow reduces the study to a local problem.

As a first application of this remark, we state that the rate of injectivity in time 1 can be seen as the area of an intersection of cubes (Proposition 2). This observation is one of the two keys of the proof of Theorem 1, the second one being the study of the action of the discretizations $\widehat{P}$ on the frequencies of differences $\rho_{\Gamma_k}(v) = D\big((\Gamma_k - v) \cap \Gamma_k\big)$. Indeed, if there exists a set $\Gamma' \subset \Gamma$ of positive density, together with a vector v such that for every $x \in \Gamma'$, we have $\widehat{P}(x) = \widehat{P}(x + v)$, then we will have $D(\widehat{P}(\Gamma)) \leq D(\widehat{P}) - D(\Gamma')$. This study of the frequencies of differences will include a Minkowski-type theorem for almost-periodic patterns (Theorem 4).

The particular problem of the discretization of linear maps has been quite little studied. To our knowledge, what has been made in this direction has been initiated by image processing. One wants to avoid phenomenons like loss of information (due to the fact that discretizations of linear maps are not injective) or aliasing (the apparition of undesirable periodic patterns in the image, due for example to a resonance between a periodic pattern in the image and the discretized map). To our knowledge, the existing studies are mostly interested in the linear maps with *rational coefficients* (see for example [8, 11] or [9]), including the specific case of *rotations* (see for example [1, 2, 12–14]). These works mainly focus on the local behaviour of the images of $\mathbf{Z}^2$ by discretizations of linear maps: given a radius R, what pattern can follow the intersection of this set with any ball of radius R? What is the number of such patterns, what are their frequencies? Are they complex (in a sense to define) or not? Are these maps bijections? In particular, the thesis [12] of B. Nouvel gives a characterization of the angles for which the discrete rotation is a bijection (such angles are countable and accumulate only on 0). Our result complements that of B. Nouvel: on the one hand it expresses that a generic sequence of discretizations is far from being a bijection, and on the other hand this remains true in any dimension.

Note that Theorem 1 will be generalized to the case of matrices of determinant 1 in [5], with more sophisticated techniques (see also [6]).

2   Almost Periodic Sets

In this section, we introduce the basic notions that we will use during the study of discretizations of isometries of $\mathbf{R}^n$.

We fix once for all an integer $n \geq 1$. We will denote by $[\![a, b]\!]$ the integer segment $[a, b] \cap \mathbf{Z}$. In this part, every ball will be taken with respect to the infinite norm; in particular, for $x = (x_1, \cdots, x_n)$, we will have

$$B(x, R) = B_\infty(x, R) = \{y = (y_1, \cdots, y_n) \in \mathbf{R}^n \mid \forall i \in [\![1, n]\!], |x_i - y_i| < R\}.$$

We will also denote $B_R = B(0, R)$. Finally, we will denote by $\lfloor x \rfloor$ the biggest integer that is smaller than x and $\lceil x \rceil$ the smallest integer that is bigger than x. For a set $B \subset \mathbf{R}^n$, we will denote $[B] = B \cap \mathbf{Z}^n$.

2.1   Almost Periodic Patterns: Definitions and First Properties

In this subsection, we define the notion of almost periodic pattern and prove that these sets possess a uniform density.

Definition 4. *Let Γ be a subset of $\mathbf{R}^n$.*

- *We say that Γ is* relatively dense *if there exists $R_\Gamma > 0$ such that each ball with radius at least R_Γ contains at least one point of Γ.*
- *We say that Γ is a* uniformly discrete *if there exists $r_\Gamma > 0$ such that each ball with radius at most r_Γ contains at most one point of Γ.*

The set Γ is called a Delone *set (see for instance [10]) if it is both relatively dense and uniformly discrete.*

Definition 5. *For a discrete set $\Gamma \subset \mathbf{R}^n$ and $R \geq 1$, we define the uniform R-density:*

$$D_R^+(\Gamma) = \sup_{x \in \mathbf{R}^n} \frac{\mathrm{Card}\left(B(x, R) \cap \Gamma\right)}{\mathrm{Card}\left(B(x, R) \cap \mathbf{Z}^n\right)}, \tag{1}$$

and the uniform upper density:

$$D^+(\Gamma) = \limsup_{R \to +\infty} D_R^+(\Gamma). \tag{2}$$

Remark that if $\Gamma \subset \mathbf{R}^n$ is Delone for the parameters r_Γ and R_Γ, then its upper density satisfies:

$$\frac{1}{(2R_\Gamma + 1)^n} \leq D^+(\Gamma) \leq \frac{1}{(2r_\Gamma + 1)^n}.$$

We can now define the notion of almost periodic pattern that we will use throughout this paper. Roughly speaking, an almost periodic pattern Γ is a set for which there exists a relatively dense set of translations of Γ, where a vector v is a translation of Γ if $\Gamma - v$ is equal to Γ up to a set of upper density smaller than ε. More precisely, we state the following definition.

Definition 6. *A Delone set* $\Gamma \subset \mathbf{Z}^n$ *is an* almost periodic pattern *if for every* $\varepsilon > 0$, *there exists* $R_\varepsilon > 0$ *and a relatively dense set* $\mathcal{N}_\varepsilon$, *called the* set of ε-translations *of* Γ, *such that*

$$\forall R \geq R_\varepsilon, \ \forall v \in \mathcal{N}_\varepsilon, \ D_R^+\big((\Gamma + v)\Delta\Gamma\big) < \varepsilon. \tag{3}$$

Of course, every lattice, or every finite union of translates of a given lattice, is an almost periodic pattern. We will see in next subsection a large class of examples of almost periodic patterns: images of $\mathbf{Z}^n$ by discretizations of linear maps.

We end this introduction to almost periodic patterns by stating that the notion of almost periodic pattern is invariant under discretizations of linear isometries: the image of an almost periodic pattern by the discretization of a linear isometry is still an almost periodic pattern.

Theorem 2. *Let* $\Gamma \subset \mathbf{Z}^n$ *be an almost periodic pattern and* $P \in O_n(\mathbf{R})$. *Then the set* $\widehat{P}(\Gamma)$ *is an almost periodic pattern.*

This implies that, given a sequence $(P_k)_{k \geq 1}$ of isometries of $\mathbf{R}^n$, the successive images $(\widehat{P_k} \circ \cdots \circ \widehat{P_1})(\mathbf{Z}^n)$ are almost periodic patterns. See Fig. 2 for an example of the successive images of $\mathbf{Z}^2$ by a random sequence of bounded matrices of $O_2(\mathbf{R})$. The proof of Theorem 2 will be done in Appendix B. Examples of sets $\widehat{P}(\mathbf{Z}^2)$ for various rotations P can be found in Fig. 4, where the almost periodicity is patent. Remark that Theorem 2 implies that the limsup in Eq. (2) is in fact a limit, which remains the same if in Eq. (1) we consider an inf instead of a sup (see [7]).

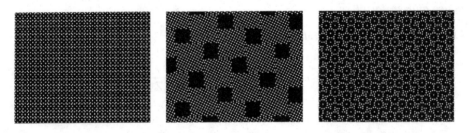

Fig. 4. Images of $\mathbf{Z}^2$ by discretizations of rotations, a point is black if it belongs to the image of $\mathbf{Z}^2$ by the discretization of the rotation. From left to right and top to bottom, angles $\pi/4$, $\pi/5$ and $\pi/6$.

2.2   Differences in Almost Periodic Patterns

We will need to understand how behave the differences in an almost periodic pattern Γ, i.e. the vectors $x - y$ with $x, y \in \Gamma$. In fact, we will study the frequency of appearance of these differences.

Definition 7. *For $v \in \mathbf{Z}^n$, we set*

$$\rho_\Gamma(v) = \frac{D\{x \in \Gamma \mid x + v \in \Gamma\}}{D(\Gamma)} = \frac{D\big(\Gamma \cap (\Gamma - v)\big)}{D(\Gamma)} \in [0,1]$$

the frequency *of the difference v in the almost periodic pattern Γ.*

Studying frequencies of differences allows to focus on the global behaviour of an almost periodic set. The function ρ_Γ is itself almost periodic in the sense given by H. Bohr (see [3]).

Definition 8. *Let $f : \mathbf{Z}^n \to \mathbf{R}$. Denoting by T_v the translation of vector v, we say that f is* Bohr almost periodic *(also called* uniformly almost periodic*) if for every $\varepsilon > 0$, the set*

$$\mathcal{N}_\varepsilon = \{v \in \mathbf{Z}^n \mid \|f - f \circ T_v\|_\infty < \varepsilon\},$$

is relatively dense.

If $f : \mathbf{Z}^n \to \mathbf{R}$ is a Bohr almost periodic function, then it possesses a *mean* $\mathcal{M}(f)$ (see for example the historical paper of H. Bohr [3, Satz VIII]), which satisfies: for every $\varepsilon > 0$, there exists $R_0 > 0$ such that for every $R \geq R_0$ and every $x \in \mathbf{R}^n$, we have

$$\left| \mathcal{M}(f) - \frac{1}{\mathrm{Card}[B(x,R)]} \sum_{v \in [B(x,R)]} f(v) \right| < \varepsilon.$$

The fact that ρ_Γ is Bohr almost periodic is straightforward.

Lemma 1. *If Γ is an almost periodic pattern, then the function ρ_Γ is Bohr almost periodic.*

In fact, we can compute precisely the mean of $\rho(\Gamma)$.

Proposition 1. *If Γ is an almost periodic pattern, then we have*

$$\mathcal{M}(\rho_\Gamma) = D(\Gamma).$$

The proof of this proposition will be done in Appendix A.

We now state a Minkowski-type theorem for the map ρ_Γ. To begin with, we recall the classical Minkowski theorem (see for example the book [4]).

Theorem 3 (Minkowski). *Let Λ be a lattice of $\mathbf{R}^n$, $k \in \mathbf{N}$ and $S \subset \mathbf{R}^n$ be a centrally symmetric convex body. If $\mathrm{Leb}(S/2) > k \, \mathrm{covol}(\Lambda)$, then S contains at least $2k$ distinct points of $\Lambda \backslash \{0\}$.*

Theorem 4. *Let $\Gamma \subset \mathbf{Z}^n$ be an almost periodic pattern of density $D(\Gamma)$. Let S be a centrally symmetric body, with $\mathrm{Leb}(S) > 4^n k$. If for every $v \in S \cap \mathbf{Z}^n$, we have $\rho_\Gamma(v) < \rho_0$, then*

$$\rho_0 \geq \frac{1}{k}\left(1 - \frac{1}{D(\Gamma)(2k+1)}\right).$$

In particular, if $k \geq \frac{1}{D(\Gamma)}$, then there exists $x \in C \cap \mathbf{Z}^n$ such that $\rho_\Gamma(x) \geq \frac{D(\Gamma)}{2}$.

Proof (of Theorem 4). Minkowski theorem (Theorem 3) asserts that $S/2$ contains at least $2k + 1$ distinct points of $\mathbf{Z}^n$, denoted by u^i. By the hypothesis on the value of ρ_Γ on S, and because the set of differences of $S/2$ is included in S, we know that the density of $(\Gamma + u^i) \cap (\Gamma + u^j)$ is smaller than $\rho_0 D(\Gamma)$. Thus,

$$D\left(\bigcup_i (\Gamma + u^i)\right) \geq \sum_i D(\Gamma) - \sum_{i<j} D\big((\Gamma + u^i) \cap (\Gamma + u^j)\big)$$

$$\geq (2k+1)D(\Gamma) - \frac{2k(2k+1)}{2}\rho_0 D(\Gamma).$$

The theorem then follows from the fact that the left member of this inequality is smaller than 1. ∎

3   Rate of Injectivity of Isometries

We now focus in more detail on the rate of injectivity of a sequence of isometries (see Definition 2).

3.1   A Geometric Viewpoint on the Rate of Injectivity

In this subsection, we present a geometric construction to compute the rate of injectivity of a generic matrix, and some applications of it.

Let $P \in O_n(\mathbf{R})$ and $\Lambda = P(\mathbf{Z}^n)$. The density of $\pi(\Lambda)$ is the proportion of $x \in \mathbf{Z}^n$ belonging to $\pi(\Lambda)$; in other words the proportion of $x \in \mathbf{Z}^n$ such that there exists $\lambda \in \Lambda$ whose distance to x (for $\|\cdot\|_\infty$) is smaller than $1/2$. remark that this property only depends on the value of x modulo Λ. If we consider the union

$$U = \bigcup_{\lambda \in \Lambda} B(\lambda, 1/2)$$

of balls of radius $1/2$ centred on the points of Λ (see Fig. 5), then $x \in \pi(\Lambda)$ if and only if $x \in U \cap \mathbf{Z}^n$. So, if we set ν the measure of repartition of the $x \in \mathbf{Z}^n$ modulo Λ, that is

$$\nu = \lim_{R \to +\infty} \frac{1}{\mathrm{Card}(B_R \cap \mathbf{Z}^n)} \sum_{x \in B_R \cap \mathbf{Z}^n} \delta_{\mathrm{pr}_{\mathbf{R}^n/\Lambda}(x)},$$

then we obtain the following formula.

Proposition 2. *For every $P \in O_n(\mathbf{R})$ (we identify U with its projection of $\mathbf{R}^n/\Lambda$),*

$$\tau(P) = D\big(\pi(\Lambda)\big) = \nu\big(\mathrm{pr}_{\mathbf{R}^n/\Lambda}(U)\big).$$

An even more simple formula holds when the matrix P is totally irrational.

Definition 9. *We say that a matrix $P \in O_n(\mathbf{R})$ is totally irrational if the image $P(\mathbf{Z}^n)$ is equidistributed[1] modulo $\mathbf{Z}^n$; in particular, this is true when the coefficients of P form a $\mathbf{Q}$-free family.*

---

[1] It is equivalent to require that it is dense instead of equidistributed.

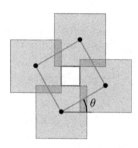

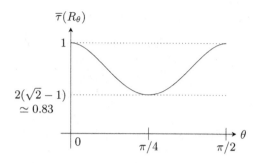

Fig. 5. Computation of the mean rate of injectivity of a rotation of $\mathbf{R}^2$: it is equal to 1 minus the area of the interior of the red square (Color figure online).

Fig. 6. Mean rate of injectivity of a rotation of $\mathbf{R}^2$ depending of the angle of the rotation.

If the matrix P is totally irrational, then the measure ν is the uniform measure. Thus, if $\mathcal{D}$ is a fundamental domain of $\mathbf{R}^n/\Lambda$, then $\tau(P)$ is the area of $\mathcal{D} \cap U$. We call the area of $\mathcal{D} \cap U$ the *mean rate of injectivity* of P and denote it by $\overline{\tau}(P)$.

With the same kind of arguments, we easily obtain a formula for $\rho_{\widehat{P}(\mathbf{Z}^n)}(v)$ (the frequency of the difference v in $\widehat{P}(\mathbf{Z}^n)$, see Definition 7).

Proposition 3. *If $P \in GL_n(\mathbf{R})$ is totally irrational, then for every $v \in \mathbf{Z}^n$,*

$$\rho_{\widehat{P}(\mathbf{Z}^n)}(v) = \mathrm{Leb}\left(B(v, 1/2) \cap U\right).$$

Proof (Sketch of proof of Proposition 3). We want to know which proportion of points $x \in \Gamma = \widehat{P}(\mathbf{Z}^n)$ are such that $x + v$ also belongs to Γ. But modulo $\Lambda = P(\mathbf{Z}^n)$, x belongs to Γ if and only if $x \in B(0, 1/2)$. Similarly, $x + v$ belongs to Γ if and only if $x \in B(-v, 1/2)$. Thus, by equirepartition, $\rho_{\widehat{P}(\mathbf{Z}^n)}(v)$ is equal to the area of $B(v, 1/2) \cap U$. $\qquad\qquad\square$

From Proposition 2, we deduce the continuity of $\overline{\tau}$. More precisely, $\overline{\tau}$ is continuous and piecewise polynomial of degree smaller than n; moreover τ coincides with a continuous function on a generic subset of $O_n(\mathbf{R})$.

It also allows to compute simply the mean rate of injectivity of some examples of matrices: for $\theta \in [0, \pi/2]$, the mean rate of injectivity of a rotation of $\mathbf{R}^2$ of angle θ is (see Figs. 5 and 6).

$$\overline{\tau}(R_\theta) = 1 - (\cos(\theta) + \sin(\theta) - 1)^2.$$

In particular, in the neighbourhoods of all the nontrivial angles on which the discrete rotation is bijective (see [12,13]), most of the rotations have a rate of injectivity bounded away from 1.

3.2 Diffusion Process

In this paragraph, we study the action of a discretization of a matrix on the set of differences of an almost periodic pattern Γ; more precisely, we study the link between the functions ρ_Γ and $\rho_{\widehat{P}(\Gamma)}$.

For $u \in \mathbf{R}^n$, we define the function φ_u, which is a "weighted projection" of u on $\mathbf{Z}^n$.

Definition 10. *Given $u \in \mathbf{R}^n$, the function $\varphi_u = \mathbf{Z}^n \to [0,1]$ is defined by*

$$\varphi_u(v) = \begin{cases} 0 & \text{if } d_\infty(u,v) \geq 1 \\ \prod_{i=1}^n (1 - |u_i + v_i|) & \text{if } d_\infty(u,v) < 1. \end{cases}$$

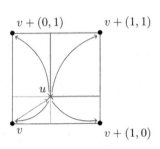

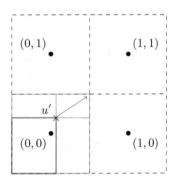

Fig. 7. The function φ_u in dimension 2: its value on one vertex of the square is equal to the area of the opposite rectangle; in particular, $\varphi_u(v)$ is the area of the rectangle with the vertices u and $v + (1,1)$ (in bold).

Fig. 8. The red vector is equal to that of Fig. 7 for $u = Pv$. If Px belongs to the bottom left rectangle, then $\pi(Px + Pv) = y \in \mathbf{Z}^2$; if Px belongs to the top left rectangle, then $\pi(Px+Pv) = y+(0,1)$ etc. (Color figure online).

In particular, the function φ_u satisfies $\sum_{v \in \mathbf{Z}^n} \varphi_u(v) = 1$, and is supported by the vertices of the integral unit cube[2] that contains[3] u. Figure 7 gives a geometric interpretation of this function φ_u.

The following property asserts that the discretization $\widehat{P}$ acts "smoothly" on the frequency of differences. In particular, when $D(\Gamma) = D(\widehat{P}\Gamma)$, the function $\rho_{\widehat{P}\Gamma}$ is obtained from the function ρ_Γ by applying a linear operator $\mathcal{A}$, acting on each Dirac function δ_v such that $\mathcal{A}\delta_u(v) = \varphi_{P(u)}(v)$. Roughly speaking, to compute $\mathcal{A}\delta_v$, we take δ_{Pv} and apply a diffusion process. In the other case where $D(\widehat{P}\Gamma) < D(\Gamma)$, we only have inequalities involving the operator $\mathcal{A}$ to compute the function $\rho_{\widehat{P}\Gamma}$.

---

[2] An integral cube has vertices with integer coordinates and its faces parallel to the canonical hyperplanes of $\mathbf{R}^n$.

[3] More precisely, the support of φ_u is the smallest integral unit cube of dimension $n' \leq n$ which contains u.

Proposition 4. *Let $\Gamma \subset \mathbf{Z}^n$ be an almost periodic pattern and $P \in O_n(\mathbf{R})$ be a generic matrix.*

(i) If $D(\widehat{P}(\Gamma)) = D(\Gamma)$, then for every $u \in \mathbf{Z}^n$,

$$\rho_{\widehat{P}(\Gamma)}(u) = \sum_{v \in \mathbf{Z}^n} \varphi_{P(v)}(u) \rho_\Gamma(v).$$

(ii) In the general case, for every $u \in \mathbf{Z}^n$, we have

$$\frac{D(\Gamma)}{D(\widehat{P}(\Gamma))} \sup_{v \in \mathbf{Z}^n} \varphi_{P(v)}(u) \rho_\Gamma(v) \le \rho_{\widehat{P}(\Gamma)}(u) \le \frac{D(\Gamma)}{D(\widehat{P}(\Gamma))} \sum_{v \in \mathbf{Z}^n} \varphi_{P(v)}(u) \rho_\Gamma(v).$$

Proof (of Proposition 4). We begin by proving the first point of the proposition. Suppose that $P \in O_n(\mathbf{R})$ is generic and that $D(\widehat{P}(\Gamma)) = D(\Gamma)$. Let $x \in \Gamma \cap (\Gamma - v)$. We consider the projection y' of $y = Px$, and the projection u' of $u = Pv$, on the fundamental domain $]-1/2, 1/2]^n$ of $\mathbf{R}^n/\mathbf{Z}^n$. We have

$$P(x + v) = \pi(Px) + \pi(Pv) + y' + u'.$$

Suppose that y' belongs to the parallelepiped whose vertices are $(-1/2, \cdots, -1/2)$ and u' (in bold in Fig. 8), then $y' + u' \in [-1/2, 1/2[^n$. Thus, $\pi(P(x+v)) = \pi(Px) + \pi(Pv)$. The same kind of results holds for the other parallelepipeds whose vertices are u' and one vertex of $[-1/2, 1/2[^n$.

We set $\Gamma = \widehat{P}(\mathbf{Z}^n)$. The genericity of P ensures that for every $v \in \mathbf{Z}^n$, the set $\Gamma \cap (\Gamma - v)$, which has density $D(\Gamma)\rho_\Gamma(v)$ (by definition of ρ_Γ), is equidistributed modulo $\mathbf{Z}^n$ (by Lemma 4). Thus, the points x' are equidistributed modulo $\mathbf{Z}^n$. In particular, the difference v will spread into the differences which are the support of the function φ_{Pv}, and each of them will occur with a frequency given by $\varphi_{Pv}(x)\rho_\Gamma(v)$. The hypothesis about the fact that the density of the sets does not decrease imply that the contributions of each difference of Γ to the differences of $\widehat{P}(\Gamma)$ add.

In the general case, the contributions to each difference of Γ may overlap. However, applying the argument of the previous case, we can easily prove the second part of the proposition. $\square$

Remark 1. We also remark that:

(i) the density strictly decreases (that is, $D(\widehat{P}(\Gamma)) < D(\Gamma)$) if and only if there exists $v_0 \in \mathbf{Z}^n$ such that $\rho_\Gamma(v_0) > 0$ and $\|Pv_0\|_\infty < 1$;
(ii) if there exists $v_0 \in \mathbf{Z}^n$ such that

$$\sum_{v \in \mathbf{Z}^n} \varphi_{P(v)}(v_0) \rho_\Gamma(v_0) > 1,$$

then the density strictly decreases by at least $\sum_{v \in \mathbf{Z}^n} \varphi_{P(v)}(v_0) \rho_\Gamma(v_0) - 1$.

3.3   Rate of Injectivity of a Generic Sequence of Isometries

We now come to the proof of the main theorem of this paper (Theorem 1). We will begin by applying the Minkowski theorem for almost periodic patterns (Theorem 4), which gives *one* nonzero difference whose frequency is positive. The rest of the proof of Theorem 1 consists in using again an argument of equidistribution. More precisely, we apply successively the following lemma, which asserts that given an almost periodic pattern Γ of density D_0, a sequence of isometries and $\delta > 0$, then, perturbing each isometry of at most δ if necessary, we can make the density of the k_0-th image of Γ smaller than $\lambda_0 D_0$, with k_0 and λ_0 depending only on D_0 and δ. The proof of this lemma involves the study of the action of the discretizations on differences made in Proposition 4

Lemma 2. *Let $(P_k)_{k \geq 1}$ be a sequence of matrices of $O_n(\mathbf{R})$ and $\Gamma \subset \mathbf{Z}^n$ an almost periodic pattern. Given $\delta > 0$ and $D > 0$ such that $D(\Gamma) \geq D$, there exists $k_0 = k_0(D)$ (decreasing in D), $\lambda_0 = \lambda_0(D, \delta) < 1$ (decreasing in D and in δ), and a sequence $(Q_k)_{k \geq 1}$ of totally irrational matrices of $O_n(\mathbf{R})$, such that $\|P_k - Q_k\| \leq \delta$ for every $k \geq 1$ and*

$$D\big((\widehat{Q_{k_0}} \circ \cdots \circ \widehat{Q_1})(\Gamma)\big) < \lambda_0 D(\Gamma).$$

We begin by proving that this lemma implies Theorem 1.

Proof (of Theorem 1). Suppose that Lemma 2 is true. Let $\tau_0 \in\,]0, 1[$ and $\delta > 0$. We want to prove that we can perturb the sequence $(P_k)_k$ into a sequence $(Q_k)_k$ of isometries, which is δ-close to $(P_k)_k$ and is such that its asymptotic rate is smaller than τ_0 (and that this remains true on a whole neighbourhood of these matrices).

Thus, we can suppose that $\tau^\infty((P_k)_k) > \tau_0$. We apply Lemma 2 to obtain the parameters $k_0 = k_0(\tau_0/2)$ (because $k_0(D)$ is decreasing in D) and $\lambda_0 = \lambda_0(\tau_0/2, \delta)$ (because $\lambda_0(D, \delta)$ is decreasing in D). Applying the lemma ℓ times, this gives a sequence $(Q_k)_k$ of isometries, which is δ-close to $(P_k)_k$, such that, as long as $\tau^{\ell k_0}(Q_0, \cdots, Q_{\ell k_0}) > \tau_0/2$, we have $\tau^{\ell k_0}(Q_1, \cdots, Q_{\ell k_0}) < \lambda_0^\ell D(\mathbf{Z}^n)$. But for ℓ large enough, $\lambda_0^\ell < \tau_0$, which proves the theorem. □

Proof (of Lemma 2). The idea of the proof is the following. Firstly, we apply the Minkowski-type theorem for almost periodic patterns (Theorem 4) to find a uniform constant $C > 0$ and a point $u_0 \in \mathbf{Z}^n \setminus \{0\}$ whose norm is "not too big", such that $\rho_\Gamma(u_0) > CD(\Gamma)$. Then, we apply Proposition 4 to prove that the difference u_0 in Γ eventually goes to 0; that is, that there exists $k_0 \in \mathbf{N}^*$ and an almost periodic pattern $\widetilde{\Gamma}$ of positive density (that can be computed) such that there exists a sequence $(Q_k)_k$ of isometries, with $\|Q_i - P_i\| \leq \delta$, such that for every $x \in \widetilde{\Gamma}$,

$$(\widehat{Q_{k_0}} \circ \cdots \circ \widehat{Q_1})(x) = (\widehat{Q_{k_0}} \circ \cdots \circ \widehat{Q_1})(x + u_0).$$

This makes the density of the k_0-th image of Γ decrease:

$$D\big((\widehat{Q_{k_0}} \circ \cdots \circ \widehat{Q_1})(\Gamma)\big) \leq D(\Gamma) - D(\widetilde{\Gamma}); \quad \cdot$$

a precise estimation of the density of $\widetilde{\Gamma}$ will then prove the lemma.

We begin by applying the Minkowski-like theorem for almost periodic patterns (Theorem 4) to a *Euclidean* ball B'_R such that (recall that $[B]$ denotes the set of integer points inside B)

$$\text{Leb}(B'_R) = V_n R^n = 4^n \left\lfloor \frac{1}{D(\Gamma)} \right\rfloor, \tag{4}$$

where V_n denotes the measure of the unit ball on $\mathbf{R}^n$. Then, Theorem 4 says that there exists $u_0 \in B'_R \cap \mathbf{Z}^n \backslash \{0\}$ such that

$$\rho_\Gamma(u_0) \geq \frac{D(\Gamma)}{2}. \tag{5}$$

We now perturb each matrix P_k into a totally irrational matrix Q_k such that for every point $x \in [B'_R] \backslash \{0\}$, the point $Q_k(x)$ is far away from the lattice $\mathbf{Z}^n$. More precisely, as the set of matrices $Q \in O_n(\mathbf{R})$ such that $Q([B'_R]) \cap \mathbf{Z}^n \neq \{0\}$ is finite, there exists a constant $d_0(R, \delta)$ such that for every $P \in O_n(\mathbf{R})$, there exists $Q \in O_n(\mathbf{R})$ such that $\|P - Q\| \leq \delta$ and for every $x \in [B'_R] \backslash \{0\}$, we have $d_\infty(Q(x), \mathbf{Z}^n) > d_0(R, \delta)$. Applying Lemma 4 (which states that if the sequence $(Q_k)_k$ is generic, then the matrices Q_k are "non resonant"), we build a sequence $(Q_k)_{k \geq 1}$ of totally irrational[4] matrices of $O_n(\mathbf{R})$ such that for every $k \in \mathbf{N}^*$, we have:

- $\|P_k - Q_k\| \leq \delta$;
- for every $x \in [B'_R] \backslash \{0\}$, we have $d_\infty(Q_k(x), \mathbf{Z}^n) > d_0(R, \delta)$;
- the set $(Q_k \circ \widehat{Q_{k-1}} \circ \cdots \circ \widehat{Q_1})(\Gamma)$ is equidistributed modulo $\mathbf{Z}^n$.

We then consider the difference u_0 (given by Eq. (5)). We denote by $\lfloor P \rfloor(u)$ the point of the smallest integer cube of dimension $n' \leq n$ that contains $P(u)$ which has the smallest Euclidean norm (that is, the point of the support of $\varphi_{P(u)}$ with the smallest Euclidean norm). In particular, if $P(u) \notin \mathbf{Z}^n$, then $\|\lfloor P \rfloor(u)\|_2 < \|P(u)\|_2$ (where $\|\cdot\|_2$ is the Euclidean norm). Then, the point (ii) of Proposition 4 shows that

$$\rho_{\widehat{Q_1}(\Gamma)}(\lfloor Q_1 \rfloor(u_0)) \geq \frac{D(\Gamma)}{D(\widehat{Q_1}(\Gamma))} \varphi_{Q_1(\lfloor Q_1 \rfloor(u_0))}(u_0)\rho_\Gamma(u_0)$$

$$\geq \frac{(d_0(R, \delta))^n}{2} D(\Gamma),$$

(applying Eq. (5)) and so on, for every $k \in \mathbf{N}^*$,

$$\rho_{(\widehat{Q_k} \circ \cdots \circ \widehat{Q_1})(\Gamma)}((\lfloor Q_k \rfloor \circ \cdots \circ \lfloor Q_1 \rfloor)(u_0)) \geq \left(\frac{(d_0(R, \delta))^n}{2} \right)^k D(\Gamma).$$

We then remark that the sequence of norms $\left\| (\lfloor Q_k \rfloor \circ \cdots \circ \lfloor Q_1 \rfloor)(u_0) \right\|_2$ is decreasing and can only take a finite number of values (it lies in $\sqrt{\mathbf{Z}}$). Then,

---

[4] See Definition 9.

there exists $k_0 \leq R^2$ such that $\left(\lfloor Q_{k_0} \rfloor \circ \cdots \circ \lfloor Q_1 \rfloor \right)(u_0) = 0$; in particular, by Eq. (4), we have

$$k_0 \leq \left(\frac{4^n}{V_n} \left\lfloor \frac{1}{D(\Gamma)} \right\rfloor \right)^{2/n}.$$

Then, point (ii) of Remark 1 applied to $v_0 = 0$ implies that the density of the image set satisfies

$$D\left((\widehat{Q_k} \circ \cdots \circ \widehat{Q_1})(\Gamma) \right) \leq \left(1 - \left(\frac{(d_0(R,\delta))^n}{2} \right)^n \right)^{k_0} D(\Gamma).$$

The conclusions of the lemma are obtained by setting $\lambda_0 = 1 - \left(\frac{(d_0(R,\delta))^n}{2} \right)^{k_0}$. $\qquad \square$

4   Conclusion

By mean of Theorem 1, we have shown why it is illusory to hope that the naive algorithm of rotation of a numerical image gives good results: applying successively the discretizations of a generic sequence of rotations leads to a complete loss of information.

A   Technical Lemmas

Let us begin by the proof of Proposition 1.

Proof (of Proposition 1). This proof lies primarily in an inversion of limits.

Let $\varepsilon > 0$. As Γ is an almost periodic pattern, there exists $R_0 > 0$ such that for every $R \geq R_0$ and every $x \in \mathbf{R}^n$, we have

$$\left| D(\Gamma) - \frac{\Gamma \cap [B(x,R)]}{\mathrm{Card}[B_R]} \right| \leq \varepsilon. \tag{6}$$

So, we choose $R \geq R_0$, $x \in \mathbf{Z}^n$ and compute

$$\frac{1}{\mathrm{Card}[B_R]} \sum_{v \in [B(x,R)]} \rho_\Gamma(v) = \frac{1}{\mathrm{Card}[B_R]} \sum_{v \in [B(x,R)]} \frac{D\left((\Gamma - v) \cap \Gamma \right)}{D(\Gamma)}$$

$$= \frac{1}{\mathrm{Card}[B_R]} \sum_{v \in [B(x,R)]} \lim_{R' \to +\infty} \frac{1}{\mathrm{Card}[B_{R'}]} \sum_{y \in [B_{R'}]} \frac{\mathbf{1}_{y \in \Gamma - v} \mathbf{1}_{y \in \Gamma}}{D(\Gamma)}$$

$$= \frac{1}{D(\Gamma)} \lim_{R' \to +\infty} \frac{1}{\mathrm{Card}[B_{R'}]} \sum_{y \in [B_{R'}]} \mathbf{1}_{y \in \Gamma} \frac{1}{\mathrm{Card}[B_R]} \sum_{v \in [B(x,R)]} \mathbf{1}_{y \in \Gamma - v}$$

$$= \frac{1}{D(\Gamma)} \lim_{R' \to +\infty} \frac{1}{\mathrm{Card}[B_{R'}]} \underbrace{\sum_{y \in [B_{R'}]} \mathbf{1}_{y \in \Gamma}}_{\text{first term}} \frac{1}{\mathrm{Card}[B_R]} \underbrace{\sum_{v' \in [B(y+x,R)]} \mathbf{1}_{v' \in \Gamma}}_{\text{second term}}.$$

By Eq. (6), the second term is ε-close to $D(\Gamma)$. Considered independently, the first term is equal to $D(\Gamma)$ (still by Eq. (6)). Thus, we have

$$\left| \frac{1}{\mathrm{Card}[B(x,R)]} \sum_{v \in [B(x,R)]} \rho_\Gamma(v) - D(\Gamma) \right| \leq \varepsilon,$$

that we wanted to prove.   □

We now state an easy lemma which asserts that for ε small enough, the set of translations $\mathcal{N}_\varepsilon$ is "stable under additions with a small number of terms".

Lemma 3. *Let Γ be an almost periodic pattern, $\varepsilon > 0$ and $\ell \in \mathbf{N}$. Then if we set $\varepsilon' = \varepsilon/\ell$ and denote by $\mathcal{N}_{\varepsilon'}$ the set of translations of Γ and $R_{\varepsilon'} > 0$ the corresponding radius for the parameter ε', then for every $k \in [\![1, \ell]\!]$ and every $v_1, \cdots, v_\ell \in \mathcal{N}_{\varepsilon'}$, we have*

$$\forall R \geq R_{\varepsilon'}, \; D_R^+\Big(\big(\Gamma + \sum_{i=1}^{\ell} v_i\big) \Delta \Gamma\Big) < \varepsilon.$$

Proof (of Lemma 3). Let Γ be an almost periodic pattern, $\varepsilon > 0$, $\ell \in \mathbf{N}$, $R_0 > 0$ and $\varepsilon' = \varepsilon/\ell$. Then there exists $R_{\varepsilon'} > 0$ such that

$$\forall R \geq R_{\varepsilon'}, \; \forall v \in \mathcal{N}_{\varepsilon'}, \; D_R^+((\Gamma + v)\Delta\Gamma) < \varepsilon'.$$

We then take $1 \leq k \leq \ell$, $v_1, \cdots, v_k \in \mathcal{N}_{\varepsilon'}$ and compute

$$D_R^+\Big(\big(\Gamma + \sum_{i=1}^{k} v_i\big)\Delta\Gamma\Big) \leq \sum_{m=1}^{k} D_R^+\Big(\big(\Gamma + \sum_{i=1}^{m} v_i\big)\Delta\big(\Gamma + \sum_{i=1}^{m-1} v_i\big)\Big)$$

$$\leq \sum_{m=1}^{k} D_R^+\Big(\big((\Gamma + v_m)\Delta\Gamma\big) + \sum_{i=1}^{m-1} v_i\Big).$$

By the invariance under translation of D_R^+, we deduce that

$$D_R^+\Big(\big(\Gamma + \sum_{i=1}^{k} v_i\big)\Delta\Gamma\Big) \leq \sum_{m=1}^{k} D_R^+((\Gamma + v_m)\Delta\Gamma)$$

$$\leq k\varepsilon'.$$

As $k \leq \ell$, this ends the proof.   □

Remark 2. In particular, this lemma implies that the set $\mathcal{N}_\varepsilon$ contains arbitrarily large patches of lattices of $\mathbf{R}^n$: for every almost periodic pattern Γ, $\varepsilon > 0$ and $\ell \in \mathbf{N}$, there exists $\varepsilon' > 0$ such that for every $k_i \in [\![-\ell, \ell]\!]$ and every $v_1, \cdots, v_n \in \mathcal{N}_{\varepsilon'}$, we have

$$\forall R \geq R_{\varepsilon'}, \; D_R^+\Big(\big(\Gamma + \sum_{i=1}^{n} k_i v_i\big)\Delta\Gamma\Big) < \varepsilon.$$

The second lemma is more technical. It expresses that given an almost periodic pattern Γ, a generic matrix $A \in O_n(\mathbf{R})$ is non resonant with respect to Γ.

Lemma 4. *Let $\Gamma \subset \mathbf{Z}^n$ be an almost periodic pattern with positive uniform density. Then the set of $A \in O_n(\mathbf{R})$ such that $A(\Gamma)$ is equidistributed modulo $\mathbf{Z}^n$ is generic. More precisely, for every $\varepsilon > 0$, there exists an open and dense set of $A \in O_n(\mathbf{R})$ such that there exists $R_0 > 0$ such that for every $R > R_0$, the projection on $\mathbf{R}^n/\mathbf{Z}^n$ of the uniform measure on $A(\Gamma \cap B_R)$ is ε-close to Lebesgue measure on $\mathbf{R}^n/\mathbf{Z}^n$.*

Proof (of Lemma 4). During this proof, we consider a distance dist on $\mathcal{P}(\mathbf{R}^n/\mathbf{Z}^n)$ which is invariant under translations, where $\mathcal{P}(\mathbf{R}^n/\mathbf{Z}^n)$ denotes the space of probability Borel measures on $\mathbf{R}^n/\mathbf{Z}^n$ endowed with weak-* topology. We also suppose that this distance satisfies the following convexity inequality: if $\mu, \nu_1, \cdots, \nu_d \in \mathcal{P}(\mathbf{R}^n/\mathbf{Z}^n)$, then

$$\text{dist}\left(\mu, \frac{1}{d}\sum_{i=1}^d \nu_i\right) \leq \frac{1}{d}\sum_{i=1}^d \text{dist}(\mu, \nu_i).$$

For the simplicity of the notations, when μ and ν have not total mass 1, we will denote by $\text{dist}(\mu, \nu)$ the distance between the normalizations of μ and ν.

We consider the set $\mathcal{U}_\varepsilon$ of matrices $A \in GL_n(\mathbf{R})$ satisfying: there exists $R_0 > 0$ such that for all $R \geq R_0$,

$$\text{dist}\left(\text{Leb}_{\mathbf{R}^n/\mathbf{Z}^n}, \sum_{x \in B_R \cap \Gamma} \bar{\delta}_{Ax}\right) < \varepsilon,$$

where $\bar{\delta}_x$ is the Dirac measure of the projection of x on $\mathbf{R}^n/\mathbf{Z}^n$. We show that for every $\varepsilon > 0$, $\mathcal{U}_\varepsilon$ contains an open dense set. Then, the set $\bigcap_{\varepsilon > 0} \mathcal{U}_\varepsilon$ will be a G_δ dense set made of matrices $A \in GL_n(\mathbf{R})$ such that $A(\Gamma)$ is well distributed.

Let $\varepsilon > 0$, $\delta > 0$, $\ell > 0$ and $A \in GL_n(\mathbf{R})$. We apply Remark 2 to obtain a parameter $R_0 > 0$ and a family $v_1, \cdots, v_n$ of ε-translations of Γ such that the family of cubes $\left(B(\sum_{i=1}^n k_i v_i, R_0)\right)_{-\ell \leq k_i \leq \ell}$ is an "almost tiling" of $B_{\ell R_0}$ (in particular, each v_i is close to the vector having $2R_0$ in the i-th coordinate and 0 in the others, see Fig. 9):

(1) this collection of cubes fills almost all $B_{\ell R_0}$:

$$\frac{\text{Card}\left(\Gamma \cap \left(\bigcup_{-\ell \leq k_i \leq \ell} B(\sum_{i=1}^n k_i v_i, R_0) \Delta B_{\ell R_0}\right)\right)}{\text{Card}(\Gamma \cap B_{\ell R_0})} \leq \varepsilon;$$

(2) the overlaps of the cubes are not too big: for all collections (k_i) and (k_i') such that $-\ell \leq k_i, k_i' \leq \ell$,

$$\frac{\text{Card}\left(\Gamma \cap \left(B(\sum_{i=1}^n k_i v_i, R_0) \Delta B(\sum_{i=1}^n k_i' v_i, R_0)\right)\right)}{\text{Card}(\Gamma \cap B_{\ell R_0})} \leq \varepsilon;$$

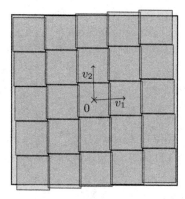

Fig. 9. "Almost tiling" of $B_{\ell R_0}$ by cubes $B(\sum_{i=1}^{n} k_i v_i, R_0)$, with $-\ell \le k_i \le \ell$.

(3) the vectors $\sum_{i=1}^{n} k_i v_i$ are translations for Γ: for every collection (k_i) such that $-\ell \le k_i \le \ell$,

$$\frac{\mathrm{Card}\left(\left(\Gamma \Delta(\Gamma - \sum_{i=1}^{n} k_i v_i)\right) \cap B_{R_0}\right)}{\mathrm{Card}(\Gamma \cap B_{R_0})} \le \varepsilon.$$

Increasing R_0 and ℓ if necessary, there exists $A' \in GL_n(\mathbf{R})$ (respectively $SL_n(\mathbf{R})$, $O_n(\mathbf{R})$) such that $\|A - A'\| \le \delta$ and that we have

$$\mathrm{dist}\left(\mathrm{Leb}_{\mathbf{R}^n/\mathbf{Z}^n}, \sum_{-\ell \le k_i \le \ell} \bar{\delta}_{A'(\sum_{i=1}^{n} k_i v_i)}\right) \le \varepsilon. \tag{7}$$

Indeed, if we denote by Λ the lattice spanned by the vectors $v_1, \cdots, v_n$, then the set of matrices A' such that $A'\Lambda$ is equidistributed modulo $\mathbf{Z}^n$ is dense in $GL_n(\mathbf{R})$ (respectively $SL_n(\mathbf{R})$ and $O_n(\mathbf{R})$).

Then, we have,

$$\mathrm{dist}\left(\mathrm{Leb}_{\mathbf{R}^n/\mathbf{Z}^n}, \sum_{\substack{-\ell \le k_i \le \ell \\ x \in \Gamma \cap B(\sum_{i=1}^{n} k_i v_i, R_0)}} \bar{\delta}_{A'x}\right)$$

$$\le \mathrm{dist}\left(\mathrm{Leb}_{\mathbf{R}^n/\mathbf{Z}^n}, \sum_{\substack{-\ell \le k_i \le \ell \\ x \in \Gamma \cap B(0, R_0)}} \bar{\delta}_{A'(\sum_{i=1}^{n} k_i v_i)+A'x}\right)$$

$$+ \mathrm{dist}\left(\sum_{\substack{-\ell \le k_i \le \ell \\ x \in \Gamma \cap B(0, R_0)}} \bar{\delta}_{A'(\sum_{i=1}^{n} k_i v_i)+A'x}, \sum_{\substack{-\ell \le k_i \le \ell \\ x \in \Gamma \cap B(\sum_{i=1}^{n} k_i v_i, R_0)}} \bar{\delta}_{A'x}\right)$$

By the property of convexity of dist, the first term is smaller than

$$\frac{1}{\mathrm{Card}\left(\Gamma \cap B(0, R_0)\right)} \sum_{x \in \Gamma \cap B(0, R_0)} \mathrm{dist}\left(\mathrm{Leb}_{\mathbf{R}^n/\mathbf{Z}^n}, \sum_{-\ell \le k_i \le \ell} \bar{\delta}_{A'(\sum_{i=1}^{n} k_i v_i)+A'x}\right);$$

by Eq. (7) and the fact that dist is invariant under translation, this term is smaller than ε. As by hypothesis, the vectors $\sum_{i=1}^{n} k_i v_i$ are ε-translations of Γ (Hypothesis (3)), the second term is also smaller than ε. Thus, we get

$$\text{dist}\left(\text{Leb}_{\mathbf{R}^n/\mathbf{Z}^n}, \sum_{\substack{-\ell \le k_i \le \ell \\ x \in \Gamma \cap B(\sum_{i=1}^{n} k_i v_i, R_0)}} \bar{\delta}_{A'x} \right) \le 2\varepsilon$$

By the fact that the family of cubes $\left(B(\sum_{i=1}^{n} k_i v_i, R_0) \right)_{-\ell \le k_i \le \ell}$ is an almost tiling of $B_{\ell R_0}$ (Hypotheses (1) and (2)), we get, for every $v \in \mathbf{R}^n$,

$$\text{dist}\left(\text{Leb}_{\mathbf{R}^n/\mathbf{Z}^n}, \sum_{x \in \Gamma \cap B_{\ell R_0}} \bar{\delta}_{A'x} \right) < 4\varepsilon.$$

Remark that we can suppose that this remains true on a whole neighbourhood of A'. We use the fact that Γ is an almost periodic pattern to deduce that A' belongs to the interior of $\mathcal{U}_\varepsilon$. □

B   Proof of Theorem 2

Definition 11. *For $A \in GL_n(\mathbf{R})$, we denote $A = (a_{i,j})_{i,j}$. We denote by $I_{\mathbf{Q}}(A)$ the set of indices i such that $a_{i,j} \in \mathbf{Q}$ for every $j \in [\![1, n]\!]$.*

The proof of Theorem 2 relies on the following remark:

Remark 3. If $a \in \mathbf{Q}$, then there exists $q \in \mathbf{N}^*$ such that $\{ax \mid x \in \mathbf{Z}\} \subset \frac{1}{q}\mathbf{Z}$. On the contrary, if $a \in \mathbf{R} \backslash \mathbf{Q}$, then the set $\{ax \mid x \in \mathbf{Z}\}$ is equidistributed in $\mathbf{R}/\mathbf{Z}$.

Thus, in the rational case, the proof will lie in an argument of periodicity. On the contrary, in the irrational case, the image $A(\mathbf{Z}^n)$ is equidistributed modulo $\mathbf{Z}^n$: on every large enough domain, the density does not move a lot when we perturb the image set $A(\mathbf{Z}^n)$ by small translations. This reasoning is formalized by Lemmas 5 and 6.

More precisely, for R large enough, we would like to find vectors w such that $D_R^+\big((\pi(A\Gamma) + w)\Delta\pi(A\Gamma)\big)$ is small. We know that there exists vectors v such that $D_R^+\big((\Gamma + v)\Delta\Gamma\big)$ is small; this implies that $D_R^+\big((A\Gamma + Av)\Delta A\Gamma\big)$ is small, thus that $D_R^+\big(\pi(A\Gamma + Av)\Delta\pi(A\Gamma)\big)$ is small. The problem is that in general, we do not have $\pi(A\Gamma + Av) = \pi(A\Gamma) + \pi(Av)$. However, this is true if we have $Av \in \mathbf{Z}^n$. Lemma 5 shows that in fact, it is possible to suppose that Av "almost" belongs to $\mathbf{Z}^n$, and Lemma 6 asserts that this property is sufficient to conclude.

The first lemma is a consequence of the pigeonhole principle.

Lemma 5. *Let $\Gamma \subset \mathbf{Z}^n$ be an almost periodic pattern, $\varepsilon > 0$, $\delta > 0$ and $A \in GL_n(\mathbf{R})$. Then we can suppose that the elements of $A(\mathcal{N}_\varepsilon)$ are δ-close to $\mathbf{Z}^n$. More precisely, there exists $R_{\varepsilon,\delta} > 0$ and a relatively dense set $\widetilde{\mathcal{N}}_{\varepsilon,\delta}$ such that*

$$\forall R \ge R_{\varepsilon,\delta}, \ \forall v \in \widetilde{\mathcal{N}}_{\varepsilon,\delta}, \ D_R^+\big((\Gamma + v)\Delta\Gamma\big) < \varepsilon,$$

and that for every $v \in \widetilde{\mathcal{N}}_{\varepsilon,\delta}$, we have $d_\infty(Av, \mathbf{Z}^n) < \delta$. Moreover, we can suppose that for every $i \in I_{\mathbf{Q}}(A)$ and every $v \in \widetilde{\mathcal{N}}_{\varepsilon,\delta}$, we have $(Av)_i \in \mathbf{Z}$.

The second lemma states that in the irrational case, we have continuity of the density under perturbations by translations.

Lemma 6. *Let $\varepsilon > 0$ and $A \in GL_n(\mathbf{R})$. Then there exists $\delta > 0$ and $R_0 > 0$ such that for all $w \in B_\infty(0, \delta)$ (such that for every $i \in I_{\mathbf{Q}}(A)$, $w_i = 0$), and for all $R \geq R_0$, we have*

$$D_R^+\big(\pi(A\mathbf{Z}^n)\varDelta\pi(A\mathbf{Z}^n + w)\big) \leq \varepsilon.$$

Remark 4. When $I_{\mathbf{Q}}(A) = \emptyset$, and in particular when A is totally irrational (see Definition 9), the map $v \mapsto \tau(A + v)$ is continuous in 0; the same proof as that of this lemma implies that this function is globally continuous.

We begin by the proofs of both lemmas, and prove Theorem 2 thereafter.

Proof (of Lemma 5). Let us begin by giving the main ideas of the proof of this lemma. For R_0 large enough, the set of remainders modulo $\mathbf{Z}^n$ of vectors Av, where v is a ε-translation of Γ belonging to B_{R_0}, is close to the set of remainders modulo $\mathbf{Z}^n$ of vectors Av, where v is any ε-translation of Γ. Moreover (by the pigeonhole principle), there exists an integer k_0 such that for each ε-translation $v \in B_{R_0}$, there exists $k \leq k_0$ such that $A(kv)$ is close to $\mathbf{Z}^n$. Thus, for every ε-translation v of Γ, there exists a $(k_0 - 1)\varepsilon$-translation $v' = (k-1)v$, belonging to $B_{k_0 R_0}$, such that $A(v + v')$ is close to $\mathbf{Z}^n$. The vector $v + v'$ is then a $k_0\varepsilon$-translation of Γ (by additivity of the translations) whose distance to v is smaller than $k_0 R_0$.

We now formalize these remarks. Let Γ be an almost periodic pattern, $\varepsilon > 0$ and $A \in GL_n(\mathbf{R})$. First of all, we apply the pigeonhole principle. We partition the torus $\mathbf{R}^n/\mathbf{Z}^n$ into squares whose sides are smaller than δ; we can suppose that there are at most $\lceil 1/\delta \rceil^n$ such squares. For $v \in \mathbf{R}^n$, we consider the family of vectors $\{A(kv)\}_{0 \leq k \leq \lceil 1/\delta \rceil^n}$ modulo $\mathbf{Z}^n$. By the pigeonhole principle, at least two of these vectors, say $A(k_1 v)$ and $A(k_2 v)$, with $k_1 < k_2$, lie in the same small square of $\mathbf{R}^n/\mathbf{Z}^n$. Thus, if we set $k_v = k_2 - k_1$ and $\ell = \lceil 1/\delta \rceil^n$, we have

$$1 \leq k_v \leq \ell \quad \text{and} \quad d_\infty\big(A(k_v v), \mathbf{Z}^n\big) \leq \delta. \tag{8}$$

To obtain the conclusion in the rational case, we suppose in addition that $v \in q\mathbf{Z}^n$, where $q \in \mathbf{N}^*$ is such that for every $i \in I_{\mathbf{Q}}(A)$ and every $1 \leq j \leq n$, we have $q\, a_{i,j} \in \mathbf{Z}$ (which is possible by Remark 2).

We set $\varepsilon' = \varepsilon/\ell$. By the definition of an almost periodic pattern, there exists $R_{\varepsilon'} > 0$ and a relatively dense set $\mathcal{N}_{\varepsilon'}$ such that Eq. (3) holds for the parameter ε':

$$\forall R \geq R_{\varepsilon'}, \ \forall v \in \mathcal{N}_{\varepsilon'}, \ D_R^+\big((\Gamma + v)\varDelta\Gamma\big) < \varepsilon', \tag{3'}$$

We now set

$$P = \big\{Av \bmod \mathbf{Z}^n \mid v \in \mathcal{N}_{\varepsilon'}\big\} \quad \text{and} \quad P_R = \big\{Av \bmod \mathbf{Z}^n \mid v \in \mathcal{N}_{\varepsilon'} \cap B_R\big\}.$$

We have $\bigcup_{R>0} P_R = P$, so there exists $R_0 > R_{\varepsilon'}$ such that $d_H(P, P_{R_0}) < \delta$ (where d_H denotes Hausdorff distance). Thus, for every $v \in \mathcal{N}_{\varepsilon'}$, there exists $v' \in \mathcal{N}_{\varepsilon'} \cap B_{R_0}$ such that

$$d_\infty(Av - Av', \mathbf{Z}^n) < \delta. \tag{9}$$

We then remark that for every $v' \in \mathcal{N}_{\varepsilon'} \cap B_{R_0}$, if we set $v'' = (k_{v'} - 1)v'$, then by Eq. (8), we have

$$d_\infty(Av' + Av'', \mathbf{Z}^n) = d_\infty(A(k_{v'}v'), \mathbf{Z}^n) \leq \delta.$$

Combining this with Eq. (9), we get

$$d_\infty(Av + Av'', \mathbf{Z}^n) \leq 2\delta,$$

with $v'' \in B_{\ell R_0}$.

On the other hand, $k_{v'} \leq \ell$ and Eq. (3') holds, so Lemma 3 (more precisely, Remark 2) implies that $v'' \in \mathcal{N}_\varepsilon$, that is

$$\forall R \geq R_{\varepsilon'}, \ D_R^+\big((\Gamma + v'')\Delta\Gamma\big) < \varepsilon.$$

In other words, for every $v \in \mathcal{N}_{\varepsilon'}$, there exists $v'' \in \mathcal{N}_\varepsilon \cap B_{\ell R_0}$ (with ℓ and R_0 independent from v) such that $d_\infty\big(A(v + v''), \mathbf{Z}^n\big) < 2\delta$. The set $\widetilde{\mathcal{N}}_{2\varepsilon, 2\delta}$ we look for is then the set of such sums $v + v''$. $\qquad\square$

Proof (of Lemma 6). Under the hypothesis of the lemma, for every $i \notin I_\mathbf{Q}(A)$, the sets

$$\left\{ \sum_{j=1}^n a_{i,j} x_j \mid (x_j) \in \mathbf{Z}^n \right\},$$

are equidistributed modulo $\mathbf{Z}$. Thus, for all $\varepsilon > 0$, there exists $R_0 > 0$ such that for every $R \geq R_0$,

$$D_R^+\left\{ v \in \mathbf{Z}^n \mid \exists i \notin I_\mathbf{Q}(A) : d\big((Av)_i, \mathbf{Z} + \frac{1}{2}\big) \leq \varepsilon \right\} \leq 2(n+1)\varepsilon.$$

As a consequence, for all $w \in \mathbf{R}^n$ such that $\|w\|_\infty \leq \varepsilon/(2(n+1))$ and that $w_i = 0$ for every $i \in I_\mathbf{Q}(A)$, we have

$$D_R^+\big(\pi(A\mathbf{Z}^n)\Delta\pi(A(\mathbf{Z}^n + w))\big) \leq \varepsilon.$$

Then, the lemma follows from the fact that there exists $\delta > 0$ such that $\|A(w)\|_\infty \leq \varepsilon/(2(n+1))$ as soon as $\|w\| \leq \delta$. $\qquad\square$

Proof (of Theorem 2). Let $\varepsilon > 0$. Lemma 6 gives us a corresponding $\delta > 0$, that we use to apply Lemma 5 and get a set of translations $\widetilde{\mathcal{N}}_{\varepsilon, \delta}$. Then, for every $v \in \widetilde{\mathcal{N}}_{\varepsilon, \delta}$, we write $\pi(Av) = Av + \big(\pi(Av) - Av\big) = Av + w$. The conclusions of Lemma 5 imply that $\|w\|_\infty < \delta$, and that $w_i = 0$ for every $i \in I_\mathbf{Q}(A)$.

We now explain why $\hat{A}v = \pi(Av)$ is a ε-translation for the set $\widehat{A}(\Gamma)$. Indeed, for every $R \geq \max(R_{\varepsilon,\delta}, MR_0)$, where M is the maximum of the greatest modulus of the eigenvalues of A and of the greatest modulus of the eigenvalues of A^{-1}, we have

$$D_R^+\Big(\pi(A\Gamma)\Delta\big(\pi(A\Gamma) + \hat{A}v\big)\Big) \leq D_R^+\Big(\pi(A\Gamma)\Delta\big(\pi(A\Gamma) + w\big)\Big)$$

$$+ D_R^+\Big(\big(\pi(A\Gamma) + w\big)\Delta\big(\pi(A\Gamma) + \hat{A}v\big)\Big)$$

(where $w = \pi(Av) - Av$). By Lemma 6, the first term is smaller than ε. For its part, the second term is smaller than

$$D_R^+\big((A\Gamma + Av)\Delta A\Gamma\big) \leq M^2 D_{RM}^+\big((\Gamma + v)\Delta\Gamma\big),$$

which is smaller than ε because $v \in \mathcal{N}_\varepsilon$.   □

References

1. Andres, E.: The quasi-shear rotation. In: Miguet, S., Ubéda, S., Montanvert, A. (eds.) DGCI 1996. LNCS, vol. 1176, pp. 307–314. Springer, Heidelberg (1996)
2. Berthé, V., Vuillon, L.: Tilings and rotations on the torus: a two-dimensional generalization of Sturmian sequences. Discrete Math. **223**(1–3), 27–53 (2000). http://dx.doi.org/10.1016/S0012-365X(00)00039-X
3. Bohr, H.: Zur theorie der fast periodischen funktionen. Acta Math. **45**(1), 29–127 (1924). I. Eine Verallgemeinerung der Theorie der Fourierreihen. http://dx.doi.org/10.1007/BF02395468
4. Gruber, P., Lekkerkerker, C.G.: Geometry of numbers, North-Holland Mathematical Library, vol. 37, 2nd edn. North-Holland Publishing Co., Amsterdam (1987)
5. Guihéneuf, P.A.: Degree of recurrence of generic diffeomorphisms (2015). arxiv:1510.00723
6. Guihéneuf, P.A.: Discrétisations spatiales de systèmes dynamiques génériques. Ph.D. thesis, Université Paris-Sud (2015)
7. Guihéneuf, P.A.: Model sets, almost periodic patterns, uniform density and linear maps, soumis (2015). arxiv:1512.00650
8. Jacob-Da Col, M.A.: Transformation of digital images by discrete affine applications. Comput. Graph. **19**(3), 373–389 (1995)
9. Jacob-Da Col, M.A.: Applications quasi-affines et pavages du plan discret. Theoret. Comput. Sci. **259**(1–2), 245–269 (2001). http://dx.doi.org/10.1016/S0304-3975(00)00006-2
10. Moody, R.V.: Meyer sets and their duals. In: Moody, R.V. (ed.) The Mathematics of Long-Range Aperiodic Order, Waterloo, ON (1995). NATO Adv. Sci. Inst. Ser. C Math. Phys. Sci., vol. 489, pp. 403–441. Kluwer Academic Publishing, Dordrecht (1997)
11. Nehlig, P.: Applications quasi affines: pavages par images réciproques. Theoret. Comput. Sci. **156**(1–2), 1–38 (1996). http://dx.doi.org/10.1016/0304-3975(95)00040-2
12. Nouvel, B.: Rotations discretes et automates cellulaires. Ph.D. thesis, Ecole normale supérieure de lyon - ENS LYON. https://tel.archives-ouvertes.fr/tel-00444088

13. Nouvel, B., Rémila, É.: Characterization of bijective discretized rotations. In: Klette, R., Žunić, J. (eds.) IWCIA 2004. LNCS, vol. 3322, pp. 248–259. Springer, Heidelberg (2004). http://dx.doi.org/10.1007/978-3-540-30503-3_19
14. Thibault, Y.: Rotations in 2D and 3D discrete spaces. Ph.D. thesis, Université Paris-Est. https://tel.archives-ouvertes.fr/tel-00596947

Discrete Tomography

A Tomographical Interpretation of a Sufficient Condition on h-Graphical Sequences

Srecko Brlek[1] and Andrea Frosini[2]([⊠])

[1] Laboratoire de combinatoire et d'informatique mathématique,
UQAM, Montreal, Canada
[2] Dipartimento di Matematica e Informatica, Università di Firenze, Firenze, Italy
andrea.frosini@unifi.it

Abstract. The notion of hypergraph generalizes that of graph in the sense that each hyperedge is a non-void subset of the set of vertices, without constraints on its cardinality.

A fundamental and widely investigated notion related both to graphs and to hypergraphs is the characterization of their degree sequences, that is the lists of their vertex degrees.

Concerning graphs, this problem has been solved in a classical study by Erdős and Gallai, while no efficient solutions are known for hypergraphs. If we restrict the (degree sequences) characterization to uniform hypergraphs, several necessary conditions are provided in the literature, but only few sufficient ones: among the latter, a recent one requires to split a sequence into suitable subsequences whose graphicality has to be recursively tested. Unfortunately, such an approach does not allow a direct efficient implementation.

We study this problem under a tomographical perspective by adapting an already known reconstruction algorithm that has been defined for regular h-uniform degree sequences to the proposed instances, providing efficiency to the sufficient condition. Furthermore, we extend the set of h-uniform degree sequences whose graphicality can be efficiently tested. This tomographical approach seems extremely promising for further developments.

Keywords: Graphic sequence · Discrete Tomography · Reconstruction problem

1 Introduction

A hypergraph $\mathcal{H}$ is defined as a couple $(Vert, \mathcal{E})$, where $Vert$ is a finite set of vertices $v_1, \ldots, v_n$, and $\mathcal{E} \subset 2^{|Vert|} \setminus \emptyset$ is a set of hyperedges, i.e. subsets of $Vert$. The notion of hypergraph naturally extends that of graph, where the edges are restricted to only couples of vertices (see [2] for preliminary notions and results on hypergraphs). In this paper, we consider *simple* hypergraphs, i.e. hypergraphs that are loopless and with distinct hyperedges. The *degree* of a vertex $v \in Vert$ is the number of hyperedges that contain v. A hypergraph is said to be *h-uniform* (simply *h-graph*), if every hyperedge has cardinality h.

© Springer International Publishing Switzerland 2016
N. Normand et al. (Eds.): DGCI 2016, LNCS 9647, pp. 95–104, 2016.
DOI: 10.1007/978-3-319-32360-2_7

A fundamental and widely investigated notion related both to graph and to hypergraph is that of degree sequence, that is the list of its vertex degrees, usually written in non-increasing order, as $d = (d_1, d_2, \ldots, d_n)$, with $d_1 \geq d_2 \geq \cdots \geq d_n$, and n being the cardinality of $Vert$.

The problem of characterizing the degree sequences for simple graphs, say graphic sequences, was solved by Erdős and Gallai (see [1,6]):

Theorem 1 (Erdős, Gallai). *A sequence $d = (d_1, d_2, \ldots, d_n)$ where $d_1 \geq d_2 \geq \cdots \geq d_n$ is graphic if and only if $\Sigma_{i=1}^n d_i$ is even and*

$$\sum_{i=1}^k d_i \leq k(k-1) + \sum_{i=k+1}^n \min\{k, d_i\}, 1 \leq k \leq n.$$

Then, other characterizations appeared in the literature: in [9], seven of them are listed and they are proved to be equivalent, one of them leading to a constructive proof of the Erdős-Gallai Theorem.

On the other hand, the problem of the characterization of the degree sequences of h-uniform hypergraphs (say h-graphic sequences) is one of the most challenging among the unsolved problems in the theory of hypergraphs even for the simplest case of $h = 3$.

In [5], Dewdney proposes the following theorem as a non-constructive characterization of an h-uniform degree sequences based on the possibility of splitting a uniform hypergraph into two uniform parts, one of them (eventually void) of smaller degree:

Theorem 2 (Dewdney). *Let $\pi = (d_1, \ldots, d_n)$ be a non-increasing sequence of non-negative integers. π is h-graphic if and only if there exists a non-increasing sequence $\pi' = (d'_2, \ldots, d'_n)$ of non-negative integers such that*

1. *π' is $(h-1)$-graphic,*
2. *$\sum_{i=2}^n d'_i = (h-1)d_1$, and*
3. *$\pi'' = (d_2 - d'_2, \ldots, d_n - d'_n)$ is h-graphic.*

The underlying idea in the characterization rests on the possibility of splitting an h-uniform hypergraph H into two parts: for each vertex v, the first one consists of the hypergraph obtained from H after deleting all the hyperedges not containing v, and then removing, from all the remaining hyperedges, the vertex v; this hypergraph is identified in the literature with $L_H(v)$, say the *link* of v, and its degree sequence the *link sequence* of v. The second hypergraph H_v^-, say the *residual* of v, is obtained from H after removing all hyperedges containing v. It is clear that H can be obtained from $L_H(v)$ and H_v^-; furthermore one can notice that $L_H(v)$ is $(h-1)$-uniform, while H_v^- preserves the h-uniformity.

Such a recursive decomposition of a uniform hypergraph into smaller parts stops when each of them either is 2-uniform, or has only one single hyperedge: in both cases the hypergraphs that realize the sequence can be efficiently reconstructed. Finally, we proceed in merging the obtained hypergraph: let $\pi' = (d'_2, \ldots, d'_n)$ be $(h-1)$-graphic and $\pi'' = (d''_2, \ldots, d''_n)$ be h-graphic, and let

$H' = (Vert', \mathcal{E}')$ and $H'' = (Vert'', \mathcal{E}'')$ be the related hypergraphs, respectively. H' and H'' can be merged into the hypergraph $H = (Vert, \mathcal{E})$ whose degree sequence is $\pi = (\frac{|\mathcal{E}|}{h-1}, d'_1 + d''_1, \ldots, d'_n + d''_n)$, as follows: identify the sets $Vert'$ and $Vert''$

- $Vert = Vert' \cup \{v\}$, v being a new vertex not in $Vert$,
- $\mathcal{E} = \mathcal{E}'_v \cup \mathcal{E}''$ being $\mathcal{E}'_v$ the set of hyperedges obtained adding to each element of $\mathcal{E}'$ the vertex v.

It is straightforward that $\mathcal{E}$ is h-uniform and it has π as degree sequence.

We point out that the process to split the degree sequence π cannot be efficiently performed, since one has to test, in general, a non-polynomial number of couple of sequences π' and π''.

So, gaining the efficiency in this step, will guarantee efficiency to the whole process, making a valuable step towards the general problem of the characterization of the degree sequences of hypergraphs. Surprisingly, many necessary conditions have been provided for a sequence to be h-uniform, most of them generalize the Erdős and Gallai Theorem, or rely on two well known theorems by Havel and Hakimi [13,14], on the other hand, few necessary ones are present. Recently, one of this latter, provided in [10], exploits Dewdney's Theorem to set a lower bound on the length of a sequence in order to be h-uniform, according to its maximum value and to the span of its elements, where span of a sequence means the maximum difference between its elements. The present paper describes and extends this result using a different perspective, providing a polynomial time strategy to determine one of the hypergraphs of the related instances. The Discrete Tomography framework we are going to consider, has valuable mathematical and statistical tools to challenge inverse problems in the form of reconstructions of discrete objects modeled as integer matrices, from the knowledge of their row and column sums, say horizontal and vertical projections. The paper is organized as follows: in Sect. 2 we introduce the main definitions and we recall some results about h-uniform degree sequences. Furthermore, after translating the h-uniformity problem in the Discrete Tomography framework, we sketch an already known hypergraph reconstruction strategy. In Sect. 3, we extend this strategy to a set of instances including those introduced in [10], and so providing for them an effective proof. Finally, in Sect. 4, we discuss future possible developments of our strategy, and we present some related open problems.

2   Definitions and Known Results

Here, we recall a sufficient condition, provided in [10], for a sequence to be h-uniform, together with an extension as corollary. This condition turns out to be non efficient in the sense that a non-polynomial number of cases has to be considered in order to test the h-uniformity. Then, we move to the Discrete Tomography environment, where we show how to embed the h-uniformity problem. A recent strategy described in [11] allows to efficiently reconstruct one of the h-uniform hypergraphs compatible with a given constant degree sequence. After recalling this result, we show how to adapt it to comprehend near-regular sequences.

2.1   A Sufficient Condition for a Sequence to be h-Graphic

A characterization of the degree sequences of simple hypergraphs is a challenging task. As a first step, we consider the subset of hypergraphs having h-uniformity, i.e. those hypergraphs whose hyperedges have the same cardinality h.

The following sufficient condition for a sequence π to be h-graphic is provided in [10] and it relies on Theorem 2; when $h = 2$, it turns out to be a simple consequence of the Erdős-Gallai Theorem for graphs. Let $\sigma(\pi)$ indicate the sum of the elements of π.

Theorem 3.   *Let π be a non-increasing sequence of length n with maximum entry Δ and t entries that are at least $\Delta - 1$. If h divides $\sigma(\pi)$ and*

$$\binom{t-1}{h-1} \geq \Delta \tag{1}$$

then π is h-graphic.

It is useful to sketch the proof of this theorem in order to underline the connection with Theorem 2 and, as a consequence, the non-efficiency of the process.

The h-graphicality of a sequence $\pi = (d_0, \ldots, d_{n-1})$ is tested after recursively split it into a series of link and residual sequences as follows: let $0 \leq i \leq \Delta - 1$,

$$s_i = \sum_{j=1}^{n-1} max\{0, d_j - i\} - (h-1)\Delta \quad \text{and} \quad c = max\{i : s_i \geq 0\}.$$

The link sequence of π is defined as the sequence $L = (l_1, \ldots, l_{n-1})$ where, for $1 \leq i \leq n - 1$,

$$l_i = \begin{cases} d_i - c - 1 & \text{if } 1 \leq i \leq s_c, \\ d_i - c & \text{otherwise.} \end{cases} \tag{2}$$

Finally, the residual sequence is defined as the sequence $R = (r_1, \ldots, r_{n-1})$, with $r_i = d_i - l_i$, for $1 \leq i \leq n - 1$. The proof proceeds by showing that the two sequences already defined are $h - 1$ and h uniform, respectively. The proof completes after noticing that a sequence of the form $\pi' = (1^{mh}, 0^{n-mh})$ is h-graphic since it is realized by a hypergraph on n vertices having m disjoint hyperedges.

Corollary 1. *Let π be a non-increasing sequence with maximum entry Δ, and let p be the minimum integer such that $\Delta \leq \binom{p-1}{h-1}$. If h divides $\sigma(\pi)$ and $\sigma(\pi) \geq (\Delta - 1)p + 1$, then π is h-graphic.*

The corollary allows to drop the near-regularity condition on the first part of the sequence, replacing it with a sufficiently large sum of the sequence.

In the sequel, we will show how one can have a better grasp on both these results once translated into the Discrete Tomography environment. Furthermore, this translation allows to use its mathematical tools to enlarge the set of sequences that can be tested be h-graphic.

2.2 Translating h-Graphicality into the Discrete Tomography Environment

The problem of checking the h-graphicality of a non-increasing integer sequence π has been related to a class of problems that are of great relevance in the field of Discrete Tomography. More precisely the aim of Discrete Tomography is the retrieval of geometrical information about a physical structure, regarded as a finite set of points in the integer lattice, from measurements, known as projections, of the number of atoms in the structure that lie on parallel lines with fixed scopes. A common simplification is to represent a finite physical structure as a binary matrix, where an entry is 1 or 0 according to the presence or absence of an atom in the structure at the corresponding point of the lattice. One of the challenging problems in the field is then to reconstruct the structure, or, at least, to detect some of its geometrical properties from a small number of projections. One can refer to the books of G.T. Herman and A. Kuba [7,8] for further information on the theory, algorithms and applications of this classical problem in Discrete Tomography.

So, let $A = (a_{i,j})$ be an $m \times n$ binary matrix; we define two vectors $H = (h_1, \ldots, h_m)$, and $V = (v_1, \ldots, v_n)$ called the horizontal and vertical projections of A, respectively, such that

$$\sum_{j=1}^{n} a_{i,j} = h_i \text{ and } \sum_{i=1}^{m} a_{i,j} = v_j \text{ with } 1 \le i \le m, 1 \le j \le n.$$

Seminal results show that the characterization, and the reconstruction of A from its two projections H and V, can be done in polynomial time; see [7] for a survey. Furthermore, in [7,8] there are applications in Discrete Tomography requiring additional constraints.

As shown in [1], Chapters 14 and 15, this problem is equivalent to the reconstruction of a bipartite graph $G = (H, V, E)$ from its degree sequences $H = (h_1, \ldots, h_m)$ and $V = (v_1, \ldots, v_n)$.

So, in this context, the problem of the characterization of the degree sequence $(d_1, d_2, \ldots, d_n)$ of an h-uniform hypergraph $\mathcal{H}$ (without parallel edges) asks whether there is a binary matrix A with non-negative projection vectors $H = (h, h, \ldots, h)$ and $V = (d_1, d_2, \ldots, d_n)$ with distinct rows, i.e., A is the incidence matrix of $\mathcal{H}$ where rows and columns correspond to hyperedges and vertices, respectively, so that the element $a_{i,j}$ has value 1 if and only if the i-th hyperedge contains the j-th vertex. We indicate with $\mathcal{E}$ the class of such matrices.

In [11], the authors consider the case of h-uniform hypergraphs that are also d-regular, i.e., each vertex has the same degree. This reflects on the vertical projection V of the related matrix by setting all its values to d.

The authors start from the following two trivial conditions that are necessary for the existence of an $m \times n$ matrix consistent with two vectors $H = (h_1, \ldots, h_m)$ and $V = (v_1, \ldots, v_n)$ of projections:

Condition 1: for each $1 \le i \le m$ and $1 \le j \le n$, it holds $h_i \le n$ and $v_j \le m$;

Condition 2: the sums of the entries of the horizontal and the vertical projections are equal, i.e., $\sum_{i=1}^{m} h_i = \sum_{j=1}^{n} v_j$,

and then they add a third one that is trivially necessary and that determines the existence of an element of $\mathcal{E}$ having projections $H = (h, \ldots, h)$ and $V = (d, \ldots, d)$:

Condition 3: the following inequality holds:

$$d \leq \frac{h}{n} \cdot \binom{n}{h}.$$

Condition 3 can be rephrased, in our setting, as follows: there does not exist a matrix in $\mathcal{E}$ having $H = (h, \ldots, h)$ and $V = (d, \ldots, d)$ as projections, and more than $\binom{n}{h}$ different rows, otherwise at least two of them have to be equal.

The characterization is obtained by showing that the three conditions are also sufficient: the authors define an efficient procedure that reconstructs an element A of $\mathcal{E}$ from its constant projections H and V, and that uses properties of combinatorics of words; we indicate this procedure by $REC(H, V, \mathcal{E})$. In particular, if we consider each row of the $m \times n$ output matrix A as a binary word, then the procedure REC inserts, inside A, a submatrix A' of dimension $m' \times n$ having the minimal admissible constant vector of vertical projections and whose rows are all the possible cyclic h-shifts of the word $u = (1)^h, (0)^{n-h}$. Here, the power notation $(x)^y$ indicates the repetition of the symbol x for y times, and the cyclic h-shift operator is defined on a generic word $w = w_1, \ldots, w_n$ as the operator $s(w)^h = w_{n-h+1}, \ldots, w_n, w_1, \ldots, w_{n-h}$. Simple computations leads to the fact that $m' = n/g.c.d.\{h, n\}$, and the vertical projections of A' have constant value $v = h/g.c.d.\{h, n\}$. The computational complexity of the reconstruction process can be obtained by observing that a CAT algorithm has been defined in [16] to generate Lyndon words of length n and given density h, and that the number of required Lyndon words is $O(m)$. Furthermore, for each Lyndon word, we require to generate the related necklace, so the total computational complexity turns out to be $O(h \cdot n^2 \cdot m^2)$.

3   A Constructive Proof of Theorem 3

Now, we present a variant of the procedure REC that includes near-regular instances, and that will be used to provide a constructive proof of Theorem 3 and Corollary 1. Such approach will allow to enlarge the set of degree sequences that can be proved to be h-graphic.

3.1   A Procedure to Reconstruct Near d-Regular, h-Uniform Hypergraphs

Let us characterize the degree sequences of h-uniform hypergraphs that are near d-regular, i.e., whose hyperedges have cardinality d or $d - 1$; we indicate the related set of binary matrices (having no equal rows, and horizontal and vertical projections $H = (h, \ldots, h)$ and $V = (d, \ldots, d, d - 1, \ldots, d - 1)$, respectively) with $\mathcal{E}_1$.

In [3], it has been proved that for these projections H and V, Conditions 1 and 2 are sufficient to ensure the existence of a compatible matrix. Adding Condition 3 and extending REC to $REC1$, we characterize the elements of $\mathcal{E}_1$. We sketch the procedure $REC1$: let $H = (h, \ldots, h)$ of length m, and $V = (d, \ldots, d, d - 1, \ldots, d - 1)$ of length n, be two vectors of projections satisfying Conditions 1, 2, and 3:

Step 1: compute the constant vectors of projection H' and V' such that $H' = (h, \ldots, h)$ has length $m' = k/h > m$, and $V' = (v', \ldots, v')$, of length n and $v' = k/n$, with k being the least integer such that it is both multiple of h and n, and greater than $h \cdot m$.

Step 2: run $REC(H', V', \mathcal{E})$, and let A be its output. Detect the submatrix A' of A whose rows are the successive h-shifts of $(1)^h (0)^{n-h}$, as defined in the previous section. Delete in A, one by one, these rows according to the order provided by the successive application of the h-shifts, till reaching the desired near-regular projections V. Give A as output.

The details of the algorithm together with the proof of its correctness can be found in [12]; its computational complexity is the same as REC procedure. In Fig. 1, (b), there is an example of the reconstruction of an element of $\mathcal{E}_1$ representing a 3-uniform hypergraph having near regular degree sequence $V = (2, 2, 2, 2, 2, 1, 1)$.

3.2   Reconstructing an h-Uniform Hypergraph Whose Degree Sequence Satisfies Inequality (1)

Let us consider a non-decreasing sequence of integers π that satisfies inequality (1); we show how to reconstruct an h-uniform hypergraph represented by an $m \times n$ matrix A, whose degree sequence is π.

The strategy, say $REC\text{-}Link$, adds efficiency to the steps in the proof of Theorem 3, and acts on each link sequence to reconstruct the related part of A from its horizontal and vertical projections. Two lemmas are needed:

Lemma 1. *Let $L = (l_1, \ldots, l_{t+1})$ be a link sequence, computed as in Theorem 3. It holds that*

$$l_{t+1} \leq \binom{t}{h - 2}.$$

Proof. Let us compute an upper bound to l_{t+1}: since $\sigma(L) = \Delta \cdot (h - 1)$, then we have $\Delta \cdot (h - 1) \leq \binom{t}{h-1} \cdot (h - 1)$. An upper bound of l_{t+1} can be set to $\binom{t}{h-1} \cdot \frac{(h-1)}{t+1}$ since L is non-increasing. The computation of the binomial coefficients leads to the thesis. □

In words, Lemma 1 states that the elements 1 in column $t + 1$ of matrix A_L are not too many, in particular they allow different configurations of the remaining $h - 2$ entries 1 from column 1 to t. Note that this inequality resembles that of Condition 3.

Remark 1. Let M be a matrix having different rows. Each permutation of its columns preserves the difference of the rows.

So, starting from $\pi = (d_0, \dots, d_{n-1})$, let Δ be its maximum element and t be the entries that are at least $\Delta - 1$, as in Theorem 3. We compute the link sequence $L = (l_1, \dots, l_{n-1})$ and define *REC-Link* as follows
REC-Link

Input: $H = (h - 1, \dots, h - 1)$ of length Δ, and $V = L$;
Output: the $\Delta \times (n - 1)$ matrix A_L compatible with H and V.
Step 1: if $t = n - 1$, then *REC1* on input (H, V) reconstructs A_L and provides
   it as output.
Step 2: if $t < n - 1$, then

   Step 2.1: place the elements 1 in A_L from column $t + 1$ to n as follows: let
      $j = 0$; for each $t + 1 \le i \le n$ place l_i elements 1 in column i, from position
      j to $j + l_i - 1 \mod (\Delta)$ and update $j = j + l_i \mod (\Delta)$.
   Step 2.2: divide A_L into $A_1, \dots, A_k$ blocks of consecutive rows according to
      their different configurations; by Step 2.1, all equal rows are grouped in
      the same block. Let h_i be the number of 1s that lie in each row of A_i, and
      r_i the number of its rows. For each block A_i run *REC1* on input (H_i, V_i),
      with $H_i = (h - 1 - h_i, \dots, h - 1 - h_i)$ of length r_i, and V_i being the
      near-regular vector such that $\sigma(V_i) = r_i \cdot (h - 1 - h_i)$, i.e. the near-regular
      vector compatible with H_i.
   Step 2.3: update in A_L the first t columns of each A_i with the matrix obtained
      as output from *REC1* on the related instance, after rearranging the first
      t columns of each block in order to let them sum up to the near-regular
      vertical projections $l_1, \dots, l_t$.

The correctness of Steps 2.1 and 2.3 directly follows from Lemma 1 and Remark 1, respectively, and its computational complexity is inherited from *REC*. Each matrix obtained from a link sequence by procedure *REC-Link* becomes part of the final matrix A, according to the proof of Theorem 3, after appending as first column a full sequence of entries 1 of length Δ. The following example clarifies the reconstruction:

Example 1. Let us consider $\pi = (14, 14, 14, 14, 13, 13, 13, 13, 12, 12, 12, 11, 11, 4)$, and check if it is 5-graphic. We compute the link sequence according to the parameters $t = 8$, $\Delta = 14$, and 34×14 being the dimension of A. The conditions of Theorem 3 are satisfied, since 5 divides $\sigma(\pi) = 170$, and $\binom{t-1}{h-1} \ge \Delta$ holds since $35 \ge 14$. The link sequence turns out to be $L = (6, 6, 6, 5, 5, 5, 5, 4, 4, 4, 3, 3, 0)$, with $c = 8$. The link sequence satisfies the equation $\sigma(L) = \Delta(h-1) = 14 \cdot 4 = 56$.

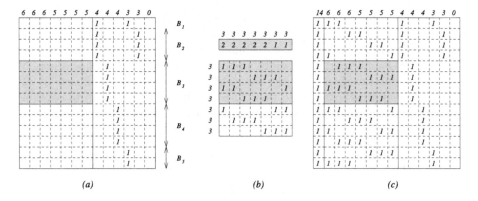

Fig. 1. (a): the link sequence L computed from π of Example 1, and the related matrix A_L to be reconstructed. (b): a call of $REC1$ on the near-regular instance (H_3, V_3). The procedure at first computes and reconstructs the matrix satisfying the minimal regular instance containing V_3, then cuts off, one after the other, those rows that are consecutive 3-shifts of $(1)^3(0)^4$ till reaching the near-regular vertical projection V_3. The darkest part of (b) is what remains. (c): all the blocks are reconstructed and their columns arranged in order to sum up to the near-regular sequence $(l_1, \ldots l_t) = (6, 6, 6, 5, 5, 5, 5)$. A first column of 1 entries is appended and the first Δ different rows of A are reconstructed.

Let us run $REC\text{-}Link$ on $H = (4, \ldots, 4)$ of dimension 14, and $V = L$. Since $t < n - 1$, then Step 2 starts: Step 2.1 place the 1s in the columns from 8 to 13 as in Fig. 1(a), and the blocks $B_1, \ldots B_5$ are detected according to Step 2.2. Step 2.3 proceeds in their reconstruction using $REC1$. Figure 1(b) shows how $REC1$ acts on B_3, one of the blocks having maximum number of rows. Lemma 1 assures that the instance (H_3, V_3) with $H_3 = (3, 3, 3, 3)$ and $V_3 = (2, 2, 2, 2, 2, 1, 1)$ can be reconstructed by $REC1$, since $\binom{7}{3} \geq 4$.

4   Conclusions and Open Problems

In this article, we consider a sufficient condition for an integer sequence π to be h-graphic, i.e. to be the degree sequence of an h-uniform hypergraph, as stated in Theorem 3, from [10]. Such a result is here approached from a different perspective, and translated into the Discrete Tomography framework where its specific tools are used to provide an effective proof. The obtained construction allows to forecast new results that constructively relax the constraint for h-graphicality stated in Theorem 3. Furthermore, this new perspective yields further challenging problems as sketched below.

First, it seems appropriate to approach the still unsolved problem of characterizing the degree sequences of 3-uniform hypergraphs. Some NP-completeness results are present in the literature mainly concerning their characterization from sets of subsumed graphs (see [4]), but they do not shed light on the general problem.

Still concerning 3-uniform hypergraphs, in [15], the authors focus on edge exchanges in order to determine how to pass from one hypergraph to another satisfying the same degree sequence. One can observe that the edge exchange has a natural counterpart in Discrete Tomography with the well known notion of switching, i.e. the changes of the values of a matrix in specific subsets of its elements that preserves the projections. It would be of some interest to develop a switching theory that also preserves the mutual difference between rows by relying on the properties of edge exchanges.

Acknowledgment. This study has been partially supported by INDAM - GNCS Project 2015.

References

1. Berge, C.: The Theory of Graphs. Dover Publications, Inc., Mineola (2001)
2. Berge, C.: Hypergraphs. North-Holland, Amsterdam (1989)
3. Brocchi, S., Frosini, A., Rinaldi, S.: The 1-color problem and the brylawski model. In: Brlek, S., Reutenauer, C., Provençal, X. (eds.) DGCI 2009. LNCS, vol. 5810, pp. 530–538. Springer, Heidelberg (2009)
4. Colbourn, C.J., Kocay, W.L., Stinson, D.R.: Some NP-complete problems for hypergraph degree sequences. Discrete Appl. Math. **14**(3), 239–254 (1986)
5. Dewdney, A.: Degree sequences in complexes and hypergraphs. Proc. Amer. Math. Soc. **53**(2), 535–540 (1975)
6. Erdősh, P., Gallai, T.: Graphs with prescribed degrees of vertices. (Hungarian) Mat. Lapok **11**, 264–274 (1960)
7. Herman, G.T., Kuba, A.: Discrete Tomography: Foundations Algorithms and Applications. Birkhauser, Boston (1999)
8. Herman, G.T., Kuba, A.: Advances in Discrete Tomography and Its Applications. Birkhauser, Boston (2007)
9. Hoogeveen, H., Sierksma, G.: Seven criteria for integer sequences being graphic. J. Graph Theor. **15**, 223–231 (1991)
10. Behrens, S., Erbes, C., Ferrara, M., Hartke, S.G., Reiniger, B., Spinoza, H., Tomlinson, C.: New results on degree sequences of uniform hypergraphs. Electron. J. Combin. **20**(4) (2013). #P14
11. Frosini, A., Picouleau, C., Rinaldi, S.: On the degree sequences of uniform hypergraphs. In: Gonzalez-Diaz, R., Jimenez, M.-J., Medrano, B. (eds.) DGCI 2013. LNCS, vol. 7749, pp. 300–310. Springer, Heidelberg (2013)
12. Frosini, A., Picouleau, C., Rinaldi, S.: A characterization of the degree sequences of uniform (almost) regular hypergraphs (2013). arXiv:1309.7832
13. Hakimi, S.: On realizability of a set of integers as degrees of the vertices of a linear graph. I. J. Soc. Indust. Appl. Math. **10**, 496–506 (1962)
14. Havel, V.: A remark on the existence of finite graphs (Czech). Časopis Pěst. Mat. **80**, 477–480 (1955)
15. Kocay, W., Li, P.C.: On 3-hypergraphs with equal degree sequences. Ars Comb. **82**, 145–157 (2007)
16. Sawada, J.: A fast algorithm to generate necklaces with fixed content. Theoret. Comput. Sci. **301**, 477–489 (2003)

Geometrical Characterization of the Uniqueness Regions Under Special Sets of Three Directions in Discrete Tomography

Paolo Dulio[1], Andrea Frosini[2], and Silvia M.C. Pagani[1(✉)]

[1] Dipartimento di Matematica "F. Brioschi", Politecnico di Milano,
Piazza Leonardo da Vinci 32, 20133 Milano, Italy
{paolo.dulio,silviamaria.pagani}@polimi.it
[2] Dipartimento di Matematica e Informatica "U. Dini", Università di Firenze,
Viale Morgagni 65, 50134 Firenze, Italy
andrea.frosini@unifi.it

Abstract. The faithful reconstruction of an unknown object from projections, i.e., from measurements of its density function along a finite set of discrete directions, is a challenging task. Some theoretical results prevent, in general, both to perform the reconstruction sufficiently fast, and, even worse, to be sure to obtain, as output, the unknown starting object. In order to reduce the number of possible solutions, one tries to exploit some a priori knowledge. In the present paper we assume to know the size of a lattice grid $\mathcal{A}$ containing the object to be reconstructed. Instead of looking for uniqueness in the whole grid $\mathcal{A}$, we want to address the problem from a *local* point of view. More precisely, given a limited number of directions, we aim in showing, first of all, which is the subregion of $\mathcal{A}$ where pixels are uniquely reconstructible, and then in finding where the reconstruction can be performed quickly (in linear time). In previous works we have characterized the shape of the region of uniqueness (ROU) for any pair of directions. In this paper we show that the results can be extended to special sets of three directions, by splitting them in three different pairs. Moreover, we show that such a procedure cannot be employed for general triples of directions. We also provide applications concerning the obtained characterization of the ROU, and further experiments which underline some regularities in the shape of the ROU corresponding to sets of three not-yet-considered directions.

Keywords: Discrete tomography · Lattice grid · Projection · Uniqueness region

1 Introduction

The retrieval of some characteristics of the internal structure of an object without direct inspection is one of the tasks of computerized tomography. Most of the data that one can achieve come out from a quantitative analysis of the spatial densities of the object, obtained by using bands of parallel high energy beams

© Springer International Publishing Switzerland 2016
N. Normand et al. (Eds.): DGCI 2016, LNCS 9647, pp. 105–116, 2016.
DOI: 10.1007/978-3-319-32360-2_8

along several directions that intersect the object itself. The attenuations of the rays are proportional to the densities of the intersected materials. The most relevant side effect that such a repeated scanning process may have on the object is the gradual changing of its internal structure, so that the object before and after the process is no longer the same. Understandably, it is desirable to obtain the maximum information about the internal structure of the object, delivering the minimum radiation dose. In the case where we consider the unknown structure as subject to a discretization process, we fall within the scope of the discrete tomography (for an overview of the topic see [12]). In this framework, an object turns out to be a finite set of points in the integer lattice, each of them having its own density value. So, an object can be modeled as a 2D or 3D matrix having integer entries, and whose dimensions are those of the minimal rectangle bounding the object, and it can be visualized in a grey-scale pixel image where each grey level corresponds to a different entry of the matrix.

For each discrete direction u, the quantitative analysis of the object's densities obtained along u can be modeled as an integer vector, say the vector of projections, summing up the values of the pixels that lie on each line parallel to u.

Practical applications require the *faithful* reconstruction of the whole unknown discrete object from projections along a set S of fixed lattice directions that are known a priori and that depend both on the accuracy and on the physics of the scanning machinery.

For only a few lattice directions, this problem is usually refereed to as *Reconstruction Problem*, and, in general, it is not quickly solvable (here quickly means in polynomial time with respect to the dimensions of the object itself) even in the case of three directions [7]. However, this result, as many similar ones in the field, relies on the possibility of solving unknown images of arbitrarily large dimensions.

Aware of these theoretical results, the physics of the scanning machinery again plays a relevant role in bounding the area where the analyzed object can lie, allowing, at least, the faithful retrieval of a subset of its pixels. As a consequence, a suitable selection of the projections' directions may allow the reconstruction of the whole object. In [5] we have characterized the shape of subsets of pixels in a generic area, whose values are uniquely determined from two projections, without explicitly computing them. These regions are called Regions of Uniqueness, briefly ROU. Such a characterization relies on the fact that the projections' values along the two directions may recursively interact in order to jointly supply information about the same subsets of pixels, till reaching their full knowledge. Furthermore, we define an algorithm to determine the shape of the ROU that performs faster than computing the density values of its internal points.

In this paper, we start to extend the above results to three directions of projections: some preliminary cases are considered and, for each of them, we define the shape of the related ROU as superposition of rectangular areas whose dimensions are strictly related to those of the directions. Then, for the remaining cases, we provide experimental evidence of the presence of strong regularities in the boundary of the ROU, which is still related to the chosen directions. This allows us to foresee the possibility to fully generalize the ROU characterization to three directions.

The paper is organized as follows. In Sect. 2 the notation is set, together with the main definitions and a summary of the algorithm proposed in [5]. In Sect. 3 the obtained results are proven, while Sect. 4 deals with some experimental results. Section 5 addresses some further investigation and concludes the paper.

2 Definition and Known Results

Assume the image we aim to reconstruct is confined in a finite $m \times n$ grid $\mathcal{A} = \{[z_1, z_2] \in \mathbb{Z} \mid 0 \le z_1 < m, 0 \le z_2 < n\}$. We agree to identify each pixel of the image with its bottom-left corner, of coordinates $[i, j]$. We define the image itself as a map $g : \mathcal{A} \longrightarrow \mathbb{Z}$, such that an integer number, corresponding to a color, is assigned to each pixel. By an abuse of notation, we will say that we reconstruct the grid instead of the contained image.

Directions are pairs (a, b) of coprime integers; the horizontal and vertical directions are $(1, 0)$ and $(0, 1)$, respectively. We say that a set $S = \{(a_k, b_k) \mid k = 1, \ldots, d\}$ of d lattice directions is *valid* for the grid $\mathcal{A}$ if

$$\sum_{k=1}^{d} |a_k| < m, \qquad \sum_{k=1}^{d} |b_k| < n.$$

We assume that m, n are large enough to guarantee that the various sets S we will consider are valid. A *lattice line* through $[i, j]$ with direction $u = (a, b)$ is the set

$$\{[i + ka, j + kb] \mid k \in \mathbb{Z}\}.$$

The *projections* along the direction u are the sums of the values of pixels on each line with direction u which meets the grid, collected in a pre-assigned order.

This is the usual way of collecting projections in the so-called grid model, which is largely employed also in other approaches to the tomographic problem, such as, for example, in the case of the Mojette transform (see for instance [8, 13]). However, the Mojette reconstruction algorithms usually move from the so-called Katz criterion ([11]), where it is required that

$$\sum_{k=1}^{d} |a_k| \ge m \quad \text{or} \quad \sum_{k=1}^{d} |b_k| \ge n.$$

When the Katz criterion is satisfied, one can ensure that the null space of the transform is empty, which means that no ghost exists, so that reconstruction is uniquely determined. This motivates the research of suitable reconstruction algorithms (see, for instance [14]). Differently, when working with sets of valid directions, uniqueness is not automatically guaranteed, so that looking for uniqueness conditions becomes a meaningful tomographic problem. This relies on the fact that, when valid directions are considered, the tomographic problem is in general ill-posed, due to the presence of *switching components*, also known as *(weakly) bad configurations*. These consist of pairs of sets (Z, W) having the same projections along the given directions. From a bad configuration (Z, W) ghosts can be easily obtained, by giving opposite sign to the weights of the pixels of Z and W, and by setting all the other pixels to value zero. Recent results concerning ghosts in Discrete Tomography can be found in [1]. A grid was proven to be uniquely determined by a set S of directions if and only if it contains no bad configurations along the directions in S (see [6]). It is true, for instance, in the case of non-valid sets of directions.

Therefore, our aim is to reconstruct the grid $\mathcal{A}$, or suitable sub-regions of $\mathcal{A}$ from its projections along a set S of valid lattice directions. Such an approach goes back to [10], where the authors characterized all the possible ghosts contained in the null space of the transform in term of polynomial factorization.

As concerns valid sets, there exist theoretical results stating that we do not need to choose large sets S to achieve uniqueness of reconstruction: as shown in [9] and then completely characterized in [2], four suitably chosen valid directions are enough to get uniqueness in a binary grid (namely, when the image is defined as $g : \mathcal{A} \longrightarrow \{0, 1\}$). This result has been extended to higher dimensions in [3]. In many cases, however, we cannot choose the set of employed directions, due to some physical or mechanical constraints. This led us to change perspective and ask a related question: Given a set of arbitrarily chosen directions, are there anyway parts of the grid which can be uniquely reconstructed? The answer is immediate: uniquely determined pixels are the ones which cannot be part of a switching component, so they are confined where switching components cannot be constructed, namely, in the corners of the grid. We then move to another question: Which pixels, among the uniquely determined ones, referred to as *(algebraic) region of uniqueness* (algebraic ROU), can be *quickly* reconstructed? In this case, quick means in linear time (we recall that the reconstruction of the whole grid is NP-hard for more than two directions). This question involves the definition of *(geometric) region of uniqueness* (geometric ROU), which has already been presented in [4,5].

Definition 1. *The* (geometric) ROU *associated to a set S of directions is the set of pixels Q such that*

(a) *Q belongs to a line, with direction in S, intersecting the grid just in Q, or*
(b) *Q lies on a line, with direction in S, whose further intersections with $\mathcal{A}$ belong to the previously determined ROU.*

For $|S| = 2$, the ROU has been completely characterized in [5]. We now aim to extend this result to sets of three directions.

Let $a, b, c, d, e, f > 0$ and consider the directions $(-a, b), (-c, d), (-e, f)$; we consider directions with negative slope in order to argue on the configuration in the bottom-left and in the upper-right corners of the grid $\mathcal{A}$. Since the configuration is symmetric in the two corners, we focus on the bottom-left one. We want to compute the ROU associated to the three directions $(-a, b), (-c, d), (-e, f)$. To do so, we exploit the Double Euclidean Division Algorithm (DEDA) shown in [5], that we summarize here for the reader's convenience.

Given a pair $(-a, b), (-c, d)$ of lattice directions, DEDA computes the shape of the ROU in the bottom-left corner of the grid. It executes the Euclidean division between a and c, and b and d, in parallel, checking at each step if one of the remainders is zero, or if the two quotients are not equal. If one of these happens, DEDA stops and returns the shape of the ROU as a list of horizontal and vertical steps. Such a zigzag path starts from the bottom-right pixel of the ROU and ends at its upper-left pixel. If the algorithm does not stop at the initial step, then of some the pixels in the algebraic ROU do not belong to the geometric ROU, and we say that there is *erosion* in the configuration.

Example 1. *Consider the pair of directions* $(-25, 17), (-11, 7)$. *DEDA computes*

	horizontal	*vertical*
level 0 :	$25 = 2 \cdot 11 + 3,$	$17 = 2 \cdot 7 + 3,$
level 1 :	$11 = 3 \cdot 3 + 2,$	$7 = 2 \cdot 3 + 1.$

At level 1 the two quotients are not equal, so DEDA stops. The final output, whose zigzag path is $\mathcal{W} = (3, 3, 3, 3, 1, 5, 10, 14, 3, 3, 3, 3, 1, 5)$, *is represented in Fig. 1.*

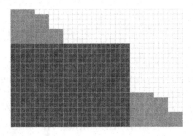

Fig. 1. The ROU associated to the pair $(-25, 17), (-11, 7)$ (depicted in the bottom-left corner of the grid).

In this paper we want to describe the shape of the ROU for three directions in some particular cases, when, for any of the three possible pairs, there is no erosion. In [4,5], the zigzag path associated to the absence of erosion for directions $(-a, b), (-c, d)$ has been proven to be

– for $a > c$, $b < d$ ([4, Corollary 1]):

$$\mathcal{W} = (b, a, d, c) \tag{1}$$

– if $a > c$, $b > d$ and moreover the quotients of the division between a and c, and b and d, are different or one of their remainders equals zero ([4, Theorem 3] and [5, Remark 11]):

$$\mathcal{W} = (d, c, b - d, a - c, d, c). \tag{2}$$

In case of erosion, the zigzag path becomes more fragmented; its detailed construction can be found in [5]. In what follows, we will study some special cases by setting $b > d > f$ and $b \geq d + f$. The strict inequalities among the first components a, c, e will determine the shape of the ROU in each case. In the following section we will refer indifferently to the absence of erosion and the corresponding zigzag path.

We will use the notation $R([i,j],[k,l])$ to refer to the rectangle whose bottom-right and upper-left corners are $[i,j]$ and $[k,l]$, respectively. Moreover, we set $\mathrm{ROU}(u_1, \ldots, u_r)$ to denote the region of uniqueness associated to the directions $u_1, \ldots, u_r$ and having its bottom-left corner in the origin of the grid.

3   Theoretical Results

Denote by $\mathcal{B}$ a minimal bad configuration associated to the set $S = \{u_1 = (-a, b), u_2 = (-c, d), u_3 = (-e, f)\}$ and placed such that its leftmost pixel is adjacent to the left side of the grid, and its rightmost pixel is in the lowest row of $\mathcal{A}$. $\mathcal{B}$ consists of eight pixels, whose coordinates are

$$B_1 = [a + c + e, 0], \; B_2 = [a + c, f], \; B_3 = [a + e, d], \; B_4 = [a, d + f],$$
$$B_5 = [c + e, b], \qquad B_6 = [c, b + f], \; B_7 = [e, b + d], \; B_8 = [0, b + d + f].$$

Note that the case where one direction is the sum of the other two cannot happen under the hypothesis of absence of erosion.

Since (algebraic) uniqueness of reconstruction is equivalent to the absence of bad configurations, we look for the ROU in the complement of the union of the rectangles having their bottom-left corner in the pixels of $\mathcal{B}$.

For two directions $(-a, b), (-c, d)$, we know from [5] that the absence of erosion corresponds to a ROU which occupies the maximum obtainable area, whose zigzag path is as in Eq. (1) or (2). We remark that, in the mentioned cases, the concepts of algebraic and geometric uniqueness coincide.

A set of pixels is said to be *horizontally and vertically convex* (briefly, hv-convex) if its rows and columns are connected sets. We recall a useful result; for the proof see [5].

Lemma 1 ([5], **Theorem** 10). *The ROU is hv-convex.*

Lemma 2. $ROU(u_1, u_2) \subseteq ROU(u_1, u_2, u_3)$.

Proof. If a pixel is uniquely determined by the pair of directions u_1, u_2, then it is uniquely reconstructed also by u_1, u_2, u_3. □

The following theorem proves the first non-erosive case.

Theorem 1. *Let* $S = \{u_1 = (-a, b), u_2 = (-c, d), u_3 = (-e, f)\}$ *be a valid set of directions for* $\mathcal{A}$, *such that* $a < c < e$ *and* $b > d > f$. *The shape of the related ROU is described by the path* (f, e, d, c, b, a).

Proof. By Lemma 2 we have that $ROU(u_1, u_2)$, defined by the path (d, c, b, a), is part of $ROU(u_1, u_2, u_3)$. Consider the rectangle $R_1 = R([a + c + e - 1, 0], [a + c, h - 1])$, adjacent to the right side of $ROU(u_1, u_2)$ and of size $e \times h$, where $h = \min(f, d - f)$ (see Fig. 2(a)). When moved along u_3, R_1 first falls in $ROU(u_1, u_2)$ and then outside the grid $\mathcal{A}$. This means that R_1 is part of $ROU(u_1, u_2, u_3)$.

Consider now a sub-rectangle of R_1, say $R_2 = R([a + 2c - 1, 0], [a + c, h - 1])$, whose length is c. A translation along u_2 moves it in $R_3 = ([a + c - 1, d], [a, d + h - 1])$, adjacent to $ROU(u_1, u_2)$ (Fig. 2(b)), and then outside $\mathcal{A}$. Therefore, also R_3 is part of $ROU(u_1, u_2, u_3)$. The sub-rectangle $R_4 = R([2a - 1, d], [a, d + h - 1])$ of R_3, of length a, when moved along u_1, ends in $R_5 = R([a - 1, b + d], [0, b + d + h - 1])$ (Fig. 2(c)) and then outside $\mathcal{A}$, so also R_5 is included in $ROU(u_1, u_2, u_3)$. If $\min(f, d - f) = f$, no further pixel is added. If $\min(f, d - f) = d - f$, now a similar argument is applied to the rectangle $R([a + c + e - 1, d - f], [a + c, f - 1])$ in order to add it and its translates to $ROU(u_1, u_2, u_3)$.

The obtained ROU has profile $\mathcal{W} = (f, e, d, c, b, a)$ and cannot be further extended. This is true since the bad configuration $\mathcal{B}$ is adjacent to $ROU(u_1, u_2, u_3)$ (Fig. 2(c)) and, by hv-convexity, all pixels in the quarters having a pixel in $\mathcal{B}$ as bottom-left corner cannot be uniquely determined. □

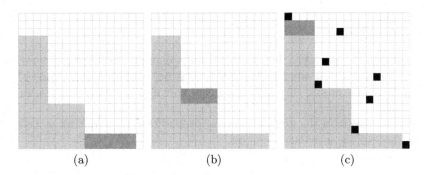

(a)   (b)   (c)

Fig. 2. (a) $ROU(u_1, u_2)$ (light grey) and R_1 (dark grey). (b) The ROU (light grey) and R_3 (dark grey). (c)The ROU (light grey), R_5 (dark grey) and $\mathcal{B}$ (black).

Following [5], we generalize the theorem above to other cases without erosion.

Theorem 2. *Let* $S = \{u_1 = (-a, b), u_2 = (-c, d), u_3 = (-e, f)\}$ *be a valid set of directions for* $\mathcal{A}$*, such that* $b > d > f$ *and* $b \geq d + f$*. Assume that the three directions do not erode pairwise. The path* $\mathcal{W}$ *delimiting the ROU is*

(a) for $a < c < e$*:* $\mathcal{W} = (f, e, d, c, b, a)$*;*
(b) for $a < e < c$*:* $\mathcal{W} = (f, e, d - f, c - e, f, e, b, a)$*;*
(c) for $c < a < e$*:* $\mathcal{W} = (f, e, d, c, b - d, a - c, d, c)$*;*
(d) for $c < e < a$*:*
    (d.1) if $a > c + e$*:* $\mathcal{W} = (f, e, d, c, b - d - f, a - c - e, f, e, d, c)$*;*

　　(d.2) if $a \le c + e$: $W = (f, e, d, c, b - d, a - c, d, c)$;
(e) for $e < a < c$: $W = (f, e, d - f, c - e, f, e, b - f, a - e, f, e)$;
(f) for $e < c < a$:
　　(f.1) if $a > c + e$: $W = (f, e, d - f, c - e, f, e, b - d - f, a - c - e, f, e, d - f, c - e, f, e)$;
　　(f.2) if $a \le c + e$, *then* $b > d + f$ *and* $W = (f, e, d - f, c - e, f, e, b - d, a - c, d - f, c - e, f, e)$.

Proof. The case (a) has already been proven in Theorem 1. We omit details of the proofs of cases $(b) - (f1)$ for brevity. These are quite similar and exploit the same strategy as in Theorem 1, even if they do not follow immediately, since the number of required steps in adding the various rectangles increases and changes from case to case. Concerning case $(f.2)$, we need a special argument. Assume $e < c < a$, $a \le c + e$, and suppose that $b = d + f$. Since $d > f$, then $d > \frac{b}{2}$, so the quotient between b and d is 1. In order to avoid erosion, the quotient between a and c has to be different from 1, or equal to 1 with remainder zero (namely, $a = c$). The last case is not possible, since $a > c$. Then $a \ge 2c$, but

$$c + e < 2c \le a,$$

a contradiction. So the case $(f.2)$ is possible only for $b > d + f$. In this case, the same argument as in Theorem 1 can be adapted to prove that $W = (f, e, d - f, c - e, f, e, b - d, a - c, d - f, c - e, f, e)$. □

In all the mentioned cases, the configuration is maximal for each pair of directions. This means that we take the whole area below the minimal bad configuration $\mathcal{B}$, whose profile corresponds to the aforementioned statements, in the various cases.

　　We will provide in the next section the profiles that come out from the non-erosive cases of the previous theorems.

4　Experimental Results

In this section, we first determine the ROU associated to given triples, and then argue on synthetic data.

　　Several triples are considered; their corresponding ROU is depicted in Fig. 3. Each triple corresponds to a case treated in Theorem 2:

　　(i)　$S = \{(-7, 13), (-8, 5), (-9, 4)\}$, corresponding to case (a) (Fig. 3(a));
　　(ii)　$S = \{(-5, 19), (-14, 13), (-7, 5)\}$, corresponding to case (b) (Fig. 3(b));
　　(iii)　$S = \{(-5, 9), (-2, 5), (-7, 2)\}$, corresponding to case (c) (Fig. 3(c));
　　(iv)　$S = \{(-7, 9), (-2, 5), (-3, 2)\}$, corresponding to case $(d.1)$ (Fig. 3(d));
　　(v)　$S = \{(-8, 9), (-2, 5), (-7, 2)\}$, corresponding to case $(d.2)$ (Fig. 3(e));
　　(vi)　$S = \{(-5, 9), (-7, 6), (-3, 2)\}$, corresponding to case (e) (Fig. 3(f));
　　(vii)　$S = \{(-12, 13), (-6, 7), (-3, 4)\}$, corresponding to case $(f.1)$ (Fig. 3(g));
　　(viii)　$S = \{(-9, 10), (-7, 5), (-5, 2)\}$, corresponding to case $(f.2)$ (Fig. 3(h)).

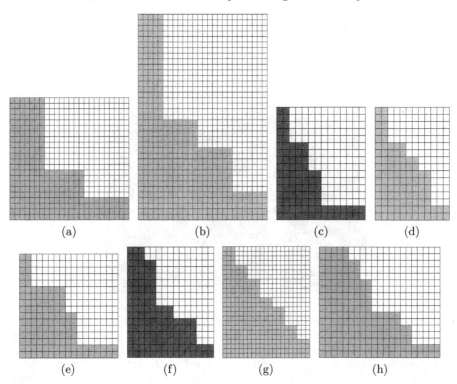

Fig. 3. The zigzag profile of the ROU determined by sets S of three directions. (a) $S = \{(-7, 13), (-8, 5), (-9, 4)\}$, corresponding to case (a) of Theorem 2. (b) $S = \{(-5, 19), (-14, 13), (-7, 5)\}$, corresponding to case (b) of Theorem 2. (c) $S = \{(-5, 9), (-2, 5), (-7, 2)\}$, corresponding to case (c). (d) $S = \{(-7, 9), (-2, 5), (-3, 2)\}$, corresponding to case $(d.1)$. (e) $S = \{(-8, 9), (-2, 5), (-7, 2)\}$, corresponding to case $(d.2)$. (f) $S = \{(-5, 9), (-7, 6), (-3, 2)\}$, corresponding to case (e). (g) $S = \{(-12, 13), (-6, 7), (-3, 4)\}$, corresponding to case $(f.1)$. (h) $S = \{(-9, 10), (-7, 5), (-5, 2)\}$, corresponding to case $(f.2)$.

The characterization provided in Theorem 2 can also be exploited to investigate possible Regions Of Interest (ROI), selected in advance. More precisely, by varying the choice of the triples, one can try to partially, or even completely, include the ROI in the ROU. This could be useful in real applications, since reconstruction in the ROU is provided in linear time, due to the uniqueness property. Therefore, one could try to adapt the ROU in order to match some ROI of the image to be reconstructed. As an example, we have considered an (80×80)-sized random phantom, consisting of 50 different colors (or grey levels), and, for a selected triangular ROI, we have explored the different portions that can be reconstructed for each one of the six triples $(-a, b), (-c, d), (-e, f)$ obtained by fixing b, d, f and considering all the possible permutations of a, c, e. Results are shown in Fig. 4. Figure 5 shows the results concerning the same six triples when applied to a (50×50)-sized random phantom, consisting of 20 different colors (or grey levels).

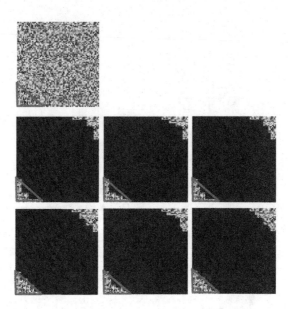

Fig. 4. Perfect and linear time reconstruction of different portions of a triangular ROI. Upper line: Original Phantom. Middle line: The ROU determined by the triples $(-5, 12), (-7, 6), (-13, 3)$ (left), $(-5, 12), (-13, 6), (-7, 3)$ (middle), $(-7, 12), (-5, 6),$ $(-13, 3)$ (right). Bottom line: The ROU determined by the triples $(-13, 12), (-5, 6),$ $(-7, 3)$ (left), $(-7, 12), (-13, 6), (-5, 3)$ (middle), $(-13, 12), (-7, 6), (-5, 3)$ (right) (Color figure online).

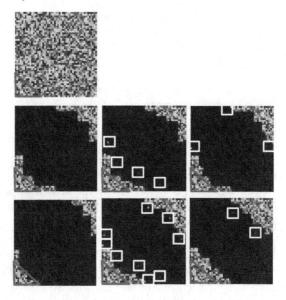

Fig. 5. The same results as in Fig. 4, for a (50×50)-sized random phantom consisting of 20 different colors. Some extra-pixels (included in the white squares) outside the ROU can sometimes be reconstructed (Color figure online).

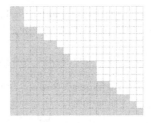

Fig. 6. The ROU associated to the triple $(-11, 8), (-7, 5), (-2, 3)$.

Note that, in this case, for some triples also a few extra-pixels not belonging to the ROU (included in white squares) can be reconstructed. This is obtained by adding to the reconstruction procedure the information that the number of employed grey levels is estimated by the number of grey levels included in the ROU. In this case, if for instance a line L has more than one intersection with $\mathcal{A}\backslash\text{ROU}$, and the projection along L equals zero, then each pixel on L gets zero value. This activates an iterative procedure which possibly allows to gain further pixels not included in the ROU.

5   Conclusions

In this paper we started extending the characterization of the region of uniqueness (ROU), obtained in [5] for pairs of directions, to sets of three directions. We have studied some special cases, when the so-called erosion does not occur, and described the profile of the ROU in the eight possible configurations. This is just a preliminary step, in view of a complete characterization of the shape of the ROU for three directions. In future work we will have to deal with profiles whose path is still not understood (see for instance Fig. 6, where the ROU has been computed by a liner-time program, just by applying Definition 1). We are confident about the fact that the shape of the ROU can be explained in terms of numerical relations among quotients and remainders of the entries of the employed directions.

References

1. Brunetti, S., Dulio, P., Hajdu, L., Peri, C.: Ghosts in discrete tomography. J. Math. Imaging Vis. **53**(2), 210–224 (2015)
2. Brunetti, S., Dulio, P., Peri, C.: Discrete tomography determination of bounded lattice sets from four X-rays. Discrete Appl. Math. **161**(15), 2281–2292 (2013)
3. Brunetti, S., Dulio, P., Peri, C.: Discrete tomography determination of bounded sets in $\mathbb{Z}^n$. Discrete Appl. Math. **183**, 20–30 (2015)
4. Dulio, P., Frosini, A., Pagani, S.M.C.: Uniqueness regions under sets of generic projections in discrete tomography. In: Barcucci, E., Frosini, A., Rinaldi, S. (eds.) DGCI 2014. LNCS, vol. 8668, pp. 285–296. Springer, Cham (2014)

5. Dulio, P., Frosini, A., Silvia, M.C.: A geometrical characterization of regions of uniqueness and applications to discrete tomography. Inverse Prob. **31**(12), 125011 (2015)
6. Fishburn, P.C., Shepp, L.A.: Sets of uniqueness and additivity in integer lattices. In: Herman, G.T., Kuba, A. (eds.) Discrete Tomography. Applied and Numerical Harmonic Analysis, pp. 35–58. Birkhäuser Boston, Boston (1999)
7. Gardner, R.J., Gritzmann, P., Prangenberg, D.: On the computational complexity of reconstructing lattice sets from their X-rays. Discrete Math. **202**(1–3), 45–71 (1999)
8. Guédon, J.-P., Normand, N.: The mojette transform: the first ten years. In: Andrès, É., Damiand, G., Lienhardt, P. (eds.) DGCI 2005. LNCS, vol. 3429, pp. 79–91. Springer, Heidelberg (2005)
9. Hajdu, L.: Unique reconstruction of bounded sets in discrete tomography. In: Proceedings of the Workshop on Discrete Tomography and its Application, vol. 20 of Electronic Notes in Discrete Mathematics, pp. 15–25 (electronic). Elsevier, Amsterdam (2005)
10. Hajdu, L., Tijdeman, R.: Algebraic aspects of discrete tomography. Journal fur die Reine und Angewandte Mathematik **534**, 119–128 (2001). Cited by 34
11. Katz, M.: Questions of Uniqueness and Resolution in Reconstruction from Projections. Springer, Heidelberg (1978)
12. Kuba, A., Herman, G.T.: Discrete tomography: a historical overview. In: Herman, G.T., Kuba, A. (eds.) Discrete Tomography. Applied and Numerical Harmonic Analysis, pp. 3–34. Birkhäuser Boston, Boston (1999)
13. Normand, N., Guédon, J.-P.: La transforme mojette: une reprsentation redondante pour l'image. Comptes Rendus de l'Acadmie des Sciences - Series I - Mathematics **326**(1), 123–126 (1998)
14. Normand, N., Kingston, A., Évenou, P.: A geometry driven reconstruction algorithm for the mojette transform. In: Kuba, A., Nyúl, L.G., Palágyi, K. (eds.) DGCI 2006. LNCS, vol. 4245, pp. 122–133. Springer, Heidelberg (2006)

A Comparison of Some Methods for Direct 2D Reconstruction from Discrete Projected Views

Preeti Gopal[1,2]([✉]), Ajit Rajwade[1], Sharat Chandran[1], and Imants Svalbe[3]

[1] Department of CSE, IIT Bombay, Mumbai, India
{preetig,ajitvr,sharat}@cse.iitb.ac.in
[2] IITB-Monash Research Academy, Mumbai, India
[3] School of Physics and Astronomy, Monash University, Clayton, Australia
imants.svalbe@monash.edu

Abstract. Tomographic acquisitions can be described mathematically as discrete projective transforms. Direct reconstruction methods aim to compute an accurate inverse for such transforms. We assemble a limited set of measurements and then apply the inversion to obtain a high-fidelity image of the original object. In this work, we compare the following direct inversion techniques for sets of discrete projections: Radon-i(inverse)Radon, a least squared error method and filtered back-projection for Mojette inversion. We observe that filtered back-projection is the best of these methods, as the reconstruction errors that arise using this method depend least strongly on the image structure. We aim to improve results for the filtered back-projection method by optimizing the design of the regularizing filter and here present work towards eliminating the regularization threshold that is used as part of this method.

1 Introduction

The classic Radon transform is based on the continuous spatial domain and requires infinitely many projections for exact reconstruction [1], which is practically infeasible. Thus, the inverse Radon transform is an ill-posed problem. This lead to the evolution of Discrete Radon Transform (DRT) in which the underlying object is viewed as a discrete entity rather than a continuous entity. Digital projections are computed along specific set of angles and if the number of angles is sufficient, then exact inversion is possible. However, in practice, the trajectory of X-rays does not follow discrete paths. The acquired measurements, which are in Radon form, can be converted to equivalent discrete measurements by solving a system of equations, as described in [2].

A type of DRT, called the Mojette transform, was developed by Guédon et al. [3]. The advantage of this transform is that, if a sufficiently large number of angles are used, then direct back-projection of discrete Mojette projections results in exact recovery of the image [4], and is reasonably robust to noise [5,6]. Each discrete projection angle is defined by a set of numbers: (p, q) where p and q are co-prime integers. Computing projection along (p, q) involves summing the intensity values of pixels that lie along the line making an angle of $\tan^{-1}\frac{p}{q}$.

© Springer International Publishing Switzerland 2016
N. Normand et al. (Eds.): DGCI 2016, LNCS 9647, pp. 117–128, 2016.
DOI: 10.1007/978-3-319-32360-2_9

This procedure is also termed as 'Dirac' Mojette transform because the underlying 2D pixel representation is a 'Dirac' field, i.e.

$$f(x, y) = \sum_k \sum_l f(k, l)\delta(x - k)\delta(y - l) \tag{1}$$

Here, there is no notion of interpolation like the Radon transform as every pixel is considered to be a weighted Dirac function [7]. The Dirac Mojette transform is formally defined as:

$$M(b, p, q) = Projection_{pq}(b)$$
$$= \sum_{k=-\infty}^{+\infty} \sum_{l=-\infty}^{+\infty} f(k, l)\delta(b + kq - pl) \tag{2}$$

where $\delta(n)$ is the kronecker delta function with value 1 when $n = 0$ and 0 everywhere else. Thus, values of all pixels that lie on the line $kq - pl = b$ are summed and put into the bin b. Both Mojette forward projection and inverse transforms of an image can be computed with the same complexity, given by $O(IN)$ [8]. Thus it is linear in I, the number of projections and N, total number of pixels in an image. The number of bins required for a projection is given by:

$$\text{Nbins}(i) = (W - 1) * |p_i| + (H - 1) * |q_i| + 1 \tag{3}$$

It is thus dependent on the angle (p_i, q_i), and the image size $H \times W$. In order to uniquely reconstruct this image, the following criterion [9] must be satisfied by the chosen angle set:

$$W \leq \sum_i |p_i| \text{ or } H \leq \sum_i |q_i| \tag{4}$$

This is called the Katz sufficiency criterion. Our aim is to perform image reconstructions below the Katz limit. When this sufficiency criterion is not met, the 'missing' projections give rise to reconstruction artefacts (called 'ghosts') [10].

Our aim here is to evaluate several methods to invert discrete projection data from a limited set of (p, q) view angles. This work is part of a study [11] reconstructing 3D and 4D image data under sparse assumptions by pooling projections across multiple 2D 'slices'. Direct reconstructions of 2D slices will be used as seeds for iterative reconstruction methods or as part of the guiding mechanism used to select and pool slices.

We seek optimal and robust 2D image reconstruction using minimal projection data. The goal of this paper is to compare three methods of direct tomographic reconstruction: classical Radon-i(inverse) Radon, filtered back-projection and a least squared error driven filtered back-projection method. We have not analyzed corner-based inversion [7], because it is noise intolerant; the Finite Radon Transform (FRT) [12], as it requires a fixed set of view angles and has strong image artefacts when the available data is under-sampled; and iterative multi-scale methods- as they work best for selected combinations of (p, q) view angles. We will include the recently published fractional Fourier method of Kingston [13] in our future work.

2   Direct Inversion Methods

We compared the performance of the following three reconstruction techniques on a set of images shown in Fig. 1:

1. Shepp-Logan inverse Radon (filtered back-projection) on Radon projected images.
2. Point Spread Function (PSF) estimation by least squared error method on Mojette projected images.
3. Filtered back-projection on Mojette projected images.

The circular mask of the images constrains the reconstructed views to have equal projection lengths for all view angles (as is common for CT where the ROI shape is framed by the edges of a rotating X-ray beam). A set of discrete angles was used for computing projections. Each angle consisted of (p, q) where p and q are co-prime integers. Throughout this paper, we refer to the 'shortest' angles as those which have small $\sqrt{p^2 + q^2}$ values. Thus, the first six shortest angles are $(0, 1)$, $(1, 0)$, $(1, 1)$, $(-1, 1)$, $(2, 1)$ and $(-2, 1)$. For all the experiments described in this section, the shortest 52 angles were used. To measure the quality of reconstructed images, we have used the Mean Squared Error (MSE) based Peak Signal to Noise Ratio (PSNR), as a metric.

2.1   Radon-iRadon

We computed the Radon projection along a fixed set of angles, each being given by $\tan^{-1} \frac{p}{q}$, corresponding to the discrete angle (p, q). Following this, the image was reconstructed using Shepp-Logan filtered back-projection for inverse Radon. The use of Shepp-Logan filter is a compromise between applying no filter (Ram-Lak) and the more heavy noise suppression of filters like Cosine or Hamming (We synthesize our projections by forward projection of the image data. Hence noise is not a major problem here). The results are shown in Fig. 2.

2.2   Least Squared Error Method

The Mojette back-projected image m and the original image im are related by the following expression [7]:

$$m_{p,q}(x, y) = im(x, y) * h_{p,q}(x, y) \tag{5}$$

where $h_{p,q}$ denotes the PSF corresponding to the Mojette angle set (p, q) along which the projections are taken and $*$ denotes convolution operator. In the frequency domain, we have

$$M(u, v) = IM(u, v) \cdot H(u, v) \tag{6}$$

We estimated $1/H(u, v)$ using a least squared error technique. The sum of $((1/H) \cdot M - IM)^2$ is minimized over all the training images. The resultant PSF

Fig. 1. Original images used for testing reconstruction algorithms throughout this paper. (a): Slice of the brain [14], (b): Cameraman, (c): House. Each image is of size 179×179

(a) PSNR = 21.24   (b) PSNR = 17.43   (c) PSNR = 19.66

Fig. 2. Reconstruction by Radon-iRadon technique with the 52 shortest projection angles

is checked for extremely small values (below a fixed threshold) and these values are replaced by the mean of their neighborhoods. The training set included 10 randomly chosen images from a dataset of images of birds and natural scenery. The size of the training set did not affect the quality of estimation. A small training set with 10 images was found to give results which were similar to those when 100 or 500 training images were used. This is because the PSF is not entirely image dependent. Its value (when estimated this way) is erroneous only in those locations where the 0/0 problem arises, because of which specific structural artefacts were observed in the reconstructed images, as shown in Fig. 3.

2.3 Filtered Back-Projection

The obvious way to reconstruct original image im is the following:

$$im(x, y) = F^{-1} \frac{M(u, v)}{H(u, v)} \tag{7}$$

where F denotes the Fourier transform. But, retrieving im as shown in Eq. 7 is unstable as H tends to be zero or very small at multiple frequencies.

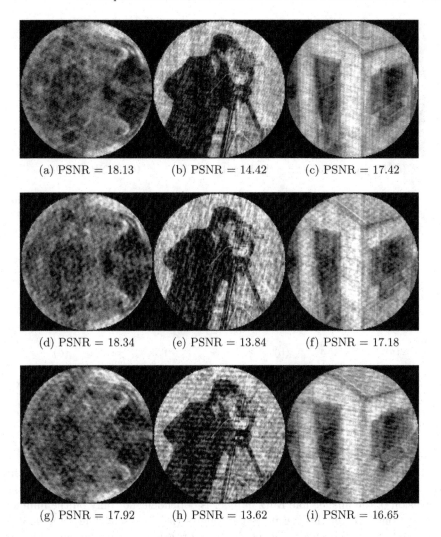

Fig. 3. Reconstruction by least squared error estimate with the 52 shortest projection angles and (a, b, c): 10 training images, (d, e, f): 100 training images, (g, h, i): 500 training images

In [15], $H_{p,q}(u, v)$ is regularized with a weighted filter so that the central region has no holes. The image is then reconstructed by direct deconvolution of the filtered PSF with the Mojette back-projected image (Fig. 4). This is equivalent to regularizing the back-projected image in the frequency domain and then taking the inverse Fourier transform. Extremely low values (below a fixed threshold) in the FFT of the filtered PSF are replaced by the mean evaluated over neighborhoods of fixed size. The PSF filtration in [15] is described by:

$$PSF_{modified} = \begin{cases} PSF. \times wnp : K > 1 \\ PSF. \times tnp \ : K < 1 \end{cases}$$

(a) PSNR = 26.32   (b) PSNR = 22.55   (c) PSNR = 23.04

Fig. 4. Reconstruction by the filtered back-projection with the 52 shortest projection angles

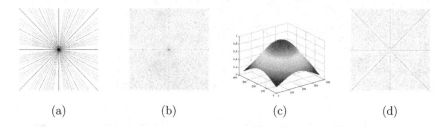

(a)   (b)   (c)   (d)

Fig. 5. (a): raw PSF (inverted); (b): FFT of PSF (center-shifted and inverted); (c): The *wnp* filter for a 179 × 179 image, measured using the 52 shortest Mojette angles; (d): absolute difference between FFT of raw PSF and filtered PSF (center-shifted and inverted)

where $.\times$ denotes 'point-to-point' multiplication and $K = k/N$; k being the Katz number given by: $k = 1 + max(\sum_i |p_i|, \sum_i |q_i|)$ [9] and $N \times N$ being the dimension of the image. The Katz number is dependent not only on the number of views but also on the specific set of angles (p, q) chosen. The filter *wnp* is computed [15] by cross-correlating the back-projected images *wp* and *wn*; *wp* being generated by back-projecting delta image along the set of angles that was used while taking measurements and *wn* being generated by back-projecting delta image along complementary set of angles.

$$wnp = (wp \star wn) \star (D \star D) \qquad (8)$$

$$tnp = (wp \star wn) \qquad (9)$$

where D is 1 within the region of interest and $\star$ denotes cross-correlation. The *wnp* filter for an image of size 179 × 179 and measurements taken along the 52 shortest angles, is shown in Fig. 5. A big advantage of this approach is that it can accommodate, with small drop in performance, angle distributions that are uniform, random or clumped.

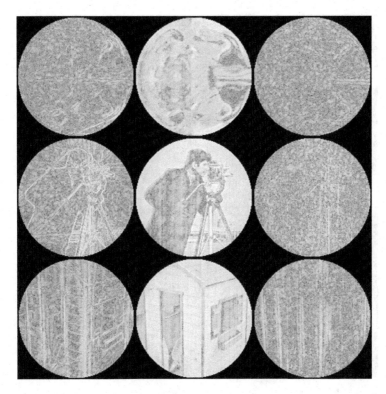

Fig. 6. Absolute error (in log scale to accentuate small intensity differences) on apply-ing different direct reconstruction techniques using the shortest 52 projection angles. Left column: Radon-iRadon; Middle column: least squared error reconstruction; Right column: filtered back-projection reconstruction for images of brain (top row), camera-man (middle row) and house (bottom row)

2.4   Comparison of Different Reconstruction Techniques

The absolute errors in reconstruction by all the above techniques are shown in Figs. 6 and 7 in logarithmic scale, while the reconstructions on rotated images are shown in Fig. 8. The errors in Radon-iRadon and more so, in least squared error based methods are strongly dependent on image structure. However, in the filtered back-projection, for a sufficient number of projection views, a significant portion of the error is image independent, indicating scope for improvement in this technique.

For example, when the number of projection views was 52, errors of all meth-ods were dependent on image structure, as shown in Fig. 6. But, when the number of views was increased to 200, the errors in the filtered back-projection method alone was weakly dependent on the image structure (Fig. 7), whereas the errors in other two methods still showed strong image dependency.

The PSF model is simple and makes no assumptions about the back-projected data. The image reconstruction is very fast and avoids any concern

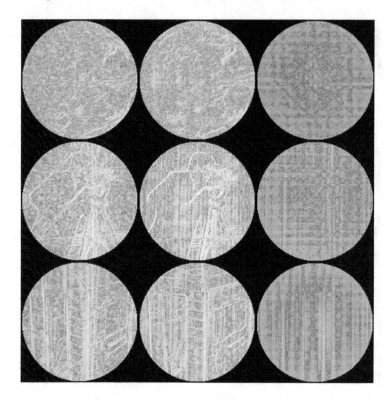

Fig. 7. Absolute error (in log scale to accentuate small intensity differences) on applying different direct reconstruction techniques using the shortest 200 projection angles. Left column: Radon-iRadon; Middle column: least squared error reconstruction; Right column: filtered back-projection reconstruction for images of brain (top row), cameraman (middle row) and house (bottom row)

over convergence rates and local minima in optimization that may constrain iterative methods.

3   Improved Filtered Back-Projection for Mojette Inversion

We wish to investigate ways to improve the filtered back-projection method, because it gives better results than the other two techniques. First, we discuss the sensitivity of this method to the value of threshold and then the strong effects of selecting particular angles as part of the projection set.

3.1   Replacing Holes in the Fourier Domain of the Filtered Point Spread Function

In filtered back-projection, the FFT of the filtered PSF is checked for values lower than a threshold. Such values are replaced by mean of their neighborhood [15].

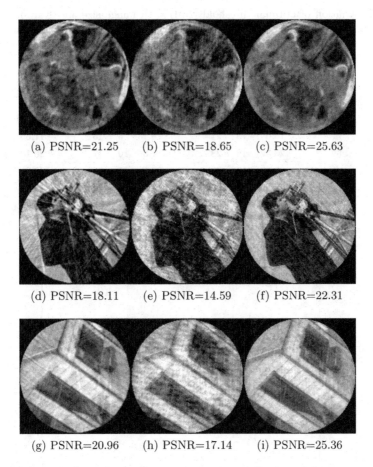

(a) PSNR=21.25   (b) PSNR=18.65   (c) PSNR=25.63

(d) PSNR=18.11   (e) PSNR=14.59   (f) PSNR=22.31

(g) PSNR=20.96   (h) PSNR=17.14   (i) PSNR=25.36

Fig. 8. Reconstruction on rotated images of brain (top row), cameraman (middle row) and house (bottom row) using Radon-iRadon (left column), least squared error method (middle column) and filtered back-projection (right column) with the 52 shortest projection angles

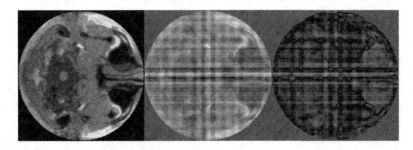

Fig. 9. Reconstruction with the shortest 50 angles excluding the angles: (0,1) and (1,0); Left: original Image; Middle: reconstructed image; Right: absolute error

Table 1. Variation of optimal threshold (fval) with image size and Katz ratio. The optimal threshold is the smallest threshold corresponding to the PSNR that lies within 0.01 of the global maximum PSNR, computed for resized versions of the cameraman image

Katz ratio	Image size: 179 × 179			Image size: 89 × 89			Image size: 43 × 43		
	# shortest angles	Peak PSNR	fval	# shortest angles	Peak Peak	fval	# shortest angles	Peak PSNR	fval
0.46	33	16.6	13	21	18.13	8	13	17.65	8
0.7	44	20.52	16	28	21.36	10	17	20.12	7
0.88	52	22.39	15	32	22.26	11	20	22.2	8
1.10	60	23.58	14	38	23.63	11	24	23.36	7
1.30	64	24.24	15	42	24.69	11	25	24.34	8
1.75	81	26.51	13	50	26.75	10	31	26.44	9

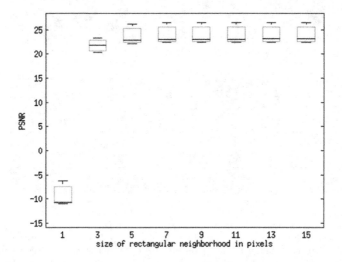

Fig. 10. Observing the quality of reconstruction as the size of neighborhood over which mean is evaluated is varied in each of the three images. The middle red line represents median and the ends of the boxes represent the 25th and 75th percentiles of PSNRs (Color figure online).

We observed the quality of reconstruction for different values of threshold. The optimal threshold is slightly dependent on image size, as seen in Table 1. This is because a larger sized PSF has higher frequency content (requiring a relatively higher threshold). However, the sensitivity of image reconstructions to the selected value of the threshold is quite weak over a wide range of image sizes and Katz values. We also observed (Fig. 10) how the size of neighborhood over which the mean is evaluated, affects the quality of reconstruction. Based on this information, we have used a 5 × 5 neighborhood.

3.2 Choice of Discrete Angles

Guillaume de Charette [16] and Matúš *et al.* [12] studied the amount of variance captured by different discrete angles. They observed that a discrete angle (p, q) captures variance (or information) proportional to $\frac{1}{p^2+q^2}$. Our observation confirms this. Figure 9 shows reconstructed image when angles $(0, 1)$ and $(1, 0)$ were omitted. The resultant reconstructed image has strong errors along $0°$ and $90°$. Hence, even when random set of angles is used, it is essential to include shorter angles like $(1, 0)$, $(0, 1)$, $(1, 1)$ and $(-1, 1)$. This effect might be compensated for by improved design of the weighting filter *wnp*.

4 Conclusion

We have compared three related techniques for direct image reconstruction from a finite set of discrete projections. We observed that as the number of projections increases, the reconstruction errors in filtered back-projection method is least dependent on the image structure and orientation. Hence, there is good scope for further improvement in this method. Ideally, we would like to modify the filter such that there is no need for a final threshold based clipping. Towards this, we observed how this optimal threshold depends on image size and Katz number.

References

1. Smith, K., Solmon, D., Wagner, S.: Practical and mathematical aspects of the problem of reconstructing objects from radiographs. Bull. Am. Math. Soc. **83**(6), 1227–1270 (1977)
2. Recur, B., Desbarats, P., Domenger, J.-P.: Radon and Mojette projections' equivalence for tomographic reconstruction using linear systems. In: 16-th International Conference in Central Europe on Computer Graphics, Visualization and Computer Vision 2008, Plzen, Czech Republic (2008). https://hal.archives-ouvertes.fr/hal-00353145
3. Guédon, J., Barba, D., Burger, N.D., Wagner, S.: Psychovisual image coding via an exact discrete radon transform. Bull. Am. Math. Soc. **2501**, 562–572 (1995)
4. Servieres, M.C.J., Normand, N., Subirats, P., Guédon, J.: Some links between continuous and discrete radon transform. In: Fitzpatrick, J.M., Sonka, M. (eds.) Medical Imaging 2004: Image Processing. Society of Photo-Optical Instrumentation Engineers (SPIE) Conference Series 1961–1971, vol. 5370, May 2004
5. Serviéres, M., Normand, N., Guédon, J.P.: Interpolation method for Mojette transform. In: SPIE-Medical Imaging: Image Processing, vol. 6142 (2006)
6. Svalbe, I., Kingston, A., Guédon, J., Normand, N., Chandra, S.: Direct inversion of mojette projections. In: 20th IEEE International Conference on Image Processing (ICIP), September 2013, pp. 1036–1040 (2013)
7. Guédon, J.: The Mojette Transform: Theory and Applications. ISTE, London; Wiley, New Jersey (2009)
8. Guédon, J.-P., Normand, N.: The Mojette transform: the first ten years. In: Andrès, É., Damiand, G., Lienhardt, P. (eds.) DGCI 2005. LNCS, vol. 3429, pp. 79–91. Springer, Heidelberg (2005)

9. Katz, M.: Questions of Uniqueness and Resolution in Reconstruction from Projections. Lecture Notes in Biomathematics, vol. 26. Springer, Heidelberg (1978)
10. Logan, B.F.: The uncertainty principle in reconstructing functions from projections. Duke Math. J. **42**(4), 661–706 (1975)
11. Gopal, P., Chandran, S., Svalbe, I., Rajwade, A.: Multi-slice tomographic reconstruction: to couple or not to couple. In: IEEE International Conference on Image Processing (ICIP) (2016, Submitted)
12. Matúš, F., Flusser, J.: Image representation via a finite radon transform. IEEE Trans. Pattern Anal. Mach. Intell. **15**(10), 996–1006 (1993)
13. Kingston, A., Li, H., Normand, N., Svalbe, I.: Fourier inversion of the Mojette transform. In: Barcucci, E., Frosini, A., Rinaldi, S. (eds.) DGCI 2014. LNCS, vol. 8668, pp. 275–284. Springer, Heidelberg (2014)
14. Cocosco, C.A., Kollokian, V., Kwan, R.K.S., Pike, G.B., Evans, A.C.: BrainWeb: online interface to a 3D MRI simulated brain database. NeuroImage **5**, 425 (1997)
15. Svalbe, I., Kingston, A., Normand, N., Der Sarkissian, H.: Back-projection filtration inversion of discrete projections. In: Barcucci, E., Frosini, A., Rinaldi, S. (eds.) DGCI 2014. LNCS, vol. 8668, pp. 238–249. Springer, Heidelberg (2014)
16. de Charette, G.: Data encryption using the finite radon transform. École Polytechnique de l'Université de Nantes, Monash University, Technical report, Département Informatique (2010)

Discrete and Combinatorial Topology

Homotopic Thinning in 2D and 3D Cubical Complexes Based on Critical Kernels

Michel Couprie[✉] and Gilles Bertrand

Université Paris-Est, LIGM, Équipe A3SI, ESIEE, Paris, France
{michel.couprie,gilles.bertrand}@esiee.fr

Abstract. We propose a symmetric thinning scheme for cubical or simplicial complexes of dimension 2 or 3. We show how to obtain, with a same generic thinning scheme, ultimate, curve or surface skeletons that are uniquely defined (no arbitrary choice is done).

Introduction

We propose a symmetric thinning scheme for cubical or simplicial complexes of dimension 2 or 3. Our motivations are listed below:

- Complexes can be used for the representation of discrete geometric objects, yielding better understanding of their structure and topological properties;
- The framework of digital topology does not permit to obtain skeletons that are provably thin, however, such a property can be proved in the framework of complexes;
- To our knowledge, there does not yet exist any symmetrical thinning algorithm in the framework of complexes. Only asymmetric algorithms, based on the collapse operation have been proposed. However, asymmetric thinning algorithms can produce, for the same object, drastically different results depending of the orientation of the object in space (see Fig. 8). On the other hand, symmetric algorithms guarantee a 90° rotation invariance.

In our previous works on critical kernels, we have proposed methods where the input and the output were "homogenous" complexes, that is, sets of pixels or sets of voxels (see *e.g.* [2,3]). The case of general complexes (made of elements of various dimensions) has never been considered in this framework.

Here, we show how to obtain, with a same generic thinning scheme, ultimate, curve or surface skeletons that are uniquely defined (no arbitrary choice is done). We also show that, if a thin skeleton is needed, it is better to use our symmetric method first and finish the thinning with a few steps of collapse.

1 Cubical Complexes

Although we focus on cubical complexes in this paper, all the notions and methods introduced from here to Sect. 5 can be readily transposed to the framework of simplicial complexes (see [1]).

© Springer International Publishing Switzerland 2016
N. Normand et al. (Eds.): DGCI 2016, LNCS 9647, pp. 131–142, 2016.
DOI: 10.1007/978-3-319-32360-2_10

Abstract complexes have been promoted in particular by V. Kovalevsky [9] in order to provide a sound topological basis for image analysis.

Intuitively, a cubical complex may be thought of as a set of elements having various dimensions (*e.g.*, vertices, edges, squares, cubes) glued together according to certain rules. In this section, we recall briefly some basic definitions on complexes, see also [2,6] for more details. We consider here n-dimensional complexes, with $0 \leqslant n \leqslant 3$.

Let S be a set. If T is a subset of S, we write $T \subseteq S$. Let $\mathbb{Z}$ denote the set of integers.

We consider the families of sets $\mathbb{F}_0^1$, $\mathbb{F}_1^1$, such that $\mathbb{F}_0^1 = \{\{a\} \mid a \in \mathbb{Z}\}$, $\mathbb{F}_1^1 = \{\{a, a+1\} \mid a \in \mathbb{Z}\}$. A subset f of $\mathbb{Z}^n$, $n \geqslant 2$, which is the Cartesian product of exactly m elements of $\mathbb{F}_1^1$ and $(n-m)$ elements of $\mathbb{F}_0^1$ is called a *face* or an *m-face* of $\mathbb{Z}^n$, m is the *dimension of* f, we write $\dim(f) = m$.

Observe that any non-empty intersection of faces is a face. For example, the intersection of two 2-faces A and B may be either a 2-face (if $A = B$), a 1-face, a 0-face, or the empty set.

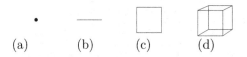

(a)   (b)   (c)   (d)

Fig. 1. Graphical representations of: (a) a 0-face, (b) a 1-face, (c) a 2-face, (d) a 3-face.

We denote by $\mathbb{F}^n$ the set composed of all m-faces of $\mathbb{Z}^n$, with $0 \leqslant m \leqslant n$. An m-face of $\mathbb{Z}^n$ is called a *point* if $m = 0$, a *(unit) interval* if $m = 1$, a *(unit) square* if $m = 2$, a *(unit) cube* if $m = 3$ (see Fig. 1).

Let f be a face in $\mathbb{F}^n$. We set $\hat{f} = \{g \in \mathbb{F}^n \mid g \subseteq f\}$ and $\hat{f}^* = \hat{f} \backslash \{f\}$.

Any $g \in \hat{f}$ is a *face of* f.

If X is a finite set of faces in $\mathbb{F}^n$, we write $X^- = \cup \{\hat{f} \mid f \in X\}$, X^- is the *closure of* X.

A set X of faces in $\mathbb{F}^n$ is a *cell* or an *m-cell* if there exists an m-face $f \in X$, such that $X = \hat{f}$. The *boundary of* a cell $\hat{f}$ is the set $\hat{f}^*$.

A finite set X of faces in $\mathbb{F}^n$ is a *complex (in $\mathbb{F}^n$)* if $X = X^-$. Any subset Y of a complex X which is also a complex is a *subcomplex of* X. If Y is a subcomplex of X, we write $Y \preceq X$. If X is a complex in $\mathbb{F}^n$, we also write $X \preceq \mathbb{F}^n$. In Fig. 2, some complexes are represented. Notice that any cell is a complex.

Let $X \subseteq \mathbb{F}^n$. A face $f \in X$ is *a facet of* X if there is no $g \in X$ such that $f \in \hat{g}^*$. We denote by X^+ the set composed of all facets of X. If X is a complex, observe that in general, X^+ is not a complex, and that $[X^+]^- = X$.

2   Collapse

In this section we recall a definition of the operation of collapse [7], which is a discrete analogue of a continuous deformation (a homotopy).

Let X be a complex in $\mathbb{F}^n$ and let $f \in X$. If there exists one face $g \in \hat{f}^*$ such that f is the only face of X which strictly includes g, then g is said to be *free for X* and the pair (f, g) is said to be a *free pair for X*. Notice that, if (f, g) is a free pair, then we have necessarily $f \in X^+$ and $\dim(g) = \dim(f) - 1$.

Let X be a complex, and let (f, g) be a free pair for X. The complex $X \backslash \{f, g\}$ is an *elementary collapse of X*.

Let X, Y be two complexes. We say that X *collapses onto Y* if $Y = X$ or if there exists a *collapse sequence from X to Y*, i.e., a sequence of complexes $\langle X_0, ..., X_\ell \rangle$ such that $X_0 = X$, $X_\ell = Y$, and X_i is an elementary collapse of X_{i-1}, $i = 1, ..., \ell$. Figure 2 illustrates a collapse sequence.

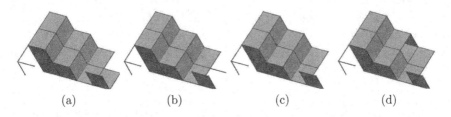

| (a) | (b) | (c) | (d) |

Fig. 2. (a): a complex $X \preceq \mathbb{F}^3$. (a–d): a collapse sequence from X.

Remark 1. *Let V be a set of 2-faces (pixels) or a set of 3-faces (voxels), and let $x \in V$. The element x is simple, in the sense of digital topology (see [6,8]) if the complex V^- collapses onto $(V \backslash \{x\})^-$.*

3   Critical Kernels

Let us briefly recall the framework introduced by one of the authors (in [1]) for thinning, in parallel, discrete objects with the warranty that we do not alter the topology of these objects. We focus here on the two- and three-dimensional cases, but in fact the results in this section are valid for complexes of arbitrary dimension. This framework is based solely on three notions: the notion of an essential face which allows us to define the core of a face, and the notion of a critical face (see illustrations in Fig. 3).

Definition 2 ([1]). *Let $X \preceq \mathbb{F}^n$ and let $f \in X$. We say that f is an* essential *face for X if f is precisely the intersection of all facets of X which contain f, i.e., if $f = \cap \{g \in X^+ \mid f \subseteq g\}$. We denote by $Ess(X)$ the set composed of all essential faces of X. If f is an essential face for X, we say that $\hat{f}$ is an* essential *cell for X. If $Y \preceq X$ and $Ess(Y) \subseteq Ess(X)$, then we write $Y \trianglelefteq X$.*

Observe that a facet of X is necessarily an essential face for X, i.e., $X^+ \subseteq Ess(X)$.

Definition 3 ([1]). *Let $X \preceq \mathbb{F}^n$ and let $f \in Ess(X)$. The* core *of $\hat{f}$ for X is the complex $Core(\hat{f}, X) = \cup \{\hat{g} \mid g \in Ess(X) \cap \hat{f}^*\}$.*

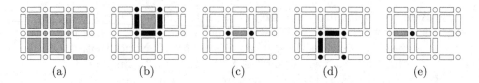

Fig. 3. (a): a complex $X \preceq \mathbb{F}^2$, the essential faces are shown in gray. (b, c, d, e): an essential face (in gray) and its core (in black). The faces in (b, e) are regular, those in (c, d) are critical.

Definition 4 ([1]). *Let $X \preceq \mathbb{F}^n$ and let $f \in X$. We say that f and $\hat{f}$ are regular for X if $f \in Ess(X)$ and if $\hat{f}$ collapses onto $Core(\hat{f}, X)$. We say that f and $\hat{f}$ are critical for X if $f \in Ess(X)$ and if f is not regular for X. If $X \preceq \mathbb{F}^n$, we set $Critic(X) = \cup\{\hat{f} \mid f$ is critical for $X\}$, we say that $Critic(X)$ is the critical kernel of X.*

If f is a pixel (resp. a voxel), then saying that f is regular is equivalent to say that f is simple in the classical sense (see Remark (1) and [6]). Thus, the notion of regular face generalizes the one of simple pixel (resp. voxel) to arbitrary facets and even to faces that are not facets.

The following theorem is the most fundamental result concerning critical kernels. We use it here in dimension 2 or 3, but notice that the theorem holds whatever the dimension.

Theorem 5 ([1]). *Let $n \in \mathbb{N}$, let $X \preceq \mathbb{F}^n$.*

(i) The complex X collapses onto its critical kernel.
(ii) If $Y \trianglelefteq X$ contains the critical kernel of X, then X collapses onto Y.
(iii) If $Y \trianglelefteq X$ contains the critical kernel of X, then any Z such that $Y \preceq Z \trianglelefteq X$ collapses onto Y.

Let n be a positive integer, let $X \preceq \mathbb{F}^n$. We define $Critic^n(X)$ as follows: $Critic^0(X) = X$, and $Critic^n(X) = Critic(Critic^{n-1}(X))$, whenever $n > 0$. If $Critic^n(X) = Critic^{n+1}(X)$, then we say that $Critic^n(X)$ is the *ultimate skeleton* of X and we write $Critic^n(X) = Critic^\infty(X)$.

From Theorem 5, we deduce immediately that for any $X \preceq \mathbb{F}^n$, the complex X collapses onto $Critic^\infty(X)$. See Fig. 4 for an illustration.

4   Symmetric Thinning Scheme

In this section, we introduce our new generic parallel thinning scheme, see Algorithm 1. It is generic in the sense that any notion of skeletal element (introduced below) may be used, for obtaining, *e.g.*, ultimate, curve, or surface skeletons.

In order to compute curve or surface skeletons, we have to keep other faces than the ones that are necessary for the preservation of the topology of the object X. In the scheme, the set K corresponds to a set of features that we want to be preserved by a thinning algorithm (thus, we have $K \subseteq X$). This set K, called *constraint set*, is updated dynamically at line 3 of the algorithm. To this aim, we will define a function $Skel_X$ from X^+ onto $\{True, False\}$, that allows us to detect some *skeletal facets* of X, e.g., some facets belonging to parts of X that are surfaces or curves. These detected facets are progressively stored in K.

Algorithm 1. SymThinningScheme$(X, Skel_X)$

Data: $X \preceq \mathbb{F}^n$, $Skel_X$ is a function from X^+ on $\{True, False\}$
Result: X
1   $K := \emptyset$;
2 **repeat**
3 | $K := K \cup \{x \in X^+ \text{ such that } Skel_X(x) = True\}$;
4 | $X := Critic(X) \cup K^-$;
5 **until** *stability* ;

Notice that, before line 4, the complex $Y = Critic(X) \cup K^-$ is such that $Y \lhd X$ and $Critic(X) \subseteq Y$. Thus, by Theorem 5(ii), the original complex X collapses onto the result of SymThinningScheme, for any X and any function $Skel_X$.

See Fig. 4 for an illustration of SymThinningScheme, using a function $Skel_X$ that yields *False* for any facet. The result of this operation is, obviously, the ultimate skeleton of the input complex X.

In order to preserve geometrical features of the original object, such as elongated or flat parts, we use two kinds of skeletal facets called isthmuses.

Intuitively, a facet f of a complex X is said to be a 1-isthmus (resp. a 2-isthmus) if the core of $\hat{f}$ for X corresponds to the one of an element belonging to a curve (resp. a surface) [3].

Let $X \subseteq \mathbb{F}^n$ be a non-empty set of faces. A sequence $(f_i)_{i=0}^{\ell}$ of faces of X is a *path in X (from f_0 to f_ℓ)* if $f_i \cap f_{i+1} \neq \emptyset$, for all $i \in [0, \ell - 1]$. We say that X *is connected* if, for any two faces f, g in X, there is a path from f to g in X.

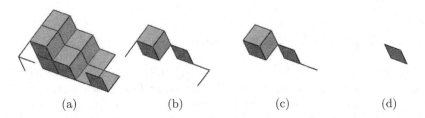

(a)   (b)   (c)   (d)

Fig. 4. (a): a complex $X \preceq \mathbb{F}^3$. (b): after one execution of the main loop of SymThinningScheme: $Critic^1(X) = Critic(X)$. (c): after two executions of the main loop: $Critic^2(X)$. (d): the final result: $Critic^3(X) = Critic^\infty(X)$.

We say that $X \preceq \mathbb{F}^n$ is a 0-*surface* if X^+ is precisely made of two facets f and g of X such that $f \cap g = \emptyset$.

We say that $X \preceq \mathbb{F}^n$ is a 1-*surface* (or a *simple closed curve*) if:

(i) X^+ is connected; and
(ii) For each $f \in X^+$, $Core(\hat{f}, X)$ is a 0-surface.

We say that $X \preceq \mathbb{F}^n$ is an *simple open curve* if:

(i) X^+ is connected; and
(ii) For each $f \in X^+$, $Core(\hat{f}, X)$ is a 0-surface or a single cell.

Definition 6. *Let $X \preceq \mathbb{F}^n$, let $f \in X^+$.*

We say that f is a 1-isthmus for X if $Core(\hat{f}, X)$ is a 0-surface.
We say that f is a 2-isthmus for X if $Core(\hat{f}, X)$ is a 1-surface.
We say that f is a 2^+-isthmus for X if f is a 1-isthmus or a 2-isthmus for X.

Our aim is to thin an object, while preserving a constraint set K that is made of faces that are detected as k-isthmuses during the thinning process. We obtain curve skeletons with $k = 1$, and surface skeletons with $k = 2^+$. These two kinds of skeletons may be obtained by using `SymThinningScheme`, with the function $Skel_X$ defined as follows:

$$Skel_X(x) = \begin{cases} True & \text{if } x \text{ is a } k\text{-isthmus for } X, \\ False & \text{otherwise,} \end{cases}$$

with k being set to 1 or 2^+.

Observe that a facet may be a k-isthmus at a given step of Algorithm 1, but not at further steps. This is why previously detected isthmuses are stored in K.

Figure 5 illustrates curve and surface skeletons. We observe that these skeletons contain faces of all dimensions: 3, 2, 1, 0. This is the counterpart of the choice of having a symmetric process, hence a 90° rotation invariance property, as illustrated in Fig. 6. We deal with the thinness issue in the next section.

Observe also that, in Fig. 6, the obtained skeletons are simple open curves, as defined above. More generally, despite the fact that they are composed of faces of various dimensions, parts of produced skeletons can be directly interpreted as pieces of curves or surfaces.

5   Asymmetric Thinning Scheme

Thinner skeletons may be obtained if we give up the symmetry. To this aim, the collapse operation may be directly used. The method described in this section corresponds to a special case of a method introduced by Liu *et al.* in [10] (see also [4]) for producing families of filtered skeletons. Here, we are interested in non-filtered skeletons obtained through parameter-free thinning methods. Besides, the filtering approach of [10] can easily be adapted to our method.

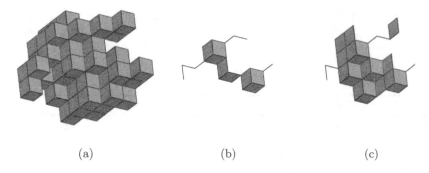

(a)   (b)   (c)

Fig. 5. (a): a complex $X \preceq \mathbb{F}^3$. (b): curve skeleton of X. (c): surface skeleton of X.

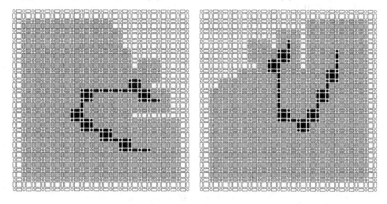

Fig. 6. Illustration of 90° rotation invariance with the symmetric thinning (algorithm SymThinningScheme).

In general, removing free pairs from a complex in parallel does not preserve topology. But under certain conditions parallel collapse of free pairs is feasible.

First, we need to define the direction of a free face. Let X be a complex in $\mathbb{F}^n$, let (f,g) be a free pair for X. Since (f,g) is free, we know that $\dim(g) = \dim(f) - 1$, and it can be easily seen that $f = g \cup g'$ where g' is the translate of g by one of the $2n$ vectors of $\mathbb{Z}^n$ with all coordinates equal to 0 except one, which is either $+1$ or -1. Let v denote this vector, and c its non-null coordinate. We define $Dir(f,g)$ as the index of c in v, it is the *direction* of the free pair (f,g). Its *orientation* is defined as $Orient(f,g) = 1$ if $c = +1$, and as $Orient(f,g) = 0$ otherwise.

Considering two distinct free pairs (f,g) and (i,j) for a complex X in $\mathbb{F}^n$ such that $Dir(f,g) = Dir(i,j)$ and $Orient(f,g) = Orient(i,j)$, we have $f \neq i$. It can easily be seen that (f,g) is free for $X \backslash \{i,j\}$, and (i,j) is free for $X \backslash \{f,g\}$. Loosely speaking, (f,g) and (i,j) may collapse in any order or in parallel. More generally, we have the following property.

Proposition 7 ([5]). *Let X be a complex in $\mathbb{F}^n$, and let $(f_1, g_1), \ldots, (f_m, g_m)$ be m distinct free pairs for X having all the same direction and the same orientation. The complex X collapses onto $X \backslash \{f_1, g_1, \ldots, f_m, g_m\}$.*

Now, we are ready to introduce Algorithm 2.

Algorithm 2. `ParDirCollapse`$(X, Skel_X)$

Data: $X \preceq \mathbb{F}^n$, $Skel_X$ is a function from X^+ on $\{\mathit{True}, \mathit{False}\}$
Result: X

1 $K := \emptyset$; $L = \{\{f, g\} \mid (f, g) \text{ is free for } X\}$;
2 **while** $L \neq \emptyset$ **do**
3 $\quad$ $K := K \cup \{x \in X^+ \text{ such that } Skel_X(x) = \mathit{True}\}$;
4 $\quad$ **for** $dir = 1 \to n$ **do**
5 $\quad\quad$ **for** $orient = 0 \to 1$ **do**
6 $\quad\quad\quad$ **for** $d = n \to 1$ **do**
7 $\quad\quad\quad\quad$ $T = \cup\{\{f, g\} \in L \mid (f, g) \text{ is free for } X \text{ and } f \notin K,$
8 $\quad\quad\quad\quad\quad$ $Dir(f, g) = dir, Orient(f, g) = orient, \dim(f) = d\}$;
9 $\quad\quad\quad\quad$ $X = X \backslash T$;

Notice that opposite orientations (*e.g.*, north and south) are treated consecutively in a same directional substep. To obtain curve or surface skeletons, we set the function $Skel_X$ as follows:

$$Skel_X(x) = \begin{cases} \mathit{True} & \text{if } \dim(x) = 1, \\ \mathit{False} & \text{otherwise.} \end{cases}$$

for curve skeletons, and

$$Skel_X(x) = \begin{cases} \mathit{True} & \text{if } \dim(x) \in \{1, 2\}, \\ \mathit{False} & \text{otherwise.} \end{cases}$$

for surface skeletons.

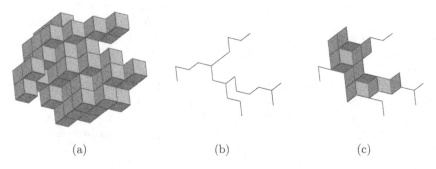

(a)   (b)   (c)

Fig. 7. (a): a complex $X \preceq \mathbb{F}^3$. (b): a curve skeleton by collapse of X. (c): a surface skeleton by collapse of X.

Figure 7 shows results of algorithm `ParDirCollapse`. Notice that the curve skeleton is only composed of 1- and 0-faces, and that the surface skeleton does not contain any 3-face. Indeed, the following property guarantees that a curve skeleton in 2D (resp. a surface skeleton in 3D) does not contain any 2-face (resp. 3-face).

Proposition 8 ([5]). *Let X be a finite complex in $\mathbb{F}^n$, with $n > 0$, that has at least one n-face. Then X has at least one free $(n-1)$-face.*

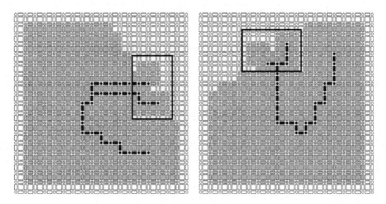

Fig. 8. Illustration of asymmetric thinning (algorithm `ParDirCollapse`). The boxed area is detailed in Fig. 9.

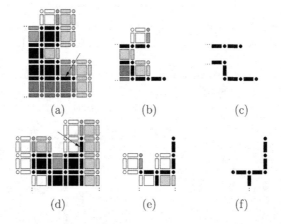

(a)   (b)   (c)

(d)   (e)   (f)

Fig. 9. Detail of the thinning by collapse (algorithm `ParDirCollapse`) of the complexes of Fig. 8. (a, d): first step. (b, e): second step. (c, f): third step. Black faces are the ones that remain at the end of the step. The order in which the faces of different directions and orientations are processed is the same in all cases: 1. horizontal, left to right (white); 2. horizontal, right to left (light gray); 3. vertical, downwards (medium gray); 4. vertical, upwards (dark gray). An arrow indicates the only 1-face that is added to the constraint set K at the beginning of the second iteration. At the beginning of the third step, all the 1-faces in black are in K. We observe the birth of two parallel branches in (c), and the merging of two branches in (f).

The price to pay for getting this thinness property is the loss of 90° rotation invariance. The example of Fig. 8 shows that differences of arbitrary size may be observed between skeletons of a same shape, depending on its position in space. On the left, we see that two parallel skeleton branches correspond to a single branch of the right image. The length of this "split branch" may be arbitrarily big, depending on the size of the whole object.

Figure 9 details the directional substeps of algorithm ParDirCollapse and shows how this algorithm may give birth to different skeleton configurations for different orientations of the same original object.

6   Experiments, Discussion and Conclusion

Skeletons are notoriously sensitive to noise, and this is major problem for many applications. Even in the continuous case, the slightest perturbation of a smooth contour shape may provoke the appearance of an arbitrarily long skeleton branch, that we will refer to as a spurious branch. A desirable property of discrete skeletonization methods is to generate as few spurious branches as possible, in response to the so-called discretization (or voxelization) noise that is inherent to any discretization process.

It would make little sense to directly compare results of SymThinningScheme with those of ParDirCollapse, as the goals of these two methods are different. On the other hand, we may compare the results of (i) ParDirCollapse with those of (ii) SymThinningScheme followed by ParDirCollapse, as both are thin skeletons.

First of all, let us take a look at Fig. 10, where the latter method is applied to the same objects as in Figs. 6 and 8. We see that the split branch artifact of Fig. 8 is avoided.

We will compare the two methods with respect to their ability to produce skeletons that are free of spurious branches. In the following, we compare how different methods behave with respect to this property.

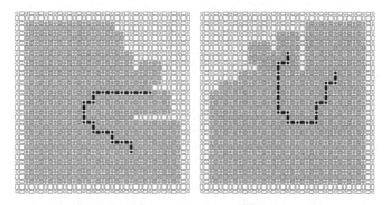

Fig. 10. Illustration of symmetric thinning (algorithm SymThinningScheme) followed by a few asymmetric thinning steps (algorithm ParDirCollapse).

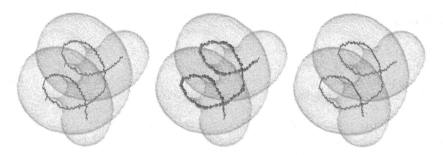

Fig. 11. Results for object X_5. Left: ParDirCollapse(X_5). Center: SymThinning Scheme(X_5). Right: ParDirCollapse(SymThinningScheme(X_5)).

Table 1. Experimental results

Object	X_1	X_2	X_3	X_4	X_5	X_6
S(ParDirCollapse(SymThinningScheme(X_i)))	4	0	0	0	0	0
S(ParDirCollapse(X_i))	16	0	0	0	8	1

In order to get ground truth skeletons, we discretized six simple 3D shapes for which the skeletons are known: a bent cylinder forming a knot (X_1), a Euclidean ball (X_2), a thickened straight segment (X_3), a torus (X_4), a thickened spiral $(X_5$, see Fig. 11), an ellipsoid (X_6). For example, a curve skeleton of a discretized torus should ideally be a simple closed discrete curve (a 1-surface). Any extra branch of the skeleton must undoubtedly be considered as spurious. Thus, a simple and effective criterion for assessing the quality of a skeletonization method is to count the number of extra branches, or equivalently in our case, the number of extra curve extremities (free faces). Notice that, even if the original objects are complexes obtained by taking the closure of sets of voxels (3-faces), the intermediate and final results are indeed general complexes, which may contain 2-facets and 1-facets.

In order to compare methods, we use the indicator $S(X) = |c(X) - c_i(X)|$, where $c(X)$ stands for the number of curve extremities for the result obtained from X after thinning, and $c_i(X)$ stands for the ideal number of curve extremities to expect with the object X. In other words, $S(X)$ counts the number of spurious branches in the skeleton of object X, a result of 0 being the best one (Table 1).

Additionally, we performed discrete rotations of the object X_4 (torus), by angles ranging from 1 to 89° by steps of 1°, and computed the values of $S(X)$ for all these rotated objects and for both methods. The mean value of $S(X)$ was 131.0 for ParDirCollapse and 69.2 for SymThinningScheme followed by ParDirCollapse, which always gave the best result.

To conclude, our symmetric parallel thinning scheme is the first one that permits to thin general 2D or 3D complexes in a symmetrical manner, avoiding any arbitrary choice. We also showed experimentally that if, however, thin skeletons are required, then it is better to use our symmetric thinning scheme first.

References

1. Bertrand, G.: On critical kernels. Comptes Rendus de l'Académie des Sciences, Série Math. **I**(345), 363–367 (2007)
2. Bertrand, G., Couprie, M.: Two-dimensional thinning algorithms based on critical kernels. J. Math. Imaging Vision **31**(1), 35–56 (2008)
3. Bertrand, G., Couprie, M.: Powerful parallel and symmetric 3D thinning schemes based on critical kernels. J. Math. Imaging Vision **48**(1), 134–148 (2014)
4. Chaussard, J.: Topological tools for discrete shape analysis. Ph.D. dissertation, Université Paris-Est (2010)
5. Chaussard, J., Couprie, M.: Surface thinning in 3D cubical complexes. In: Wiederhold, P., Barneva, R.P. (eds.) IWCIA 2009. LNCS, vol. 5852, pp. 135–148. Springer, Heidelberg (2009)
6. Couprie, M., Bertrand, G.: New characterizations of simple points in 2D, 3D and 4D discrete spaces. IEEE Trans. Pattern Anal. Mach. Intell. **31**(4), 637–648 (2009)
7. Giblin, P.: Graphs, Surfaces and Homology. Chapman and Hall, London (1981)
8. Yung Kong, T., Rosenfeld, A.: Digital topology. Comput. Vision Graph. Image Process. **48**, 357–393 (1989)
9. Kovalevsky, V.A.: Finite topology as applied to image analysis. Comput. Vision Graph. Image Process. **46**, 141–161 (1989)
10. Liu, L., Chambers, E.W., Letscher, D., Ju, T.: A simple and robust thinning algorithm on cell complexes. Comput. Graph. Forum **29**(7), 2253–2260 (2010)

P-Simple Points and General-Simple Deletion Rules

Kálmán Palágyi$^{(\boxtimes)}$

Department of Image Processing and Computer Graphics,
University of Szeged, Szeged, Hungary
`palagyi@inf.u-szeged.hu`

Abstract. Reductions transform binary pictures only by changing some black points to white ones. Parallel reductions can alter a set of black points simultaneously, while a sequential reduction traverses the black points of a picture, and changes the actually visited single point if the considered deletion rule is satisfied. Two reductions are called equivalent if they produce the same result for each input picture. General-simple deletion rules yield pairs of equivalent topology-preserving parallel and sequential reductions in arbitrary binary pictures. This paper bridges P-simple points and general-simple deletion rules: we show that some deletion rules that delete P-simple points are general-simple, and each point deleted by a general-simple deletion rule is P-simple.

Keywords: Topology preservation · Equivalent reductions · P-simple points · General-simple deletion rules

1 Introduction

Let $\mathcal{V}$ be a *digital space* (e.g., $\mathbb{Z}^2$ or $\mathbb{Z}^3$), where the elements of $\mathcal{V}$ are called *points*. A *digital binary picture* [12] on $\mathcal{V}$ is a mapping that assigns only two possible values, *black* and *white* to each point. A *reduction* [7] transforms a binary picture only by changing some black points to white ones, which is referred to as *deletion*. Reductions play a key role in some topological algorithms, e.g., *thinning* [7] or *reductive shrinking* [8].

Parallel reductions can delete a set of black points simultaneously, while a *sequential reduction* traverses the black points of a picture, and considers the actually visited point for possible deletion at a time. These two strategies are illustrated in Algorithms 1 and 2. Note that Algorithm 2 (i.e., the sequential approach) can specify a unique mapping of pictures to pictures with the help of the input parameter Π (i.e., the order in which the points are selected by the **foreach** loop). For practical purposes we assume that all input pictures are *finite* (i.e., they contain finitely many black points).

Algorithms 1 and 2 classify the set of black points B in the input picture into two (disjoint) subsets: points in the *constraint set* C are not taken into consideration (i.e., each element in C is absolutely protected), and the *deletion rule* R associated with these algorithms is evaluated only for the elements of

© Springer International Publishing Switzerland 2016
N. Normand et al. (Eds.): DGCI 2016, LNCS 9647, pp. 143–153, 2016.
DOI: 10.1007/978-3-319-32360-2_11

Algorithm 1.   Parallel reduction

Input: set of black points $B \subseteq \mathcal{V}$, constraint set $C \subseteq B$, and deletion rule R
Output: set of black points $PB \subseteq B$
// *specifying the set points X to be investigated*
$X = B \setminus C$
// *determining the set of deletable points D*
$D = \{ \, p \mid p \in X \text{ and } R(p, B, C) = \textbf{true} \, \}$
// *deletion of the set of points D*
$PB \; = \; B \setminus D$

Algorithm 2.   Sequential reduction

Input: set of black points $B \subseteq \mathcal{V}$, constraint set $C \subseteq B$, deletion rule R, and
permutation Π of points in $B \setminus C$
Output: set of black points $SB \subseteq B$
// *specifying the set points X to be investigated*
$X = B \setminus C$
$SB = B$
// *traversal of X according to permutation Π*
foreach $p \in X$ **do**
  if $R(p, SB, C) = \textbf{true}$ **then**
    // *deletion of the single point p*
    $SB \; = SB \setminus \{p\}$

---

the set $X = B \backslash C$. Note that thinning algorithms can delete some *border points* [12], hence their constraint sets contains all *interior points* [12] and some types of *endpoints* (i.e., points that provide relevant geometrical information with respect to the shape of the objects) [7,16] or accumulated *isthmuses* (i.e., generalization of curve/surface interior points) [3].

An investigated black point p is *deletable* by the deletion rule R, if $R\,(\,p,\,A,\,C\,) = \textbf{true}$, where $A = B$ in parallel reductions (see Algorithm 1), and $A = SB \subseteq B$ in the sequential case (see Algorithm 2). Thus parallel reductions consider the initial set of black points B when the deletion rule is evaluated. On the contrary, the set of black points is dynamically altered when a sequential reduction is performed. We lay stress upon the fact that both the constraint set C and the set of points $X = B \backslash C$ are not modified by Algorithms 1 and 2.

Sequential reductions (with the same deletion rule) suffer from the drawback that different visiting orders (raster scans) of points in $X = B \backslash C$ may yield various results. A deletion rule R is called *order-independent* [21] if the sequential reductions with R produce the same output picture for all the possible visiting orders. By extending this concept: a sequential reduction is said to be *order-independent* if its deletion rule is order-independent.

The concept of *simple point* is fundamental in topological algorithms. A (black or white) point is called a simple point if its *alteration* (i.e., changing its 'color') *preserves topology* [12]. A reduction is said to be *topology-preserving* if the set of black points in the output picture can be obtained from the set of black points in the input picture by sequential deletions of simple points [12].

Unfortunately, simultaneous deletion of a (sub)set of simple points may alter the topology. That is why Bertrand has proposed the notion of *P-simple point* [1].

Two reductions are called *equivalent* if they produce the same result for each input picture [18]. With the help of the notion of *general-simple* deletion rule, the author showed that general-simple deletion rules yield pairs of equivalent and topology-preserving parallel and (order-independent) sequential reductions [18].

This paper bridges *P*-simple points and general-simple deletion rules. In Sect. 3, we prove that some deletion rules that delete *P*-simple points are general-simple, and Sect. 4 shows that each point deleted by a general-simple deletion rule is *P*-simple.

2 Basic Notions and Results

In this paper we consider the traditional paradigm of digital topology as reviewed by Kong and Rosenfeld [12].

Let $(k, \bar{k})$ be a pair of *adjacency* relations on *digital space* $\mathcal{V}$. A $(k, \bar{k})$ *digital picture* is a quadruple $(\mathcal{V}, k, \bar{k}, B)$, where each point in $B \subseteq \mathcal{V}$ is called a *black point*, each point in $\mathcal{V} \backslash B$ is said to be a *white point*, k-adjacency/connectivity is used for black points, and $\bar{k}$-adjacency/connectivity is assigned to white points.

A black point p is called an *interior point* if each point that is $\bar{k}$-adjacent to p is black. A black point is said to be a *border point* if it is not an interior point.

The concept of *simple* points is well established in digital topology: A point is called a *simple point* in a picture if its alteration *preserves topology* [12].

There are several useful characterizations of simple points in 2D [5,9–12], 3D [5,10,12,13,22], 4D [5,11], and higher dimensional [15] pictures. Since all examples presented in this paper assume 2D digital pictures sampled on $\mathbb{Z}^2$ (that is dual to the regular square grid), we recall the following characterization of simple points on $(8, 4)$ pictures.

Theorem 1. [9] *A (black or white) point $p \in \mathbb{Z}^2$ is simple in an $(8, 4)$ picture if and only if p is matched by at least one of the base templates depicted in Fig. 1 or their rotated and reflected versions.*

Fig. 1. Base matching templates for characterizing $(8, 4)$-simple (black or white) points. Notations: each black element matches a black point; each white element matches a white point; each element depicted in gray matches either a black or a white point; each central element marked a '★' is coincident with the investigated point. Note that interior points are non-simple points.

In order to construct topology-preserving parallel reductions and provide a verification method, Bertrand has proposed the notion of *P-simple point* [1]. Let us consider a set of points $B \subseteq V$ and a set of points $P \subseteq B$. Then a point $p \in P$ is *P-simple* in B if for each set of points $Q \subseteq P\backslash\{p\}$, point p is simple in $B\backslash Q$.

Bertrand showed that a parallel reduction that deletes only *P*-simple points is topology-preserving [1], and he gave a local characterization of *P*-simple points on $(26, 6)$ 3D pictures [2]. Note that Bertrand and Couprie linked *critical kernels* to *minimal non-simple sets* and *P*-simple points [4].

Figure 2 classifies black points of an $(8, 4)$ picture into (non-simple) interior points, non-simple border points, *P*-simple points, and additional simple points (that are not *P*-simple points).

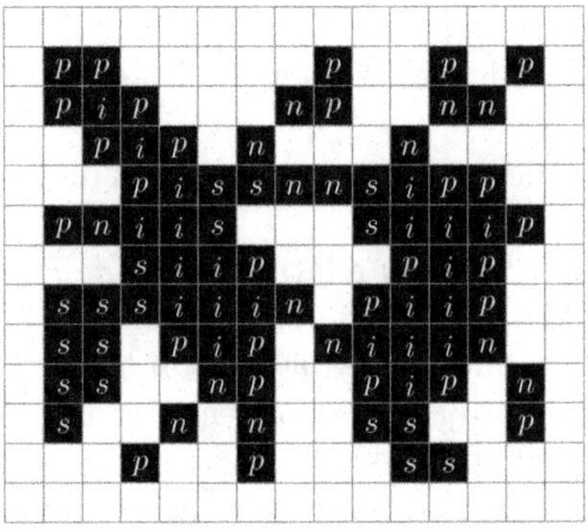

Fig. 2. Classifying black points in an $(8, 4)$ picture. Notations: (non-simple) interior points are marked '*i*', non-simple border points are marked '*n*', *P*-simple points are marked '*p*', simple (border) points that are not *P*-simple points are marked '*s*'.

The author established a new sufficient condition for arbitrary pictures with the help of *general-simple* deletion rules [18]. Let us recall and rephrase the corresponding concepts and results.

Definition 1. *Deletion rule R (with respect to the constraint set $C \subseteq B$) is general if $R(p, B, C) = \textbf{true}$ for a point $p \in B$, then $R(q, B, C) = R(q, B\backslash\{p\}, C)$ for each point $q \in B\backslash\{p\}$.*

Definition 2. *Deletion rule R is* general-simple *if R is general and it deletes only simple points.*

Theorem 2. *A deletion rule R yields an order-independent sequential reduction if and only if R is general.*

Theorem 3. *General deletion rules specify pairs of equivalent parallel and order-independent sequential reductions.*

Theorem 4. *A parallel reduction is topology-preserving if its deletion rule is general-simple.*

Let us summarize Theorems 2–4: if a deletion rule deletes only simple points, and the 'deletability' of any point does not depend on the 'color' of any 'deletable' point, then that deletion rule specifies a pair of topology-preserving equivalent parallel and (order-independent) sequential reductions.

3   From P-Simple Points to General-Simple Deletion Rules

In this section a special deletion rule is constructed that deletes P-simple points, and we verify that our deletion rule is general-simple. Throughout this section, we assume that set B contains all black points in the input picture (see Algorithms 1 and 2).

Let us define constraint set C as

$$C = \{\, p \mid p \text{ is not a simple point in } B \,\}. \tag{1}$$

Let Y be a set of black points in a picture such that $Y \subseteq B$ and $Y \supseteq C$. Then consider the following set of points

$$\mathcal{P}_{C,Y} = \{\, p \mid p \in Y \text{ and } p \text{ is a simple point in } Y \,\}. \tag{2}$$

The deletion rule $\mathcal{PS}$ (with respect to the constraint set C) is defined as

$$\mathcal{PS}\,(p,\ Y,\ C) = \begin{cases} \textbf{true} & \text{if } p \text{ is } P\text{-simple in } Y \\ \textbf{false} & \text{otherwise} \end{cases}, \tag{3}$$

where $P = \mathcal{P}_{C,Y}$.

Since a P-simple point is necessarily a simple point, $\mathcal{PS}$ deletes only simple points. We show that deletion rule $\mathcal{PS}$ is general-simple.

Let us introduce a relation marked '$\leftrightsquigarrow$' between two sets of black points. We say that $B_1 \leftrightsquigarrow B_2$ if B_2 can be derived from B_1 by sequential alteration of simple points. It can readily be seen that '$\leftrightsquigarrow$' is an equivalence relation. We can write that $B \leftrightsquigarrow B\backslash\{p\}$ if $p \in B$ is a simple black point, and $B \leftrightsquigarrow B \cup \{p\}$ if $p \notin B$ is a simple white point.

Lemma 1. *Assume that $\mathcal{PS}(p,B,C) = \textbf{true}$. Then $\mathcal{P}_{C,B\backslash\{p\}} = \mathcal{P}_{C,B}\backslash\{p\}$.*

Proof.

It is obvious that $\mathcal{P}_{C,B\backslash\{p\}} \subseteq \mathcal{P}_{C,B}\backslash\{p\}$. Then we prove that $\mathcal{P}_{C,B} \subseteq \mathcal{P}_{C,B\backslash\{p\}} \cup \{p\}$.

Let q be a point, such that $q \in \mathcal{P}_{C,B}\backslash\{p\}$. In consequence, q is a simple point in B. Since $\mathcal{PS}(p,B,C) = \textbf{true}$ for a point $p \in B\backslash C$, p is a simple point in B. As point p is $\mathcal{P}_{C,B}$-simple in B, p is a simple point in $B\backslash\{q\}$. Hence we can write

$$\begin{aligned} B &\leftrightsquigarrow B\backslash\{q\}, \\ B &\leftrightsquigarrow B\backslash\{p\}, \\ B\backslash\{q\} &\leftrightsquigarrow (B\backslash\{q\}) \backslash \{p\}. \end{aligned} \tag{4}$$

Since '$\leftrightarrow$' is an equivalence relation,

$$B\backslash\{p\} \; \leftrightarrow \; B \; \leftrightarrow \; B\backslash\{q\} \; \leftrightarrow \; (B\backslash\{q\}) \backslash \{p\} = (B\backslash\{p\})\backslash\{q\}. \tag{5}$$

Consequently, $B\backslash\{p\} \leftrightarrow (B\backslash\{p\})\backslash\{q\}$, thus q is a simple point in $B\backslash\{p\}$. Hence $q \in \mathcal{P}_{\mathcal{C},B\backslash\{p\}} \cup \{p\}$. $\qquad\square$

Lemma 1 states that deletion of a point by $\mathcal{PS}$ does not alter the simpleness of the (black) simple points that are not in the constraint set $\mathcal{C}$. Thus we can state that

$$\mathcal{P}_{\mathcal{C},B} = B\backslash\mathcal{C}, \tag{6}$$

and since $\mathcal{PS}(p, B, \mathcal{C}) = \mathbf{true}$, then

$$\mathcal{P}_{\mathcal{C},B\backslash\{p\}} = (B\backslash\{p\}) \backslash \mathcal{C}. \tag{7}$$

We are now ready to state the key theorem.

Theorem 5. *Deletion rule $\mathcal{PS}$ (with respect to constraint set $\mathcal{C}$) is general.*

Proof.

It is to be proved that if $\mathcal{PS}(p, B, \mathcal{C}) = \mathbf{true}$, then $\mathcal{PS}(q, B, \mathcal{C}) = \mathcal{PS}(q, B\backslash\{p\}, \mathcal{C})$ for any two points $p \in B$ and $q \in B\backslash\{p\}$.

Since $\mathcal{PS}(p, B, \mathcal{C}) = \mathbf{true}$, $p \in \mathcal{P}_{\mathcal{C},B}$ (see Eqs. 1–3, 6). If $q \in \mathcal{C}$, then $\mathcal{PS}(q, B, \mathcal{C}) = \mathcal{PS}(q, B\backslash\{p\}, \mathcal{C}) = \mathbf{false}$ (see Eq. 3). Hence, by Eqs. 6 and 7, it is sufficient to consider a point q such that $q \in \mathcal{P}_{\mathcal{C},B\backslash\{p\}}$.

- First we prove that if $\mathcal{PS}(q, B, \mathcal{C}) = \mathbf{true}$, then $\mathcal{PS}(q, B\backslash\{p\}, \mathcal{C}) = \mathbf{true}$.
  We give an indirect proof. Let us assume that $\mathcal{PS}(q, B, \mathcal{C}) = \mathbf{true}$ and $\mathcal{PS}(q, B\backslash\{p\}, \mathcal{C}) = \mathbf{false}$ for a point $q \in \mathcal{P}_{\mathcal{C},B}$. Thus there is a set $Q \subseteq \mathcal{P}_{\mathcal{C},B\backslash\{p\}} \backslash \{q\}$ such that q is not a simple point in $(B\backslash\{p\})\backslash Q = B\backslash(Q \cup \{p\})$. Hence we can write

$$B\backslash(Q \cup \{p\}) \; \not\leftrightarrow \; (B\backslash(Q \cup \{p\}))\backslash\{q\}. \tag{8}$$

Since, by Lemma 1, $Q \cup \{p\} \subseteq \mathcal{P}_{\mathcal{C},B}\backslash\{q\}$ holds, and $\mathcal{PS}(q, B, \mathcal{C}) = \mathbf{true}$, q is simple in $B\backslash(Q \cup \{p\})$. Hence

$$B\backslash(Q \cup \{p\}) \; \leftrightarrow \; (B\backslash(Q \cup \{p\}))\backslash\{q\}. \tag{9}$$

Thus Eq. 9 contradicts Eq. 8.
- Then we prove that if $\mathcal{PS}(q, B\backslash\{p\}, \mathcal{C}) = \mathbf{true}$, then $\mathcal{PS}(q, B, \mathcal{C}) = \mathbf{true}$.
  Proving indirectly, let us assume that $\mathcal{PS}(q, B\backslash\{p\}, \mathcal{C}) = \mathbf{true}$ and $\mathcal{PS}(q, B, \mathcal{C}) = \mathbf{false}$. Consequently, there is a set $Q \subseteq \mathcal{P}_{\mathcal{C},B}\backslash\{q\}$ such that q is not a simple point in $B\backslash Q$. Thus

$$B\backslash Q \; \not\leftrightarrow \; (B\backslash Q)\backslash\{q\}. \tag{10}$$

There are two cases to be investigated:

- $p \in Q$:

Consider the set $Q' = Q \backslash \{p\}$. Then by Lemma 1, $Q' \subseteq \mathcal{P}_{\mathcal{C},B \backslash \{p\}}$ holds. Since $\mathcal{PS}(q, B \backslash \{p\}, \mathcal{C}) = \textbf{true}$, point q is simple in $(B \backslash \{p\}) \backslash Q'$. Thus

$$(B \backslash \{p\}) \backslash Q' \; \backsimeq \; ((B \backslash \{p\}) \backslash Q') \backslash \{q\}. \tag{11}$$

Since $Q = Q' \cup \{p\}$, Eq. 11 can be rewritten as

$$B \backslash Q \; \backsimeq \; (B \backslash Q) \backslash \{q\}. \tag{12}$$

Thus Eq. 12 contradicts Eq. 10.

- $p \notin Q$:

In this case $Q \subseteq \mathcal{P}_{\mathcal{C},B}$ and $(Q \cup \{q\}) \subseteq \mathcal{P}_{\mathcal{C},B}$.

Since we assumed that $\mathcal{PS}(p, B, \mathcal{C}) = \textbf{true}$, point p is simple in sets $B \backslash Q$ and $B \backslash (Q \cup \{q\})$. Hence the following two equations hold

$$B \backslash Q \; \backsimeq \; (B \backslash Q) \backslash \{p\} \; = \; (B \backslash \{p\}) \backslash Q, \tag{13}$$

$$(B \backslash Q) \backslash \{q\} \; = \; B \backslash (Q \cup \{q\}) \; \backsimeq \; B \backslash (Q \cup \{q\}) \backslash \{p\} \; = \; B \backslash Q \backslash \{p, q\}. \tag{14}$$

Since, by Lemma 1, $Q \subseteq \mathcal{P}_{\mathcal{C},B \backslash \{p\}} \backslash \{q\}$ holds, and $\mathcal{PS}(q, B \backslash \{p\}, \mathcal{C}) = \textbf{true}$, point q is simple in $(B \backslash \{p\}) \backslash Q$. Thus

$$(B \backslash \{p\}) \backslash Q \; \backsimeq \; ((B \backslash \{p\}) \backslash Q) \backslash \{q\} \; = \; B \backslash Q \backslash \{p, q\}. \tag{15}$$

Since '$\backsimeq$' is an equivalence relation, by Eqs. 13–15, we can write

$$B \backslash Q \; \backsimeq \; (B \backslash \{p\}) \backslash Q \; \backsimeq \; B \backslash Q \backslash \{p, q\} \; \backsimeq \; (B \backslash Q) \backslash \{q\}. \tag{16}$$

Thus Eq. 16 contradicts Eq. 10.

We proved that if $\mathcal{PS}(p, B, \mathcal{C}) = \textbf{true}$, then $\mathcal{PS}(q, B, \mathcal{C}) = \mathcal{PS}(q, B \backslash \{p\}, \mathcal{C})$ for any two points $p \in B$ and $q \in B \backslash \{p\}$. $\qquad\square$

As deletion rule $\mathcal{PS}$ is general by Theorem 5, and it deletes only simple points, $\mathcal{PS}$ is general-simple (see Definition 2). Thus the following theorem is an easy consequence of Theorems 2–5.

Theorem 6. *The followings hold for deletion rule $\mathcal{PS}$ (that deletes P-simple points, and considers constraint set $\mathcal{C}$):*

1. *The sequential reduction (see Algorithm 2) with deletion rule $\mathcal{PS}$ is order-independent and topology-preserving.*
2. *The sequential and parallel reductions specified by $\mathcal{PS}$ are equivalent (i.e., they produce the same result for each input picture).*

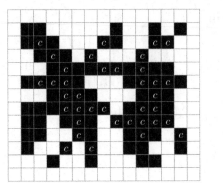 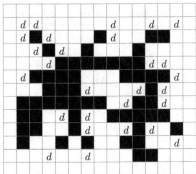

Fig. 3. Example of the pair of equivalent topology-preserving parallel and (order-independent) sequential reductions that use deletion rule $\mathcal{PS}$ with respect to constraint set $\mathcal{C}$. Elements in the constraint set (i.e., non-simple points) are marked 'c' in the input $(8,4)$ picture (left), and deleted points are marked 'd' in the output picture (right).

3. *The parallel reduction (see Algorithm 1) with deletion rule $\mathcal{PS}$ is topology-preserving.*

It can readily be seen that if we combine deletion rule $\mathcal{PS}$ with an arbitrary constraint set $C \supset \mathcal{C}$, then Lemma 1, Theorems 5 and 6 hold. Hence we can get various pairs of equivalent and topology-preserving parallel and (order-independent) sequential reductions by considering different geometrical constraints (e.g., characterizations of endpoints [7,16] and types of isthmuses [3]).

Figure 3 gives an example of a pair of equivalent topology-preserving parallel and order-independent sequential reductions with deletion rule $\mathcal{PS}$.

4   From General-Simple Deletion Rules to P-Simple Points

In this section we show that each general-simple deletion rule deletes P-simple points.

Theorem 7. *If a deletion rule R (with respect to a constraint set $C \subseteq B$) is general-simple, then each point that can be deleted by R is P-simple in B, where $P = \{ p \mid p \in B \backslash C$ and $R(p, B, C) = \textbf{true} \}$.*

Proof. Let $p \in P$ be a point such that $R(p, B, C) = \textbf{true}$, and consider a set $Q \subseteq P \backslash \{p\}$. It is to be proved that point p is simple in $B \backslash Q$.

There are two cases to be investigated:

– $Q = \emptyset$: Since deletion rule R is general-simple, it deletes only simple points. Hence deletable point p is simple in $B = B \backslash \emptyset = B \backslash Q$.

– $Q \neq \emptyset$: The general-simple deletion rule R is general as well (see Definitions 1 and 2), hence it is order-independent by Theorem 2. It means that 'deletable' point p remains 'deletable' after deletion of some previously visited 'deletable' points. Thus $R(p, B \backslash Q, C) = \mathbf{true}$.

Since the general-simple deletion rule R deletes only simple points, point p is simple in $B \setminus Q$. □

Let us recall the 2D parallel thinning algorithm proposed by Manzanera et al. [14]. That algorithm falls into the category of *fully parallel thinning* [7] since it uses the same reduction in each thinning phase (i.e., iteration step). The deletion rule $\mathcal{M}$ of that algorithm was given by three classes of matching templates. The base templates α_1, α_2, and β are depicted in Fig. 4. All their rotated versions are templates as well, where the rotation angles are 90°, 180°, and 270°. All elements of α_1-type and α_2-type are *removing templates*, while β and its rotated versions are *preserving templates*. A black point is designated to be deleted if at least one removing template matches it, but it is not matched by any preserving template. The constraint set $\mathcal{I}$ comprises all interior points in the input picture of the actual iteration step.

In [17] the author proved that the deletion rule of the thinning algorithm proposed by Manzanera et al. [14] is general-simple. Thus that algorithm is topology-preserving, and it is equivalent to an order-independent sequential algorithm that uses the same deletion rule (see Theorems 2–4). By Theorem 7, we can state that the existing thinning algorithm proposed by Manzanera et al. [14] deletes P-simple points from $(8, 4)$ input pictures.

Figure 5 is to illustrate one iteration step of the thinning algorithm in question. It is easy to check that each deletable point by $\mathcal{M}$ is P-simple, where P is the set of deletable points.

Note that Palágyi, Németh, and Kardos proposed a pair of equivalent 4-subiteration 2D sequential and parallel thinning algorithms [19], and four pairs of equivalent 6-subiteration 3D sequential and parallel surface-thinning algorithms [20]. In addition, they showed in [20] that the deletion rule of the 3D parallel surface-thinning algorithm of Gong and Bertrand [6] is general-simple. Since the deletion rules of the 2D and 3D thinning algorithm mentioned above are all general simple, all of these algorithms delete P-simple points by Theorem 7.

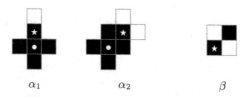

α_1 $\qquad$ α_2 $\qquad$ β

Fig. 4. The three base templates associated with the deletion rule $\mathcal{M}$. Notations: each black element matches a black point; each white element matches a white point; black elements marked '★' are the central positions of the templates; black elements with white bullets match interior points (i.e., elements in the constraint set $\mathcal{I}$).

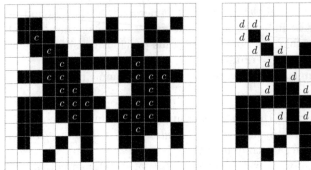

Fig. 5. Example of the pair of equivalent topology-preserving parallel and (order-independent) sequential reductions that use deletion rule $\mathcal{M}$ with respect to constraint set $\mathcal{I}$. These reductions are associated to one iteration step of the fully parallel thinning algorithm proposed by Manzanera et al. [14]. Elements in the constraint set $\mathcal{I}$ (i.e., interior points) are marked 'c' in the input $(8, 4)$ picture (left), and deleted points by $\mathcal{M}$ are marked 'd' in the output picture (right).

5 Conclusions

This work bridges P-simple points and general-simple deletion rules that specify pairs of equivalent topology-preserving parallel and (order-independent) sequential reductions. On the one hand, we showed that deletion rules that delete P-simple points (with respect to constraint sets containing all non-simple points) are general-simple. Hence parallel reductions with these deletion rules are equivalent to topology-preserving and order-independent sequential reductions. On the other hand, we proved that each point deleted by a general-simple deletion rule is P-simple. Note that the results presented in this work are all valid for digital pictures in arbitrary dimensions.

Acknowledgements. This work was supported by the grant OTKA K112998 of the National Scientific Research Fund.

References

1. Bertrand, G.: On P-simple points. Compte Rendu de l'Académie des Sciences de Paris, Série Math. **I**(321), 1077–1084 (1995)
2. Bertrand, G.: Sufficient conditions for 3D parallel thinning algorithms. In: Proceedings of SPIE - Vision Geometry IV, vol. 2573, pp. 52–60 (1995)
3. Bertrand, G., Couprie, M.: Transformations topologiques discrètes. In: Coeurjolly, D., Montanvert, A., Chassery, J. (eds.) Géométrie discrète et images numériques, pp. 187–209. Hermès Science Publications (2007)
4. Bertrand, G., Couprie, M.: On parallel thinning algorithms: minimal non-simple sets, P-simple points and critical kernels. J. Math. Imaging Vision **35**, 637–648 (2009)

5. Couprie, M., Bertrand, G.: New characterizations of simple points in 2D, 3D, and 4D discrete spaces. IEEE Trans. Pattern Anal. Mach. Intell. **31**, 23–35 (2009)
6. Gong, W.X., Bertrand, G.: A simple parallel 3d thinning algorithm. In: Proceedings of 10th IEEE International Conference Pattern Recognition, ICPR 1990, pp. 188–190 (1990)
7. Hall, R.W.: Parallel connectivity-preserving thinning algorithms. In: Kong, T.Y., Rosenfeld, A. (eds.) Topological Algorithms for Digital Image Processing, pp. 145–179. Elsevier Science B.V. (1996)
8. Hall, R.W., Kong, T.Y., Rosenfeld, A.: Shrinking binary images. In: Kong, T.Y., Rosenfeld, A. (eds.) Topological Algorithms for Digital Image Processing, pp. 31–98. Elsevier Science B.V. (1996)
9. Kardos, P., Palágyi, K.: On topology preservation in triangular, square, and hexagonal grids. In: Proceedings of 8th International Symposium on Image and Signal Processing and Analysis, IEEE/EURASIP, ISpPA 2013, pp. 782–787 (2013)
10. Kong, T.Y.: On topology preservation in 2-d and 3-d thinning. Int. J. Pattern Recognit. Artif. Intell. **9**, 813–844 (1995)
11. Kong, T.Y.: Topology-preserving deletion of 1's from 2-, 3- and 4-dimensional binary images. In: Ahronovitz, E., Fiorio, C. (eds.) DGCI 1997. LNCS, vol. 1347, pp. 1–18. Springer, Heidelberg (1997)
12. Kong, T.Y., Rosenfeld, A.: Digital topology: introduction and survey. Comput. Vision Graph. Image Process. **48**, 357–393 (1989)
13. Malandain, G., Bertrand, G.: Fast characterization of 3d simple points. In: International Conference on Pattern Recognition, ICPR 1992, pp. 232–235 (1992)
14. Manzanera, A., Bernard, T.M., Pretêux, F., Longuet, B.: n-dimensional skeletonization: a unified mathematical framework. J. Electron. Imaging **11**, 25–37 (2002)
15. Niethammer, M., Kalies, W.D., Mischaikow, K., Tannenbaum, A.: On the detection of simple points in higher dimensions using cubical homology. IEEE Trans. Image Process. **15**, 2462–2469 (2006)
16. Palágyi, K., Németh, G., Kardos, P.: Topology preserving parallel 3D thinning algorithms. In: Brimkov, V.E., Barneva, R.P. (eds.) Digital Geometry Algorithms, pp. 165–188. Springer-Verlag (2012)
17. Palágyi, K.: Equivalent 2D sequential and parallel thinning algorithms. In: Barneva, R.P., Brimkov, V.E., Šlapal, J. (eds.) IWCIA 2014. LNCS, vol. 8466, pp. 91–100. Springer, Heidelberg (2014)
18. Palágyi, K.: Equivalent sequential and parallel reductions in arbitrary binary pictures. Int. J. Pattern Recognit. Artif. Intell. **28**, 1460009-1–1460009-16 (2014)
19. Palágyi, K., Németh, G., Kardos, P.: Topology-preserving equivalent parallel and sequential 4-subiteration 2D thinning algorithms. In: Proceedings of 9th International Symposium on Image and Signal Processing and Analysis, IEEE/EURASIP, ISPA 2015, pp. 306–311 (2015)
20. Palágyi, K., Németh, G., Kardos, P.: Equivalent sequential and parallel subiteration-based surface-thinning algorithms. In: Barneva, R.P., Bhattacharya, B.B., Brimkov, V.E. (eds.) IWCIA 2015. LNCS, vol. 9448, pp. 31–45. Springer, Heidelberg (2015)
21. Ranwez, V., Soille, P.: Order independent homotopic thinning for binary and grey tone anchored skeletons. Pattern Recogn. Lett. **23**, 687–702 (2002)
22. Saha, P.K., Chaudhuri, B.B.: Detection of 3-d simple points for topology preserving transformations with application to thinning. IEEE Trans. Pattern Anal. Mach. Intell. **16**, 1028–1032 (1994)

Two Measures for the Homology Groups of Binary Volumes

Aldo Gonzalez-Lorenzo[1,2]([✉]), Alexandra Bac[1], Jean-Luc Mari[1], and Pedro Real[2]

[1] Aix-Marseille University, CNRS, LSIS UMR 7296, Marseille, France
`aldo.gonzalez-lorenzo@univ-amu.fr`
[2] Institute of Mathematics IMUS, University of Seville, Seville, Spain

Abstract. Given a binary object (2D or 3D), its Betti numbers characterize the number of holes in each dimension. They are obtained algebraically, and even though they are perfectly defined, there is no unique way to display these holes. We propose two geometric measures for the holes, which are uniquely defined and try to compensate the loss of geometric information during the homology computation: the *thickness* and the *breadth*. They are obtained by filtering the information of the persistent homology computation of a filtration defined through the signed distance transform of the binary object.

1 Introduction

Homology computation is a very useful tool for the classification and the understanding of binary objects in a rigorous way. It provides a class of descriptors summarizing the basic structure of the considered shape.

A binary object is a set $X \subset \mathbb{Z}^d$ together with a connectivity relation. We will assume through this paper that X is a volume ($d = 3$) and the elements of X will be called *voxels*. However, the generalization to higher dimensions is direct.

Roughly speaking, homology deals with the "holes" of an object. Holes can be classified by dimensions: 0-holes correspond to connected components, 1-holes to tunnels or handles and 2-holes to cavities. When computing homology, we usually expect to find the number of these holes (called the Betti numbers) and a representative of each hole (called representative cycle of a homology generator). Nevertheless, the homology computation works at an abstract level (the chain complex) which ignores the embedding of the object. Thus, the geometry of the object is in some way neglected.

Moreover, holes are difficult to visualize. We can know how many they are, but not where they are. This is due to the fact that in homology computation, a choice for determining a linearly independent set of holes must be done. This is something difficult to apprehend: for instance, we remind that a cube with its interior and faces removed (as depicted in Fig. 1) contains five 1-holes instead of six.

© Springer International Publishing Switzerland 2016
N. Normand et al. (Eds.): DGCI 2016, LNCS 9647, pp. 154–165, 2016.
DOI: 10.1007/978-3-319-32360-2_12

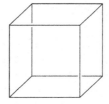

Fig. 1. There are only five 1-holes in this object, and there is no natural way to state where they are.

The aim of this paper is to endow Betti numbers with an additional information containing a geometric interpretation. We define two measures—*thickness* and *breadth*—which, unlike homology generators, are uniquely determined by the object and can be used as concise topology descriptors. They are obtained through the persistent homology of the filtration induced by the signed distance transform of the volume.

Previous efforts have been made to combine homology and geometry. We cite herein some works that are related to our problem. Erickson and Whittlesey founded an algorithm in [EW05] that computes the shortest base for the first homology group for oriented 2-manifolds. Dey et al. [DFW13] developed a similar work, but also classifying the 1-holes into tunnels and handles. Chen and Freedman [CF10] measured the 1-holes of a complex by the "length" of the homology generators and they also gave an algorithm to compute the smallest set of homology generators. Zomorodian and Carlsson introduced the *localized homology* in [ZC08], which allows to locate each homology class in a subset of a given cover. This cover, or collection of subcomplexes whose union contains the complex, successfully give a geometric sense to the Betti numbers.

Contributions

We define two geometric measures that enrich the Betti numbers of a binary volume. They are uniquely defined up to a choice of connectivity relation and distance. They can be computed through a distance transform and persistent homology, so it has matrix multiplication complexity over the number of voxels in the bounding box of the volume. These measures can be considered as pairs of numbers associated to each homology generator but also visualized as three-dimensional balls on the volume.

2 Preliminaries

2.1 Cubical Complexes and Homology

Cubical Complex — This section is derived from [KMM04]. For a deeper understanding of these concepts, the reader can refer to it. An *elementary interval* is an interval of the form $[k, k+1]$ or a degenerate interval $[k, k]$, where $k \in \mathbb{Z}$.

An *elementary cube* in $\mathbb{R}^n$ is the Cartesian product of n elementary intervals, and the number of non-degenerate intervals in this product is its *dimension*. An elementary cube of dimension q will be called *q-cube*.

Given two elementary cubes σ and τ, we say that σ is a *face* of τ if $\sigma \subset \tau$. It is a *primary face* if the difference of their dimensions is 1. Similarly, σ is a *coface* of τ if $\sigma \supset \tau$. A *cubical complex* is a set of elementary cubes with all of their faces. The *boundary* of an elementary cube is the collection of its primary faces.

Chain Complex — A *chain complex* (C_*, d_*) is a sequence of groups $C_0, C_1, \ldots$ (called *chain groups*) and homomorphisms $d_1 : C_1 \to C_0$, $d_2 : C_2 \to C_1, \ldots$ (called *differential* or *boundary operators*) such that $d_{q-1}d_q = 0, \forall q > 0$.

Given a cubical complex, we define its chain complex as follows:

– C_q is the free group generated by the q-cubes of the complex;
– d_q gives the "algebraic" boundary, which is the linear operator that maps every q-cube to the sum of its primary faces.

The elements of the chain group C_q, which are formal sums of q-cubes with coefficients in $\mathbb{Z}_2$, are called *q-chains*. They can be seen as sets of cubes of the same dimension.

Homology Groups — A q-chain x is a *cycle* if $d_q(x) = 0$, and a *boundary* if $x = d_{q+1}(y)$ for some $(q+1)$-chain y. By the property $d_{q-1}d_q = 0$, every boundary is a cycle, but the reverse is not true: a cycle which is not a boundary contains a "hole". The q-th homology group of the chain complex (C_*, d_*) contains the q-dimensional "holes": $H(C)_q = \ker(d_q)/\operatorname{im}(d_{q+1})$. This set is a finitely generated group, so there is a "base" typically formed by the holes of the cubical complex. Since our ring of coefficients is $\mathbb{Z}_2$, this group is isomorphic to $\mathbb{Z}^{\beta_q}$ and β_q is the q-th *Betti number*.

Cubical Complex Associated to a Binary Volume — Given a binary volume, we can define two different associated cubical complexes encoding the 6 or the 26-connectivity relation.

– Primal associated cubical complex (26-connectivity): every voxel $x = (x_1, x_2, x_3)$ generates the 3-cube $[x_1, x_1 + 1] \times [x_2, x_2 + 1] \times [x_3, x_3 + 1]$ and all its faces.
– Dual associated cubical complex (6-connectivity): let us first adapt the notion of *clique* to our context. A *d-clique* is a maximal (in the sense of inclusion) set of voxels such that their intersection is a d-cube. First, for every voxel (in fact 3-clique) $x = (x_1, x_2, x_3)$ of the volume, we add the 0-cube $\sigma = [x_1] \times [x_2] \times [x_3]$. Then, for every d-clique ($d < 3$) in the volume, we add to the cubical complex a $(3 - d)$-cube such that its vertices are the voxels of the d-clique.

These cubical complexes can be defined for any dimension. Figure 2 illustrates both complexes.

Fig. 2. Left: a binary volume. Center: its primal associated cubical complex. Right: its dual cubical complex

2.2 Persistent Homology

This section introduces the persistent homology. A more rigorous presentation can be found in [EH08] and the references therein.

A *filtration* is a finite (or countable) sequence of nested cubical complexes $X_1 \subset X_2 \subset \cdots X_n$. It can also be described by a function f on the final complex X_n, which assigns to each q-cube the first index at which it appears in the complex. Since it is a sequence of cubical complexes, a cube cannot appear strictly before its faces, so the function f must verify

$$f(\sigma) \leq f(\tau), \forall \sigma \subset \tau \tag{1}$$

Therefore, a function $f : X \to \mathbb{R}$ defined over a cubical complex is a *filtration function* if its image is a finite (or countable) set and if it satisfies Eq. (1). Given such a function, its filtration is the sequence $\mathcal{F}_f(X) = \{X_i\}_{i=0}^{n}$, where $a_0 < a_1 < \cdots < a_n$ are the images of f and $X_i = f^{-1}(]-\infty, a_i])$.

As illustrated in Fig. 3, the homology groups of the complex can evolve as "time" goes on. The *persistence diagram* [EH08, p. 3] records these changes: a q-hole being born in X_i and dying (or vanishing) in X_j is represented in the persistence diagram as the point (i, j). This is also called a *P-interval* in [ZC05]. A homology generator of X_n, which never dies, is represented by the point (i, ∞). The reader can find some persistence diagrams in Sect. 4. Persistent homology can also be visualized in terms of barcodes [Ghr08], where each point (i, j) is represented as an interval in the real line.

We can assign a pair of cubes (σ_i, σ_j) to each P-interval (i, j). The first cube creates the hole (e.g., a point in a new connected component or the edge that closes a handle) while the second one merges the hole into the set of boundaries (e.g., the edge that connects two connected components, or the square that fills a handle). The reader could try to find these pairs of cubes in the filtration described in Fig. 3.

There has been an extensive research in the computation of persistent homology. An algorithm in matrix multiplication time was introduced in [MMS11]. The algorithms present in [ZC05, MN13] have cubical worst-time complexity, but they seem to behave better in practice. An algorithm adapted for cubical complexes was developed in [WCV12].

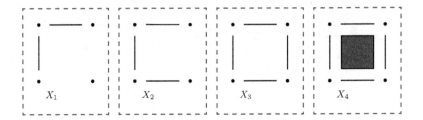

Fig. 3. A filtration. X_1: there are two 0-holes (connected components). X_2: one 0-hole dies. X_3: a 1-hole is born. X_3: that 1-hole dies.

Note that in this article we do not use Čech, Vietoris-Rips nor Alpha complexes, as it is usually done in the persistent homology literature. The filtrations considered are based on a function defined over a cubical complex.

2.3 Distance Transform

Given a binary volume $X \subset \mathbb{Z}^3$, its *distance transform* DT_X is the map that sends every voxel $x \in X$ to $DT_X(x) = \min_{y \in \mathbb{Z}^3 \setminus X} d(x, y)$. It can be seen as the radius of the maximal ball centered at x and contained in X. Note that we must consider one specific distance for the volume: L_p metrics such as the Manhattan distance (L_1), the Euclidean distance (L_2) or the chessboard distance (L_∞); distances based on chamfer masks [Mon68,Bor84] or sequences of chamfer masks [MDKC00,NE09], etc.

There exist linear time algorithms for computing the Euclidean distance on nD binary objects [BGKW95,Hir96,MRH00,MQR03].

We can then consider the signed distance transform, which maps every voxel $x \in X$ to $-DT_X(x)$ and $x \notin X$ to $DT_{\mathbb{Z}^3 \setminus X}(x)$.

3 The Two Measures

3.1 Motivation

The main motivation for this work comes from the desire of comparing binary volumes obtained from real acquisition of complex structures. Using mere Betti numbers for this objective can be ineffective, so we thought of enhancing this basic homological information by defining two geometry-aware measures for the holes.

In the following we give an intuitive introduction to the two new measures. We consider two binary images which are repeatedly eroded and dilated respectively, and we comment what happens to their homology groups. Then we figure out how we can treat this problem with persistent homology.

When we erode the image in Fig. 4-(a), we observe that a 1-hole disappears (Fig. 4-(b)), a 0-hole is created (Fig. 4-(c)) and a 0-hole disappears (Fig. 4-(d)).

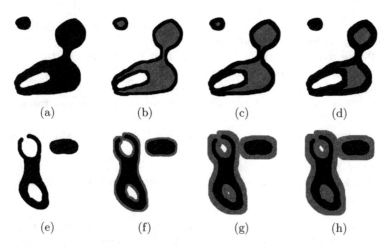

Fig. 4. Two images being eroded and dilated. We can notice a change in its homology at every step.

On the other hand, when we dilate the image in Fig. 4-(e), a new 1-hole appears (Fig. 4-(f)), a 0-hole vanishes (Fig. 4-(g)) and a 1-hole disappears (Fig. 4-(h)).

Thanks to persistent homology, we can record these events. Given a binary volume $X \subset [0, m_1] \times [0, m_2] \times [0, m_3] \subset \mathbb{Z}^3$, we build the associated primal (or dual) cubical complex K of the bounding box $[0, m_1] \times [0, m_2] \times [0, m_3]$. Then we compute a filtration function f associated to the signed distance transform of X: the 3-cubes (0-cubes) are mapped to the value of their associated voxels (cf. Sect. 2.1). The image of the rest of the cubes is coherently assigned in order to produce a filtration, that is, each q-cube, for $0 \le q \le 2$ ($1 \le q \le 3$), takes the minimum (maximum) value of its 3-dimensional cofaces (0-dimensional faces).

Let $a_0 < a_1 < \cdots < a_i = 0 < \cdots < a_n$ be the different values of f over the volume. Thus, we can consider the filtration $F_f(K)$:

$$X_0 = f^{-1}(]-\infty, a_0]) \subset \cdots \subset X \subset \cdots \subset X_n = f^{-1}(]-\infty, a_n])$$

Let us now see how the previously described phenomena are encoded in the persistent homology of the filtration $F_f(K)$. Note that the eroded images are now seen in reversed time:

- Figure 4-(b): a 1-hole is born in negative time (1b–);
- Figure 4-(c): a 0-hole dies in negative time (0d–);
- Figure 4-(d): a 0-hole is born in negative time (0b–);
- Figure 4-(f): a 1-hole is born in positive time (1b+);
- Figure 4-(g): a 0-hole dies in positive time (0d+);
- Figure 4-(h): a 1-hole dies in positive time (1d+);

Since we are interested in the homology groups of the given volume, we only consider the persistent intervals that start at negative time and finish at positive time. We can give an intuitive and physical interpretation to those events.

0b– A 0-hole being born at time $t_0 < 0$ means that the maximal ball included in that connected component has radius $-t_0$;

0d+ A 0-hole dying at time $t_0 > 0$ means that the shortest distance between that connected component and any other is $2 \cdot t_0$;

1b– A 1-hole being born at time $t_0 < 0$ means that $-t_0$ is the smallest radius needed for a ball included in X that, once removed, breaks or vanishes that hole;

1d+ A 1-hole dying at time $t_0 > 0$ means that the maximal ball that can pass through that hole has radius t_0;

2b– Same as 1b–;

2d+ A 2-hole dying at time $t_0 > 0$ means that the maximal ball that can fit inside that hole has radius t_0;

Thus, for $q = 1, 2$, the time t at which a q-hole is born can be seen as the thickness of the hole, or how far it is from being destructible. We will call it *thickness*. Also, the time t at which a q-hole dies can be interpreted as a kind of size or *breadth*. We choose this second term, since the size of a 2-hole is better understood as the volume it covers or the area of its boundary. These two terms are not suitable for dimension 0, so we use the terms *breadth* (again) and *separability* respectively.

3.2   Definitions

Let $X \subset [0, m_1] \times [0, m_2] \times [0, m_3] \subset \mathbb{Z}^3$ be a binary volume and $d : \mathbb{Z}^3 \to \mathbb{R}^+$ a distance. Consider the *signed distance transform*

$$sDT_X(x) = \begin{cases} -\min\{d(x, y) : y \notin X\}, & x \in X \\ \min\{d(x, y) : y \in X\}, & x \notin X \end{cases}$$

Depending on which connectivity relation we use for the volume, we build its associated cubical complex K (cf. Sect. 2.1) and the filtration function f_X induced by sDT_X (cf. Sect. 3.1).

The *thickness* and *breadth* of the q-holes of X are the pairs $\{(-i_k, j_k)\}_{k=1}^{\beta_q}$, where $\{(i_k, j_k)\}_{k=1}^{\beta_q}$ are the P-intervals of the filtration f_X such that $i_k < 0$ and $j_k > 0$ for all $1 \leq k \leq \beta_q$. We can represent these pairs as points in $\mathbb{R}^2$ in the *thickness-breadth diagram*.

The final step of the filtration is the full cubical complex associated to the bounding box of the volume. If the volume is not empty, as the bounding box has one connected component there exists a P-interval $(-i, \infty)$ of dimension 0. We will represent this thickness-breadth pair as $(-i, -1)$ in the thickness-breadth diagram. As a consequence, a volume has $\beta_0 - 1$ separation values. This is coherent with its interpretation, as there are only $n - 1$ shortest distances between n objects.

Note that the thickness-breadth pairs are uniquely defined for the filtration and they only depend on the connectivity relation of the volume and the distance considered.

3.3 Visualization

The breadth and thickness values of a volume can be visualized in terms of balls, as it was suggested in Sect. 3.1. Given a thickness-breadth pair (i, j) with its (non-unique) associated pair of cubes (σ, τ) (cf. Sect. 2.2), we can also assign a pair of voxels (p, q) to them. For the primal (dual) associated cubical complex, according to its construction (cf. Sect. 2.1), we choose any coface of dimension 3 (face of dimension 0) with the same image under f for each cube and we take its associated voxel in the original volume. We call these voxels, which are not unique, $v(i)$ and $v(j)$.

Therefore, for each hole we can visualize its thickness by the ball centered at $v(i)$ with radius i, and similarly for its breadth. Figure 5 illustrates this procedure.

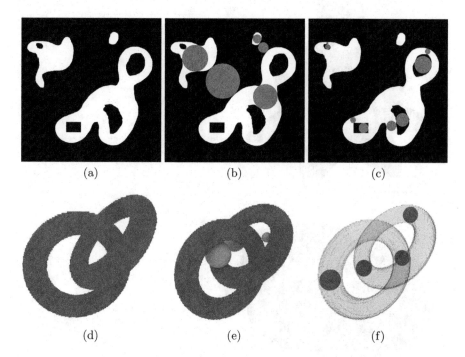

(a)   (b)   (c)

(d)   (e)   (f)

Fig. 5. Example of visualization of the thickness and the breadth through balls. Top: a binary image (a) with the balls of H_0 (b) and H_1 (c). Bottom: a binary volume (d) which contains two chained volumetric tori, its breadth (e) and its thickness balls (f) of dimension 0 and 1.

4 Results

Our implementation for computing the breadth and the thickness uses the DGtal library [DGt] for the distance transform and the Perseus software [Nan] for the persistent homology. In order to obtain the centers of the balls, we developed a specific software. The following volumes were voxelized with the Binvox software [Min, NT03].

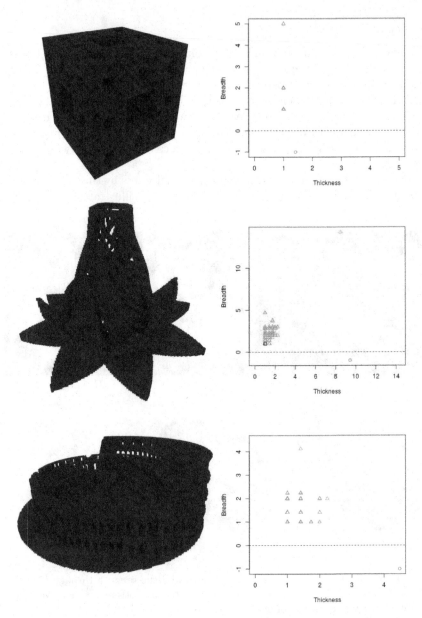

Fig. 6. Left: three binary volumes. Right: their thickness-breadth diagrams. 0-holes are represented by red circles, 1-holes by green triangles and 2-holes by blue squares (Color figure online).

Figure 6 shows some examples of thickness-breadth diagrams. For each binary volume, we show the diagram containing the thickness-breadth pairs of the homology groups H_0 (red circles), H_1 (green triangles) and H_2 (blue squares).

Note that we represent the pairs (i, ∞), associated to the "broadest" connected component of the volume, as $(i, -1)$.

We can appreciate the fractal structure of the Menger sponge after three iterations in its diagram. There are only three possible values for the breadth and the thickness: $(1, \frac{3^0+1}{2})$, $(1, \frac{3^1+1}{2})$ and $(1, \frac{3^3+1}{2})$. For the lamp [Rey15], we can easily recognize a lot of similar holes and a bigger one, which traverses the volume along the z-axis. The small 2-hole follows from a discretization error. The diagram of the Colosseum volume [Gas15] presents a regular shape. All the 1-holes (i.e., the doors) have similar measures, except for one which has greater thickness. It corresponds to the ground floor concentric corridor.

Figure 7 illustrates the pertinence of the visualization described in Sect. 3.3. More examples are available on [GL].

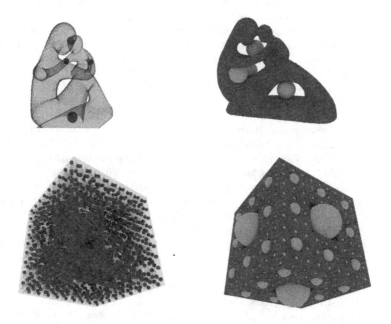

Fig. 7. Examples of thickness (red) and breadth (green) balls. Top: Fertility volume, available by AIM@SHAPE with the support of M. Couprie. Bottom: Menger sponge, available on [GL] (Color figure online).

5 Conclusion and Future Work

This paper introduces a concise and rigorous geometrical and topological information for binary volumes that extends the Betti numbers. This could be used for a statistical analysis of the volume or a better understanding of its topological features.

An interesting issue that should be addressed in the near future is the stability of this definition. Are the breadth and thickness stable under small perturbations of the volume? How much do these values change when we consider different connectivity relations or distances?

In addition, it seems that the intersection of the thickness and breadth balls with the volume could provide a heuristic for computing geometry-aware homology and cohomology generators.

References

[BGKW95] Breu, H., Gil, J., Kirkpatrick, D., Werman, M.: Linear time Euclidean distance transform algorithms. IEEE Trans. Pattern Anal. Mach. Intell. **17**, 529–533 (1995)

[Bor84] Borgefors, G.: Distance transformations in arbitrary dimensions. Comput. Vis. Graph. Image Process. **27**(3), 321–345 (1984)

[CF10] Chen, C., Freedman, D.: Measuring and computing natural generators for homology groups. Comput. Geom. **43**(2), 169–181 (2010). Special Issue on the 24th European Workshop on Computational Geometry (EuroCG 2008)

[DFW13] Dey, T.K., Fan, F., Wang, Y.: An efficient computation of handle and tunnel loops via Reeb graphs. ACM Trans. Graph. **32**(4), 32:1–32:10 (2013)

[DGt] DGtal: Digital geometry tools and algorithms library. http://dgtal.org

[EH08] Edelsbrunner, H., Harer, J.: Persistent homology - a survey. In: Pach, J., Goodman, J.E., Pollack, R. (eds.) Surveys on Discrete and Computational Geometry: Twenty Years Later. Contemporary Mathematics, vol. 453, pp. 257–282. American Mathematical Society, Providence (2008)

[EW05] Erickson, J., Whittlesey, K.: Greedy optimal homotopy and homology generators. In: Proceedings of the Sixteenth Annual ACM-SIAM Symposium on Discrete Algorithms, SODA 2005, pp. 1038–1046. Society for Industrial and Applied Mathematics, Philadelphia (2005)

[Gas15] Gaspard, J.: Roman colosseum completely detailed see the world, August 2015. Creative Commons License. https://www.thingiverse.com/thing: 962416

[Ghr08] Ghrist, R.: Barcodes: the persistent topology of data. Bull. Am. Math. Soc. **45**, 61–75 (2008)

[GL] Gonzalez-Lorenzo, A.: http://aldo.gonzalez-lorenzo.perso.luminy.univ-amu.fr/measures.html. Accessed 20 Oct 2015

[Hir96] Hirata, T.: A unified linear-time algorithm for computing distance maps. Inf. Process. Lett. **58**(3), 129–133 (1996)

[KMM04] Kaczynski, T., Mischaikow, K., Mrozek, M.: Computational Homology, (chapters 2 and 7), vol. 157, pp. 255–258. Springer, New York (2004)

[MDKC00] Mukherjee, J., Das, P.P., Aswatha Kumar, M., Chatterji, B.N.: On approximating Euclidean metrics by digital distances in 2D and 3D. Pattern Recogn. Lett. **21**(6–7), 573–582 (2000)

[Min] Min, P.: Binvox, 3D mesh voxelizer. http://www.cs.princeton.edu/~min/binvox/. Accessed 7 Oct 2015

[MMS11] Milosavljević, N., Morozov, D., Skraba, P.: Zigzag persistent homology in matrix multiplication time. In: Proceedings of the Twenty-Seventh Annual Symposium on Computational Geometry, SoCG 2011, pp. 216–225. ACM, New York (2011)

[MN13] Mischaikow, K., Nanda, V.: Morse theory for filtrations and efficient computation of persistent homology. Discrete Comput. Geom. **50**(2), 330–353 (2013)

[Mon68] Montanari, U.: A method for obtaining skeletons using a quasi-Euclidean distance. J. ACM **15**(4), 600–624 (1968)

[MQR03] Maurer Jr., C.R., Qi, R., Raghavan, V.: A linear time algorithm for computing exact Euclidean distance transforms of binary images in arbitrary dimensions. IEEE Trans. Pattern Anal. Mach. Intell. **25**(2), 265–270 (2003)

[MRH00] Meijster, A., Roerdink, J., Hesselink, W.H.: A general algorithm for computing distance transforms in linear time. In: Goutsias, J., Vincent, L., Bloomberg, D.S. (eds.) Mathematical Morphology and its Applications to Image and Signal Processing. Computational Imaging and Vision, vol. 18, pp. 331–340. Springer, New York (2000)

[Nan]   Nanda, V.: Perseus, the persistent homology software. http://www.sas. upenn.edu/~vnanda/perseus. Accessed 7 Oct 2015

[NE09] Normand, N., Évenou, P.: Medial axis lookup table and test neighborhood computation for 3D chamfer norms. Pattern Recogn. **42**(10), 2288–2296 (2009). Selected papers from the 14th IAPR International Conference on Discrete Geometry for Computer Imagery 2008

[NT03] Nooruddin, F.S., Turk, G.: Simplification and repair of polygonal models using volumetric techniques. IEEE Trans. Vis. Comput. Graph. **9**(2), 191–205 (2003)

[Rey15] Reynolds, A.: Radiant blossom, August 2015. Creative Commons License. https://www.thingiverse.com/thing:978768

[WCV12] Wagner, H., Chen, C., Vuçini, E.: Efficient computation of persistent homology for cubical data. In: Peikert, R., Hauser, H., Carr, H., Fuchs, R. (eds.) Topological Methods in Data Analysis and Visualization II. Mathematics and Visualization, pp. 91–106. Springer, Heidelberg (2012)

[ZC05] Zomorodian, A., Carlsson, G.: Computing persistent homology. Discrete Comput. Geom. **33**(2), 249–274 (2005)

[ZC08] Zomorodian, A., Carlsson, G.: Localized homology. Comput. Geom. **41**(3), 126–148 (2008)

Shape Classification According to LBP Persistence of Critical Points

Ines Janusch$^{(\boxtimes)}$ and Walter G. Kropatsch

Pattern Recognition and Image Processing Group,
Institute of Computer Graphics and Algorithms, TU Wien, Vienna, Austria
{ines,krw}@prip.tuwien.ac.at

Abstract. This paper introduces a shape descriptor based on a combination of topological image analysis and texture information. Critical points of a shape's skeleton are determined first. The shape is described according to persistence of the local topology at these critical points over a range of scales. The local topology over scale-space is derived using the local binary pattern texture operator with varying radii. To visualise the descriptor, a new type of persistence graph is defined which captures the evolution, respectively persistence, of the local topology. The presented shape descriptor may be used in shape classification or the grouping of shapes into equivalence classes. Classification experiments were conducted for a binary image dataset and the promising results are presented. Because of the use of persistence, the influence of noise or irregular shape boundaries (e.g. due to segmentation artefacts) on the result of such a classification or grouping is bounded.

Keywords: Shape descriptor · Shape classification · Local topology · Persistence · LBP · Local features

1 Introduction

We present in this paper a shape descriptor based on local topology and persistence in order to classify shapes. In prior work [7] we presented an approach to derive Reeb graphs using skeletons and local binary patterns (LBPs). One crucial decision when applying this approach is the choice of radius for the LBP computation. The degree of detail respectively noise (e.g. segmentation artefacts) that is captured by the representation depends on this radius. In order to analyse this influence of the parameter we conducted the following experiments: For the critical points of a skeleton (branching and end positions) all possible radii were tested. We can restrict the LBP analysis to these critical points as the parts of a skeleton in between critical points form a continuous curve representing a ribbon-like shape. These skeleton segments in between critical points correspond to edges in a Reeb graph and therefore do not represent any change in topology.

The experiments we conducted not only allow us insight into the impact of a change in the radius parameter but it also provides the basis for a new shape

© Springer International Publishing Switzerland 2016
N. Normand et al. (Eds.): DGCI 2016, LNCS 9647, pp. 166–177, 2016.
DOI: 10.1007/978-3-319-32360-2_13

descriptor. According to this descriptor a shape is represented by the evolution of LBP types for critical points of a skeleton and for varying LBP radii. This approach is therefore related to persistence as introduced by Edelsbrunner et al. [4]. In persistence, those features which persist for a parametrised family of spaces over a range of parameters are considered signals of interest. Short-lived features are treated as noise [5]. Such a range of parameters is for example given by succeeding scales as it is done for scale-space representations [10]. The persistence of a feature is given as its lifetime, that is the range of scales for which the feature is present. The approach presented is based on persistence over scales. We consider LBPs for ranges of radii and use the persistence of LBP types according to these radii as shape descriptors. Similar shape descriptors based on topological persistence are for example barcodes [1] or persistence diagrams [4]. The shape descriptor presented in this paper was inspired by a method to compute Reeb graphs. Nevertheless, for shape classification using a graph representation, graph matching is needed, which is not trivial and may be computationally expensive. A classification of shapes using the shape descriptor presented in this paper is done by comparison of feature vectors using the edit distance. As these feature vectors are only computed for a small number of characteristic positions within the shape, the number of feature vectors for each shape is limited. Thus, a shape classification based on the new descriptor can be done efficiently. These feature vectors are built as a scale-space representation. The sampling of the scale-space may of course limit the degree of detail of the representation. In the discrete space a complete sampling over all possible scales can be done. However, as the shape descriptor depends on this sampling of the scale-space it is not invariant to the size of the shape represented. For shape classification using shapes of varying sizes, a normalisation is needed beforehand.

The rest of the paper is structured as follows: Sect. 2 gives a short summary of the theory of LBPs. The influence of the LBP parameters on our approach is discussed in Sect. 3. In Sect. 4 a new shape descriptor based on the variation of one such parameter is presented. Experimental results obtained for this shape descriptor are given in Sect. 5. Section 6 concludes the paper.

2 Introduction to LBPs

LBPs were first introduced for texture classification [11] and since became popular texture operators: To determine the LBP for a pixel, this pixel is compared to the subsampled neighbourhood around it. The according position in a bit pattern is set to 1 if the value of a neighbouring pixel is larger than or equal to the value of the center pixel and to 0 otherwise (Fig. 1a and b). The two parameters P and R may be adjusted for the LBP operator. P fixes the number of sampling points along a circle of radius R around the center pixel, for which the LBP operator is computed [12]. Figure 1c shows different parameter configurations.

The bit pattern encodes the local topology. According to the LBP, the center pixel can be classified as:

– (local) maximum (the bit pattern contains only 0s),

$x_1 \geq c$	$x_2 < c$	$x_3 < c$
$x_8 \geq c$	c	$x_4 < c$
$x_7 \geq c$	$x_6 \geq c$	$x_5 \geq c$

1	0	0
1	c	0
1	1	1

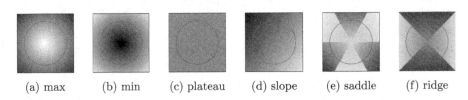

(a)   compar-
ison   with
neighbours

(b) bit pattern

(c) (P,R) = (8,1); (P,R) = (16,2); (P,R)
= (8,2) - according to [12].

Fig. 1. (a) and (b) LBP computation for center pixel c and (c) variations of the parameters P (sampling points) and R (radius).

(a) max	(b) min	(c) plateau	(d) slope	(e) saddle	(f) ridge

Fig. 2. Local topology encoded by an LBP at the radius marked in red (Color figure online).

- (local) minimum (the bit pattern contains only 1s),
- plateau (the bit pattern contains only 1s, but all pixels of the region have the same gray value),
- slope (the bit pattern of the region contains one connected component of 1s and one connected component of 0s - compare uniform patterns [12]),
- saddle point otherwise [6].

We further define a special case of saddles, the ridge to contain two connected components of 1s and two connected components of 0s. Figure 2 shows all these neighbourhood configurations.

3   Impact of the LBP Radius

In [7] we used a combination of LBPs and the medial axis to derive a graph representation of a binary image. The shape is first represented using a skeleton. Each skeleton pixel is then considered the center pixel of an LBP computation and the LBP type is determined. The LBP is computed for a radius R' that is larger than the radius r of the maximally inscribed circle stored for every medial axis pixel, in order to consider also the shape boundary: $R' = r + \epsilon$.

The graph representations obtained using this method depend strongly on the chosen LBP parameters. While adapting the number of sampling points P results in no or only minor changes of the graph, if the sampling density is high enough to capture the finest details, variations of the radius R have a strong influence on the resulting graph since many details inside the circle may be ignored.

The medial axis radius r of a certain skeleton pixel needs to be enlarged for the radius R' used in the LBP computation in order to take the shape boundaries into account (this also compensates for possible segmentation artefacts). Therefore, $R' = r + \epsilon$. In this process ϵ controls the radius and thus the level of detail that is captured or discarded. Enlarging ϵ may be compared to a smoothing operation along the shape's boundary. The impact on the resulting graph can be best described as graph pruning. Branches shorter than $r + \epsilon$ are discarded by this approach. However, how to choose factor ϵ is a crucial decision.

The choice of ϵ in general depends on the dataset. Moreover, ϵ should not be set to a fixed number of pixels for all skeleton pixels of the shape or even of the dataset. It should rather be defined as an adapting factor dependent on the medial axis radii. When varying the LBP radius, the LBP type of a skeleton pixel changes: For $\epsilon = 0$ the LBP sampling points are located along the maximally inscribed circle of the medial axis. Therefore, for a binary segmented image the according LBP is of type plateau. For a slightly increased radius the LBP in general equals the connectivity of the skeleton pixel (branching point, end point or ridge). For further increasing ϵ the LBP may change its type [2]. LBPs of the following types: plateau, slope, saddle and ridge may occur. The largest analysed radius is reached once an LBP of type maximum is observed. This configuration appears, once the whole shape is covered by the circle spanned by radius R'. For this maximal radius R'_{max} the following condition holds:

$$R'_{max} \leq \begin{cases} \frac{diameter}{2} & \text{for center of shape} \\ diameter & \text{everywhere} \end{cases} \tag{1}$$

Some examples of maximally inscribed circles with radius r and circles according to an enlarged radius R' are shown in Fig. 3. While the LBPs for maximally inscribed circles encode a plateau, the local topology according to the LBP of the larger circle shows a saddle (Fig. 3a) and a slope (Fig. 3b).

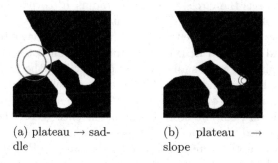

(a) plateau → saddle

(b) plateau → slope

Fig. 3. LBP for increasing radius captures evolution of local topology.

4   Shape Descriptor: Persistence of Critical Points

We further analyse the evolution of the LBP types for increasing radii R' starting with the medial axis radius r:

$$r \leq R' \leq diameter. \tag{2}$$

The evolution, respectively persistence of the LBP types of critical points of a skeleton (e.g. branching points and end points) can be described using a vector (LBP persistence vector) and a graph visualisation. We compute the LBP type for a range of radii and store it for every radius analysed. The persistence of such an LBP type is defined as its lifetime. In case of our persistence vector the lifetime is given as the length of an uninterrupted sequence of identical LBP type entries. We use the graph and the underlying LBP persistence vector as a shape descriptor and as a tool for shape classification.

For a binary shape, the approach presented in this paper proceeds as follows:

1. obtain topology preserving skeleton (using morphological thinning)
2. derive medial axis radii (Euclidean distance from boundary)
3. locate branching points of the skeleton
4. compute LBP persistence vector for each branching point
5. shape matching: compare LBP persistence vectors using the edit distance [9]

More details and properties of this approach are discussed in the following subsections.

4.1   LBP Persistence Vectors

We iteratively analyse the local topology captured by the LBP operator for the critical points using increasing LBP radii R'. We start with the radius r of the medial axis $R'_0 = r$ and increase the radius in each iteration. This computation terminates as soon as the circle spanned by radius R' covers the entire shape. Therefore, we start with an LBP of type plateau $R'_0 = r$ and stop at an LBP of type maximum. In between these two states the LBP type may alternate between ridge, saddle, and slope. An LBP of type minimum can not appear, as for this the LBP center pixel as well as all sampling points along the LBP radius would need to be in the background region, which is not possible since the LBP center pixel is a branching point of a skeleton and therefore always inside a shape.

For every critical point we store in the persistence vector the LBP type of every radius analysed. Thus, an LBP type of high persistence is captured as sequence of uninterrupted, repeated entries of the same LBP type. Alternatively the interval of radii for which one LBP type stays the same may be stored. In the persistence graph we assign the LBP types to one axis and plot the according radii along the second axis. Figure 4 shows a small toy example for a simple shape.

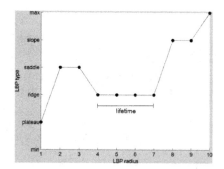

$v = \{pl, sa, sa, ri, ri, ri, ri, sl, sl, max\}$

(a) Simple shape and some LBP radii marked in red.

(b) Persistence graph and LBP persistence vector v.

Fig. 4. Toy example for LBP persistence vector plotted as graph (Color figure online).

4.2 Critical Points of the Skeleton

Critical points of the skeleton (branching and end positions) represent changes in topology: the number of connected components changes when walking along a skeleton and passing a critical point. The skeleton segments in between these critical points correspond to topological persistence, the number of connected components does not change when moving along these segments. For these skeleton segments the LBP centered at a skeleton pixel is always of type ridge, upon discarding intersections of parts of the shape that are not in the same connected component as the center pixel for the LBP computation. Branching points thus serve as suitable locations within a shape to compute LBP persistence vectors. However, for some shapes no branching points will exist. In this case the skeleton is given as single curve with two endpoints. A skeleton given as a single point (medial axis of a perfect circle) or a closed curve (medial axis of a perfect ring) are special cases which are not considered here. For skeletons without branching points a representation of the shape is possible using LBP persistence vectors computed for the end points. The LBP persistence vectors then allow to deduct information about the curvature of the shape's boundary. Taking the sequence of LBP types for increasing radii along a branchless medial axis into consideration, bounds for the degree of curvature of the medial axis can be estimated (Fig. 5a and c). Furthermore, changes in thickness are detectable using the medial axis radius stored for the skeleton pixels (Fig. 5b). Shocks as introduced by [13] present a suitable alternative to analyse and represent such shapes.

4.3 Shape Classification

Since we compute the proposed shape descriptor (the LBP persistence vector) centered at the branching points of the skeleton, one for each branching point, we derive several shape descriptors representing one shape. We tested different

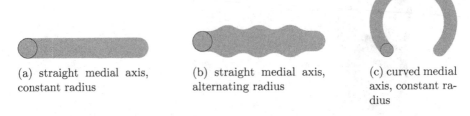

(a) straight medial axis, constant radius

(b) straight medial axis, alternating radius

(c) curved medial axis, constant radius

Fig. 5. Examples for shapes with no branching points in the skeleton.

approaches on how to choose from them or how to combine them in order to classify a shape in our experiments and present them in Sect. 5. However, the LBP persistence vectors corresponding to the different branching points of a single shape all individually cover the whole shape. Therefore, it is sufficient to choose just one of the branching points and its corresponding persistence vector as shape descriptor. Nevertheless, some branching points may be better suited as center of a shape descriptor (e.g. because they require less iteration to cover the whole shape). Future research may therefore be focused on methods and properties to judge the suitability of a branching point.

For classification of shapes the LBP persistence vectors are compared using the Levenshtein distance (edit distance) [9] as a measure of similarity which is also used by Sebastian et al. when comparing shock graphs [13]. Since every shape descriptor starts with an LBP of type plateau and ends with an LBP of type maximum, edit operations only alter the LBP types in between. These are of type saddle, slope or ridge. LBPs of these three types may alternate in an LBP persistence vector, therefore every edit operation generates a valid LBP persistence vector. The LBP type is stored as entry in the LBP persistence vector for every radius analysed, a sequence of identical entries represents an LBP type of high persistence. To transform such a vector to a sequence of alternating LBP types (therefore low persistence) many edit operations are needed. Thus, the information about the lifetime of one LBP type over a range of radii is taken into account when using the edit distance. In addition, this shape descriptor is robust in regard to small changes of the shape (e.g. small perturbations along the shape's contour), since such details do not persist over a large range of scales (radii) and thus have low persistence. These small changes therefore lead to changes of short sequences in the LBP persistence vector. Regarding a comparison using the edit distance, differences in such few entries are inexpensive and therefore the two shapes (with one of them showing small perturbations) will show high similarity.

5   Experiments

We tested the proposed shape descriptor on a subset of the Kimia 99 dataset [13]. For the subset used in our evaluation of shape classification we chose shape

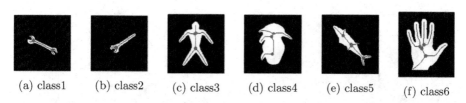

(a) class1   (b) class2   (c) class3   (d) class4   (e) class5   (f) class6

Fig. 6. Representative image for each class in the dataset together with its skeleton and branching points marked in red (Color figure online).

classes for which at least four images are in the class. Furthermore, we did not use the partially occluded shapes of the dataset. Following this rules, our dataset contains 38 images, which are grouped into six classes holding four to eleven images each. Figure 6 shows one representative image together with its skeleton and branching points of the skeleton (marked in red) for all classes of the dataset[1].

To evaluate the classification efficiency of our shape descriptor we performed a leave-one-out cross validation for each of the 38 shapes in the dataset and for the 105 LBP persistence vectors corresponding to the 105 branching points in the dataset. For every shape several descriptors (LBP persistence vectors), one corresponding to each branching point of the skeleton, may exist. The shape can be classified using one of these descriptors or a combination of them. We tested the following approaches:

1. according to the LBP persistence of only one branching point,
2. according to the average LBP persistence of all branching points of the shape,
3. according to the combined LBP persistence of all branching points of the shape using a majority vote,
4. according to the combined LBP persistence of all branching points of the shape using a weighted majority vote.

5.1 Classification Using Single LBP Persistence

Every image is classified by a comparison of the LBP persistence vector of only one of its branching points to all other LBP persistence vectors. For the dataset used in this experiment 105 branching points and respectively 105 LBP persistence vectors represent the 38 images. Figure 7a shows the confusion matrix for the distances between all 105 vectors. For every LBP persistence vector the class with the smallest edit distance, is chosen as resulting class. We evaluated this classification using leave-on-out cross validation for the 105 vectors. Only 15 out of 105 LBP persistence vectors were wrongly classified, 86 % were correctly matched. The confusion matrix in Fig. 7a shows that smallest distances correspond mostly to comparisons within one class for class 1, 3 and 6. This is

---

[1] The complete dataset with marked skeletons and branching points can be found at: prip.tuwien.ac.at/staffpages/ines/docs/dataset_dgci16.pdf.

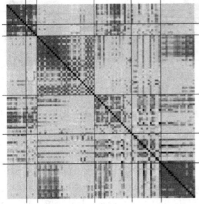 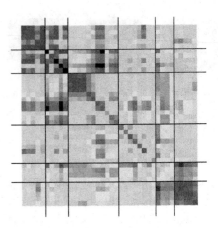

(a) distances between all LBP persis-   (b) average distances of the LBP persistence vectors   tence vectors of all images

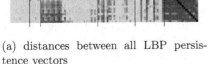

Fig. 7. Confusion matrix: distances of the (a) 105 LBP persistence vectors and (b) 38 average LBP persistence vectors. Blue: low distances; red: high distances. The lines separate the individual classes: class1 to class6 from left to right and top to bottom (Color figure online).

well visible by the dark blue blocks along the diagonal. Especially for class 5 distances are in general high both for comparisons within and outside the same class. This may be due to the high variation of the shapes within this class.

5.2 Classification Using Average LBP Persistence

Here we compute the distance of one image of the dataset to all other images as the average of the edit distances of all LBP persistence vectors representing one image. We then classify every image using leave-one-out cross validation. Again the class of the image with the smallest distance is taken as classification result The confusion matrix for this experiment is shown in Fig. 7b. For classes 1, 3, 4 and 6 the smallest distances are found within the same class. This is again visible by the blue blocks along the diagonal. 8 of the 38 images were wrongly classified, 79 % of the images are correctly classified. The classification error is slightly higher in comparison with the method presented in Sect. 5.1.

5.3 Classification Using a Majority Vote for Combined LBP Persistence

For this experiment we do not combine the distances. We first determine for all LBP persistence vectors of one image the class of smallest distance (as it is done in Sect. 5.1). The final classification is obtained using a majority vote on these classes. In case no majority can be found a choice is made randomly. We again classified all 38 images of the dataset using a leave-one-out cross validation:

Table 1. Classification results using a majority vote (result1) and a weighted vote (result2). Wrongly classified samples are highlighted in bold font.

Image	img1	img2	img3	img4	img5	img6	img7	img8	img9	img10	img11	img12	img13
Class	1	1	1	1	1	2	2	2	2	2	3	3	3
result1	1	1	1	1	1	**1**	**1**	**1**	**1**	**1**	3	3	3
result2	1	1	1	1	1	2	**1**	**1**	**1**	**1**	3	3	3
Image	img14	img15	img16	img17	img18	img19	img20	img21	img22	img23	img24	img25	img26
Class	3	3	3	3	3	3	3	3	4	4	4	4	4
result1	3	3	3	3	3	3	3	3	4	4	4	4	4
result2	3	3	3	3	3	3	3	3	4	4	4	4	4
Image	img27	img28	img29	img30	img31	img32	img33	img34	img35	img36	img37	img38	
Class	4	4	4	5	5	5	5	6	6	6	6	6	
result1	4	**1**	**2**	5	5	5	5	6	6	6	6	6	
result2	4	**1**	**2**	5	5	5	5	6	6	6	6	6	

7 images were wrongly classified, thus 82 % of the dataset were correctly classified. Table 1 shows the classification results in row *result1*. All samples of class 2 were wrongly classified as class 1. This may be due to the quite high similarity of the shapes apart from symmetry (see Fig. 6). However, all samples of class 5, which has a high variation within the class, are correctly classified in contrast to the two classification procedures presented in Sects. 5.1 and 5.2.

5.4 Classification Using a Weighted Vote for Combined LBP Persistence

This classification procedure works similar to the one presented in Sect. 5.3. However, here the LBP persistence vectors of one image are not equally involved in the voting but have a weighted vote according to their distance values. We compute the weight as the ratio of the distance of one LBP vector to the sum of distances of all vectors belonging to one image. Smallest distances are associated with the highest weights. The final class of an image is determined as the class with the highest vote when summing up the weights. The 38 images of the dataset were classified using leave-one-out cross validation. The results are shown in Table 1 in row *result2*. 6 images were wrongly classified, a correct classification was obtained for 84 % of the dataset. This result is very similar to the one obtained using a majority vote, except for one shape, which was wrongly classified using the majority vote, but is correctly classified using the weighted vote.

5.5 Discussion of the Experiments

Because of multiple branching points in a skeleton, several shape descriptors may exist for one shape when using the presented approach. We tested four methods to choose one or to combine these shape descriptors when classifying a shape. Choosing just one shape descriptor based on a branching point of the shape randomly yields the best results as the shape descriptor of one branching point in any case takes the whole shape into account. Combinations of the descriptors seem to introduce noise and therefore reduce the representational power.

The experiments show the classification capability of the presented approach: 79 % to 86 % of the dataset (subset of the Kimia 99 dataset [13] - occluded shapes and single shapes were not considered) are correctly classified. The authors of [13] report a 100 % (respectively 87 %) correct shape classification when using their shock graphs and considering the top 3 (respectively top 10) matches. Using only the best match and a variation of the dataset, 86 % accuracy in shape classification are reached by our algorithm. However, the authors of [13] also mention the high computational time needed to classify a shape. The classification presented in this paper is very efficient as it is computed only for a small number of locations or even only one location, within the shape. Furthermore the approach presented does not analyse all points of a shape, but employs the efficient LBP approach and computes the shape descriptor only on a subset of the shape's points (given as subsampling of a circle with origin inside the shape).

6   Conclusion and Outlook

The presented approach uses the persistence of LBP types computed at characteristic positions within the shape to classify it. It therefore derives information about local topology and persistence using a texture operator at increasing scales. The feature vectors used for classification are given as the evolution respectively persistence of the LBP types as LBP persistence vectors. For each shape the computation is guided by a skeleton representation of the shape. LBP persistence vectors are only computed for a very limited number of locations - the branching points of the skeleton.

For future work we would like to further test our shape classification using partially occluded shapes. Besides the edit distance, used in the experiments of this paper, further distance measurements need to be tested. For the decision on critical points that are used in the shape descriptor computation, we will evaluate methods to judge the suitability of the critical points. We may for example consider the diameter of the shape that is based on the maximum eccentricity of a shape [8]. Branching points of the skeleton close to the boundary mainly represent small irregularities of the shape boundary, e.g. segmentation artefacts. This estimation of the position of branching points in the shape can be used to rate the importance of branching points according to Eq. 1. Thus, the set of critical points can be reduced to the most important ones.

Moreover, as LBPs are not limited to binary images, we would like to test an application of the presented approach on gray-scale images. An initial segmentation may still be needed to derive the skeleton which guides all further operations. However, to reduce the impact of segmentation artefacts the LBP computations may then be done on the original gray-scale data. Couprie et al. presented grayscale skeletons [3] that may be used as an alternative to an initial segmentation. Moreover, a gray-level image can be interpreted as a landscape according to the gray-values. Critical points of this landscape as they are computed by Cerman et al. [2], may be used as an alternative and a segmentation and skeleton representation may no longer be needed.

References

1. Carlsson, G., Zomorodian, A., Collins, A., Guibas, L.J.: Persistence barcodes for shapes. Int. J. Shape Model. **11**(02), 149–187 (2005)
2. Cerman, M., Gonzalez-Diaz, R., Kropatsch, W.: LBP and irregular graph pyramids. In: Azzopardi, G., Petkov, N., Effenberg, A.O. (eds.) CAIP 2015. LNCS, vol. 9257, pp. 687–699. Springer, Heidelberg (2015). doi:10.1007/978-3-319-23117-4_59
3. Couprie, M., Bezerra, F.N., Bertrand, G.: Topological operators for grayscale image processing. J. Electron. Imaging **10**(4), 1003–1015 (2001)
4. Edelsbrunner, H., Letscher, D., Zomorodian, A.: Topological persistence and simplification. Discrete Comput. Geom. **28**(4), 511–533 (2002)
5. Ghrist, R.: Barcodes: the persistent topology of data. Bull. Am. Math. Soc. **45**(1), 61–75 (2008)
6. Gonzalez-Diaz, R., Kropatsch, W.G., Cerman, M., Lamar, J.: Characterizing configurations of critical points through LBP. In: SYNASC 2014 Workshop on Computational Topology in Image Context (2014)
7. Janusch, I., Kropatsch, W.G.: Reeb graphs through local binary patterns. In: Liu, C.-L., Luo, B., Kropatsch, W.G., Cheng, J. (eds.) GbRPR 2015. LNCS, vol. 9069, pp. 54–63. Springer, Heidelberg (2015)
8. Kropatsch, W.G., Ion, A., Haxhimusa, Y., Flanitzer, T.: The eccentricity transform (of a digital shape). In: Kuba, A., Nyúl, L.G., Palágyi, K. (eds.) DGCI 2006. LNCS, vol. 4245, pp. 437–448. Springer, Heidelberg (2006)
9. Levenshtein, V.I.: Binary codes capable of correcting deletions, insertions, and reversals. Soviet Physics Doklady **10**, 707–710 (1966)
10. Lindeberg, T.: Scale-space theory: a basic tool for analyzing structures at different scales. J. Appl. Stat. **21**(1–2), 225–270 (1994)
11. Ojala, T., Pietikäinen, M., Harwood, D.: A comparative study of texture measures with classification based on featured distributions. Pattern Recogn. **29**(1), 51–59 (1996)
12. Pietikäinen, M., Hadid, A., Zhao, G., Ahonen, T.: Computer Vision Using Local Binary Patterns. Computational Imaging and Vision, vol. 40. Springer, London (2011)
13. Sebastian, T., Klein, P., Kimia, B.: Recognition of shapes by editing shock graphs. In: IEEE International Conference on Computer Vision, vol. 1, p. 755. IEEE Computer Society (2001)

Shape Descriptors

Signature of a Shape Based on Its Pixel Coverage Representation

Vladimir Ilić[1]([⊠]), Joakim Lindblad[2,3], and Nataša Sladoje[2,3]

[1] Faculty of Technical Sciences, University of Novi Sad, Novi Sad, Serbia
`vlada.mzsvi@uns.ac.rs`
[2] Centre for Image Analysis, Uppsala University, Uppsala, Sweden
[3] Mathematical Institute, Serbian Academy of Sciences and Arts, Belgrade, Serbia

Abstract. Distance from the boundary of a shape to its centroid, a.k.a. signature of a shape, is a frequently used shape descriptor. Commonly, the observed shape results from a crisp (binary) segmentation of an image. The loss of information associated with binarization leads to a significant decrease in accuracy and precision of the signature, as well as its reduced invariance w.r.t. translation and rotation. Coverage information enables better estimation of edge position within a pixel. In this paper, we propose an iterative method for computing the signature of a shape utilizing its pixel coverage representation. The proposed method iteratively improves the accuracy of the computed signature, starting from a good initial estimate. A statistical study indicates considerable improvements in both accuracy and precision, compared to a crisp approach and a previously proposed approach based on averaging signatures over α-cuts of a fuzzy representation. We observe improved performance of the proposed descriptor in the presence of noise and reduced variation under translation and rotation.

Keywords: Shape signature · Centroid distance function · Pixel coverage representation · Sub-pixel accuracy · Precision

1 Introduction

A shape signature is any one-dimensional function which represents a two-dimensional shape. Several versions are proposed in the literature, such as centroid distance, centroidal profile, complex coordinates, tangent angle, cumulative angle; overviews of shape descriptors can be found in [3,10]. Among them, signature based on centroid distance is popular due to its simple and intuitive definition, as well as its suitability for further processing and, *e.g.*, derivation of Fourier or wavelet descriptors of the shape. As a boundary based descriptor, however, centroid distance signature suffers from noise sensitivity. Additionally, the signature of a discrete shape representation is often altered by translation and/or rotation of a shape, which is an undesired property of a shape descriptor.

We suggest, in this paper, to improve accuracy and precision (i.e., reduce bias and variance) of the centroid distance signature by computing it from a

© Springer International Publishing Switzerland 2016
N. Normand et al. (Eds.): DGCI 2016, LNCS 9647, pp. 181–193, 2016.
DOI: 10.1007/978-3-319-32360-2_14

pixel coverage representation of the shape [7,8], instead of following the traditional approach and utilize a crisp shape representation. Previous results on shape representation based on coverage of pixels by the imaged object have confirmed that such an information rich representation may be utilized to increase accuracy and precision of several shape descriptors, compared to binary (crisp) approaches [6,7]. In a coverage representation every pixel is assigned an intensity value proportional to the relative coverage of the pixel by the imaged shape. This intensity value can be utilized to determine sub-pixel position of the shape boundary, which can then be used to produce more precise and accurate shape signature. In this paper, we propose a novel method for estimating the shape signature of a starshaped object utilizing a coverage representation of the shape. We present an algorithm which iteratively improves accuracy of the computed signature, alternating between estimation of boundary position and shape normal direction. Evaluation results confirm a number of advantages of the proposed approach. The signature derived for a coverage representation is computed by a simple and fast procedure, and exhibits increased accuracy and precision, and invariance to rotation and translation, as well as decreased sensitivity to noise, compared to previously used crisp approach.

2   Background

Usually, an object is extracted from an image by crisp segmentation, providing a crisp representation of the shape, such that each pixel is either completely included, or completely excluded, in/from it. This process is associated with a loss of information about the original shape. Shape descriptors are then computed utilizing this crisp representation. The reduced amount of information preserved in the crisp representation reduces the descriptors' accuracy and precision, as well as their invariance under translation and rotation. One way to improve performance of shape descriptors is to compute them from a coverage representation of the shape. Advantages of such an approach, compared to utilizing a crisp representation, are observed in, *e.g.*, [2,6,7]. These results motivated our work on development of an estimator for centroid distance shape signature which utilizes coverage representation of the shape. Before presenting the algorithm for signature computation in Sect. 3, and showing the results of its evaluation in Sect. 4, we introduce the main concepts and notions we utilize.

2.1   Fuzzy Set Theory

Fuzzy set theory is a generalization of the classical (crisp) set theory, allowing partial belongingness of elements of some reference set to a fuzzy set.

Definition 1. [9] *A fuzzy subset S of a reference set X is a set of ordered pairs $S = \{(x, \mu_S(x)) \mid x \in X\}$, where $\mu_S : X \to [0,1]$ is the membership function of S in X.*

A crisp set is a special case of a fuzzy set and its characteristic function, allowing only values $\{0,1\}$, is a special case of a fuzzy membership function.

Important notions related to fuzzy set are those of α-cut and support:

Definition 2. *[9] For $\alpha \in (0,1]$, the $\alpha-$cut of S, defined by μ_S, is the crisp set $S_\alpha = \{x \in S \mid \mu_S(x) \geq \alpha\}$.*

Definition 3. *[9] The support of a fuzzy set S is the crisp set $Supp(S) = \{x \in X \mid \mu_S(x) > 0\}$.*

Definition 4. *For a fuzzy set $S = \{((x,y), \mu(x,y)) \mid (x,y) \in X\}$, defined on a discrete reference set $X \subset R^2$, the coordinates of its centroid (x_c, y_c) are*

$$x_c = \frac{\sum_x \sum_y x \cdot \mu(x,y)}{\sum_x \sum_y \mu(x,y)}, \qquad y_c = \frac{\sum_x \sum_y y \cdot \mu(x,y)}{\sum_x \sum_y \mu(x,y)}.$$

Definition 4 can be used to compute the centroid of a crisp shape as well, utilizing the characteristic function of the crisp shape as its membership function.

2.2   Shape Signature

The centroid distance signature of a continuous shape is a continuous function which assigns distance from the shape centroid to every point on the continuous shape boundary, parametrized in some appropriate way. For a starshaped object, it holds that every straight half-line starting from the centroid intersects the boundary of the shape in exactly one point; therefore there exists a one-to-one mapping from the interval $[0, 2\pi)$ of angles to the set of boundary points, which we use for parametrization.

Definition 5. *The signature of a continuous starshaped object S is a continuous function $r(\theta)$ which assigns the length of the vector determined by the shape centroid (x_c, y_c) and the boundary point $(x(\theta), y(\theta))$, to every angle $\theta \in [0, 2\pi)$, where $\theta = \arctan\left(\frac{y(\theta) - y_c}{x(\theta) - x_c}\right)$:*

$$r(\theta) = \sqrt{(x(\theta) - x_c)^2 + (y(\theta) - y_c)^2}. \tag{1}$$

Throughout the paper we assume that $\arctan\left(\frac{y}{x}\right)$ computes the four-quadrant inverse tangent, appropriately handling the case $x = 0$ and providing an output in the range $[0, 2\pi)$.

The signature of a discrete representation of a continuous shape is a restriction of the continuous shape signature to some appropriate sample values θ_i of the interval $[0, 2\pi)$. The sample values θ_i can be determined beforehand, or be derived as a part of the process of signature computation. Often used sample points/angles are those of a 4-connected, or an 8-connected boundary.

Definition 6. *The signature of a discrete representation of a (continuous) starshaped object S is a collection of pairs $(r(\theta_i), \theta_i)$, where $r(\theta_i)$ assigns the length of the vector determined by the shape centroid (x_c, y_c) and the boundary point $(x(\theta_i), y(\theta_i))$, to a set of values $\theta_i \in [0, 2\pi)$:*

$$r(\theta_i) = \sqrt{(x(\theta_i) - x_c)^2 + (y(\theta_i) - y_c)^2}, \qquad i = 1, 2, \ldots, N. \tag{2}$$

In absence of any information about the boundary location within a pixel, the centre point of the pixel is a reasonable sample point. It is, however, rather unlikely that the centre of a pixel belongs to the boundary of the continuous object, which introduces errors in computation of the signature, and leads to translation and rotation variance of this descriptor.

An approach to compute the signature of a fuzzy shape representation, and by that reduce negative effects of crisp segmentation, is proposed in [1]. Following one of the main fuzzification principles, the signature of a fuzzy shape is computed as the average of the signatures of all its α-cuts. This approach is shown to outperform crisp approach in terms of achieved precision of the estimator. Bias of the estimator is not addressed/corrected, which affects its accuracy.

2.3   Coverage Model

Let the Voronoi region in $\mathbb{R}^2$ of a grid point $x \in \mathbb{Z}^2$, i.e., a pixel, be denoted by $\sigma(x)$. *Coverage digitization* in $\mathbb{R}^2$ is defined in [8] as:

Definition 7. *For a given continuous object $S \subset \mathbb{R}^2$, inscribed into an integer grid $\mathbb{Z}^2$, the pixel coverage digitization of S is*

$$\mathcal{D}_C(S) = \left\{ (x, a(x)) \mid x \in \mathbb{Z}^2 \right\}, \qquad a(x) = \frac{|\sigma(x) \cap S|}{|\sigma(x)|}, \qquad (3)$$

where $|S|$ denotes area of a set S.

For a real image, the pixel coverage values can be estimated by coverage segmentation methods, [4,5,7].

Coverage representation of a continuous crisp shape is characterized by pixel values equal to one, for pixels completely covered by the shape, pixel values equal to zero for pixels completely covered by the background, and pixel values strictly between 0 and 1, for partly covered pixels that typically form a one pixel thick discrete shape boundary.

The coverage function a is a particular type of fuzzy membership function, suitable for representation of imaged objects. Definitions and properties derived for (spatial) fuzzy sets are thus directly applicable to the coverage representation.

3   Shape Signature with Sub-pixel Precision

We present a novel approach for estimating shape signature based on a pixel coverage representation of the object. If we approximate the boundary of a shape within a pixel with a straight line segment, we can estimate the edge position with sub-pixel accuracy from the pixel (coverage) value, assuming that we know the edge direction. We can then compute a discrete signature of the shape by selecting a sample point on the estimated edge. Any point on the edge segment is a possible choice. To have a well defined inner point of the pixel, we use the midpoint of the estimated edge as a sample point for the proposed signature computation. Based on the above, we propose the following Pixel Coverage Shape Signature (PCSS), in accordance with Defintion 6.

The pixel coverage signature of a starshaped continuous object S is a set of pairs $(r(\theta_i), \theta_i)$, where $r(\theta_i)$ assigns the length of the vector given by the shape centroid (x_c, y_c) and the point $(x_{\text{mid}}(\theta_i), y_{\text{mid}}(\theta_i))$, to every angle $\theta_i \in [0, 2\pi)$, such that $\theta_i = \arctan\left(\frac{y_{\text{mid}}(\theta_i) - y_c}{x_{\text{mid}}(\theta_i) - x_c}\right)$, where $(x_{\text{mid}}(\theta_i), y_{\text{mid}}(\theta_i))$ is the midpoint of the edge in the i-th pixel along the object boundary and

$$r(\theta_i) = \sqrt{(x_{\text{mid}}(\theta_i) - x_c)^2 + (y_{\text{mid}}(\theta_i) - y_c)^2}, \quad i = 1, 2, \ldots, N. \qquad (4)$$

3.1   Computation of Midpoint of Edge Segment Within a Pixel

Assuming that the object boundary within a pixel centred at (x, y) can be approximated by a straight edge, let us denote the normal direction of the edge by $\boldsymbol{n} = (\cos\varphi, \sin\varphi)$, and the coverage value of the observed pixel by a.

The coordinates of the midpoint of the edge inside the observed pixel can be expressed as

$$\left(x + e_x(a, \varphi),\, y + e_y(a, \varphi)\right), \qquad (5)$$

where, for angles $\varphi \in [0, \pi/4]$, it holds that

$$(e_x(a, \varphi), e_y(a, \varphi)) = \begin{cases} \left(\sqrt{\frac{a\tan\varphi}{2}} - \frac{1}{2}, \sqrt{\frac{a}{2\tan\varphi}} - \frac{1}{2}\right), & 0 \le a \le \frac{1}{2}\tan\varphi, \\ \left(a - \frac{1}{2}, 0\right), & \frac{1}{2}\tan\varphi \le a \le 1 - \frac{1}{2}\tan\varphi, \\ \left(\frac{1}{2} - \sqrt{\frac{(1-a)\tan\varphi}{2}}, \frac{1}{2} - \sqrt{\frac{1-a}{2\tan\varphi}}\right), & 1 - \frac{1}{2}\tan\varphi \le a \le 1. \end{cases}$$

$$(6)$$

Due to (anti)symmetry of the above expressions, it is enough to consider values $0 \le a \le 0.5$ and $\varphi \in [0, \pi/4]$; other cases can be obtained by changing the sign and/or swapping the coordinates of the computed midpoint.

3.2   Algorithm for Shape Signature Estimation

To compute PCSS using (4) and (5), we need to estimate the normal direction $\boldsymbol{n} = (\cos\varphi, \sin\varphi)$ of the edge within each boundary pixel. We propose an iterative procedure which utilizes the current signature estimate to derive the normal direction of the edge, and then to improve the signature itself. This method is simple and fast, utilizes only the values of the observed pixel and its two neighbours, and does not require convolution masks for gradient estimation.

An initial signature is computed as a sequence of distances from the shape centroid to midpoints of the edges in each boundary pixel. We consider boundary pixels to be pixels with $a > 0$ which are 8-connected to the pure background pixels, having coverage $a = 0$. For an object without holes, such boundary consists of a thin 4-connected sequence of pixel which is easily detected and indexed. For computation of initial midpoints, normal directions are initialized as $\varphi_i = \theta_i$.

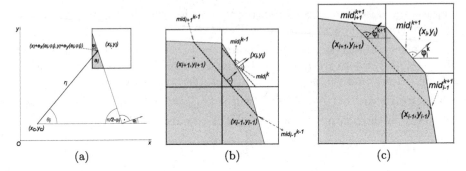

Fig. 1. (a) Shape signature based on pixel coverage, PCSS, computed to the mid-point of the edge segment of a pixel (x_i, y_i), for coverage a_i and normal direction φ_i. (b) and (c) Illustration of iterative estimation of boundary position (midpoint $\text{mid}_i = (x_i + e_x(a_i, \varphi_i), y_i + e_y(a_i, \varphi_i))$) and edge normal direction φ_i, in two consecutive iterations (k and $k+1$) of Algorithm 1, by using estimated midpoints of the edges in two neighbouring boundary pixels.

In each iteration the estimate of the edge position in a pixel is updated by computing the normal of the line connecting the midpoints of the edges in the two boundary pixels neighbouring the observed one. These midpoints are determined based on the signature estimate in the previous iteration. The new edge position (and its new midpoint) is computed using the new normal estimate and the given coverage value. Figure 1 illustrates this process. The procedure terminates when the difference between two consecutive signature estimates stays below a given tolerance, or when some predefined number of iterations is reached.

The complete algorithm is presented below. Each line in the algorithm assumes parallel/vectorized computation for all $i \in \{1, 2, \ldots, N\}$. Cyclic indexing is assumed when index $i \pm 1$ reaches the end values of the sequence.

Algorithm 1

Input: Pixel centers (x_i, y_i) counterclockwise along the boundary of S and corresponding coverage values a_i, $i = 1, \ldots, N$. Tolerance $\varepsilon > 0$, maxiter> 0.
Output: Pixel coverage signature $PCSS$ of S, consisting of N pairs (r_i, θ_i).

$$\varphi_i \leftarrow \arctan\left(\frac{y_i - y_c}{x_i - x_c}\right)$$
for $k = 1, 2, \ldots, \text{maxiter}$
    compute $(e_x(a_i, \varphi_i), e_y(a_i, \varphi_i))$ according to (6)
$$r_i^k \leftarrow \sqrt{(x_i + e_x(a_i, \varphi_i) - x_c)^2 + (y_i + e_y(a_i, \varphi_i) - y_c)^2}$$
$$\theta_i^k \leftarrow \arctan\left(\frac{y_i + e_y(a_i, \varphi_i) - y_c}{x_i + e_x(a_i, \varphi_i) - x_c}\right)$$
    if $k > 1 \wedge \left\|r^k - r^{k-1}\right\|_\infty < \varepsilon$ then

exit the for loop

$$\tilde{x}_i \leftarrow r_i^k \cos(\theta_i^k), \quad \tilde{y}_i \leftarrow r_i^k \sin(\theta_i^k)$$
$$\varphi_i \leftarrow \arctan\left(\frac{\tilde{x}_{i-1} - \tilde{x}_{i+1}}{\tilde{y}_{i+1} - \tilde{y}_{i-1}}\right)$$

end for

$$PCSS = \{(r_i^k, \theta_i^k)\}, \text{ for } i = 1, 2, \dots, N$$

For poorly resolved non-smooth objects, there may be cases where one boundary pixel is intersected several times by the shape boundary, *i.e.*, where a single straight edge segment poorly models the boundary position. In such cases the above algorithm may produce suboptimal edge position estimates. We therefore adjust utilization of coverage values for edge direction estimation if a pixel is visited more than once while the boundary is traversed. We appropriately split its coverage and utilize it for each of the edge estimates separately. This is conveniently achieved by a simple update of the coverage values to be used by the algorithm, as follows:

---

for all pixels j appearing $n_j > 1$ times in the traced boundary
    if $a_j < \frac{1}{2}$, let $a_j \leftarrow \frac{a_j}{n_j}$
    else, let $a_j \leftarrow 1 - \frac{1 - a_j}{n_j}$

This pre-processing step results in improved boundary estimation at sharp corners, as illustrated in Fig. 2.

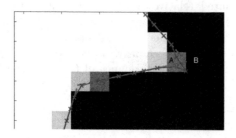

Fig. 2. Example showing the need for adjusting coverage of pixels intersected by more than one edge. Pixel A (with coverage 0.52) is intersected twice by the true object boundary (red dashed line), passing pixel B (with a small positive coverage) in between. Without adjustment, both edge midpoints of A would incorrectly be placed very close to the pixel centre. If we instead assume that each intersecting edge cuts away half of the background and utilize a coverage value (for each edge individually) of 0.76, a better result (blue line, the crosses indicate computed edge midpoints) is achieved (Color figure online).

4 Evaluation

We evaluate performance of the proposed method for computation of the signature from a coverage representation of a shape, by comparing it with methods utilizing crisp shape representation, according to Definition 6, and utilizing α-cuts of a fuzzy representation of the shape, as proposed in [1]. The approaches are denoted by BSS (Binary Shape Signature), ACSS (α-cut Shape Signature) and PCSS (Pixel Coverage Shape Signature). Our evaluation is focused on accuracy, precision, noise sensitivity, and invariance to translation and rotation.

The observed test objects are disks, 6-cornered stars, and rectangles, for which the true signatures used in comparisons are analytically derived. Crisp shape representations of the observed objects are obtained by Gauss centre point digitization of continuous shapes. The coverage representations of these objects are analytically derived, and therefore exact (to the level of floating point operations). The fuzzy representation considered is the coverage representation, *i.e.*, ACSS and PCSS are applied to the same shape representations. Tolerance used for the exit criterion in Algorithm 1 is set to $\varepsilon = 0.0001$ and maxiter $= 100$.

We first show an illustrative example motivating our proposed approach. Figure 3 presents centroid distance signatures estimated by BSS, ACSS, and PCSS, of a disk with radius $r = 66.64$ pixels. Comparing with the true signature of the disk (constant function $y = 66.64$), we observe very high accuracy and precision of the proposed PCSS method, compared to rather precise, but non-accurate (biased) ACSS, and comparably imprecise and also biased BSS.

The distributions of errors for the different methods are presented in Fig. 4, estimated for a disk and a 6-cornered star, both of a radius $r = 66.64$. Each object is digitized 100 times, at random positions (and rotations) in the grid, and errors are computed relative to the resp. true continuous shape. Both 4- and 8-connected boundaries are considered for BSS, whereas only 8-connected boundary is observed for ACSS (per recommendation in [1]). For PCSS it is most natural to use the 4-connected boundary, which gives more sample points than the 8-connected one. Excellent performance of the proposed PCSS in terms of

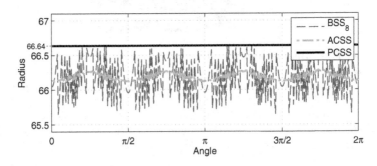

Fig. 3. Shape signatures computed by BSS (8-connected boundary observed), ACSS and PCSS for a disk with radius of 66.64 pixels. At this scale, errors of the PCSS method are too small to be visible.

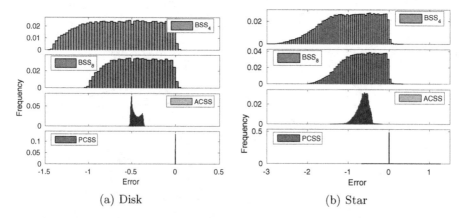

Fig. 4. Histograms of errors of BSS, ACSS and PCSS applied to a disk (a) and a star (b), of radius of 66.64 pixels. Errors larger than 0.01 pixel are very rare for PCSS.

accuracy is clearly visible and confirmed by the significantly smaller range of errors, compared to the range of errors of both BSS (4- and 8-connected boundary), and ACSS. The biases of BSS and ACSS are also clearly visible.

We further present results of a statistical study of the estimation error, for the three observed signature estimators for objects of increasing sizes. We observe the root-mean-square error RMSE $= \sqrt{\frac{1}{N} \sum_{i=1}^{N} (\hat{x}_i - x_i)^2}$, where N is the number of boundary pixels, x_i are true values of the shape signature and $\hat{x}_i$ are estimated shape signature values. We also analyse the maximal absolute error (MaxErr) of the methods. For each radius in the range 0 to 100 pixels, 100 disks (100 stars) with centres randomly translated within one pixel, are observed and their signatures compared with the ground truth (true signatures of the corresponding continuous objects). For the case of stars, random rotations of the objects are included as well.

To compare and evaluate performance of the estimators in terms of their precision, we correct for the bias of BSS and ACSS. We estimate their resp. bias by averaging signatures of 100 randomly positioned disks of a radius 1000, and subtract it from estimated signatures observed in the test.

Figure 5 presents plots of obtained RMSE and MaxErr. Outstanding performance of the proposed PCSS estimator on the disks can better be appreciated in the log-log plots. Linear decrease of both RMSE and MAE is observed for PCSS for disks, with increase of object size (or image resolution). The proposed method clearly outperforms the competitors also when applied to the more complex objects (stars) with non-smooth boundaries.

Rotation invariance of the proposed PCSS estimator is illustrated in Fig. 6. Three rotations of a rectangle (9×11 pixels), shown in Fig. 6(b) by their coverage representations, are observed. Corresponding signatures of these shapes, estimated by the three compared methods, are shown in Fig. 6(a). Each plot represents signatures of the three shapes, with differently marked points (triangles, squares, and stars), estimated by one of the three observed methods.

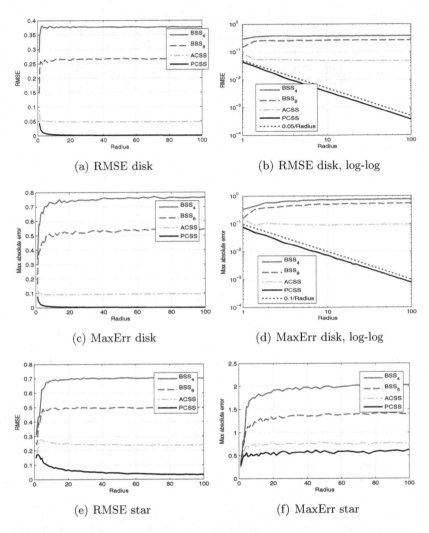

Fig. 5. RMSE and MaxErr (in pixels) of BSS (4- and 8-connected), ACSS (8-conn) and PCSS, applied to disks (first and second row) and stars (third row) of increasing radii.

The blue continuous line indicates the true signature of the observed rectangle. The signatures are shifted to the same starting angle so that they, in the ideal case, should coincide. Translation and rotation invariance is however violated for BSS, but also ACSS, due to discretization. The cloud of points clearly deviates from the blue line (lack of accuracy) for both BSS and ACSS as well as between observations (lack of precision). For the PCSS estimate, the points are consistently correctly positioned, exhibiting both high accuracy and high precision.

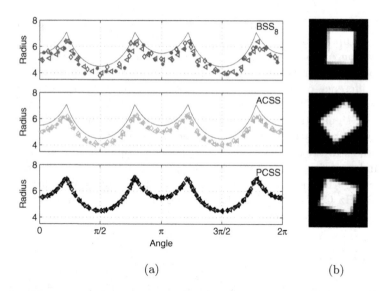

(a)   (b)

Fig. 6. Signatures of three rotations of a rectangle, estimated by BSS, ACSS, and PCSS. Points of the different signatures are marked by triangles, squares, and stars. Coverage representations of the displaced rectangles are shown to the right.

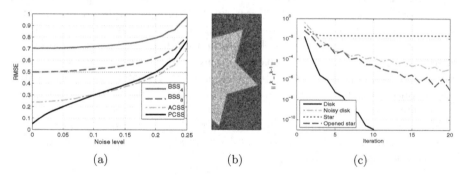

(a)   (b)   (c)

Fig. 7. RMSE for the methods for increasing levels of Gaussian noise. Both PCSS and ACSS perform better when degraded by up to 20 % noise than what a noise free crisp approach (blue dotted line) does. (b) Part of the observed object with 20 % noise added. (c) Asymptotic behavior of the sequence $\left\| r^k - r^{k-1} \right\|_\infty$, $k \in \mathbb{N}$ for different cases (Color figure online).

Sensitivity to noise of PCSS is evaluated in comparison with the performance of BSS and ACSS, observing RMSE of the estimated signatures of a star, in presence of increasing levels of added zero mean Gaussian noise with standard deviation in the range $[0, 0.25]$ (where a white pixel has level 1). Each object is segmented by either Otsu thresholding (for crisp approaches) or with the fast soft thresholding technique proposed in [7] (for ACSS and PCSS). We observe, in Fig. 7, that both PCSS and ACSS perform better than what a *noise free* crisp approach does (blue dotted line), even when the coverage based approaches are degraded by up to 20 % of noise.

Illustration of convergence of the proposed algorithm is given in Fig. 7(c). Results of the empirical evaluation indicate that the sequence $\left\| r^k - r^{k-1} \right\|_\infty$ showing difference of two consecutive signature estimates (non-monotonically) decreases for shapes which do not have sharp corners. Convergence of the algorithm when signature of a star is estimated is violated, so the algorithm terminates with the reached maximal number of iterations. However, if the sharp peaks of the star are smoothed by a morphological opening, the algorithm converges, and so it does when applied to both noisy and noise-free disks.

Processing time is around 0.01 s for a 600 pixel long signature (Star) on a single 3 GHz CPU in Matlab. The algorithm is easy to parallelize if higher processing speed is required.

5   Conclusion

We propose a novel method for estimating the centroid distance signature of a shape utilizing its pixel coverage representation. Pixel value, representing fraction of the pixel area covered by the imaged object, is used to estimate the edge position of the object inside the pixel. Midpoint of the estimated edge is used to define the signature. An iterative procedure is suggested to improve edge direction estimation, and by that to improve the signature estimate in each iterative step. Evaluation of the proposed method confirms its superiority regarding accuracy, precision, robustness to noise, and invariance w.r.t. rotation and translation, compared to the shape signatures estimated from crisp shape representation, or by averaging the signatures of the α-cuts of the coverage representation. The proposed iterative algorithm is fast, intuitive and simple.

We have, on the example of shape signature, again confirmed that the information rich coverage representation enables estimation of shape descriptors with high accuracy and precision.

Acknowledgments. Authors are supported by the Ministry of Science of the Republic of Serbia through Project ON174019 (V. Ilić) and Projects ON174008 and III44006 (J. Lindblad, N. Sladoje). N. Sladoje is supported by the Swedish Governmental Agency for Innovation Systems (VINNOVA).

References

1. Chanussot, J., Nyström, I., Sladoje, N.: Shape signatures of fuzzy star-shaped sets based on distance from the centroid. Patt. Recogn. Lett. **26**, 735–746 (2005)
2. Gustavson, S., Strand, R.: Anti-aliased Euclidean distance transform. Patt. Recogn. Lett. **32**, 252–257 (2011)
3. Kindratenko, V.: On using functions to describe the shape. J. Math. Imag. Vis. **18**, 225–245 (2003)

4. Lindblad, J., Sladoje, N.: Coverage segmentation based on linear unmixing and minimization of perimeter and boundary thickness. Patt. Recogn. Lett. **33**(6), 728–738 (2012)
5. Malmberg, F., Lindblad, J., Sladoje, N., Nyström, I.: A graph-based framework for sub-pixel image segmentation. Theor. Comput. Sci. **412**, 1338–1349 (2011)
6. Sladoje, N., Lindblad, J.: Estimation of moments of digitized objects with fuzzy borders. In: Roli, F., Vitulano, S. (eds.) ICIAP 2005. LNCS, vol. 3617, pp. 188–195. Springer, Heidelberg (2005)
7. Sladoje, N., Lindblad, J.: High-precision boundary length estimation by utilizing gray-level information. IEEE Trans. Patt. Anal. Mach. Intell. **31**, 357–363 (2009)
8. Sladoje, N., Lindblad, J.: The coverage model and its use in image processing. In Selected Topics on Image Processing and Cryptology, Zbornik Radova (Collection of Papers), vol. 15, pp. 39–117 (2012). Math. Inst. of Serbian Academy of Sciences and Arts
9. Zadeh, L.: Fuzzy sets. Inf. Control **8**, 338–353 (1965)
10. Zhang, D., Lu, G.: Review of shape representation and description techniques. Pattern Recogn. **37**, 1–19 (2004)

Computation of the Normal Vector to a Digital Plane by Sampling Significant Points

Jacques-Olivier Lachaud[1], Xavier Provençal[1], and Tristan Roussillon[2(✉)]

[1] Université Savoie Mont Blanc, LAMA, UMR5127,
73376 Le Bourget-du-Lac, France
{jacques-olivier.lachaud,xavier.provencal}@univ-smb.fr
[2] Université de Lyon, CNRS INSA-Lyon, LIRIS, UMR5205,
69622 Villeurbanne, France
tristan.roussillon@liris.cnrs.fr

Abstract. Digital planes are sets of integer points located between two parallel planes. We present a new algorithm that computes the normal vector of a digital plane given only a predicate "is a point x in the digital plane or not". In opposition with the algorithm presented in [7], the algorithm is fully local and does not explore the plane. Furthermore its worst-case complexity bound is $O(\omega)$, where ω is the arithmetic thickness of the digital plane. Its only restriction is that the algorithm must start just below a Bezout point of the plane in order to return the exact normal vector. In practice, our algorithm performs much better than the theoretical bound, with an average behavior close to $O(\log \omega)$. We show further how this algorithm can be used to analyze the geometry of arbitrary digital surfaces, by computing normals and identifying convex, concave or saddle parts of the surface.

Keywords: Digital geometry · Digital plane · Recognition · Normal vector estimation · Lattice reduction

1 Introduction

The study of the linear geometry of digital sets has raised a considerable amount of work in the digital geometry community. In 2D, digital straightness has been extremely fruitful. Indeed, properties of digital straight lines have impacted the practical analysis of 2D shapes, especially through the notion of *maximal segments* [3,4], which are unextensible pieces of digital straight lines along digital contours.

In 3D, the main problem is that there is no more an implication between "being a maximal plane" and "being a tangent plane" as it is in 2D. This was highlighted in [2], where maximal planes were then defined as planar extension

---

This work has been partly funded by DIGITALSNOW ANR-11-BS02-009 research grant.

N. Normand et al. (Eds.): DGCI 2016, LNCS 9647, pp. 194–205, 2016.
DOI: 10.1007/978-3-319-32360-2_15

of maximal disks. To sum up, the problem is not so much to recognize a piece of plane, but more to group together the pertinent points onto the digital shape.

A first algorithm was proposed by the authors in [7]. Given only a predicate "is point x in plane $\mathbf{P}$?" and a starting point in $\mathbf{P}$, this algorithm extracts the exact characteristics of $\mathbf{P}$. The idea is to deform an initial unit tetrahedron based at the starting point with only unimodular transformations. Each transformation is decided by looking only at a few points around the tetrahedron. This step is mostly local but may induce sometimes a dichotomic exploration. At the end of this iterative process, one face of the tetrahedron is parallel to $\mathbf{P}$, and thus determines its normal. The remarkable idea is that the algorithm decides itself on-the-fly which points have to be considered for computing the local plane geometry, in opposition to usual plane recognition algorithms [1,5,6,8]. This approach was thus also interesting for analysing 3D shape boundaries.

This paper proposes another algorithm that extracts the characteristics of plane $\mathbf{P}$ given only this predicate and a starting configuration. This new algorithm is also an iterative process that deforms an initial tetrahedron and stops when one face is parallel to $\mathbf{P}$. However, this new algorithm both complements and enhances the former approach. It complements it since it extracts triangular facets onto the plane whose Bezout point is above the facet, while the former algorithm extracts the ones whose Bezout point is not above the facet. See Fig. 2 for an example of execution of both algorithms with the same input. It is also a geometrical algorithm using convex hull and Delaunay circumsphere property, while the former was mostly arithmetic. It enhances it for several reasons. First, its theoretical complexity is slightly better (it drops a log factor) and it is faster also in practice. Second, we control the position of the evolving tetrahedron, which always stays around the starting point. Third, each step is purely local and tests the predicate on only six points. Fourth, it can detect planarity defects onto digital surfaces that are not digital planes. Last, a variant of the proposed algorithm *almost always* produce directly a reduced basis of the lattice of upper leaning points, without lattice reduction. Of course, this algorithm presents one disadvantage with respect to the former one: the starting configuration must lie at a reentrant corner of the plane, more precisely below the Bezout point of the plane. If it starts at another corner, then the algorithm will stop sooner and outputs only an approximation of the normal of $\mathbf{P}$.

The paper is organized as follows. First, we give basic definitions and present our new algorithm. Second we show its correctness and exhibit worst-case upperbound for its time complexity. Third we study how often this algorithm extracts a reduced basis of the lattice of upper leaning points and present a variant — with the same properties — that returns almost always a reduced basis. Afterwards we exploit this algorithm to determine the linear geometry of digital surfaces, and we show how to deal with starting configurations that are not under the Bezout point. Finally the pros and cons of this algorithm are discussed and several research directions are outlined.

Algorithm 1. Extracts a triangle aligned with the digital plane **P** by successive convex hull computations.

Input: a predicate "$x \in \mathbf{P}$?", an initial triangle $\mathbf{T}^{(0)}$
1 $\mathbf{T} \leftarrow \mathbf{T}^{(0)}$;
2 **while** $\Sigma_{\mathbf{T}} \cap \mathbf{P} \neq \emptyset$ **do**
3 $\quad\quad$ Compute the convex hull of $\mathbf{T} \cup (\Sigma_{\mathbf{T}} \cap \mathbf{P})$;
4 $\quad\quad$ Find $\mathbf{T}'$, defined as the upper triangular facet intersected by $[\boldsymbol{pq}]$;
5 $\quad\quad$ $\mathbf{T} \leftarrow \mathbf{T}'$;
6 **return** $\mathbf{T}$;

---

2 Notations and Algorithm

In this section, we introduce a few notations before presenting our new algorithm. A digital plane is defined as the set

$$\mathbf{P} = \{x \in \mathbb{Z}^3 | \mu \leq x \cdot \mathbf{N} < \mu + \mathbf{s} \cdot \mathbf{N}\},$$

where $\mathbf{N} \in \mathbb{Z}^3$ is the *normal vector* whose components (a, b, c) are relatively prime, $\mu \in \mathbb{Z}$ is the *intercept*, $\mathbf{s}$ is the *shift vector*, equal to $(\pm1, \pm1, \pm1)$ in the standard case.

By translation, we assume w.l.o.g. that $\mu = 0$. Moreover, by symmetry, we assume w.l.o.g. that the normal of the plane lies in the first octant, i.e. its components are positive. We also exclude cases where a component is null since then it falls back to a 2D algorithm. Thus, $a, b, c > 0$ and $\mathbf{s} = (1, 1, 1)$. Finally, we denote by $\omega := \mathbf{s} \cdot \mathbf{N} = a+b+c$ the *thickness* of the standard digital plane **P**.

The above definition of digital plane suggest to see the space as partitionned into layers of coplanar points, orthogonal to $\mathbf{N}$. The value $x \cdot \mathbf{N}$, called *height*, is a way of sorting these layers in direction $\mathbf{N}$. Points of height 0 (resp. $\omega-1$), which are extreme in **P**, are called *lower leaning points* (resp. *upper leaning points*). Points of height ω, the closest ones above **P**, are called *Bezout points*.

Algorithm 1 finds $\mathbf{N}$ given a predicate "is $x \in \mathbf{P}$?". The algorithm starts at any reentrant corner as follows: the *corner point* $\boldsymbol{p}$ is in **P**; the *shifted point* $\boldsymbol{q} := \boldsymbol{p}+\boldsymbol{s}$ is not in **P** since $\boldsymbol{s} \cdot \mathbf{N}$ is the thickness of **P**; the initial *triangle* $\mathbf{T}^{(0)} := (\boldsymbol{v}_k^{(0)})_{k \in \{0,1,2\}}$ is such that $\forall k, \boldsymbol{v}_k^{(0)} := \boldsymbol{p} + \boldsymbol{e}_k + \boldsymbol{e}_{k+1}$ (Fig. 1.c). It is easy to check that $\mathbf{T}^{(0)} \subset \mathbf{P}$ for any $\boldsymbol{p}$ such that $0 \leq \boldsymbol{p} \cdot \mathbf{N} < \min\{a, b, c\}$, which corresponds exactly to reentrant corners onto a standard digital plane. The algorithm then updates this initial triangle and iteratively aligns it with the plane by calling the above predicate for well-chosen points, until the solution equals $\mathbf{N}$.

At step $i \in \mathbb{N}$, the solution is described by a triangle denoted by $\mathbf{T}^{(i)}$. Algorithm 1 is designed so that $\forall i, \mathbf{T}^{(i)}$ is included in **P** and is intersected by segment $[\boldsymbol{pq}]$. Segment $[\boldsymbol{pq}]$ thus forces the exploration to be local.

Let k be an integer taken modulo 3, i.e. $k \in \mathbb{Z}/3\mathbb{Z}$. The three counterclockwise oriented vertices of $\mathbf{T}^{(i)}$ are denoted by $\boldsymbol{v}_k^{(i)}$ (Fig. 1.a). For sake of clarity,

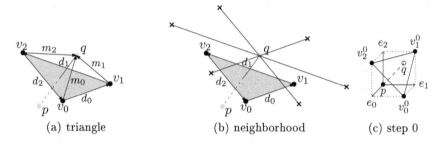

(a) triangle   (b) neighborhood   (c) step 0

Fig. 1. Main notations are illustrated in (a) and (b) (iteration number as exponent (i) is dropped for sake of clarity). The starting triangle is illustrated in (c).

we will write $\forall k$ instead of $\forall k \in \mathbb{Z}/3\mathbb{Z}$. We also introduce the following vectors (Fig. 1.a):

$$\forall k, \ \boldsymbol{m}_k^{(i)} := \boldsymbol{q} - \boldsymbol{v}_k^{(i)} \tag{1}$$

$$\forall k, \ \boldsymbol{d}_k^{(i)} := \boldsymbol{v}_{k+1}^{(i)} - \boldsymbol{v}_k^{(i)} = \boldsymbol{m}_k^{(i)} - \boldsymbol{m}_{k+1}^{(i)}. \tag{2}$$

The normal of $\mathbf{T}^{(i)}$, denoted by $\hat{\mathbf{N}}(\mathbf{T}^{(i)})$, is merely defined by the cross product between two consecutive edge vectors of $\mathbf{T}^{(i)}$, i.e. $\hat{\mathbf{N}}(\mathbf{T}^{(i)}) = \boldsymbol{d}_0^{(i)} \times \boldsymbol{d}_1^{(i)}$.

In order to improve the guess at step i, the algorithm checks if the points of a small neighborhood around $\boldsymbol{q}$, parallel and above $\mathbf{T}^{(i)}$, belong to $\mathbf{P}$ or not. This neighborhood is defined as follows (Fig. 1.b):

$$\Sigma_{\mathbf{T}^{(i)}} := \left\{ \boldsymbol{q} \pm \boldsymbol{d}_k^{(i)} \right\}_{k \ \in \{0,1,2\}}. \tag{3}$$

The algorithm then computes $\Sigma_{\mathbf{T}^{(i)}} \cap \mathbf{P}$ by making six calls to the predicate. In Algorithm 1, the new triangle $\mathbf{T}^{(i+1)}$ is simply defined as the other triangular facet of the convex hull of $\mathbf{T}^{(i)} \cup (\Sigma_{\mathbf{T}^{(i)}} \cap \mathbf{P})$ which intersects $[\boldsymbol{pq}]$. See Fig. 2 for an example. The algorithm stops when $\Sigma_{\mathbf{T}^{(n)}} \cap \mathbf{P}$ is empty. We show in Theorem 2 that, when $\boldsymbol{p}$ is a lower leaning point (its height is 0), the vertices of the last triangle $\mathbf{T}^{(n)}$ are upper leaning points of $\mathbf{P}$ (their height is $\omega - 1$). Corollary 1 implies that the normal to $\mathbf{T}^{(n)}$ is the normal to $\mathbf{P}$ and Corollary 2 implies that triangle edges form a basis of the lattice of upper leaning points to $\mathbf{P}$.

3 Validity and Complexity

In this section, we first prove the two invariants of Algorithm 1 and fully characterize its main operation (lines 3–5 of Algorithm 1). Then we prove that if $\boldsymbol{p}$ is a lower leaning point, Algorithm 1 retrieves the true normal $\mathbf{N}$ in less than $\omega - 3$ steps.

Let us assume that the algorithm always terminate in a finite number of steps whatever the starting point and let n be the last step (the proof is postponed, see Theorem 1).

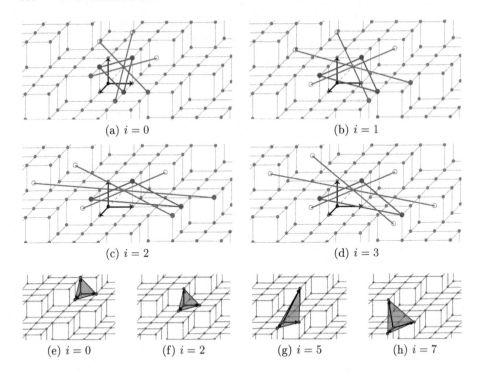

(a) $i = 0$ (b) $i = 1$

(c) $i = 2$ (d) $i = 3$

(e) $i = 0$ (f) $i = 2$ (g) $i = 5$ (h) $i = 7$

Fig. 2. Illustration of the running of Algorithm 1 and algorithm from [7] on a digital plane of normal vector $(1, 2, 5)$. Images (a) to (d) shows the four iterations of Algorithm 1 starting from the origin. For $i \in \{0, 1, 2, 3\}$, triangles $\mathbf{T}^{(i)}$ are in blue, whereas neighborhood points are depicted with red disks (resp. circles) if they belong (resp. do not belong) to $\mathbf{P}$. In (d), the normal of the last triangle is $(1, 2, 5)$. Images (e) to (h) shows iterations 0 (initial), 2, 5 and 7 (final) of the algorithm from [7]. The initial tetrahedron (a) is placed at the origin and the final one (h) has an upper triangle with normal vector $(1, 2, 5)$ (Color figure online).

3.1   Algorithm Invariants

The following two invariants are easy to prove by induction since by construction, each new triangle is chosen as a triangular facet of the convex hull of some subset of $\mathbf{P}$, which is intersected by $]\boldsymbol{pq}]$.

Invariant 1. $\forall i \in \{0, \ldots, n\}, \; \forall k, \; \boldsymbol{v}_k^{(i)} \in \mathbf{P}$.

Invariant 2. $\forall i \in \{0, \ldots, n\}$, the interior of $\mathbf{T}^{(i)}$ is intersected by $]\boldsymbol{pq}]$.

The only difficulty in Invariant 2 is to show that the triangle boundary is never intersected by $]\boldsymbol{pq}]$. Below, for lack of space, we only show by contradiction that $]\boldsymbol{pq}]$ does not intersect any edge of $\mathbf{T}^{(i+1)}$, if Invariant 2 is assumed to be true for $i \in \{0, \ldots, n-1\}$.

Proof. If $]pq]$ and some edge $[xy]$ of $\mathbf{T}^{(i+1)}$ intersect, points p, q, x and y are coplanar. But, since the boundary of $\mathbf{T}^{(i)}$ is not intersected by $]pq]$ and since points of $\Sigma_{\mathbf{T}^{(i)}}$ are located on lines that are parallel to the sides of $\mathbf{T}^{(i)}$, x and y must be opposite points in $\Sigma_{\mathbf{T}^{(i)}}$, i.e. $x := q + d_l^{(i)}$ for some $l \in \{0, 1, 2\}$ and $y = q - d_l^{(i)}$.

However, since $q \notin \mathbf{P}$, x and y cannot be both in $\mathbf{P}$ by linearity and thus cannot be the ends of an edge of $\mathbf{T}^{(i+1)}$, which is included in $\mathbf{P}$ by Invariant 1. □

3.2   Operation Characterization

The two following lemmas characterize the operation that transforms a triangle into the next one (lines 3–5 of Algorithm 1).

Lemma 1. $\forall i \in \{0, \ldots, n-1\}$, $\mathbf{T}^{(i)}$ and $\mathbf{T}^{(i+1)}$ *share exactly one or two vertices.*

Proof. Let γ be the number of common vertices between $\mathbf{T}^{(i)}$ and $\mathbf{T}^{(i+1)}$. Since $\mathbf{T}^{(i)}$ and $\mathbf{T}^{(i+1)}$ have three vertices each, $0 \leq \gamma \leq 3$, but we prove below that (i) $\gamma \neq 3$ and (ii) $\gamma \neq 0$.

Inequality (i) is trivial. Indeed, since the volume of the convex hull of $\mathbf{T}^{(i)} \cup (\Sigma_{\mathbf{T}^{(i)}} \cap \mathbf{P})$ is not empty for $i \in \{0, \ldots, n-1\}$ (see Algorithm 1, l. 2), we necessarily have $\mathbf{T}^{(i)} \neq \mathbf{T}^{(i+1)}$, which implies that $\gamma \neq 3$.

In order to prove inequality (ii), let $\mathcal{P}$ be the plane passing by the points of $\Sigma_{\mathbf{T}^{(i)}}$ (points of $\Sigma_{\mathbf{T}^{(i)}}$ are coplanar by (2) and (3)). Note that $q \in \mathcal{P}$ by (3).

Let us assume that $\mathbf{T}^{(i+1)}$ is included in $\mathcal{P}$. By invariant 1, $\mathbf{T}^{(i+1)} \subset \mathbf{P}$. However, $q \notin \mathbf{P}$ by definition of q. As a consequence, the upper leaning plane of $\mathbf{P}$, i.e. $\{x \in \mathbb{R}^3 | x \cdot \mathbf{N} = \omega - 1\}$, strictly separates $\mathbf{T}^{(i+1)}$ from q in the plane $\mathcal{P}$. In addition, p is located strictly below $\mathbf{T}^{(i)}$ by Invariant 2 and is thus also below $\mathcal{P}$, which is parallel and above $\mathbf{T}^{(i)}$ by definition. As a consequence, $\mathbf{T}^{(i+1)}$ is clearly not intersected by $]pq]$, which contradicts Invariant 2. We conclude that $\mathbf{T}^{(i+1)}$ is not included in $\mathcal{P}$, which means that $\gamma \neq 0$. □

The following lemma fully characterizes the main operation of Algorithm 1:

Lemma 2. $\forall i \in \{0, \ldots, n-1\}$, $\forall k$, $\begin{cases} either \ v_k^{(i+1)} = v_k^{(i)}, \\ or \ v_k^{(i+1)} = v_k^{(i)} + m_l^{(i)}, l \neq k. \end{cases}$

We know by Lemma 1 that $\mathbf{T}^{(i)}$ and $\mathbf{T}^{(i+1)}$ share one or two vertices. The proof is thus in two parts. In each part however, we use infinite cones of apex q to check whether Invariant 2 is true or not. Let us define $\forall i \in \{0, \ldots, n\}$, $\mathcal{T}_\infty^{(i)}$ as the set of infinite rays emanating from q and intersecting the triangular facet $\mathbf{T}^{(i)}$. Invariant 2 is equivalent to the following

Invariant 2': $\forall i \in \{0, \ldots, n\}$, the interior of $\mathcal{T}_\infty^{(i)}$ contains p.

Proof. We first assume that $\mathbf{T}^{(i)}$ and $\mathbf{T}^{(i+1)}$ share two vertices. Let us assume w.l.o.g. that the index of the vertex of $\mathbf{T}^{(i)}$ that is not in $\mathbf{T}^{(i+1)}$ is 0.

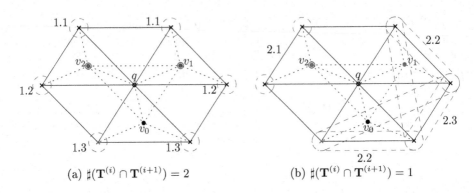

(a) $\sharp(\mathbf{T}^{(i)} \cap \mathbf{T}^{(i+1)}) = 2$ (b) $\sharp(\mathbf{T}^{(i)} \cap \mathbf{T}^{(i+1)}) = 1$

Fig. 3. Illustration of the proof of Lemma 2. Case where $\mathbf{T}^{(i)}$ and $\mathbf{T}^{(i+1)}$ share $v_1^{(i)}$ and $v_2^{(i)}$ in (a), but only $v_2^{(i)}$ in (b) (exponent (i) is omitted in the figures).

Since $v_0^{(i+1)} \in \Sigma_{\mathbf{T}^{(i)}} \cap \mathbf{P}$, there are six cases for $v_0^{(i+1)}$, which are grouped below two by two (see Fig. 3.a). Invariant 1 is true for all cases. We show however by contradition that the first four cases (items 1.1 and 1.2) are not possible because otherwise, Invariant 2' is not true:

1.1 Let us assume that $v_0^{(i+1)} = v_l^{(i)} + m_0^{(i)}, l \in \{1,2\}$. In these cases, $\mathcal{T}_\infty^{(i+1)}$ is adjacent to $\mathcal{T}_\infty^{(i)}$ along facet $(q, v_1^{(i)}, v_2^{(i)})$. We conclude that the interior of $\mathcal{T}_\infty^{(i+1)}$ and $\mathcal{T}_\infty^{(i)}$ are disjoint and cannot both contains p, which raises a contradiction.

1.2 If we assume now that $v_0^{(i+1)} = v_1^{(i)} + m_2^{(i)}$ or $v_0^{(i+1)} = v_2^{(i)} + m_1^{(i)}$. In these cases, $v_0^{(i+1)}$ lies on the plane passing by q, $v_1^{(i)}$ and $v_2^{(i)}$. We conclude that the interior of $\mathcal{T}_\infty^{(i+1)}$ is empty and cannot contains p, a contradiction.

1.3 Hence, $v_0^{(i+1)} = v_0^{(i)} + m_l^{(i)}, l \in \{1,2\}$. In this case, $\mathcal{T}_\infty^{(i+1)}$, which contains $\mathcal{T}_\infty^{(i)}$, contains p.

We now assume that $\mathbf{T}^{(i)}$ and $\mathbf{T}^{(i+1)}$ share one vertex. Let us assume w.l.o.g. that its index is 2.

By definition, $v_0^{(i+1)}, v_1^{(i+1)} \in \Sigma_{\mathbf{T}^{(i)}} \cap \mathbf{P}$. There are $\binom{6}{2} = 15$ cases to consider for $v_0^{(i+1)}$ and $v_1^{(i+1)}$, which are grouped below into four different items. See Fig. 3.b for the last three items. Invariant 1 is true for all cases. We show however that Invariant 2 is true for only the three cases described at item 2.3:

2.0 (3 cases) Let us assume that $v_0^{(i+1)}$ and $v_1^{(i+1)}$ are opposite with respect to q, i.e. one is equal to $q + d_k^{(i+1)}$, $k \in \{0,1,2\}$, whereas the other is equal to $q - d_k^{(i+1)}$. Since $q \notin \mathbf{P}$, $q + d_k^{(i+1)}$ and $q - d_k^{(i+1)}$ cannot be both in $\mathbf{P}$ by linearity, which raises a contradiction.

2.1 (7 cases) Let us assume that $v_0^{(i+1)} = v_2^{(i)} + m_l^{(i)}, l \in \{1,2\}$ or $v_1^{(i+1)} = v_2^{(i)} + m_l^{(i)}, l \in \{1,2\}$. In all cases, $\mathcal{T}_\infty^{(i+1)}$ is adjacent to $\mathcal{T}_\infty^{(i)}$ by edge $(q, v_2^{(i)})$, by facet $(q, v_0^{(i)}, v_2^{(i)})$ or by facet $(q, v_1^{(i)}, v_2^{(i)})$. As a consequence, the interior of $\mathcal{T}_\infty^{(i+1)}$ cannot contains p.

2.2 (2 cases) Let us assume that $v_0^{(i+1)} = v_l^{(i)} + m_{l+1}^{(i)}$ and $v_1^{(i+1)} = v_l^{(i)} + m_{l+2}^{(i)}$, for $l \in \{0,1\}$. In both cases, $\mathcal{T}_\infty^{(i+1)}$ is adjacent to $\mathcal{T}_\infty^{(i)}$ by edge $(q, v_l^{(i)})$. As a consequence, the previous argument also applies.

2.3 (3 cases) $\mathcal{T}_\infty^{(i+1)}$ contains $\mathcal{T}_\infty^{(i)}$ and thus contains p in the following cases: $v_0^{(i+1)} = v_0^{(i)} + m_{l_0}^{(i)}, l_0 \in \{1,2\}$ and $v_1^{(i+1)} = v_1^{(i)} + m_{l_1}^{(i)}, l_1 \in \{0,2\}$, with $(l_0, l_1) \neq (1,0)$. □

Lemma 3. *Let $\mathbf{M}^{(i)}$ be the 3×3 matrix that consists of the three row vectors $(m_k^{(i)})_k \in \{0,1,2\}$. Then, $\forall i \in \{0, \ldots, n\}$, $\det(\mathbf{M}^{(i)}) = 1$.*

Proof. We can easily check that $\det(\mathbf{M}^{(0)}) = 1$.

We now prove that if $\det(\mathbf{M}^{(i)}) = 1$ for $i \in \{0, \ldots, n-1\}$, then $\det(\mathbf{M}^{(i+1)}) = 1$. By Lemmas 1 and 2, we have $v_k^{(i+1)} = v_k^{(i)} + m_l^{(i)} \Leftrightarrow m_k^{(i+1)} = m_k^{(i)} - m_l^{(i)}, l \neq k$ for at most two rows over three, while the remaining one or two rows correspond to identity. Such matrix operations do not change the determinant, which concludes. □

3.3 Termination

In the following proofs, we compare the position of the points along direction $\mathbf{N}$. For the sake of simplicity, we use the bar notation $\bar{\cdot}$ above any vector x to denote its height relative to $\mathbf{N}$. Otherwise said, $\bar{x} := x \cdot \mathbf{N}$. Even if $\mathbf{N}$ is not known, $\bar{q} \geq \omega$ by definition and for all $x \in \mathbf{P}$, $0 \leq \bar{x} < \omega$.

Theorem 1. *The number of steps in Algorithm 1 is bounded from above by $\omega - 3$.*

Proof. First, it is easy to see that $\forall k, m_k^{(0)} = e_{k+2}$. Thus, $\sum_k \overline{m}_k^{(0)} = \omega$.

Let us recall that $\forall i \in \{0, \ldots, n\}, \forall k, m_k^{(i)} = q - v_k^{(i)}$. We have $\bar{q} \geq \omega$ by definition of q and $\forall i \in \{0, \ldots, n\}, \forall k, 0 \leq \overline{v}_k^{(i)} < \omega$ by Invariant 1 and by definition of $\mathbf{P}$. As a consequence, $\forall i \in \{0, \ldots, n\}, \forall k, \overline{m}_k^{(i)} > 0$ and $\sum_k \overline{m}_k^{(i)} \geq 3$.

Moverover, by Lemma 2 and since any $\overline{m}_k^{(i)}$ is strictly positive $\forall i \in \{0, \ldots, n\}$, we clearly have $\forall i \in \{0, \ldots, n-1\}, \sum_k \overline{m}_k^{(i)} > \sum_k \overline{m}_k^{(i+1)}$.

The sequence $(\sum_k \overline{m}_k^{(i)})_{i=0,\ldots,n}$ is thus a strictly decreasing sequence of integers between ω and 3. □

Remark that this bound is reached when running the algorithm on a plane with normal $\mathbf{N}(1,1,r)$. We now give the main result when the starting point p is a lower leaning point, i.e. $\bar{p} = 0$. Note that in this case $\bar{q} = \omega$. Since we focus below on the last step n, we omit the exponent (n) in the proofs to improve their readability.

Theorem 2. *If p is a lower leaning point (i.e. $\bar{p} = 0$ and thus $\bar{q} = \omega$), the vertices of the last triangle are upper leaning points, i.e. $\forall k, \overline{v}_k^{(n)} = \omega - 1$.*

Proof. If there exists $k \in \{0, 1, 2\}$ such that $\overline{d}_k \neq 0$, then either (i) $\overline{d}_k < 0$ or (ii) $\overline{d}_k > 0$. Since $\overline{q} = \omega$ and $|\overline{d}_k| < \omega$, either (i) $q + d_k \in \mathbf{P}$ or (ii) $q - d_k \in \mathbf{P}$. This implies that $\Sigma_{\mathbf{T}} \cap \mathbf{P} \neq \emptyset$, which is a contradiction because $\Sigma_{\mathbf{T}} \cap \mathbf{P} = \emptyset$ at the last step (see Algorithm 1, l. 2). As a consequence, $\forall k, \overline{d}_k = 0$ and $\forall k, \overline{m}_k = \alpha$, a strictly positive integer.

Let us denote by $\mathbf{1}$ the vector $(1, 1, 1)^T$. We can write the last system as $\mathbf{M}\mathbf{N} = \alpha\mathbf{1}$. Since $\mathbf{M}$ is invertible (because $\det(\mathbf{M}) = 1$ by Lemma 3), $\mathbf{N} = \mathbf{M}^{-1}\alpha\mathbf{1}$ and as a consequence $\alpha = 1$ (because the components of $\mathbf{N}$ are relatively prime and $\mathbf{M}^{-1}$ is unimodular).

We conclude that $\forall k, \overline{m}_k = 1$ and, straightforwardly, $\overline{v}_k = \omega - 1$. $\square$

The following two corollaries are easily derived from Lemma 3 and Theorem 2.

Corollary 1. *If p is a lower leaning point, the normal of the last triangle is equal to $\mathbf{N}$, i.e. $\hat{\mathbf{N}}(\mathbf{T}^{(n)}) = \mathbf{N}$.*

Corollary 2. *If p is a lower leaning point, $(d_0^{(n)}, d_1^{(n)})$ is a basis of the lattice of upper leaning points $\{x \in \mathbf{P} | \overline{x} = \omega - 1\}$.*

4   Lattice Reduction and Delaunay Condition

Corollary 2 implies that the edge vectors of the last triangle form a basis of the lattice of upper leaning points to $\mathbf{P}$.

Let us recall that a basis (x, y) is reduced if and only if $\|x\|, \|y\| \leq \|x - y\| \leq \|x + y\|$. Given (x, y) a basis of a two dimensional lattice, there is an iterative algorithm to compute a reduced basis from (x, y). This algorithm consists in replacing the longest vector among (x, y) by the shortest one among $x + y$ and $x - y$, if it is smaller.

We have found many examples for which the basis returned by Algorithm 1 is not reduced, even if we take the two shortest vectors in $\{d_k^{(n)}\}_{k \in \{0,1,2\}}$ (see Fig. 4.a for an example). To cope with this problem, we can either apply the above reduction algorithm to the returned basis, or keep triangles as "small" as possible so that the last triangle leads to a reduced basis, without any extra reduction steps. To achieve this aim, we use a variant of Algorithm 1; see Algorithm 2.

Note that in Algorithm 2, we compute the convex hull of $T \cup S^\star$, but since $S^\star \subset \Sigma_{\mathbf{T}} \cap \mathbf{P}$ and since $\Sigma_{\mathbf{T}} \cap \mathbf{P} \neq \emptyset \Rightarrow S^\star \neq \emptyset$, all the results of Sect. 3 remain valid.

In Fig. 4, the results of the two algorithms for the digital plane of normal $(2, 3, 9)$ are compared. The triangles returned by Algorithm 2 are more compact and closer to segment $[pq]$. The last triangle returned by Algorithm 2 leads to a reduced basis, but this is not true for Algorithm 1.

In practice, Algorithm 2 drastically reduces the cases of non-reduced basis. We have compared the two variants for normal vectors ranging from $(1, 1, 1)$ to $(200, 200, 200)$. There are 6578833 vectors with

Algorithm	Algorithm 1	Algorithm 2
avg. nb. steps	21.8	20.9
nb. non-reduced	4803115 (73 %)	924 (0.01 %)
avg. nb. reductions	5.5	1

Algorithm 2. Variant of Algorithm 1 based on a Delaunay condition.

Input: a predicate "$x \in \mathbf{P}$?", an initial triangle $\mathbf{T}^{(0)}$

1 $\mathbf{T} \leftarrow \mathbf{T}^{(0)}$;

2 **while** $\Sigma_{\mathbf{T}} \cap \mathbf{P} \neq \emptyset$ **do**

3   Compute the set S^* of points $s^* \in (\Sigma_{\mathbf{T}} \cap \mathbf{P})$ such that the circumsphere of $\mathbf{T} \cup \{s^*\}$ does not include any point $s \in (\Sigma_{\mathbf{T}} \cap \mathbf{P})$ in its interior ;

4   Compute the convex hull of $\mathbf{T} \cup S^*$;

5   Find $\mathbf{T}'$, defined as the upper triangular facet intersected by $[\boldsymbol{pq}]$;

6     $\mathbf{T} \leftarrow \mathbf{T}'$;

7 **return** T;

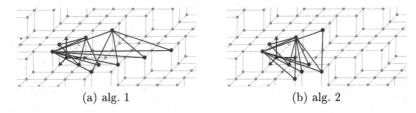

(a) alg. 1   (b) alg. 2

Fig. 4. Triangles returned by Algorithm 1 in (a) and Algorithm 2 in (b) for a digital plane with normal $(2, 3, 9)$ (the corner point $\boldsymbol{p}$ is set to the origin). In (b), triangles returned by Algorithm 2 are smaller and the last one defines a reduced basis.

relatively prime components in this range. It turns out that less than $0.01\,\%$ of the basis returned by Algorithm 2 are non-reduced against about $73\,\%$ for Algorithm 1.

5   Applications to Digital Surfaces

In this section, we consider a set of voxels, Z, where voxels are seen as unit cubes whose vertices are in $\mathbb{Z}^3$. The *digital boundary,* $\mathrm{Bd}Z$, is defined as the topological boundary of Z. Since a digital boundary looks locally like a digital plane, it is natural to run our algorithm at each reentrant corner of the digital boundary with predicate "Is $\boldsymbol{x}$ in $\mathrm{Bd}Z$" in order to estimate the local normal vector to the volume Z (see Fig. 5).

Although we do not go into details here due to space limitations, we can mention the following points:

- Our algorithm works well on digitally convex shapes Z. The set of points $\boldsymbol{x} \in \mathrm{Bd}Z$ located below some facet $\mathcal{F}$ of $\mathrm{Conv}(\mathrm{Bd}Z)$ whose normal vector is in the first octant and such that points $\boldsymbol{x} + \boldsymbol{s}$ are strictly above $\mathcal{F}$, is a piece of a digital plane, $\mathbf{P}$. If the starting reentrant corner is a lower leaning point of $\mathbf{P}$, the last triangle computed by our algorithm is included in $\mathcal{F}$ (see Fig. 5). We call such triangles *patterns* of $\mathbf{P}$.

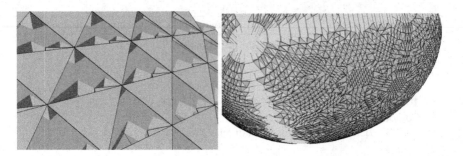

Fig. 5. Our algorithm has been runned at each reentrant corner of a digital plane (left) and an ellipse (right). The last triangle of each run is printed.

- When starting from a corner that is not a lower leaning point of a facet of $\mathrm{Conv}(\mathrm{Bd}Z)$, the algorithm returns a triangle called *reduced pattern* of **P** which is approximately aligned with the plane.
- It is possible to detect and remove reduced patterns with a simple algorithm. Put in queue every point q of all computed triangles, together with their three vectors m_k. Loop on the queue while it is not empty by popping them in sequence. For each popped point q and vectors m_k, check if q translated by any of m_k is a shifted point q' of another triangle. In this case, mark q' as reduced pattern and put it back in queue but now with the vectors m_k. Otherwise, do nothing. When the algorithm stops, since reduced patterns have shifted points above the shifted point of the lower leaning point, all reduced patterns are marked.
- If the shape is not convex, the algorithm can be adapted to detect planarity defects. The idea is that $\Sigma_{\mathbf{T}} \cap \mathrm{Bd}Z$ should contain at most three elements if it is locally a piece of plane. Moreover these elements must always be consecutive neighbors around q in $\Sigma_{\mathbf{T}}$. We thus stop the algorithm whenever at least one of the two previous conditions fails, meaning that the surface is locally non convex. We also check that each new triangle is separating onto the digital boundary: for any point $x \in \mathrm{Bd}Z$ below a triangle of shift s, the point $x + s$ must be above the triangle.

6   Conclusion and Perspectives

In this paper, we proposed a new algorithm that computes the parameters of a digital plane. In opposition to usual plane recognition algorithms, this algorithm greedily decides on-the-fly which points have to be considered like in [7]. Compared to [7], this algorithm is however simpler because it consists in iterating one geometrical operation, which is fully characterized in Lemma 2. In addition, the returned solution, which is described by a triangle parallel to the sought plane, is close to the starting point. On one hand, the starting point must projects onto the triangle and on the other hand, the two shortest edge vectors of the triangle almost always form a reduced basis (in the second version of our algorithm).

For the future, we would like to find a theoretically sound way of always getting a reduced basis, without any extra reduction operation. Moreover, we would like to find a variant of our algorithm in order to retrieve complement triangles whose Bezout point is not above the triangle. For sake of completeness, we are also interested in degenerate cases, where at least one component is null. After having achieved these goals, we would have a complete working tool for the analysis of digitally convex surfaces. In order to correctly process concave or saddle parts however, we must precisely associate to a given triangle a piece of digital plane containing the triangle vertices and having the same normal vector than the triangle one. Although there are many such pieces of digital plane for a given triangle, we hope that this work will provide a canonical hierarchy of pieces of digital plane.

References

1. Charrier, E., Buzer, L.: An efficient and quasi linear worst-case time algorithm for digital plane recognition. In: Coeurjolly, D., Sivignon, I., Tougne, L., Dupont, F. (eds.) DGCI 2008. LNCS, vol. 4992, pp. 346–357. Springer, Heidelberg (2008)
2. Charrier, E., Lachaud, J.-O.: Maximal planes and multiscale tangential cover of 3D digital objects. In: Aggarwal, J.K., Barneva, R.P., Brimkov, V.E., Koroutchev, K.N., Korutcheva, E.R. (eds.) IWCIA 2011. LNCS, vol. 6636, pp. 132–143. Springer, Heidelberg (2011)
3. de Vieilleville, F., Lachaud, J.-O., Feschet, F.: Maximal digital straight segments and convergence of discrete geometric estimators. J. Math. Image Vis. 27(2), 471–502 (2007)
4. Doerksen-Reiter, H., Debled-Rennesson, I.: Convex and concave parts of digital curves. In: Klette, R., Kozera, R., Noakes, L., Weickert, J. (eds.) Geometric Properties for Incomplete Data. Computational Imaging and Vision, vol. 31, pp. 145–160. Springer, The Netherlands (2006)
5. Gérard, Y., Debled-Rennesson, I., Zimmermann, P.: An elementary digital plane recognition algorithm. Discrete Appl. Math. 151(1), 169–183 (2005)
6. Kim, C.E., Stojmenović, I.: On the recognition of digital planes in three-dimensional space. Pattern Recogn. Lett. 12(11), 665–669 (1991)
7. Lachaud, J.-O., Provençal, X., Roussillon, T.: An output-sensitive algorithm to compute the normal vector of a digital plane. Theoretical Computer Science (2015, to appear)
8. Veelaert, P.: Digital planarity of rectangular surface segments. IEEE Trans. Pattern Anal. Mach. Intell. 16(6), 647–652 (1994)

Finding Shortest Triangular Path in a Digital Object

Apurba Sarkar[1]([✉]), Arindam Biswas[2], Shouvick Mondal[1],
and Mousumi Dutt[3]

[1] Department of Computer Science and Technology,
Indian Institute of Engineering Science and Technology, Shibpur, India
as.besu@gmail.com, shouvick.mondal.cemk@gmail.com
[2] Department of Information Technology,
Indian Institute of Engineering Science and Technology, Shibpur, India
barindam@gmail.com
[3] Department of Computer Science and Engineering,
International Institute of Information Technology, Naya Raipur, India
duttmousumi@gmail.com

Abstract. A combinatorial algorithm to find a shortest triangular path (STP) between two points inside a digital object imposed on triangular grid is designed having $O(\frac{n}{g} \log \frac{n}{g})$ time complexity, n being the number of pixels on the contour of the object and g being the grid size. First the inner triangular cover of the given digital object is constructed which maximally inscribes the object. Certain combinatorial rules are formulated based on the properties of triangular grid and are applied whenever necessary to obtain a shortest triangular path, where the path lies entirely in the digital object and moves only along the grid edges. The length of STP and number of monotonicity may be two useful parameters to determine shape complexity of the object. Experimental results show the effectiveness of the algorithm.

Keywords: Shortest path · Shortest triangular path · Monotone path · Shape analysis · Shape complexity · Digital geometry

1 Introduction

The shortest path problem is a well-studied problem in graphs (directed and undirected). It enquires the shortest path between two vertices in a graph such that the sum of the weights of its constituent edges is minimized. The weights of the edges may vary depending on the problem being studied. Shortest path problem is a broadly useful problem solving model in robot navigation, texture mapping, typesetting in TeX, urban traffic planning, optimal pipe-lining of VLSI chip, subroutine in advanced algorithms, telemarketer operator scheduling, routing of telecommunications messages, approximating piecewise linear functions, network routing protocols, and optimal truck routing through given traffic congestion pattern [1]. The complete history of shortest path problem can

© Springer International Publishing Switzerland 2016
N. Normand et al. (Eds.): DGCI 2016, LNCS 9647, pp. 206–218, 2016.
DOI: 10.1007/978-3-319-32360-2_16

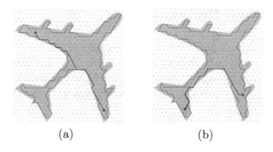

(a)   (b)

Fig. 1. A digital object with inner cover (blue), and shortest paths (a) in (red, green) and (b) in (red, green, purple) (Color figure online)

be found in [13]. The first reported algorithm on shortest path is by Shimbel in 1954 [14], where he had reported few observations on calculating distance and proposed a method which later became known as the Bellman-Ford method.

In 1958, Bellman proposed a dynamic programming based approach for solving shortest path problem in [3] which runs in $O(n^3)$ time, where n is the number of vertices in the graph. In 1957, Moore proposed another algorithm on shortest paths [11]. In 1959, Dijkstra presented a simpler algorithm which runs in $O(n^2)$ time in [5]. Most of the works on shortest paths reported so far mainly find paths between two vertices in a graph. However, we are proposing a combinatorial algorithm on shortest paths which finds a path between two points inside a digital object in triangular grid. M. Dutt et al. [6,7] proposed a combinatorial algorithm to find a shortest isothetic path between two grid points inside a digital object without any holes. In [12], B. Nagy analysed some properties of hexagonal and triangular grid (considering cell model), where distance based neighborhood sequence is defined and an algorithm to calculate shortest distance between two arbitrary points is proposed. In [2], chain-code representation in triangular grid is discussed. There exists another work on finding shortest distance on the hexagonal grid [10], based on the distance function and the neighboring relations. A shortest path in triangular grid is a series of digital straight lines [8].

This paper focuses on finding a shortest path between two grid points inside a digital object imposed on a background triangular grid. First, the inner triangular cover of the given object is computed using the algorithm presented in [4]. This is done to ensure that the computed path does not go outside the inner cover and hence outside the object. An appropriate parallelogram is considered keeping the points at diagonally opposite corners. Combinatorial rules are applied on the intersection points generated on intersection of the inner cover with the parallelogram. Two triangular shortest paths are shown in Fig. 1(a) and (b).

The rest of the paper is organized as follows. All the required definitions and preliminaries are presented in Sect. 2. The method to obtain the shortest path is elaborated in Sect. 3. Estimation of running time of the proposed algorithm is explained in Sect. 4. Section 5 presents the experimental results with analysis and the conclusion is presented in Sect. 6.

2   Definitions and Preliminaries

A *digital object* (henceforth referred as an object A) is a finite subset of $\mathbb{Z}^2$, which consists of one or more 8-connected components. In this paper, a connected hole-free object is considered. A *triangular grid* (henceforth simply referred as grid) $\mathbb{T} := (\mathbb{L}_{60}, \mathbb{L}_0, \mathbb{L}_{120})$ consists of three sets of parallel grid lines, which are inclined at $60°$, $0°$, and $120°$ (w.l.o.g) w.r.t. x-axis [9].The grid lines in $\mathbb{L}_{60}$, $\mathbb{L}_0$, $\mathbb{L}_{120}$ correspond to three distinct coordinates, namely α, β, γ. Three grid lines, one each from $\mathbb{L}_{60}$, $\mathbb{L}_0$, $\mathbb{L}_{120}$, intersect at a (real) grid point. The distance between two consecutive grid points along a grid line is termed as *grid size, g*. A line segment of length g connecting two consecutive grid points on a grid line is called *grid edge*. The smallest-area triangle formed by three grid edges, one each from $\mathbb{L}_{60}$, $\mathbb{L}_0$, and $\mathbb{L}_{120}$, is called *unit grid triangle* (UGT). For a given grid point, p, there are six neighboring UGTs, given by $\{T_i : i = 0, 1, \ldots, 5\}$ as shown in Fig. 2. A portion of the triangular grid is shown in Fig. 2 along with direction codes. It has six distinct regions called sextants, each of which is well-defined by two rays starting from $(0, 0, 0)$. For example, Sextant 1 is defined by the region $\alpha_+ \cap \beta_+$, Sextant 2 is defined by the region $\alpha_- \cap \gamma_-$, and so on.

The *triangular distance* (d_t) between two points $p(\alpha_p, \beta_p, \gamma_p)$ and $q(\alpha_q, \beta_q, \gamma_q)$ is defined by $d_t(p, q) = max(|\alpha_p\ \alpha_q|, |\beta_p\ \beta_q|, |\gamma_p\ \gamma_q|)$.

The 6-neighborhood of a point (α, β, γ) is given by $N_6(\alpha, \beta, \gamma) = \{(\alpha', \beta', \gamma') : max(|\alpha - \alpha'|, |\beta - \beta'|, |\gamma - \gamma'|) = 1\}$.

A (finite) polygon P imposed on the grid $\mathbb{T}$ is termed as a *triangular polygon* if its sides are collinear with lines in $\mathbb{L}_{60}$, $\mathbb{L}_0$, and $\mathbb{L}_{120}$. It consists of a set of UGTs, and is represented by the (ordered) sequence of its vertices, which are grid points. Its interior is defined as the set of points with integer coordinates lying inside it. An *inner triangular polygon* (or simply *inner polygon*) tightly inscribes

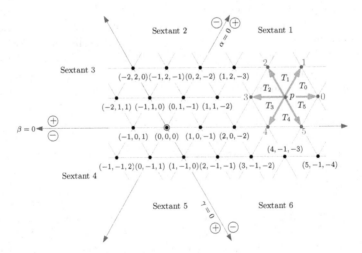

Fig. 2. Portion of a triangular canvas, the UGTs $\{T_0, T_1, \ldots, T_5\}$ incident at a grid point p, and the direction codes $\{0, 1, \ldots, 5\}$ of neighboring grid points of p.

A such that its border is a subset of A and the number of its constituting UGTs is maximum. An *inner triangular hole polygon* (or *inner hole polygon*) tightly circumscribes a hole and its border is a subset of A. The *inner triangular cover* (ITC), $\underline{P}$, is the set of inner polygons and inner hole polygons, such that the region given by the union of the inner polygons minus the union of the interiors of the inner hole polygons, contains a UGT if and only if it is a subset of A. In this paper, ITC containing one inner polygon is considered.

A (simple) *triangular path* π from a grid point p to a grid point q is a sequence of n distinct points $p_1, p_2, \ldots, p_n$ with $p_1 = p$ and $p_n = q$ such that $p_i \in N_6(p_{i+1})$, for $1 < i < n$. The length of a given triangular path is the sum of distances traversed along each axis. A triangular path π is said to be shortest if it is of minimum length. A path π in triangular grid is *monotone* if it consists of only one direction or two consecutive directions. Otherwise π is said to be non-monotone. In Fig. 1(a), the triangular path contains two monotone sub-paths whereas in Fig. 1(b), the triangular path contains three monotone sub-paths.

Deriving the Inner Triangular Cover (ITC): The inner triangular cover of A, $\underline{P}$, is constructed using the same method as outer triangular cover as explained in [4], but in the reverse manner. A grid point q is classified as a vertex of the inner cover, if and only if at least one (and at most five) of the six $UGTs$ incident at q is fully occupied by the object A i.e., $T_i^q \cap A = T_i^q$ where $i \in \{0, 1, 2, 3, 4, 5\}$. The *object occupancy vector*, $A_q = \langle a_0 a_1 \ldots a_5 \rangle$, where $a_i = 1$ if T_i^q is fully occupied else it is 0, is used to determine the type of the vertex. Let k denote the number of fully occupied $UGTs$, then for $k = 0$: q is an exterior point of $\underline{P}$, $k = 6$: interior point, $k = 1$: a 60^0 vertex (included angle is 60^0), and $k = 5$: 300^0 vertex. For other cases, $k = 2, 3, 4$, **Type** of q is derived based on the incoming (d) and outgoing direction (d') at q. If the incoming direction is d, then $a_j = 1$ and $a_{(j+1) \bmod 6} = 1$ where $j = (d + 2) \bmod 6$. Now, j is incremented until the next 1-bit in A_q, say at j', $a_{j'} = 1$, then the outgoing direction, $d' = j'$. A **Type 3** vertex is considered as edge point. **Type 1, 2** vertices are considered as convex vertices and **Type 4, 5** vertices as concave vertices. The construction of $\underline{P}$ keeps A' (background) to the right during the traversal. The polygon is traced to the next grid point q_n, type of q_n is determined and the direction of traversal from q_n is computed. The traversal continues until the start vertex, v_s is reached. During the construction of $\underline{P}$, 4 lists are maintained L, L_α, L_β, and L_γ, where, L is a doubly linked list of vertices (corner points) of $\underline{P}$ and L_α, L_β, and L_γ simultaneously contain vertices as well as edge points of $\underline{P}$ in lexicographically sorted order with their respective primary and secondary keys. The primary key for L_α is α and secondary key is β, similarly the primary and secondary keys for L_β and L_γ can be defined. An index (in increasing order) is assigned to each vertex of $\underline{P}$ in order of their occurrence in $\underline{P}$.

3   Finding Shortest Path

To find a shortest path between two points p and q, an appropriate parallelogram, B, is constructed keeping p and q at diagonally opposite corners and then a traversal is made along one of the semi-perimeters. Throughout this work, the object is assumed to lie in sextant 5 and 6, point p is assumed to be above q and the left semi-perimeter is traversed to find shortest path. Three different types of parallelogram are considered depending on the positions of point $q(\alpha_q, \beta_q, \gamma_q)$. Regions are separated by red lines as shown in Fig. 3 and are determined as follows, q is in *Region 1* if $\beta_q < \beta_p$ and $\alpha_q < \alpha_p$; q is in *Region 2* if $\alpha_q > \alpha_p$ and $\gamma_q > \gamma_p$ and finally q is in *Region 3* if $\beta_q < \beta_p$ and $\gamma_q < \gamma_p$. If q lies on the region separator, then there will be only one shortest path (as object does not have holes) and determining the shortest path is straight-forward. The reason behind the construction of parallelogram is that the semi-perimeters are the shortest distance between the two points (if semi-perimeters lie completely within the object). So, to construct a shortest path the traversal is made along one of the semi-perimeters. During this traversal if the semi-perimeter does not lie completely inside the object, intersection points between the semi-perimeter and the inner cover of the object are determined. The traversal is then guided through those intersection points possibly applying the reduction rules to shorten the path in such a way that the path lies inside the object. It is to be noted that not all intersection points will be important to guide the traversal and are eliminated using few combinatorial rules as explained in Sect. 3.1. The traversal continues this way applying the reduction rules whenever necessary until it reaches q. The reduction rules are explained in Sect. 3.2.

3.1   Finding Intersection Points

W.l.o.g, let p be always above q in the bounding parallelogram B and c_1, q, c_2 are the vertices in order, to the left of p in anti-clockwise direction. Then the left semi-perimeter of B is defined by $\overline{pc_1}$ and $\overline{c_1 q}$ (Fig. 3). The procedure CONTROL-POINTS in (Algorithm 1) finds out the points which guide the traversal via the semi-perimeter. If the semi-perimeter $\overline{pc_1}, \overline{c_1 q}$ lies entirely within the object, then the semi-perimeter itself will be a shortest path. Otherwise, the intersection points of $\overline{pc_1}$ with the inner cover of the object are found out by searching $L_x, x \in \{\alpha, \beta, \gamma\}$ depending on the orientation of $\overline{pc_1}$ and stored in M_1 and those of $\overline{c_1 q}$ stored in M_2. If an intersection point w_i lies on the edge $v_j v_{j+1}$ of $\underline{P}$, then its index is set to $j + 0.5$ to maintain the order that it appears after v_j. The lists are further examined and some of the points are removed as they will not be important to find the shortest path. The points in

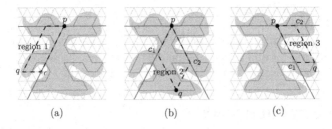

(a)   (b)   (c)

Fig. 3. Three regions and corresponding orientations of bounding parallelogram (a, b, c) (Color figure online)

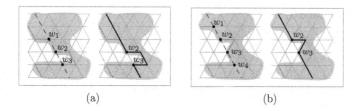

(a)   (b)

Fig. 4. Illustration of removal of unimportant intersection points. Dashed line represents one side of the parallelogram. (a) A concave vertex, w_1, is followed by a convex vertex w_2, w_1 is removed because it has lower index. (b) w_4, a concave vertex with higher index, is removed as it is preceded by a convex vertex, w_3.

two appropriate lists among $M_\alpha, M_\beta, M_\gamma$ are considered in pairs and if a convex (concave) vertex is followed by a concave (convex) vertex, then the vertex with greater (lower) index is discarded using REMOVE-POINTS (Steps 1 and 2) (as shown in Algorithm 1). In Steps 3–9, the final list of intersection points M, is formed by concatenating p, M_1, c_1 (if it is inside A), M_2 and q. In steps 16–22, the indices of the pairs of intersection points are checked to find whether they are in increasing or decreasing order and whether the index falls within indices of the extreme two points in M, namely $M[1]$ and $M[k+k']$, where k is the total number of intersection points. The value of k' indicates whether c_1 has been included in $M(k' = 1(\text{Step } 5))$ or not $(k' = 0$ (Step 8)) (Fig. 4).

If the test in Step 16 succeeds, then three consecutive intersection points in M are removed when c_1 is the second next point from $M[i]$ (Step 18); else next two consecutive points are removed (Step 20). Finally M is the required list. Let $M = \langle p, w_1, w_2, \ldots, w_k, q \rangle$. When the left semi-perimeter of B lies inside $\underline{P}$, (semi-perimeter of)B is traversed; otherwise, $\underline{P}$. So pw_1 is traversed along B, then w_1w_2 along $\underline{P}$, next w_2w_3 along B, again w_3w_4 along $\underline{P}$ and so on. Such an alternate traversal is made possible by reordering the vertices in Steps 16–22 if the index ordering does not hold.

Algorithm 1. CONTROL-POINTS

```
Input: M₁, M₂, p, q, c₁                11  while M[i] ≠ q do
Output: M                               12     if M[i] = c₁ then
1  REMOVE-POINTS(M₁);                   13        i ← i + 1;
2  REMOVE-POINTS(M₂);                   14        continue
3  if c₁ ∈ A then                       15     else
      /* c₁ is the corner point on left 16        while ∼ (((index[M[i]] <
         semi-perimeter            */                index[M[i + 1]]) ∧ (index[M[i + 1]] ≤
4     M ← CONCAT(p, M₁, c₁, M₂, q);                  index[M[k + k′]]))∨ ((index[M[i]] >
5     k′ ← 1;                                        index[M[i + 1]]) ∧ (index[M[i + 1]] ≥
6  else                                              index[M[k + k′]]))) do
7     M ← CONCAT(p, M₁, M₂, q);         17           if M[i + 2] = c₁ then
8     k′ ← 0;                           18              DELETE(M[i + 1], M[i +
9  end                                                  2], M[i + 3])
10 i ← 1;                               19           else
                                        20              DELETE(M[i + 1], M[i + 2])
                                        21           end
                                        22        end
                                        23     end
                                        24     i ← i + 2;
                                        25  end
                                        26  return M
```

3.2   Reduction Rules

While traversing from p via the semi-perimeter of B, the intersection points in M are also encountered to reach q. This path may include convexities which are to be removed to shorten the path. The rules are discussed here. The convexities are detected when the turn at a vertex or the sum of the turns at two consecutive vertices is equal to or greater than $120°$. A clockwise (anticlockwise) change in direction at a vertex is considered as a positive (negative) turn by the corresponding angle. All possible cases are depicted in Fig. 5. A **Type 1** vertex makes a turn of $120°$ so it is a convex vertex. Similarly turn at two consecutive vertices of types **22**, **21** creates a turn of $120°$ or more and hence create convexities. Pattern **12** and **11** also create convexity.

It is to be noted that although a **Type 2** vertex is treated as a convex vertex, unlike **Type 1** vertex, it alone cannot create a convexity. The proposed algorithm maintains with each vertex, its **Type** (t), length (l), and the outgoing direction (d). Removal of convexity sometimes requires removal of some or all vertices that are involved in the convexity and the deletion of vertex needs adjustment of those information with vertex that precedes or follows convexity. If the convexity is created by a **Type 1** vertex, four consecutive vertices, $v_0 v_1 v_2 v_3$, where v_2 is the vertex of **Type 1** and v_3 is the most recently visited vertex, are considered to apply the rule to remove convexity. On the other hand, if the convexity is created by two consecutive convex vertices (of **Type 22** or **21**) then five consecutive vertices $v_0 v_1 v_2 v_3 v_4$, where v_1 and v_2 are convex vertices and v_4 is the most recently visited vertex; are considered to apply the rule to remove convexity. The type of the start (p) and end (q) vertices are set to **6** since in general the path is found between two points that lie inside the cover. The rules are explained as follows.

Pattern $< t_1 1 t_3 >$ $t_1, t_3 \in \{4, 5, 6\}$

This pattern implies a convex region created by a single **Type 1** vertex and it is preceded or followed by concave vertices. We consider four most recently traversed vertices, $v_0(t_0, l_0) v_1(t_1, l_1) v_2(t_2, l_2) v_3(t_3, l_3)$, v_3 being the most recent. Depending on the lengths l_1 and l_2, three rules are as follows.

R11 $(l_1 < l_2)$: $\langle v_0(t_0, l_0) v_1(t_1, l_1) v_2(t_2, l_2) v_3(t_3, l_3) \rangle \rightarrow$
$\langle v_0(t_0, l_0) v_1(t_1 - 1, l_1) v_2(t_2 + 1, l_2 - l_1) v_3(t_3, l_3) \rangle$
R12 $(l_1 = l_2)$: $\langle v_0(t_0, l_0) v_1(t_1, l_1) v_2(t_2, l_2) v_3(t_3, l_3) \rangle \rightarrow$
$\langle v_0(t_0, l_0) v_1(t_1 - 1, l_1) v_3(t_3 - 1, l_3) \rangle$
R13 $(l_1 > l_2)$: $\langle v_0(t_0, l_0) v_1(t_1, l_1) v_2(t_2, l_2) v_3(t_3, l_3) \rangle \rightarrow$

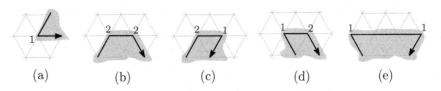

(a)   (b)   (c)   (d)   (e)

Fig. 5. Types of convexities present in $\underline{P}$

$\langle v_0(t_0, l_0)v_1(t_1, l_1 - l_2)v_2(t_2 + 1, l_2)v_3(t_3 - 1, l_3)\rangle$. After application of the rule whichever necessary, if type of any of the vertices is **3** then its length is added to the vertex that precedes it and the vertex is deleted. For example, if the type of vertex v_3 becomes **3** then l_3 is added to length of v_2 and v_3 is deleted. The illustration of rule **R1** is shown in Fig. 6. **Pattern** $< t_1 t_2 t_3 t_4 >$ where $t_2, t_3 \in \{1, 2\}$ and $t_1, t_4 \in \{4, 5, 6\}$.

This pattern implies two consecutive convex vertices followed and preceded by concave vertices. There will be three possible cases depending on the length of v_2 and v_3 as explained in the following rules (illustrated in Fig. 7).

R21: $(l_1 < l_3)$ $\langle v_0(t_0, l_0)v_1(t_1, l_1)v_2(t_2, l_2)v_3(t_3, l_3)v_4(t_4, l_4)\rangle \rightarrow$
$\langle v_0(t_0, l_0)v_1(t_1 - 1, (t_2 + t_3 - 3)l_1 + l_2)v_3(t_3, l_3 - l_1)v_4(t_4, l_4)\rangle$
R22: $(l_1 = l_3)$ $\langle v_0(t_0, l_0)v_1(t_1, l_1)v_2(t_2, l_2)v_3(t_3, l_3)v_4(t_4, l_4)\rangle \rightarrow$
$\langle v_0(t_0, l_0)v_1(t_1 - 1, (t_2 + t_3 - 3)l_1 + l_2)v_4(t_4 - 1, l_4)\rangle$
R23: $(l_1 > l_3)$ $\langle v_0(t_0, l_0)v_1(t_1, l_1)v_2(t_2, l_2)v_3(t_3, l_3)v_4(t_4, l_4)\rangle \rightarrow$
$\langle v_0(t_0, l_0)v_1(t_1, l_1 - l_3)v_2(t_2, (t_2 + t_3 - 3)l_3 + l_2)v_4(t_4 - 1, l_4)\rangle$

Pattern $< t_1 t_2 t_3 t_4 >$ where $t_2, t_3, t_4 \in \{1, 2\}$ and $t_1 \in \{4, 5, 6\}$.

This pattern implies a concave vertex (t_1) is followed by 3 consecutive convex t_2, t_3, t_4 vertices. The total turn at these three consecutive convex vertices may be more than $180°$ and if $l_1 > l_3$ the traversal may enter into a convoluted region which should be avoided to obtain shorter path. This is explained with the help of a sample case shown in Fig. 8. Consider the line h along $v_2 v_1$ and the line h' in the direction of $v_2 v_3$ projected at $v' v_4$ (h and h' meet at v'). To avoid the convoluted region, a traversal is made from v_4 and every time it reaches a new vertex, a check is made to determine whether it has entered the free region

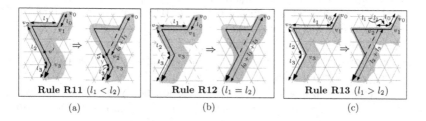

| Rule R11 ($l_1 < l_2$) | Rule R12 ($l_1 = l_2$) | Rule R13 ($l_1 > l_2$) |
| (a) | (b) | (c) |

Fig. 6. Illustration of rules: (a) **R11**, (b) **R12**, (c) **R13**

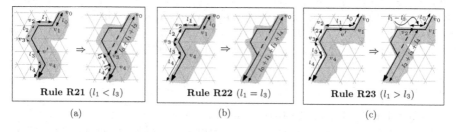

| Rule R21 ($l_1 < l_3$) | Rule R22 ($l_1 = l_3$) | Rule R23 ($l_1 > l_3$) |
| (a) | (b) | (c) |

Fig. 7. Illustration of rules (a) **R21**, (b) **R22**, (c) **R23**

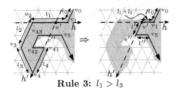

Rule 3: $l_1 > l_3$

Fig. 8. Illustration of Rule 3

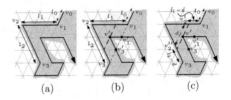

(a)　　　　(b)　　　　(c)

Fig. 9. Nearest concavity line

defined by v_1, v', v_4. For example when the traversal is at vertex v_{41} or v_{42} or v_{43}, it is on or to the right of the line h' and below (right of) the line h and hence they are within the convoluted region. When the traversal reaches v_5, it is on the left of h' and below h which is the free region. The vertices starting from v_2 to v_{43} are deleted, length l' from $v'v''$ and l'' from v'' to v_5 are determined and set accordingly. The type of v'' i.e. t'' is calculated from incoming and outgoing direction at v'' by the formula $(d_{in} - complement(d_{out}) + 6) \mod 6$. The rule is given below.

R3: $\langle v_0(t_0, l_0)v_1(t_1, l_1)v_2(t_2, l_2)v_3(t_3, l_3)v_4(t_4, l_4)\rangle \rightarrow$
$\langle v_0(t_0, l_0)v_1(t_1, l_1 - l_3)v'(t_2, l')v''(t'', l'')\rangle$

However, if $l_1 \leq l_3$, the convexities can be avoided using **Rule 2**. One or more concavity can be intruded inside a convex region. In a convex region, we have to find the concavity which is mostly intruded. This checking has to be performed while applying reduction rules. For example, the convexity created by vertices v_1, v_2, v_3, shown in Fig. 9(a) is of type $< t_1 l t_3 >$ with $l_1 = l_2$. Within the convex region a concave portion is there. To keep the shortest path inside the object the path should pass along the concavity line which is mostly intruded in the convex region (dashed line via v' as shown in Fig. 9(b)). To locate the required concavity line, the intersection points with line $v_1 v_3$ (dotted line in Fig. 9(b)) are found out by searching the appropriate list $(L_\alpha, L_\beta, L_\gamma)$. If there is no intersection point, then appropriate reduction rule is applied directly. On the other hand, if there are intersection points then a traversal is made in the portion of $\underline{P}$ starting from one intersection point to the next intersection point to find out the nearest concavity line. For example, in Fig. 9(b) line $v_1 v_3$ intersects $\underline{P}$ at v'_1 and v'_3. A traversal is made from v'_3 to v'_1, finding distance from every new vertex it meets to v_2 and choosing the one with minimum distance, v' in this case. So the reduction is made upto v' via the dashed line. This introduces two new vertices v'_2 and v''_2 and their length and types are adjusted as shown in Fig. 9(c). If the distance of v_2 from v' is d, then length of v_1 and v''_2 is set to $l_1 - d$ and $l_2 - d$ respectively. Length of v'_2 is set to d and v_2 is deleted.

3.3 Algorithms

The algorithm FIND-STP (Algorithm 2) takes the inner cover $\underline{P}$, the lists L_α, L_β, and L_γ, source and destination points, p and q, as input. The point of intersections of B with the semi-perimeter $\overline{pc_1}$, $\overline{c_1q}$ are obtained (Steps 3–4) and non-essential points are removed using the procedure CONTROL-POINTS (Algorithm 1) (Step 5). p is appended to the shortest path, π. In the while loop (Steps 8–23), each point in M is considered until it reaches q. If $M[i]$ is the corner point c_1, then it is appended to π and then reduced (Steps 10–11). The procedure REDUCE uses the reduction rules to remove the convexity in π. In Steps 15–16, $M[i]$ is added and reduction rules are applied if needed. In Step 17, the portion of $\underline{P}$ between $M[i]$ a 19, $M[i+1]$ is added to π and reduced if needed.

In the procedure TRAVERSE in Algorithm 3, if the index of $M[i]$ is less than that of $M[i+1]$, then $\underline{P}$ is traversed in an anticlockwise manner (Steps 1–11); otherwise, $\underline{P}$ is traversed clockwise (Steps 12–22). In Steps 2–3, l' and l'' indicate the pointers to the neighbor vertices of $M[i]$ and $M[i+1]$ in $\underline{P}$, taken appropriately. After adding $\underline{P}[l']$ to path π (Step 4 or 15), each vertex on the path is appended to π in the while loop (Steps 7–11 or 18-22) until the vertex $\underline{P}[l'']$ is reached. Appropriate reduction rules are applied by calling REDUCE in Steps 5 and 16, 9 and 20 as and when necessary. Procedure REDUCE is explained in Sect. 3.2 with reduction rules, and procedure SEARCH is used to search intersection points of the boundary of $\underline{P}$ with the semi-perimeter of the bounding parallelogram.

Algorithm 2. FIND-STP

Input: $\underline{P}, L_\alpha, L_\beta, L_\gamma, p, q$
Output: π
1 $c_1 \leftarrow$ corner point on left semi-perimeter;
2 $\theta_1, \theta_2 \leftarrow$ Orientation of segment $\overline{pc_1}, \overline{c_1q}$;
3 $M_1 \leftarrow$ SEARCH(p, c_1, θ_1);
4 $M_2 \leftarrow$ SEARCH(c_1, q, θ_2);
5 $M \leftarrow$
   CONTROL-POINTS(M_1, M_2, p, q, c_1);
6 $i \leftarrow 1, \pi \leftarrow \phi$;
7 APPEND(π, p);
8 **while** $M[i] \neq q$ **do**
9   **if** $M[i] = c_1$ **then**
10   APPEND(π, c_1);
11   REDUCE(π);
12       $i \leftarrow i + 1$;
13   **continue**
14   **end**
15   APPEND$(\pi, M[i])$;
16   REDUCE(π);
17   TRAVERSE$(\underline{P}, M[i], M[i+1], \pi)$;
18   APPEND$(\pi, M[i+1])$;
19   REDUCE(π);
20    $i \leftarrow i + 2$;
21   APPEND(π, q);
22   REDUCE(π);
23 **end**
24 **return** π

Algorithm 3. TRAVERSE

Input: $(\underline{P}, M[i], M[i+1], \pi)$
1 **if** $index[M[i]] < index[M[i+1]]$ **then**
2     $l' \leftarrow \lfloor (index[M[i]] + 1) \rfloor$;
3     $l'' \leftarrow \lceil (index[M[i+1]] - 1) \rceil$;
4   APPEND$(\pi[m], \underline{P}[l'])$;
5   REDUCE(π);
6     $j \leftarrow l' + 1$;
7   **while** $j \leqslant l''$ **do**
8   APPEND$(\pi[m], \underline{P}[j])$;
9   REDUCE(π);
10       $j \leftarrow j + 1$;
11   **end**
12 **else**
13    $l' \leftarrow \lceil (index[M[i]] - 1) \rceil$;
14    $l'' \leftarrow \lfloor (index[M[i+1]] + 1) \rfloor$;
15   APPEND$(\pi[m], \underline{P}[l'])$;
16   REDUCE(π);
17    $j \leftarrow l' - 1$;
18   **while** $j \geqslant l''$ **do**
19   APPEND$(\pi[m], \underline{P}[j])$;
20   REDUCE(π);
21       $j \leftarrow j - 1$;
22   **end**
23 **end**

---

4 Time Complexity

To compute the running time of the proposed algorithm let us look at the steps involved and the cost of each step. Initially the inner cover of the object is computed by the algorithm presented in [4] which costs $O(n/g)$ time, n being the number pixels in the perimeter of the object and g is the grid size. During the construction of inner cover three sorted lists L_α, L_β and L_γ are also constructed in $O(n/g \log n/g)$ time. The intersection points on the inner cover of the object with the semi-perimeter of the parallelogram are found by searching L_α or L_β or L_γ in $O(\log n/g)$ time. The algorithm to find shortest path uses control points to reach the destination and applies reduction rules whenever necessary. Reductions can be performed in $O(1)$ time. So, the overall running time of the algorithm amounts to $O(n/g) + O(n/g \log n/g) + O(\log n/g) + O(1) \simeq O(n/g \log n/g)$.

5   Experimental Results and Analysis

The proposed algorithm is implemented in C in Ubuntu 12.04, 64-bit, kernel version 3.5.0-43-generic, the processor being Intel i5-3570, 3.4 GHz FSB and tested exhaustively to show the efficacy and correctness of the algorithm. Two instances of shortest paths for two different objects along with the bounding parallelogram (purple) through which the shortest path is calculated are shown with $g = 8$ in Fig. 10. The number of monotone sub-paths (m) with different

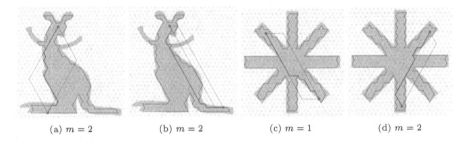

(a) $m = 2$ (b) $m = 2$ (c) $m = 1$ (d) $m = 2$

Fig. 10. Shortest Path of three different objects with $g = 8$ and # monotone paths, m, {(a), (b)} Kangaroo, {(c), (d)} Device

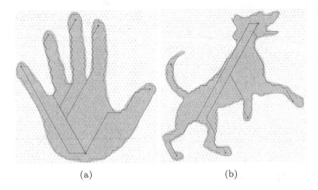

(a)   (b)

Fig. 11. Shortest Paths of two different objects with $g = 8$ from single source (black) to multiple destinations (red) (Color figure online)

colors are also shown under each object in the results. Figure 11 also shows shortest paths from a single source to multiple destinations for two different objects. It is evident from the results that the reported shortest path is not only the shortest path between the two points but also there exists a set of shortest paths having same path-length and our algorithm reports one of the paths between the two points.

6   Conclusions

A combinatorial algorithm to find a shortest triangular path between two points inside a digital object is presented here, which is not unique. Our algorithm reports one of the shortest triangular paths. Thus in future, this work can be extended to determine all shortest paths between two points. The number of monotone triangular sub-paths depends on the position of the two points inside the digital object and also on the shape of the object. The number of monotone triangular sub-paths and other related properties, e.g., length of the path, distance between two points, can be used to determine shape complexity of the object. These metrics will also be useful for determining shape signatures.

References

1. Ahuja, R.K., Magnanti, T.L., Orlin, J.B.: Network Flows: Theory, Algorithms and Applications. Prentice Hall, Upper Saddle River (1993)
2. Balint, G.T., Nagy, B.: Finiteness of chain-code picture languages on the triangular grid. In: Image and Signal Processing and Analysis (ISPA), pp. 310–315 (2015)
3. Bellman, R.: On a routing problem. Q. Appl. Math. **16**, 87–90 (1958)
4. Das, B., Dutt, M., Biswas, A., Bhowmick, P., Bhattacharya, B.B.: A combinatorial technique for construction of triangular covers of digital objects. In: Barneva, R.P., Brimkov, V.E., Šlapal, J. (eds.) IWCIA 2014. LNCS, vol. 8466, pp. 76–90. Springer, Heidelberg (2014)
5. Dijkstra, E.: A note on two problems in connexion with graphs. Numer. Math. **1**, 269–271 (1959)
6. Dutt, M., Biswas, A., Bhowmick, P., Bhattacharya, B.B.: On finding a shortest isothetic path and its monotonicity inside a digital object. Ann. Math. Artif. Intell. **75**, 27–51 (2015)
7. Dutt, M., Biswas, A., Bhowmick, P., Bhattacharya, B.B.: On finding shortest isothetic path inside a digital object. In: Barneva, R.P., Brimkov, V.E., Aggarwal, J.K. (eds.) IWCIA 2012. LNCS, vol. 7655, pp. 1–15. Springer, Heidelberg (2012)
8. Freeman, H.: Algorithm for generating a digital straight line on a triangular grid. IEEE Trans. Comput. **28**, 150–152 (1979)
9. Her, I.: Geometric transformation on the hexagonal grid. IEEE Trans. Image Process. **4**, 1213–1222 (1995)
10. Luczak, E., Rosenfeld, A.: Distance on a hexagonal grid. IEEE Trans. Comput. **25**(5), 532–533 (1976)
11. Moore, E.: The shortest path through a maze. In: Proceedings of an International Symposium on the Theory of Switching, 25 April 1957, pp. 285–292. Harvard University Press, Cambridge (1959)

12. Nagy, B.: Shortest paths in triangular grids with neighbourhood sequences. J. Comput. Inf. Technol. **11**(2), 111–122 (2003)
13. Schrijver, A.: On the history of the shortest path problem. Doc. Math. 155–167 (2012)
14. Shimbel, A.: Structural parameters of communication networks. Bull. Math. Biophys. **15**(4), 501–507 (1953)

A Measure of Q-Convexity

Péter Balázs[1] and Sara Brunetti[2(✉)]

[1] Department of Image Processing and Computer Graphics,
University of Szeged, Árpád tér 2, Szeged 6720, Hungary
`pbalazs@inf.u-szeged.hu`
[2] Dipartimento di Ingegneria dell'Informazione e Scienze Matematiche,
Via Roma, 56, 53100 Siena, Italy
`sara.brunetti@unisi.it`

Abstract. We define new measures of convexity for binary images. The convexity considered here is the so called Q-convexity, that is, convexity by quadrants. This kind of convexity has been mostly studied in Discrete Tomography for its good properties, and permits to generalize h-convexity to any two or more directions. Moreover convex binary images are also Q-convex, and for these two classes similar properties hold. Here we present two measures based on the geometrical properties of "Q-convex shape" which have the following features: (1) their values range from 0 to 1; (2) their values equal 1 if and only if the binary image is Q-convex; (3) their efficient computation can be easily implemented.

1 Introduction

The measure of convexity is one of the most important shape descriptors used in digital image analysis [12]. Various continuous and discrete convexity measures have been proposed in image processing which can be grouped into different categories. Area based measures form one popular category [3,16,17], while boundary-based ones [18] are also frequently used. Other methods use simplification of the contour [13] or a probabilistic approach [14,15] to solve the problem. In discrete geometry, and especially in discrete tomographic reconstruction a straightforward alternative of the continuous convexity concept is the horizontal and vertical convexity (or shortly, h-convexity), arising inherently from the pixel-based representation of the digital image (see, e.g., [2,8,9]). A measure of horizontal (or vertical) convexity was introduced in [1], showing also that the aggregation of the measure in two dimensions can be a difficult task. In this paper we propose an immediate two-dimensional convexity measure, based on the concept of Q-convexity [5,6]. This kind of convexity has been mostly studied in Discrete Tomography for its good properties, and permits to generalize h-convexity to any two or more directions. Moreover convex binary images are also Q-convex, and for these two classes similar properties hold.

The notion of salient points of a Q-convex image has been introduced in [10,11] as the analogue of extremal points of a convex set. They have similar features, and in particular a Q-convex image is characterized by its salient points. Therefore, salient points have been employed for the random generation of

© Springer International Publishing Switzerland 2016
N. Normand et al. (Eds.): DGCI 2016, LNCS 9647, pp. 219–230, 2016.
DOI: 10.1007/978-3-319-32360-2_17

Q-convex images [7]. Further, as salient points can be generalized for any binary image, they have been studied to model the "complexity" of a binary image. In this paper, we focus on estimators of shape descriptors, in particular convexity estimators. The idea is to exploit the geometrical description of a binary image provided by salient and generalized salient points. The advantages of this approach are that normalization is straightforward; the measures are equal to 1 if and only if the binary image is Q-convex; the efficient computation of the measures can be easily implemented; and the measures are easy to generalize to any two or more directions.

2   Preliminaries

In this section we introduce the necessary notation and definitions. Any binary image F is a $m \times n$ binary matrix, and it can be represented by a set of cells/pixels (unit squares) or, equivalently, by a finite subset of $\mathbb{Z}^2$ contained in a lattice grid $\mathcal{G}$ (rectangle of size $m \times n$) up to a translation. Throughout the paper, we are going to use both representations as notation for the latter one is more suitable to describe geometrical properties whereas the images are illustrated as sets of cells. For our convenience, we use F for both the image and its representation.

Let us consider the horizontal and vertical directions, and denote the coordinate of any point M of the grid $\mathcal{G}$ by (x_M, y_M). Then, M and the directions determine the following four quadrants:

$$Z_0(M) = \{N \in \mathcal{G} : 0 \le x_N \le x_M, \ 0 \le y_N \le y_M\}$$
$$Z_1(M) = \{N \in \mathcal{G} : x_M \le x_N < m, \ 0 \le y_N \le y_M\}$$
$$Z_2(M) = \{N \in \mathcal{G} : x_M \le x_N < m, \ y_M \le y_N < n\}$$
$$Z_3(M) = \{N \in \mathcal{G} : 0 \le x_N \le x_M, \ y_M \le y_N < n\}.$$

Definition 1. *A binary image F is Q-convex with respect to the horizontal and vertical directions if $Z_p(M) \cap F \ne \emptyset$ for all $p = 0, \ldots, 3$ implies $M \in F$.*

Figure 1 illustrates two Q-convex images.

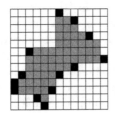 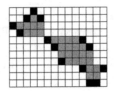

Fig. 1. Two Q-convex images. Salient points are indicated by black cells.

The Q-convex hull of F can be defined as follows:

Definition 2. *The Q-convex hull $\mathcal{Q}(F)$ of a binary image F is the set of points $M \in \mathcal{G}$ such that $Z_p(M) \cap F \neq \emptyset$ for all $p = 0, \ldots, 3$.*

Therefore, if F is Q-convex then $F = \mathcal{Q}(F)$. Differently, if F is not Q-convex, then $\mathcal{Q}(F) \backslash F \neq \emptyset$ (see Fig. 2). Denote the cardinality of F and $\mathcal{Q}(F)$ by α_F and $\alpha_{\mathcal{Q}(F)}$, respectively.

Definition 3. *For a given binary image F, its Q-convexity measure $\Theta(F)$ is defined to be $\Theta(F) = \frac{\alpha_F}{\alpha_{\mathcal{Q}(F)}}$.*

Since $\Theta(F)$ holds 1, if F is Q-convex, and $\alpha_{\mathcal{Q}(F)} \geq \alpha_F$, it ranges in $(0, 1]$. This Q-convexity measure corresponds to the classical convexity measure defined as the ratio between the area of the considered shape and the area of its convex hull. Similarly, $\Theta(F)$ is easy to compute but presents similar defects as does not detect defects in the shape which does not impact on the sizes α_F and $\alpha_{\mathcal{Q}(F)}$ (see Sect. 4 about the experiments with intrusions/protrusions images). Indeed, in the first case both α_F and $\alpha_{\mathcal{Q}(F)}$ differ by n so that Θ is close to 1, whereas in the second case α_F is close to $2/3\alpha_{\mathcal{Q}(F)}$ for n big.

Therefore, we are going to define a new measure based on the geometrical properties of the "shape".

Definition 4. *Let F be a binary image. A point $M \in F$ is a salient point of F if $M \notin Q(E \backslash \{M\})$.*

Denote the set of salient points of F by $\mathcal{S}(F)$ (or simply $\mathcal{S}$): of course $\mathcal{S}(\emptyset) = \emptyset$. In particular, it can be proven [10] that the salient points of F are the salient points of the Q-convex hull $\mathcal{Q}(F)$ of F. This means that if F is Q-convex, its salient points completely characterize F [6]. If it is not, there are other points belonging to the Q-convex hull of F but not in F that "track" the non-Q-convexity of F. These points are called generalized salient points. The generalized salient points $\mathcal{S}_g(F)$ of F are obtained iterating the definition of salient points on the sets obtained each time by discarding the points of the set from its Q-convex hull, i.e. using the set notation:

Definition 5. *If F is a binary image, then the set of its generalized salient points $\mathcal{S}_g(F)$ is defined by $\mathcal{S}_g(F) = \cup_k S(F_k)$, where $F_0 = F$, $F_{k+1} = \mathcal{Q}(F_k) \backslash F_k$.*

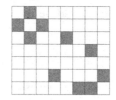 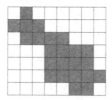

Fig. 2. Left: A non Q-convex binary image (all dark cells are salient points). Right: its Q-convex hull.

By definition, $\mathcal{S}(F) \subseteq S_g(F)$ and the equality holds when F is Q-convex. Moreover, as the points of $\mathcal{S}_g(F)$ are chosen among the points of subsets of $\mathcal{Q}(F)$, then $\mathcal{S}_g(F) \subseteq \mathcal{Q}(F)$.

We are now in the position to define two Q-convexity measures:

Definition 6. *For a given binary image F, its Q-convexity measure $\Psi_1(F)$ is defined by*

$$\Psi_1(F) = \alpha_{\mathcal{S}(F)}/\alpha_{\mathcal{S}_g(F)},$$

where $\mathcal{S}(F)$ and $\mathcal{S}_g(F)$ denote the sets of its salient and generalized salient points, respectively.

$\Psi_1(F)$ measures the Q-convexity of F in terms of proportion between salient points and generalized salient points. Indeed, if the generalized salient points are many with respect to salient points, then F is far to be Q-convex. This measure is purely qualitative because is independent from the size of the image.

Definition 7. *For a given binary image F, its Q-convexity measure $\Psi_2(F)$ is defined by*

$$\Psi_2(F) = \frac{\alpha_{\mathcal{Q}(F)} - \alpha_{\mathcal{S}_g(F)}}{\alpha_{\mathcal{Q}(F)} - \alpha_{\mathcal{S}(F)}},$$

where $\mathcal{Q}(F)$ denotes its Q-convex hull and $\mathcal{S}(F)$ and $\mathcal{S}_g(F)$ denote the sets of its salient and generalized salient points, respectively.

$\Psi_2(F)$ takes salient, and generalized salient points with respect to the Q-convex hull of the image into account. In other words, it measures the non-Q-convexity of F as the proportion of generalized salient points which are not salient and the points of the Q-convex hull which are not salient.

Since $\mathcal{S}(F) \subseteq \mathcal{S}_g(F) \subseteq \mathcal{Q}(F)$, both Q-convexity measures satisfy the following properties:

– the Q-convexity measure ranges from 0 to 1;
– the Q-convexity measure equals 1 if and only if F is Q-convex.

In particular, for $\Psi_1(F)$, since there are examples where $\mathcal{S}_g(F) = \mathcal{Q}(F)$ (for instance in the chessboard configuration), the ratio decreases with the inverse of the size of $\mathcal{Q}(F)$. For $\Psi_2(F)$, in the same case, we get exactly 0. We point out that we must have $\mathcal{S}(F) \subset \mathcal{Q}(F)$ in the definition of $\Psi_2(F)$, but $\mathcal{S}(F) = \mathcal{Q}(F)$ holds only for binary images of size smaller or equal to two. In these cases of course we also have $\mathcal{S}_g(F) = \mathcal{Q}(F)$, so that $\Psi_2(F)$ would be undefined.

3   Implementation

The measures Ψ_1 and Ψ_2 can be computed efficiently. Let us associate a boolean variable $V_p(M)$ to each point $M \in \mathcal{G}$, $p = 0, \ldots, 3$, such that $V_p(M) = 1$ if

$Z_p(M) \cap F \neq \emptyset$, else $V_p(M) = 0$. The variables can be easily computed by iteration:

$$V_0(x_M, y_M) = V_0(x_M - 1, y_M) \vee V_0(x_M, y_M - 1) \vee \text{"}(x_M, y_M) \in F\text{"}$$
$$V_1(x_M, y_M) = V_1(x_M + 1, y_M) \vee V_1(x_M, y_M - 1) \vee \text{"}(x_M, y_M) \in F\text{"}$$
$$V_2(x_M, y_M) = V_2(x_M + 1, y_M) \vee V_2(x_M, y_M + 1) \vee \text{"}(x_M, y_M) \in F\text{"}$$
$$V_3(x_M, y_M) = V_3(x_M - 1, y_M) \vee V_3(x_M, y_M + 1) \vee \text{"}(x_M, y_M) \in F\text{"}.$$

Finally the Q-convex hull of F is obtained by the formulas:

$$Q(F) = \{M \in \mathcal{G} : V_0(x_M, y_M) \wedge V_1(x_M, y_M) \wedge V_2(x_M, y_M) \wedge V_3(x_M, y_M)\},$$

and the computation requires a number of operations equal to the size of $\mathcal{G}$. The same variables permit to compute $\mathcal{S}(F)$ and, hence $\mathcal{S}_g(F)$, as follows. It can be proven [10] that M is a salient point of F if and only if there exists p such that $Z_p(M) \cap F = \{M\}$. So let $\mathcal{S}_p(F) = \{M \in F : Z_p(M) \cap F = \{M\}\}$; we get $\mathcal{S}(F) = \mathcal{S}_0(F) \cup \mathcal{S}_1(F) \cup \mathcal{S}_2(F) \cup \mathcal{S}_3(F)$. The set of salient points can be easily computed as we have:

$$\mathcal{S}_0(F) = \{M \in F : \neg V_0(x_M - 1, y_M) \wedge \neg V_0(x_M, y_M - 1)\}$$
$$\mathcal{S}_1(F) = \{M \in F : \neg V_1(x_M + 1, y_M) \wedge \neg V_1(x_M, y_M - 1)\}$$
$$\mathcal{S}_2(F) = \{M \in F : \neg V_2(x_M + 1, y_M) \wedge \neg V_2(x_M, y_M + 1)\}$$
$$\mathcal{S}_3(F) = \{M \in F : \neg V_3(x_M - 1, y_M) \wedge \neg V_3(x_M, y_M + 1)\},$$

where $\neg V_p$ is the negation of V_p. By definition, the set of generalized salient points is computed by iterating the computation of salient points on $\mathcal{Q}(F_k) \backslash F_k$ until the set reduces to the empty set. Therefore this computation depends on the size of $\mathcal{Q}(F)$ which bounds the number of iterations.

4   Case Study

We report on the behavior of the previous Q-convexity measures on the following representative configurations (F is a $n \times n$ binary image):

Chessboard. It is easy to check that if the items of F are arranged to form a chessboard, then $\mathcal{S}_g(F) = \mathcal{Q}(F)$. As a consequence, $\Psi_1(F) = \alpha_{\mathcal{S}(F)}/\alpha_{\mathcal{Q}(F)}$ depends on the size of $\mathcal{Q}(F)$ since $\alpha_{\mathcal{S}(F)} \in \{4, 6, 8\}$. This means that $\Psi_1(F)$ assigns different values to the same configuration decreasing for increasing sizes of F. In other words, bigger matrices are "less" Q-convex than smaller ones. Differently, $\Psi_2(F) = 0$ by definition. Therefore, the chessboard configuration is the "least" Q-convex configuration (for the Ψ_2 measure) (Fig. 3).

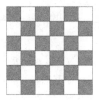

Fig. 3. $\Psi_1(F) = 0.176470$, $\Psi_2(F) = 0$.

Stripe Pattern. By counting reasoning, it is easy to see that if the items of F are arranged to constitute a stripe pattern, then $\alpha_{\mathcal{S}(F)} = 6$ whereas $\alpha_{\mathcal{S}_g(F)} = 6\lfloor\frac{n-1}{2}\rfloor+4$, where $n \times n$ is the matrix size. Therefore $\Psi_1(F)$ decreases for n which increases. The behavior is similar to the chessboard case, but for the same size n, a chessboard configuration is "less" Q-convex than a stripe pattern configuration (in the first case the decrement is quadratic, in the second linear w.r.t n). For the Ψ_2 measure, the $\alpha_{\mathcal{Q}(F)}$ term dominates, so that its value is close to 1 for n big (for $n = 10$ $\Psi_2 = 0.58$, for $n = 50$ $\Psi_2 = 0.93$, for $n = 100$ $\Psi_2 = 0.965$). Hence, for this configuration the two measures give two different (opposite) "responses" (Fig. 4).

Fig. 4. $\Psi_1(F) = 0.375$, $\Psi_2(F) = 0.583333$.

Frame. If the items of F are arranged to form a frame, $\alpha_{\mathcal{S}(F)} = 4$ and $\alpha_{\mathcal{S}_g(F)} = 8$ or 5 if there is only one 0 item in F. Moreover, $\alpha_{\mathcal{Q}(F)} = n^2$. Therefore, $\Psi_1 = 1/2$ or $4/5$ in the latter case (constant), whereas Ψ_2 tends to 1 for increasing values of n. In both cases, the value is independent from α_F (from the "thickness" of the frame) (Fig. 5).

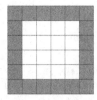

Fig. 5. $\Psi_1(F) = 0.5$, $\Psi_2(F) = 0.875$.

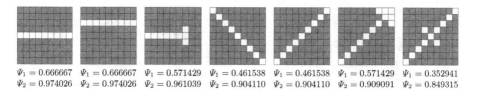

$\Psi_1 = 0.666667$ $\Psi_1 = 0.666667$ $\Psi_1 = 0.571429$ $\Psi_1 = 0.461538$ $\Psi_1 = 0.461538$ $\Psi_1 = 0.571429$ $\Psi_1 = 0.352941$
$\Psi_2 = 0.974026$ $\Psi_2 = 0.974026$ $\Psi_2 = 0.961039$ $\Psi_2 = 0.904110$ $\Psi_2 = 0.904110$ $\Psi_2 = 0.909091$ $\Psi_2 = 0.849315$

Fig. 6. Q-convexity values measured on intrusions images.

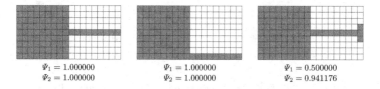

$\Psi_1 = 1.000000$ $\Psi_1 = 1.000000$ $\Psi_1 = 0.500000$
$\Psi_2 = 1.000000$ $\Psi_2 = 1.000000$ $\Psi_2 = 0.941176$

Fig. 7. Q-convexity values measured on protrusions images.

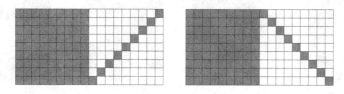

Fig. 8. Q-convexity values measured on protrusions images. $\Psi_1 = 0.550000$, $\Psi_2 = 0.921739$

"Bad" Configurations. The following figures illustrate examples of the application of the proposed Q-convexity measures for simple synthetic polygons: the values show the effects of rotation and translation of intrusions/protrusions images. In particular, consider the fourth image in Fig. 6. Note that $\alpha_{\mathcal{Q}(F)} = O(n^2)$; $\mathcal{S}(F)$ does not depend on the size of F; all (except the two on the first and last rows) the 0's item belong to $\mathcal{S}_g(F)$, and so $\alpha_{\mathcal{S}_g(F)} = O(n)$. We point out in addition that $\mathcal{S}_g$ does not depend on the "thickness" of the diagonal intrusion. Therefore, $\Psi_1(F)$ decreases when n increases; differently $\Psi_2(F)$ depends on $\mathcal{Q}(F)$ so that it is equal to 1 for $n = 2$ (and in fact F is Q-convex) and it is close to 1 for n big even if the diagonal intrusion is "thicker".

Consider the images illustrated in Fig. 8 (of course they have same measures by symmetry). Similar considerations hold with the difference that $\alpha_{\mathcal{S}(F)} = O(n)$ and the "thickness" of the protrusion affects $\alpha_{\mathcal{S}_g(F)}$ since it reduces items in $\mathcal{S}_g$. Therefore when n is big, Ψ_1 is greater or equal to $1/2$, and tends to 1 when the "protrusion" becomes very "thick", and Ψ_2 is close to 1.

In Fig. 10 we present the images used in [1] for the analysis of the horizontal and vertical convexity measure. For comparison, we also calculated the values of Q-convexity according to the measures Ψ_1 and Ψ_2. We deduce that the Q-convexity measures are in accordance with the horizontal and vertical convexity measures of [1] in the sense that higher Q-convexity values correspond to higher

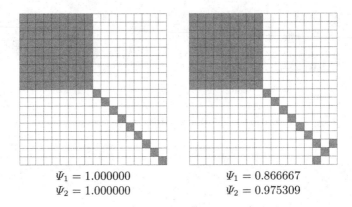

$$\Psi_1 = 1.000000 \qquad\qquad \Psi_1 = 0.866667$$
$$\Psi_2 = 1.000000 \qquad\qquad \Psi_2 = 0.975309$$

Fig. 9. Q-convexity values measured on protrusions images.

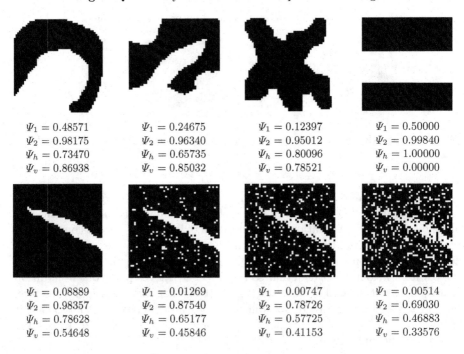

$\Psi_1 = 0.48571$	$\Psi_1 = 0.24675$	$\Psi_1 = 0.12397$	$\Psi_1 = 0.50000$
$\Psi_2 = 0.98175$	$\Psi_2 = 0.96340$	$\Psi_2 = 0.95012$	$\Psi_2 = 0.99840$
$\Psi_h = 0.73470$	$\Psi_h = 0.65735$	$\Psi_h = 0.80096$	$\Psi_h = 1.00000$
$\Psi_v = 0.86938$	$\Psi_v = 0.85032$	$\Psi_v = 0.78521$	$\Psi_v = 0.00000$

$\Psi_1 = 0.08889$	$\Psi_1 = 0.01269$	$\Psi_1 = 0.00747$	$\Psi_1 = 0.00514$
$\Psi_2 = 0.98357$	$\Psi_2 = 0.87540$	$\Psi_2 = 0.78726$	$\Psi_2 = 0.69030$
$\Psi_h = 0.78628$	$\Psi_h = 0.65177$	$\Psi_h = 0.57725$	$\Psi_h = 0.46883$
$\Psi_v = 0.54648$	$\Psi_v = 0.45846$	$\Psi_v = 0.41153$	$\Psi_v = 0.33576$

Fig. 10. Example binary images of size 50×50, with the Ψ_1 and Ψ_2 convexity values shown. Black pixels indicate object points. For comparison the horizontal (Ψ_h) and vertical (Ψ_v) convexity values are also given. Bottom row: same image without, and with 5 %, 10 %, and 20 % salt-and-pepper noise, from left to right, respectively.

h-convexity (v-convexity) values. Moreover, our measures visibly integrate the two one-directional measures (see especially the fourth image in the top row of Fig. 10). We also found that our measures scale well on noisy images (see bottom row of Fig. 10, for example). Finally let us notice that the values of the

Θ measure are $0.609828, 0.750000, 0.737407$ and 0.660000 for the binary images from left to right, respectively, in the top row, showing that the Θ measure does not "classify" the images in accordance with Ψ_1 and Ψ_2: for instance, according to Θ the first image is "less" Q-convex than the second one, whereas according to Ψ_1 and Ψ_2 the opposite holds.

5   Strongly Q-Convexity

The notion of Q-convexity can be extended to more than two directions [4,7]. Consider the horizontal, vertical and diagonal (i.e. $\boldsymbol{h} = (1,0)$, $\boldsymbol{v} = (0,1)$, $\boldsymbol{d} = (-1,1)$) directions. Let

$$s_h^+(M) = \{N \in \mathcal{G} \; : \; y_M = y_N \text{ and } \boldsymbol{h} \cdot \boldsymbol{ON} \geq \boldsymbol{h} \cdot \boldsymbol{OM}\}$$
$$s_h^-(M) = \{N \in \mathcal{G} \; : \; y_M = y_N \text{ and } \boldsymbol{h} \cdot \boldsymbol{ON} \leq \boldsymbol{h} \cdot \boldsymbol{OM}\}$$
$$s_v^+(M) = \{N \in \mathcal{G} \; : \; x_M = x_N \text{ and } \boldsymbol{v} \cdot \boldsymbol{ON} \geq \boldsymbol{v} \cdot \boldsymbol{OM}\}$$
$$s_v^-(M) = \{N \in \mathcal{G} \; : \; x_M = x_N \text{ and } \boldsymbol{v} \cdot \boldsymbol{ON} \leq \boldsymbol{v} \cdot \boldsymbol{OM}\}$$
$$s_d^+(M) = \{N \in \mathcal{G} \; : \; x_M + y_M = x_N + y_N \text{ and } \boldsymbol{d} \cdot \boldsymbol{ON} \geq \boldsymbol{d} \cdot \boldsymbol{OM}\}$$
$$s_d^-(M) = \{N \in \mathcal{G} \; : \; x_M + y_M = x_N + y_N \text{ and } \boldsymbol{d} \cdot \boldsymbol{ON} \leq \boldsymbol{d} \cdot \boldsymbol{OM}\}$$

where O is the origin and $\cdot$ denotes the scalar product.

Definition 8. *An almost-semi-plane (or ASP) $\Pi(M)$ along the horizontal, vertical and diagonal directions is a zone $Z_p^{(i,j)}(M)$ ($i \neq j \in \{h, v, d\}$, $p \in \{0, \ldots, 3\}$) such that for each direction k only one of the two semi-lines s_k^+ $s_k^-(M)$ is contained in $\Pi(M)$ for $k = h, v, d$.*

Let $\Pi_0(M)$ be the ASP containing $s_k^+(M)$ for each $k = h, v, d$. We denote the other almost-semi-planes encountered clockwise around M from $\Pi_1(M), \ldots, \Pi_5(M)$. In particular,

$$\Pi_0(M) = Z_2^{(dh)}(M) = \{N \in \mathcal{G} : x_M + y_M \leq x_N + y_N, y_M \leq y_N\}$$
$$\Pi_1(M) = Z_2^{(vd)}(M) = \{N \in \mathcal{G} : x_M + y_M \leq x_N + y_N, x_M \leq x_N\}$$
$$\Pi_2(M) = Z_1^{(h)}(M) = \{N \in \mathcal{G} : x_M \leq x_N, y_M \geq y_N\}$$
$$\Pi_3(M) = Z_0^{(dh)}(M) = \{N \in \mathcal{G} : x_M + y_M \geq x_N + y_N, y_M \geq y_N\}$$
$$\Pi_4(M) = Z_0^{(vd)}(M) = \{N \in \mathcal{G} : x_M + y_M \geq x_N + y_N, x_M \geq x_N\}$$
$$\Pi_5(M) = Z_3^{(h)}(M) = \{N \in \mathcal{G} : x_M \geq x_N, y_M \leq y_N\}.$$

Definition 9. *A binary image F is strongly Q-convex around the horizontal, vertical and diagonal directions if $\Pi_p(M) \cap F \neq \emptyset$ for all $p = 0, \ldots, 5$ implies $M \in F$.*

The definitions of Q-convex hull, salient and generalized salient points can be easily extended by replacing quadrants with ASP's. In order to compute them note that we can associate a boolean variable $V_p(M)$ to each point $M \in \mathcal{G}$ $p = 0, \ldots, 5$ such that $V_p(M) = 1$ if $\Pi_p(M) \cap F \neq \emptyset$, else $V_p(M) = 0$. Similar formulas to the two direction case can be derived to iteratively compute the variables. By their computation we can also derive the set of salient points $\mathcal{S}_p(M)$ $p = 0, \ldots, 5$ and finally the set of generalized salient points. Indeed, we can extend the proof in [10] to show that $M \in \mathcal{S}(F)$ iff there exists p such that $\Pi_p(M) \cap F = \{M\}$ and $\mathcal{S}(F) = \mathcal{S}(\mathcal{Q}(F))$.

We conducted the same experiments as in Sect. 4 for measuring the strongly Q-convexity of the images. We do not report all the results for space limits, but they are very similar with few differences.

For the intrusions test, as expected, the fourth image of Fig. 6 receives a better value, whereas the third and sixth are worse. For the protrusions test, values for images in Fig. 7 are worse because they are not strongly Q-convex, whereas values for images in Fig. 9 are exactly the same. The only significant difference concerns the second image in Fig. 8 for which $\Psi_1 = 0.571429$, $\Psi_2 = 0.975410$: indeed, $\mathcal{S} = 4$ and $\mathcal{S}_g = 7$ are constant (whereas they depend on n in case of Q-convexity w.r.t. horizontal and vertical directions), and, moreover, they score better values since the protrusion is in the diagonal direction.

Regarding the images of Fig. 10 we observed that the three directional Q-convexity values were more or less the same as in the two directional case, except the third image of the first row ($\Psi_1 = 0.08108$ and $\Psi_2 = 0.93621$ for the strongly Q-convexity measures) and the first image of the second row ($\Psi_1 = 0.11111$ and $\Psi_2 = 0.98717$). The reason is that the two images are "almost" convex in the diagonal direction.

6   Conclusion and Discussion

In this paper, we proposed two measures of convexity based on the notions of generalized salient points and Q-convex hull. First, we considered convexity in the horizontal and vertical directions and then we extended also the measures to a third diagonal direction. The experiments we conducted are promising as the measures correctly incorporate the convexity along the considered directions. In particular, measure Ψ_1 is more sensible to small modification of the shape, while Ψ_2 is more robust.

As a short discussion, let us notice here that a correcting factor could be used which takes into account also quantitative information. For example look at the images illustrated in Fig. 11, where Ψ_1 is constant ($1/2$) and Ψ_2 is close to 1, since the geometry is preserved. But it would be desirable that the Q-convexity measures assign a bigger value to the configuration on the left of Fig. 11 and smaller to the configuration to the right. This can be obtained by multiplying the previous measures to the factor $\Theta(F) = \alpha_F / \alpha_{\mathcal{Q}(F)}$. For the image on the left $\alpha_{\mathcal{S}(F)} = 4$, $\alpha_{\mathcal{S}_g(F)} = 8$, $\alpha_{\mathcal{Q}(F)} = 64$, and $\alpha_F / \alpha_{\mathcal{Q}(F)} = 0.84345$, while for the image on the right $\alpha_{\mathcal{S}(F)} = 4$, $\alpha_{\mathcal{S}_g(F)} = 8$, $\alpha_{\mathcal{Q}(F)} = 64$, $\alpha_F / \alpha_{\mathcal{Q}(F)} = 0.34375$ which underpins

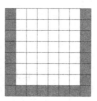

Fig. 11. Images to show the importance of the correcting factor. For the image on the left $\alpha_F/\alpha_{Q(F)} = 0.84345$ while for the image on the right $\alpha_F/\alpha_{Q(F)} = 0.34375$.

the above argument. Further work can be conducted to investigate the role of the correcting factor and to extend the measure to any two or more directions.

Acknowledgements. The collaboration of the authors was supported by the EXTREMA COST Action MP1207 "Enhanced X-ray Tomographic Reconstruction: Experiment, Modeling, and Algorithms". The research of Péter Balázs was supported by the OTKA K112998 grant of the National Scientific Research Fund.

References

1. Balázs, P., Ozsvár, Z., Tasi, T.S., Nyúl, L.G.: A measure of directional convexity inspired by binary tomography. Fundamenta Informaticae **141**(2–3), 151–167 (2015)
2. Barcucci, E., Del Lungo, A., Nivat, M., Pinzani, R.: Medians of polyominoes: a property for the reconstruction. Int. J. Imag. Syst. Technol. **9**, 69–77 (1998)
3. Boxter, L.: Computing deviations from convexity in polygons. Pattern Recognit. Lett. **14**, 163–167 (1993)
4. Brunetti, S., Daurat, A., Del Lungo, A.: An algorithm for reconstructing special lattice sets from their approximate X-rays. In: Borgefors, G., Nyström, I., Sanniti di Baja, G. (eds.) DGCI 2000. LNCS, vol. 1953, pp. 113–125. Springer, Heidelberg (2000)
5. Brunetti, S., Daurat, A.: An algorithm reconstructing convex lattice sets. Theor. Comput. Sci. **304**(1–3), 35–57 (2003)
6. Brunetti, S., Daurat, A.: Reconstruction of convex lattice sets from tomographic projections in quartic time. Theor. Comput. Sci. **406**(1–2), 55–62 (2008)
7. Brunetti, S., Daurat, A.: Random generation of Q-convex sets. Theor. Comput. Sci. **347**(1–2), 393–414 (2005)
8. Brunetti, S., Del Lungo, A., Del Ristoro, F., Kuba, A., Nivat, M.: Reconstruction of 4- and 8-connected convex discrete sets from row and column projections. Linear Algebra Appl. **339**, 37–57 (2001)
9. Chrobak, M., Dürr, C.: Reconstructing hv-convex polyominoes from orthogonal projections. Inform. Process. Lett. **69**(6), 283–289 (1999)
10. Daurat, A.: Salient points of Q-convex sets. Int. J. Pattern Recognit. Artif. Intell. **15**, 1023–1030 (2001)
11. Daurat, A., Nivat, M.: Salient and reentrant points of discrete sets. Electron. Notes Discrete Math. **12**, 208–219 (2003)

12. Gonzalez, R.C., Woods, R.E.: Digital Image Processing, 3rd edn. Prentice Hall, Harlow (2008)
13. Latecki, L.J., Lakamper, R.: Convexity rule for shape decomposition based on discrete contour evolution. Comput. Vis. Image Und. **73**(3), 441–454 (1999)
14. Rahtu, E., Salo, M., Heikkila, J.: A new convexity measure based on a probabilistic interpretation of images. IEEE Trans. Pattern Anal. **28**(9), 1501–1512 (2006)
15. Rosin, P.L., Zunic, J.: Probabilistic convexity measure. IET Image Process. **1**(2), 182–188 (2007)
16. Sonka, M., Hlavac, V., Boyle, R.: Image Processing, Analysis, and Machine Vision, 3rd edn. Thomson Learning, Toronto (2008)
17. Stern, H.: Polygonal entropy: a convexity measure. Pattern Recognit. Lett. **10**, 229–235 (1998)
18. Zunic, J., Rosin, P.L.: A new convexity measure for polygons. IEEE Trans. Pattern Anal. **26**(7), 923–934 (2004)

Models for Discrete Geometry

Symmetric Masks for In-fill Pixel Interpolation on Discrete p:q Lattices

Imants Svalbe[1(✉)] and Arnaud Guinard[2]

[1] School of Physics and Astronomy, Monash University, Melbourne, Australia
imants.svalbe@monash.edu
[2] PolytechNantes, Nantes, France
arnaud.guinard@etu.univ-nantes.fr

Abstract. A 2D p:q lattice contains image intensity entries at pixels located at regular, staggered intervals that are spaced p rows and q columns apart. Zero values appear at all other intermediate grid locations. We consider here the construction, for any given p:q, of convolution masks to smoothly and uniformly interpolate values across all of the intermediate grid positions. The conventional pixel-filling approach is to allocate intensities proportional to the fractional area that each grid pixel occupies inside the boundaries formed by the p:q lines. However these area-based masks have asymmetric boundaries, flat interior values and may be odd or even in size. We ask here if smoother, symmetric versions of such convolution masks exist and, if so, is their structure unique for each p:q lattice? The answer appears to be yes on both counts. The coefficients of the masks constructed here have simple integer values whose distribution is derived purely from symmetry considerations. We have application for these symmetric interpolation masks as part of a precise image rotation algorithm, as well as to smooth back-projected values when performing discrete tomographic image reconstruction.

Keywords: Discrete Haar interpolation · Rational angle rotation · Mojette discrete back-projection

1 Introduction

We consider here the problem of distributing regularly-spaced but sparse image intensities to fill the gaps between known values. The known values are located at discrete pixel locations that fall on straight lines oriented at an angle θ, whose tangents are rational fractions, for example $\theta = tan(q/p)$, where p and q are co-prime integers. When p or q is negative, θ becomes $-\theta$.

The values at all other grid locations are set to zero. We call this 2D set of sparse image points a p:q lattice. These lattices occur as a result of affine rotation of discrete images [1,7] and also when discrete 1D projected views of a 2D image are back-projected across a 2D space when reconstructing images from projections [4].

© Springer International Publishing Switzerland 2016
N. Normand et al. (Eds.): DGCI 2016, LNCS 9647, pp. 233–243, 2016.
DOI: 10.1007/978-3-319-32360-2_18

The way we choose to perform pixel in-filling on a p:q lattice is important. Interpolation changes the spectral content of the original image through the process of convolving the distribution of interpolating mask values with the original image values that are now spread across a p:q lattice.

Masks for pixel in-filling are traditionally defined on the basis of area intersection: the value interpolated at each vacant pixel position depends on the fraction of each pixel area that lies inside the bounding lines of the p:q lattice [7].

This paper presents a method to construct symmetric in-fill masks. These masks provide an alternative to conventional asymmetric, area-based masks. We compare the effects of mask symmetry on in-filled images when they are subsequently down-scaled as part of an image rotation scheme. We also examine the use of symmetric masks to smooth discrete projection data.

An affine rotation of discrete image data, at an angle θ, remaps the content of the original image pixels located at (x, y) to a new image with pixels positioned at (x', y') where (using Matlab notation), $[x'; y'; 1] = [q\ -p\ 0;\ p\ q\ 0;\ 0\ 0\ 1]*[x; y; 1]$ in homogeneous image coordinates. This rotation also expands a square of unit area to have area $A = p^2 + q^2$, leaving all the intermediate pixel values as zero. Pixel in-filling 'joins the dots' between known values in the up-scaled and rotated data, producing a smooth grey-scale image.

Discrete projection of an image involves summing pixel values along parallel rays to form a 1D sum of the 2D image data at orientation p:q. Collecting views at a set of different p:q angles permits reconstruction of the 2D image from its 1D projections (discrete tomography, [4]). One method by which images can be reconstructed is through back-projection of the 1D data across the 2D image plane, usually with a filtration step to reduce the anomalies that occur when mapping polar samples back across a Cartesian grid. The process of back-projecting 1D data generates p:q lattices at each view angle, for which these interpolation masks again prove useful.

Where many neighbouring p:q lattice values have the same intensity, these interpolating masks must all produce an exactly uniform response for all pixels that lie between those lattice points. An example of a 1:5 lattice image is shown in Fig. 1, together with the image the results after convolving with the area-based mask shown in Fig. 2.

Fig. 1. Left: a 1:5 lattice image. Right: the same image after interpolation using the area-based mask shown in Fig. 2

1	3	5	7	9	1
9	10	10	10	10	3
7	10	10	10	10	5
5	10	10	10	10	7
3	10	10	10	10	9
1	9	7	5	3	1

0	0	0	4	21	9	0	0
0	1	15	29	30	26	1	0
9	26	30	30	30	30	15	0
21	30	30	30	30	30	29	4
4	29	30	30	30	30	30	21
0	15	30	30	30	30	26	9
0	1	26	30	29	15	1	0
0	0	9	21	4	0	0	0

Fig. 2. Area-based interpolation mask coefficients. Left for a 1:5 lattice, right for a 3:5 lattice

The area-based interpolation masks (Fig. 2) have asymmetric edges (along which the area fractions change quite rapidly) with very flat interiors since the areas of most 'internal' pixels are fully included. These masks can be odd or even in size and hence have an ambiguously-positioned centre for interpolation. The values in the mask for a p:q lattice need to be transposed for use on a q:p lattice. The transpose of the masks shown in Fig. 2 must be used to in-fill pixels on a 5:1 and 5:3 lattice, respectively. 1D projection of these 2D area-based masks is equivalent to forming the corresponding Haar interpolation filter [4].

This paper is organised as follows: Sect. 2 outlines a method to construct symmetric pixel in-filling masks for any p:q. Section 3 presents examples of in-filled lattice structures, and discusses the uniqueness properties of these masks and compares results obtained using symmetric and area-based masks. Section 4 presents image rotation examples obtained using symmetric and the conventional masks. Section 5 examines filters built as 1D projection of the 2D area-based and symmetric masks, whilst Sect. 6 looks at future extensions of this work.

2   Construction of Symmetric Interpolation Masks

To construct a symmetric mask for any p:q lattice, we require just that the mask coefficients have fourfold symmetry (assuming we use a square grid) and that the four nearest neighbours on a p:q lattice (call them A, B, C and D) contribute equally to form the value of the single pixel (or group of pixels) that lie at the geometric centre of these 4 lattice points.

To contribute equal proportions to the intensity at the centre pixel(s) we expect A, B, C and D to add as $[1, 1, 1, 1]$, i.e. each adds 25 % of their value to the final central value. Off-centre pixels are then expected to add as $[2, 1, 1, 0]$, $[2, 2, 0, 0]$, $[3\,1, 0, 0]$ and finally $[4, 0, 0, 0]$ for pixels that lie at or close to lattice points A, B, C or D. The symmetric mask coefficients thus only contain integer values of 0, 1, 2, 3, or 4, since we combine intensities in multiples of 25 % portions.

When $s = p + q + 1$ is odd, there is a single pixel for which A, B, C and D combine equally as $[1, 1, 1, 1]$ and the mask has size s*s. When s is even, there will be 4 central pixels located exactly between A, B, C and D, each of which will be summed as $[1, 1, 1, 1]$. The mask size is then $(s + 1) * (s + 1)$. Note that the full mask size, S, is now always odd. Let the mask have coefficients $M(x, y)$, where $-w \leq x$, $y \leq +w$, and where $w = (S-1)/2$. All coefficients of M that fall outside the S*S area are assumed to have value zero.

2.1 Rules for Constructing Symmetric Masks to In-fill p:q Lattice Pixels

(1) The coefficients for contributions from points A, B, C and D are set to [1, 1, 1, 1] at the position of the single centre pixel when s is odd. The same values are applied at the four centremost pixels when s is even.

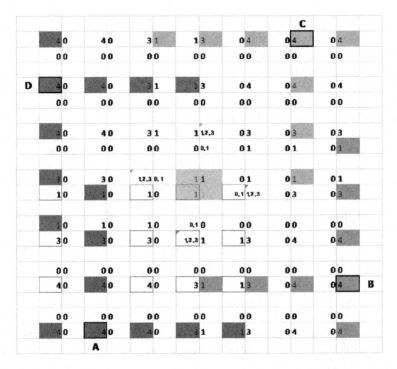

Fig. 3. Construction of 1:5 symmetric mask values from 4 points (labelled A (red), B (green), C (grey) and D (blue) on a 1:5 lattice. Point D is located 1:5 steps from point A. Point B is located 5:1 steps from point A. The 4 numbers clustered around each location denote the relative contribution summed to that point by each of the 4 lattice pixels. The four numbers, starting anticlockwise from bottom left, come from lattice points A, B, C and D, respectively. The same symmetric 7*7 mask $M(x, y)$ is placed with its centre at each of the 4 lattice points. The fourfold symmetry means we only need to find one octant of the coefficients in M. The yellow square marks the initial centre pixel (it is equidistant to A, B C and D). It is initialised with equal contributions [1, 1, 1, 1] from A, B, C and D as the starting point and from which we find all of the remaining coefficients of the mask M. Where two options to fill a pixel are possible (shown in smaller font), we choose the maximum of those two values (the resulting values that get selected are shown here as bold) (Color figure online)

(2) If $M(x,y) = v$, then $M(x,-y) = M(-x,\pm y) = M(\pm y, \pm x) = v$, i.e. four-fold symmetry, $\{v = 0, 1, 2, 3 \text{ or } 4\}$

(3) Lattice point $D = 4 - v$ when A, B, C = $[0, 0, v]$, $[0, v, 0]$ or $[v, 0, 0]$.

(4) If $M(x \pm 1, y) = v$ or $M(x, y \pm 1) = v$, then $M(x,y) \geq v$, monotonicity

(5) If $M(x,y) = u \| v$, then $M(x,y) = max(u,v)$ when a paired choice possibility arises

Figure 3 provides an example of the construction method used to find the coefficients for the 7*7 symmetric interpolation mask M for $p{:}q = 1{:}5$. Figure 4 summarises the final coefficients of M as obtained from Fig. 3.

It seems that simply imposing four-fold symmetry and requiring equal sharing at the centre pixel location(s) is sufficient to constrain the values for all of the coefficients of $M(x,y)$ for any p:q. It is however tedious to work manually thorough the symmetry points to establish values for each of the coefficients for large p:q. It would be gratifying to have a theoretical proof that the conditions imposed above are necessary and sufficient for all p:q.

Figure 5 shows example mask coefficients of M for the lattice directions 1:7 and 3:7. A symmetric mask can be found for each p:q that we have tried. To date, we have no counter examples.

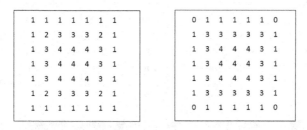

Fig. 4. Symmetric coefficients M for a 1:5 lattice. Left: using the maximum value (Rule 5) when forced to choose. Right: resulting mask from the alternate choice

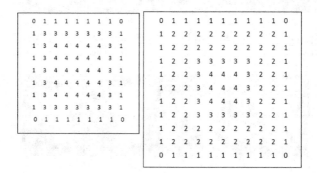

Fig. 5. Symmetric coefficients for 1:7 (9*9 mask) and 3:7 (11*11 mask) lattices

3   Mask Performance for Pixel In-filling

For a p:q lattice that has constant intensity values, the interpolated response must be exactly flat for all interior pixels of a finite lattice of any size. Figure 6 shows in-filled lattices for $p{:}q = 1{:}5$ and 3:7 for a small triangular area containing 15 lattice pixels. The matching result for the same traditional area-based filter is shown for comparison. As expected, by construction, the interior region for all masks is exactly flat.

The borders of the in-filled regions provide an indication of how the masks perform when there are different values at the nearest lattice points. In-filled pixels in the interior of these binary shapes will be interpolated from four know values and will (by design) result in perfectly smooth values. Pixels near the edge of the binary shape will have in-filled values interpolated from one, two or three known values. Thus the edge pixel intensities depend strongly on the design of the mask coefficients.

The size of the symmetric interpolating masks is always larger than the area masks (because these masks extend across the centre pixel region) so the symmetric interpolated edges are wider, but the interpolated edge values are better aligned with the row and column axes.

If the rules for symmetric filling are relaxed (in particular Rule 5), then alternate versions of these filters can be constructed, as was also shown in Fig. 4. Figure 7 shows the resultant in-filling for a small triangle of points on a 3:5 lattice. The position of each original lattice pixels is shown here as a black point inside the triangle. Results for three different symmetric masks as well as the area-based mask are shown. The mask values are given in Fig. 8. Mask (d) of Fig. 8 adheres to the full rules for symmetric mask generation as it has the broadest

Fig. 6. A triangular area of uniform valued lattice pixels filled by, left to right: $p{:}q = 1{:}5$ area filled then symmetric filled, $p{:}q = 3{:}7$, area filled then symmetric filled

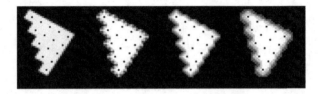

Fig. 7. Interpolated images for 3:5 lattice points inside a triangular area containing 15 points using (left to right) the four mask values shown in Fig. 8(a) to (d) respectively. The original unit-intensity lattice pixel positions are shown as the black points inside the triangular area

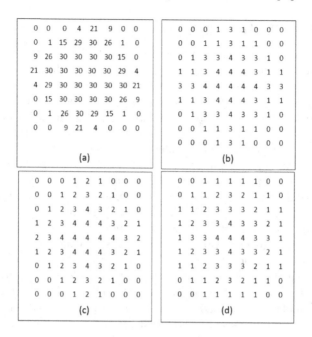

0	0	0	4	21	9	0	0
0	1	15	29	30	26	1	0
9	26	30	30	30	30	15	0
21	30	30	30	30	30	29	4
4	29	30	30	30	30	30	21
0	15	30	30	30	30	26	9
0	1	26	30	29	15	1	0
0	0	9	21	4	0	0	0

(a)

0	0	0	1	3	1	0	0	0
0	0	1	1	3	1	1	0	0
0	1	3	3	4	3	3	1	0
1	1	3	4	4	4	3	1	1
3	3	4	4	4	4	4	3	3
1	1	3	4	4	4	3	1	1
0	1	3	3	4	3	3	1	0
0	0	1	1	3	1	1	0	0
0	0	0	1	3	1	0	0	0

(b)

0	0	0	1	2	1	0	0	0
0	0	1	2	3	2	1	0	0
0	1	2	3	4	3	2	1	0
1	2	3	4	4	4	3	2	1
2	3	4	4	4	4	4	3	2
1	2	3	4	4	4	3	2	1
0	1	2	3	4	3	2	1	0
0	0	1	2	3	2	1	0	0
0	0	0	1	2	1	0	0	0

(c)

0	0	1	1	1	1	1	0	0
0	1	1	2	3	2	1	1	0
1	1	2	3	3	3	2	1	1
1	2	3	3	4	3	3	2	1
1	3	3	4	4	4	3	3	1
1	2	3	3	4	3	3	2	1
1	1	2	3	3	3	2	1	1
0	1	1	2	3	2	1	1	0
0	0	1	1	1	1	1	0	0

(d)

Fig. 8. Alternate forms for 3:5 lattice masks. (a) Traditional area-based, (b) to (d) using variations of Rule 5 to select between tied values for coefficients of symmetric masks $M(x, y)$. (b) Minimum, (c) intermediate, (d) maximum

spread, which is ensured by requiring that the maximum coefficients occur along the periphery of the mask.

4 Rotation of Images by p:q and q:p

4.1 Discrete Rotation Angles

Whilst it is common to use simple units of degrees for angular measurements, it turns out that discrete angles defined by p:q ratios provide good coverage of uniformly-spaced angles when using reasonably small values of p and q (naturally any angle can be approximated with arbitrary precision by $atan(q/p)$ using arbitrarily large integers p and q). For example, of the 44 angles from 1° to 44° in 1° steps, 27 of these angles can be approximated to within ±0.5° as discrete p:q ratios with $|p|$ and $|q| \leq 11$.

Another nice feature of discrete rational angles is the way they can be combined to reach multiples of 90°. For example, a rotation of $p:q = \theta$ followed by $q : p = 90° - \theta$ is a 90° rotation for any choice of p:q. Similarly, we can find sets of distinct p:q rotations that cumulatively effect a net rotation by a multiple of 90°. For example, the net rotations {7:4, 7:6, 1:5, 1:7, 3:4, 1:4} is exactly 180°.

4.2   Measuring the Quality of Rotated Images

Rigorous, perceptually-based measures of image quality to assess digital image rotation algorithms are difficult to design and use. The fidelity of a rotated image depends on the rotation angle as well as the image content. There is no way to evaluate a rotated image without recourse to a 'golden standard'.

We then resort to using rough but simple physical measures such as mean squared error (MSE) and peak signal to noise ratio (PSNR). The choice of angle at which to compare the efficacy of different rotation algorithms is again somewhat arbitrary. Many algorithm evaluations have used multiple rotations by the same angle to arrive at a cumulative rotation such as $90°$ or $180°$, for which we can make an exact comparison copy for any image data (for example, $6 * 15°$ rotations $= 90°$). This strategy is better than using rotations of $+\theta$ followed by $-\theta$ to effect a net zero degree rotation, because the positive and negative rotation errors of a given algorithm can contrive to cancel exactly, giving the illusion that each θ rotation was perfect. However the interpretation that can be assigned to the quality of a single rotation of $16°$ is unclear, when given an MSE obtained for 6 successive such rotations (the interpolation effects of rotations are not linearly cumulative, especially when the same angle is used repeatedly).

Here we choose to use sets of 6 different p:q rotations and then apply the 6 rotations many times in randomised order to get a mean quality measure. We can then apply the same test using 5 or 7 sets of similar angles to judge the level of distortion contributed by adding or subtracting one extra rotation.

4.3   Rotation Algorithms

Algorithms for efficient discrete image rotation date from the beginnings of computer vision (for example [3]). The three-pass skew algorithm [2,5,6] that was developed more recently seems to perform exceptionally well. Another means to accurately rotate images is by extending (by reflection) the sinogram of a high precision conventional $180°$ Radon transform to cover projection angles $0°$ to $360°$ and then indexing the 'zero' starting angle to the required rotation angle for precise image reconstruction (by filtered back-projection, for example).

Here we perform an exact, lossless p:q affine image rotation, then apply the p:q in-fill masks $M(x, y)$ to interpolate the image that is now up-scaled in area by $p^2 + q^2$. The (potentially large) increase in the up-scaled image size is not a problem if the computer memory capacity permits. We then downsize the rotated image by factor $f = 1/\sqrt{p^2 + q^2}$ along both column and row axes to obtain a rotated image at the original image scale. The quality of the final rotated image then depends on only the pixel in-fill method applied and the method used for image downsizing (although we apply Gaussian smoothing to the up-scaled and in-filled image before downsizing to reduce the aliasing).

Surprisingly, the supposedly simple process of downsizing images causes a lot of variation in the rotated image. We tried to stabilize these results by pre-rotating a test image. The test image, of the same size as the original data, is comprised of a unit-valued pixel located at the rotation centre, with between 7

Fig. 9. Six successive rotations, by {1:7, 1:5, 3:4, 7:4, 7:6, 1:4}, of a test image (a 509 × 509 portion of boats.jpg) for which the final accumulated angle is 180°. Top row: symmetric masks, middle row: area-base masks, bottom row: normalised difference images. The resulting net PSNR (measured over a circular region with a diameter just inside the final aperture), remains about 33 ± 3 for the same set of angles for the same set of rotations performed in random order

to 10 unit-valued pixels, placed at nearly equal-spaced angular intervals, around the perimeter of a circle drawn about the image centre.

The test image is then up-scaled by the selected p:q rotation and convolved with the appropriate p:q in-fill mask. We then shift and zero-pad this test image until the downsizing by the scale factor f preserves the intensity of the rotated test image centre pixel and perimeter pixels to better than 95 % (where possible). We then downsize the rotated up-scaled and in-filled original image data, applying the optimal shift and pad values that best preserved the test image pixels.

Figure 9 shows a typical result obtained for the set of rotations {7:4, 7:6, 1:5, 1:7, 3:4, 1:4}. The PSNR of the final result has a mean value of around 33 ± 3, depending on the order in which the rotations are performed. This result is consistent with the values we obtained using traditional bi-linear or bi-cubic interpolation algorithms, but poorer than for the three-pass skew rotation method [2,5,6], which averaged a PSNR closer to 40.

The symmetric masks can perform better than the area-based masks because of the enforced alignment of the interpolated values along the image rows and columns, providing that the downsizing step can be well synchronised.

We observed 'perfect' results (precisely zero MSE) using the symmetric masks for a net 90° rotation done as 3:4 followed by 4:3 (provided no anti-alias smoothing is applied). The downsizing factor for 3:4 is an exact integer, $f = \sqrt{3^2 + 4^2} = 5$. The intermediate image is however not perfect. It is clear that here the visible 3:4 rotation errors are exactly undone by the 4:3 rotation.

5   Projections of 2D Symmetric Masks for 1D Haar Interpolation

The in-filling of pixels on a p:q lattice also has relevance for back-projecting data on a grid at discrete angles. The summation of discrete pixels along a line (under the Dirac model) can be converted to area-based integrals (that represent rays of finite width) by convolving the 1D Dirac projections with Haar filters. The appropriate Haar filter for projection p:q is the 1D projection of the 2D area-based p:q mask [4].

Figure 10 displays the 1D projection at 3:5 for the 3:5 area-based and symmetric masks. The projection profiles are notably distinct and reflect the very different underlying assumptions behind their construction. Haar filtered back-projected images for direction 3:5 that arise from a point are also shown for comparison. Because the symmetric filters all have odd size, the 2D filled back-projection can always be replicated exactly by 1D convolution of the projected data for any p:q. For area-based masks where $p+q$ is even, the 1D back-projection is misaligned at the edges of the in-filled rays.

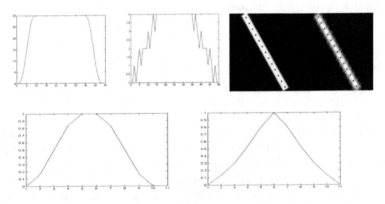

Fig. 10. Top row, left to right: normalised 3:5 projection of the 2D 3:5 mask for area-filled and symmetric masks, then area-based and symmetric in-filled back-projection of a single image point at angle 3:5. Black points mark the position of the original un-filtered back-projected pixels. Bottom row: normalised 1D column (or row) sums of the 2D 3:5 masks, left, area-based mask, right symmetric mask.

6   Conclusions and Future Work

We considered here the construction of convolution masks to uniformly in-fill pixels on p:q lattices using simple symmetry constraints, rather than the traditional area-based assumptions. We tested the efficiency of these interpolation masks when used as part of a discrete rotation algorithm. The quality of the image rotations obtained using the symmetric masks was at least as good and usually better than that when using the area-based versions.

We have yet to examine how these symmetric masks perform in discrete image reconstruction when used as 1D filters to smooth discrete back-projected data. The construction of 2D symmetric masks should also be able to be extended to higher dimensions, for example to $M(x, y, z)$, using, as far as possible, the same symmetry rules as listed here in Sect. 2. It would be interesting to examine the rules that constrain the construction of six-fold symmetric 2D masks for hexagonal p:q lattices.

Acknowledgements. The School of Physics and Astronomy at Monash University provided partial funding to IS for this research. It also supported the residence of AG during an internship at Monash University Clayton in 2015, as part of his M.Sc. studies at Polytech Nantes in Nantes, France. AG also received funding assistance from the Region Pays de Loire in France. IS acknowledges ongoing collaboration with members of the IVC group at Polytech Nantes.

References

1. Blot, V., Coeurjolly, D.: Quasi-affine transformation in higher dimension. In: Brlek, S., Reutenauer, C., Provençal, X. (eds.) DGCI 2009. LNCS, vol. 5810, pp. 493–504. Springer, Heidelberg (2009)
2. Condat, L., Ville, D.D., Forster-Heinlein, B.: Reversible, fast and high-quality grid conversions. IEEE Trans. Image Process. **17**(5), 679–693 (2008)
3. Fraser, D.: Comparison of high spatial frequencies of two-pass and one-pass geometric transformation algorithms. Computer Vis. Graph. Image Process. **46**, 267–283 (1989)
4. Guedon, J.: The Mojette Transform: Theory and Applications. ISTE-Wiley, Chichester (2009)
5. Unser, M., Thévenaz, P., Yaroslavsky, L.: Convolution-based interpolation for fast, high-quality rotation of images. IEEE Trans. Image Process. **4**(10), 1371–1381 (1995)
6. Owen, C.B.: High quality alias free image rotation. In: Conference Record of the Asilomar Conference on Signals, Systems and Computers. IEEE, Pacific Grove, November 1996
7. Svalbe, I.: Exact, scaled image rotations in finite radon space. Pattern Recogn. Lett. **32**, 1415–1420 (2011)

Digital Surfaces of Revolution Made Simple

Eric Andres[✉] and Gaelle Largeteau-Skapin

Université de Poitiers, Laboratoire XLIM, SIC, UMR CNRS 7252, BP 30179,
Futuroscope, Chasseneuil, 86962 Poitiers, France
{eric.andres,gaelle.largeteau.skapin}@univ-poitiers.fr

Abstract. In this paper we present a new, simple, method for creating digital 3D surfaces of revolution. One can choose the topology of the surface that may have 0, 1 or no tunnels. The definition we propose is not limited to circles as curves of revolution but can be extended to any type of implicit curves.

Keywords: Digital surfaces · Implicit functions · Surface of revolution

1 Introduction

Pottery wheels made the surface of revolution one of the first types of complex surfaces that man has created. In Computer Graphics [1], surfaces of revolution, as special case of swept surfaces, have always represented a simple way of constructing surfaces that appear quite frequently in real life: glasses, chess pieces, lamps, flower pots, etc. They are based on a 2D curve profile called *generatrix* and a *revolution curve* which is classically a 2D circle. In digital geometry, there are not many papers that have dealt specifically with the problem of generating digital surfaces of revolution. There is the works of Nilo Stolte that indirectly looked at such problems as part of research on the visualization of implicit surfaces with cylindrical coordinates [2,3]. In two more recent papers, G. Kumar and P. Bhowmick [4,5] proposed a virtual pottery design tool based on digital surfaces of revolution. Their idea is to generate a digital surface of revolution by superposing 2D digital annuli. The horizontal section of a continuous surface of revolution is, by definition, the curve of revolution and thus, in general, a circle. A vertical slice of height one can of course be very complicated but rasterized, it corresponds to a 2D digital annulus. The authors work with a classical notion of circle close to the one proposed by Bresenham, defined only for integer radii and center, and this creates some difficulties. While the center of the annulus for z is the point of integer coordinates $(0, 0, z)$, there is no particular reason for the interior or exterior radii to be an integer. Moreover, Bresenham circles do not fill space (i.e. concentric Bresenham circles of increasing radii leave points that do not belong to any circle). G.Kumar and P.Bhowmick solve this by determining the missing (also sometimes called *absentee*) voxels that would otherwise leave 6-connected holes in the revolution surface. Their method is quite general since it does not require the generatrix to be functional in z (only one ordinate value

© Springer International Publishing Switzerland 2016
N. Normand et al. (Eds.): DGCI 2016, LNCS 9647, pp. 244–255, 2016.
DOI: 10.1007/978-3-319-32360-2_19

per z, if we consider the generatrix to be in the yz-space). The method for filling the holes with absentee voxels is however quite complicated and limited to circular curves of revolution. One way around this problem could be to use Andres circles [6–8] that, do, fill space and are defined for arbitrary center and radii.

In this paper we are going to adopt a different approach based on the digitization of implicit curves and surfaces proposed at the last DGCI conference [9] with the participation of the authors of the present paper. We showed that it is possible to analytically define digital implicit surfaces in dimension n with controlled topology. Under some regularity conditions [9], one can ensure that the analytically defined digital implicit surface is a k-tunnel free $(n-1)$-dimensional digital surface in dimension n, with $0 \leq k < n$. In this paper, we show how this can be used to create surfaces of revolution with a 2D explicit function as generatrix and a 2D implicit curve as curve of revolution. The generatrix and the curve of revolution are combined in order to obtain a 3D implicit surface that is then digitized into 0, 1 or 2-tunnel free digital surfaces of revolution. The control of the topology allows us to choose the type of surface we need for a given application. For instance, for visualization purposes, a thinner 2-tunnel free surface might suffice. If we want to use a 3D printer however, we need 2-connectivity (voxel-face connectivity) and thus 0 or 1-tunnel free surfaces of revolution. The curve of revolution is, in our model, not limited to a circle. One can use any implicit 2D curve as long as it separates space into a positive and a negative side. Of course, if the curve is infinite, one has to consider a generation window in which to consider the revolution curve or simply be satisfied with an analytical description of the surface of revolution. By considering only an explicit function generatrix of type $y = f(z)$, our generatrix are less general than [4,5]. There are however solutions and we will give an example in the conclusion. There remains some work to be done in this regard. The method presented in this paper is very simple and general but it comes with some limitations when the curve of revolution does not respect some regularity conditions [9]. We will give an example of a surface of revolution that has holes.

In Sect. 2, we present basic notations and the analytical implicit surface digitization method. In Sect. 3, we present our digital surface of revolution, show some results and extensions. We finish the section with the limitations of our generation method. In Sect. 4, we conclude and present short and long term perspectives for this work.

2   Basic Notions and Notations

Let $\{e_1, \ldots, e_n\}$ denote the canonical basis of the n-dimensional Euclidean vector space. Let $\mathbb{Z}^n$ be the subset of $\mathbb{R}^n$ that consists of all the integer coordinate points. A *digital (resp. Euclidean) point* is an element of $\mathbb{Z}^n$ (resp. $\mathbb{R}^n$). We denote by x_i the i-th coordinate of a point or a vector x, that is its coordinate associated to e_i. A *digital (resp. Euclidean) object* is a set of digital (resp. Euclidean) points.

For all $k \in \{0, \ldots, n-1\}$, two integer points v and w are said to be k-*adjacent* or k-*neighbors*, if for all $i \in \{1, \ldots, n\}$, $|v_i - w_i| \leq 1$ and $\sum_{j=1}^{n} |v_j - w_j| \leq n-k$. In the

2-dimensional plane, the 0- and 1-neighborhood notations correspond respectively to the classical 8- and 4-neighborhood notations. In the 3-dimensional space, the 0-, 1- and 2-neighborhood notations correspond respectively to the classical 26-,18- and 6-neighborhood notations.

A k-*path* is a sequence of integer points such that every two consecutive points in the sequence are k-adjacent. A digital object E is k-connected if there exists a k-path in E between any two points of E. A maximum k-connected subset of E is called a k-*connected component*. Let us suppose that the complement of a digital object E, $\mathbb{Z}^n \backslash E$ admits exactly two k-connected components F_1 and F_2, or in other words that there exists no k-path joining integer points of F_1 and F_2, then E is said to be k-*separating*, or k-*tunnel free*, in $\mathbb{Z}^n$. If there is no path from F_1 to F_2 then E is said to be 0-separating or simply *separating*.

Let $\oplus$ be the Minkowski addition, known as dilation, such that $\mathcal{A} \oplus \mathcal{B} = \cup_{b \in \mathcal{B}} \{a + b : a \in \mathcal{A}\}$.

2.1 Implicit Surface Digitization

In this paper we are considering analytical digital implicit surfaces as defined in [9]. Let $\mathcal{S}$ be an implicit surface $\mathcal{S} = \{x \in \mathbb{R}^n : f(x) = 0\}$ which separates space into one (or several) region(s) where $f(x) < 0$ and one (or several) region(s) where $f(x) > 0$. Our digitization method is based on a morphological type digitization method for implicit surfaces based on k-flakes (introduced in [10]). Those adjacency flakes can be described as the union of a finite number of straight segments centered on the origin (see Fig. 1).

Definition 1. *Let* $0 \leq k < 3$*. The minimal* k*-adjacency flake,* $F_k(\rho)$ *with radius* $\rho \in \mathbb{R}^+$ *is defined by:*

$$F_k(\rho) = \left\{ \lambda u : \lambda \in [0, \rho], u \in \{-1, 0, 1\}^3, \sum_{i=1}^{3} |u_i| = 3 - k \right\}.$$

The Flake-digitization of a $(n-1)$-dimensional surface S in $\mathbb{R}^n$ is defined by $\mathcal{F}_k(S) = \{v \in \mathbb{Z}^n : \left(F_k\left(\frac{1}{2}\right) \oplus v\right) \cap S \neq \varnothing\}$. Under some conditions [9], the flake digitization can be analytically characterized by considering only the vertices of the k-flake. We are now simply going to consider that this analytical characterization corresponds to a proper analytical digitization method defined by:

$$\mathcal{A}_k(S) = \left\{ v \in \mathbb{Z}^n : \begin{array}{l} min\{f(x) : x \in (v \oplus F_k(1/2))\} \leq 0 \\ \text{and } max\{f(x) : x \in (v \oplus F_k(1/2))\} \geq 0 \end{array} \right\}.$$

When the regularity conditions are verified then $\mathcal{A}_k(S) = \mathcal{F}_k(S)$ otherwise there are some differences that may in some cases create topological problems (see Sect. 3.6) but this is largely compensated by the fact that $\mathcal{A}_k(S)$ is easy to construct while $\mathcal{F}_k(S)$ may not. The regularity conditions as presented in [9] are the following: S should be r-regular [11] nD surface with $r > \left(\sqrt{n-k} + \sqrt{n}\right)/2$. An r-regular set [11] is closed set such that for all $x \in \delta E$, it is possible to find

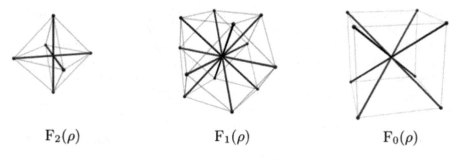

$F_2(\rho)$ $F_1(\rho)$ $F_0(\rho)$

Fig. 1. Adjacency flakes in 3D.

two osculating open balls of radius r, one lying entirely inside E and the other lying entirely outside E. These conditions ensures the k-tunnel freeness of the digital implicit surface. They also preserve the connected components between the complement of the implicit surface and the complement of its digitization with regard to the k-adjacency relationship. These conditions are sufficient but not necessary. A necessary condition is still an open (and difficult) question. For short, $\mathcal{A}_k(S)$ is different from $\mathcal{F}_k(S)$ in places where the details of the curve S are small compared to the grid size which is not completely surprising. What is important is that the flake digitization $\mathcal{F}_k(S)$ is k-separating in $\mathbb{Z}^n$ if it, basically, defines a positive and a negative region (for f defining S) greater than a pixel [9,10]. This remains true for $\mathcal{A}_k(S)$ under the restrictions we mentioned.

3   Digital Surface of Revolution

3.1   Implicit Surface of Revolution Definition

The surfaces of revolution are usually defined by rotating a curve (called the generatrix) around a straight line (the axis). In this paper, we generalise the notion: we use an explicit function $y = g(z)$ as a generatrix and instead of rotating it around an axis, we are using a *revolution* curve. In our work, the revolution function can be any 2D implicit continuous curve $r(x, y) = 0, (x, y) \in [-1, 1]^2 \subset \mathbb{R}^2$, that separates the window $[-1, 1]^2$ into two regions (one where the function r is positive, one where it is negative). If the curve of revolution is a circle then we obtain the classical surface of revolution. Note that the y of the generatrix definition $y = g(z)$ is not the same y that appears in the curve of revolution implicit equation $r(x, y) = 0$.

The surfaces we define use the generatrix as an homothetic factor for the revolution curve:

Definition 2

$$S(g, r) = \left\{ (x, y, z) \in \mathbb{R}^3,\ r\left(\frac{x}{g(z)}, \frac{y}{g(z)} \right) = 0 \right\}$$

There is therefore a constraint on the generatrix: $\forall z \in \mathbb{R}$, $g(z) \neq 0$. Actually, if one considers surfaces or revolution, then one can restrict the definition to $g(z) >= 1$. For $g(z) = 1$, with $r(x, y)$ defined on $[-1, 1]^2$, the slice of the digital surface of revolution corresponding to z will only be, at most, one voxel big. Smaller values of $g(z)$ may only make the slice disappear because it is smaller than what the analytical flake digitization can detect. Also, this constraint can be lifted in many cases by simply developing the expression r depending on how it is defined. For instance, when $r(x, y) = x^2 + y^2 - 1 = 0$ is the unit circle c, the resulting surface is the classical surfaces of revolution: $S(g, c) = \{x, y, z \in \mathbb{R}^3, \ x^2 + y^2 - g(z)^2 = 0\}$.

In all cases, we obtain an implicit surface that can be digitized using the $\mathcal{A}_k$-digitization.

3.2   Digital Generatrix

Sometimes, the generatrix we would like to consider is not defined by an explicit function but by a digital curve given as a set of digital points (such as an hand drawn generatrix for instance). In this case, one can easily recreate an explicit function generatrix.

There are several simple ways to transform a set of discrete points into an explicit function. The point to look out for is that the generatrix must be defined in such a way that it allows to compute values for $z \pm \frac{1}{2}$ (see Sect. 3.3). A first method consist in creating an explicit function by interpolation of the discrete points that form your digital curve. A second method consists in simply consider that each integer z value is valid for the interval $[z - 1/2, z + 1/2]$.

The last method consists in decomposing the digital curve into digital line segments and then taking the continuous analog of the digital line: for a digital

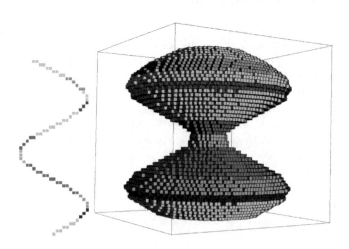

Fig. 2. A digital generatrix decomposed into digital straight segments (left) and the resulting revolution surface (right) using the unit implicit circle as revolution curve.

line $0 \le ax - by + c < b$, we will consider the continuous line $ax - by - c - b/2 = 0$ or $ax - by - c - a/2 = 0$ depending on the orientation. A little care has to be taken for the end points of the different digital lines. If the intersections of two consecutive line segments does not fall in the pixel of the digital point corresponding to the common end point then a little patch function has to be added (see [12]) or one has to use an adapted line recognition where this problem does not occur [13]. This creates a piecewise defined explicit function. Figure 2 shows a digitized sinusoid that we have decomposed into digital line segments. In this example, the curve of revolution is a circle.

3.3 Algorithm

The digitization algorithm has been implemented in Mathematica. It is decomposed into two main functions: the first function tests if a point is in the digitization; the second scans the 3D digital space, applies the previous test and adds the valid points to the digitization result. We present here the three functions (one for each 3D flake digitization) that tests if a voxel $v(x, y, z)$ belongs to the digitization of the surface S defined by its implicit equation. Then we present the scanning function. Since the code for those functions is very simple, we present the functions as they appear in the mathematica notebook.

$\mathcal{A}_2$-**digitization test:** This will define a surface that is 2-separating and 1-connected. This is what is classically called a *naive* surface in the digital geometry community. S is the implicit 3D surface. The result is a Boolean expression.

inNaif3D$[S_-, x_-, y_-, z_-] :=$
Module$[\{f1, f2, f3, f4, f5, f6\},$
$f1 = 1.0 * S[x + 0.5, y, z]; f2 = 1.0 * S[x - 0.5, y, z];$
$f3 = 1.0 * S[x, y - 0.5, z]; f4 = 1.0 * S[x, y + 0.5, z];$
$f5 = 1.0 * S[x, y, z + 0.5]; f6 = 1.0 * S[x, y, z - 0.5];$
And$[\textbf{Min}[f1, f2, f3, f4, f5, f6] <= 0, \textbf{Max}[f1, f2, f3, f4, f5, f6] >= 0]];$

$\mathcal{A}_1$-**digitization test:** This will define a surface that is 1-separating and 0-connected. This is an intermediary surface type between the naive and the supercover types.

inInter3D$[S_-, x_-, y_-, z_-] :=$
Module$[\{f1, f2, f3, f4, f5, f6, f7, f8, f9, f10, f11, f12\},$
$f1 = 1.0 * S[x + 0.5, y + 0.5, z]; \quad f2 = 1.0 * S[x - 0.5, y + 0.5, z];$
$f3 = 1.0 * S[x + 0.5, y - 0.5, z]; \quad f4 = 1.0 * S[x - 0.5, y - 0.5, z];$
$f5 = 1.0 * S[x + 0.5, y, z + 0.5]; \quad f6 = 1.0 * S[x + 0.5, y, z - 0.5];$
$f7 = 1.0 * S[x - 0.5, y, z + 0.5]; \quad f8 = 1.0 * S[x - 0.5, y, z - 0.5];$
$f9 = 1.0 * S[x, y + 0.5, z + 0.5]; \quad f10 = 1.0 * S[x, y + 0.5, z - 0.5];$
$f11 = 1.0 * S[x, y - 0.5, z + 0.5]; f12 = 1.0 * S[x, y - 0.5, z - 0.5];$
And$[\textbf{Min}[f1, f2, f3, f4, f5, f6, f7, f8, f9, f10, f11, f12] <= 0,$
Max$[f1, f2, f3, f4, f5, f6, f7, f8, f9, f10, f11, f12] >= 0]];$

$\mathcal{A}_0$-**digitization test:** This will define a surface that is 0-separating and 0-connected. This is close to the classical supercover type surfaces [14–17].

inSuper3D$[S_-, x_-, y_-, z_-] :=$
Module$[\{f1, f2, f3, f4, f5, f6, f7, f8\},$
$f1 = 1.0 * S[x + 0.5, y + 0.5, z + 0.5];$ $f2 = 1.0 * S[x + 0.5, y + 0.5, z - 0.5];$
$f3 = 1.0 * S[x + 0.5, y - 0.5, z + 0.5];$ $f4 = 1.0 * S[x + 0.5, y - 0.5, z - 0.5];$
$f5 = 1.0 * S[x - 0.5, y + 0.5, z + 0.5];$ $f6 = 1.0 * S[x - 0.5, y + 0.5, z - 0.5];$
$f7 = 1.0 * S[x - 0.5, y - 0.5, z + 0.5];$ $f8 = 1.0 * S[x - 0.5, y - 0.5, z - 0.5];$
And$[$**Min**$[f1, f2, f3, f4, f5, f6, f7, f8] <= 0,$
Max$[f1, f2, f3, f4, f5, f6, f7, f8] >= 0]];$

Scan Algorithm: In the following module, (x0,x1), (y0,y1) and (z0,z1) are the intervals on which to generate the surface. A list of elements in mathematica, for instance a, b, c, is defined by $\{a, b, c\}$. The operation 'AppendTo' adds a value at the end of the list. The result of this module will be a list of points (a point is defined as a list of three coordinates). 'TopoTest' stands for one of the above mentioned modules inNaif3D, inInter3D, inSuper3D.

Revolution $[S_-, x0_-, x1_-, y0_-, y1_-, z0_-, z1_-] :=$
Module$[\{x, y, z, listpoint\}, listpoint = \{\};$
For$[x = x0, x <= x1, x + +,$
For$[y = y0, y <= y1, y + +,$
For$[z = z0, z <= z1, z + +,$
If$[TopoTest[S, x, y, z],$
AppendTo$[listpoint, \{x, y, z\}]];$
$]]];$
$listpoint];$

In order to be concise, the scan algorithm we present here is a very basic one. It is not very difficult to design one that works by propagation with a seed point and this for all three types of surfaces. A seed point is also not difficult to determine when considering surfaces of revolution.

3.4   Results

In this part, we present some examples of digital surfaces of revolution. Figure 3 presents the digitization of a generalized surface of revolution with a sinusoid function $g(z) = Sin[z/9] * 12 + 17$ as generatrix and the lemniscate function $r(x, y) = (0.25 * x^2 + 0.15 * y^2)^2 - 0.2 * (0.25 * x^2 - 0.15 * y^2)$ as revolution curve.

Figure 4 presents similar objects than in Fig. 3. Between two objects, the lemniscate (the revolution curve) has been rotated and the phase of the sinusoid (the generatrix) has been changed[1]. The view point of the camera is the same for the four objects.

Figure 5 presents the three 3D digitizations of a surface: the naive $\mathcal{A}_2$-digitization on the top right, the $\mathcal{A}_1$-digitization on the bottom left and the $\mathcal{A}_0$-digitization on the bottom right. This surface is built using the same generatrix as in Fig. 3 and $r(x, y) = (2x^2 + 2y^2)^3 - 48x^2 y^2 = 0$ as a revolution

---

[1] An animated version of these results can be seen on the following web page: http://xlim-sic.labo.univ-poitiers.fr/themes/ig/ig_axe_3.php.

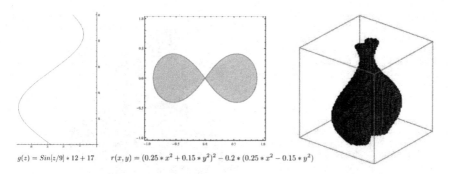

$$g(z) = Sin|z/9| * 12 + 17 \qquad r(x,y) = (0.25 * x^2 + 0.15 * y^2)^2 - 0.2 * (0.25 * x^2 - 0.15 * y^2)$$

Fig. 3. Digital Generalized Surface of Revolution. The generatrix (left), the revolution curve (center) and the resulting digital surface (right).

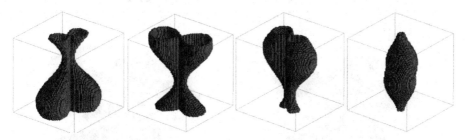

Fig. 4. Several examples of digital surfaces obtained with lemniscate revolution curves and sinusoidal generatrices.

curve (top left of the figure). Voxels are represented smaller in those pictures to emphasize the topological properties (connectivity and tunnels).

3.5 Extensions

The method we have presented can easily be extended to generate various, more general, types of surfaces. One simple extension is to use the generatrix, not as an homothetic function but as central axis for the revolution curve:

Definition 3

$$S(g,r) = \left\{ x, y, z \in \mathbb{R}^3, \ r\left(x, y - g(z)\right) = 0 \right\}$$

In an other example, we combine a homothetic function $h(z)$ and two functions for translations $t(z), u(z)$ of the center of the revolution curve:

Definition 4

$$S(h,t,r) = \left\{ x, y, z \in \mathbb{R}^3, \ r\left(\frac{x}{h(z)} - u(z), \frac{y}{h(z)} - t(z) \right) = 0 \right\}$$

Fig. 5. The three different digitizations ($\mathcal{A}_2$,$\mathcal{A}_1$ and $\mathcal{A}_0$) with $r(x,y) : (2x^2 + 2y^2)^3 - 48x^2y^2 = 0$ as a revolution curve and the generatrix $g(z) = Sin[z/9] * 12 + 17$.

On the left of Fig. 6 we can see an example where a sinusoid generatrix is considered as a translation function for the revolutions function (here the unit circle). On the right of Fig. 6, the digital surface is obtained with an homothetic factor $h(z) = |z|/5$, two translation functions $t(z) = sin(z/5)$ and $u(z) = cos(z/5)$ and the unit circle $r(x,y) = x^2 + y^2 - 1 = 0$ as a revolution function. The implicit equation of the surface is the one of Definition 4.

3.6   Limitations

The main limitation of this digital revolution surface generation method is that the analytical digitization can miss some points. Figure 7 (left) shows that a

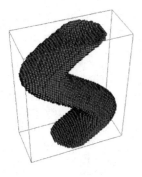

Fig. 6. Two examples of some other surface we can build with our algorithm.

curve can cross a voxel but the vertices of the flake are on only one side of the curve. The voxel is therefore wrongly discarded from the digitization result and disconnections appear. This problem is classically dealt with interval arithmetics [18]. The problem with an interval arithmetic approach is that it is easy to use to replace $\mathcal{A}_0$ but more complicated for both other analytical digitizations. This could be an interesting problems for the future.

Figure 7 presents an example where this limitation can be seen: there are some missing voxels in the surface. Those missing voxels correspond to the points where the revolution curve (middle of Fig. 7) crosses itself.

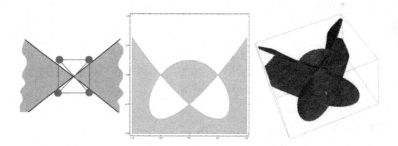

Fig. 7. $\mathcal{A}_2$-digitization with $r(x,y) = 16y^3 + 12y^2 - (4x^2 - 1)^2 = 0$ as a revolution curve and the generatrix $g(z) = Sin[z/9] * 12 + 17$. Visible holes in the surface.

4 Conclusion

In this paper, we have presented a very simple generation algorithm for digital surfaces of revolution. We have proposed some extensions that correspond to more general surfaces such as swept surfaces. We need now to investigate that and see if it is possible to create all/most types of digital swept surfaces. The algorithm leads to topologically controlled digital surfaces (0, 1 or 2 connected) upto some special cases where the surface might have unexpected holes. Let us note that this problem seems to appear quite rarely. We had to run a lot of tests in order to get such a figure. Compared to previous methods [4,5], our method appears to be simpler and more general. Our perspectives are firstly to combine a parametric function as generatrix with implicit revolution curves, the control of the topology in this case could be ensured by the implicit expression of the surface. Secondly, we want to investigate the digitization of full parametric surfaces to allow a hand-drawn generatrix and hand-drawn revolution curve. This should solve some of the problems we have with some configurations where we had holes in the surfaces. The problem in Fig. 7, for instance, would not occur anymore. We already have some interesting results (see Fig. 8 for an example) which need to be consolidated.

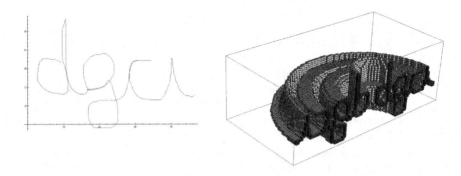

Fig. 8. Hand-drawn generatrix with an implicit circle as revolution curve.

Acknowledgement. This work has been supported by the CPER 2015-2020, NUMERIC Program and FEDER-FSE Project MODEGA of the Poitou-Charentes Region, France.

References

1. Salomon, D.: Curves and Surfaces for Computer Graphics. Springer, New York (2006)
2. Stolte, N., Kaufman, A.E.: Novel techniques for robust voxelization and visualization of implicit surfaces. Graph. Models **63**(6), 387–412 (2001)
3. Yongsheng, L., Stolte, N.: Robust voxelization based ray tracing of implicit surfaces. In: Proceedings of the 6th IASTED, Honolulu, Hawaii, USA, pp. 177–180 (2003)
4. Bhowmick, P., Bera, S., Bhattacharya, B.B.: Digital circularity and its applications. In: Wiederhold, P., Barneva, R.P. (eds.) IWCIA 2009. LNCS, vol. 5852, pp. 1–15. Springer, Heidelberg (2009)
5. Kumar, G., Sharma, N.K., Bhowmick, P.: Wheel-throwing in digital space using number-theoretic approach. IJART **4**(2), 196–215 (2011)
6. Andres, E.: Discrete circles, rings and spheres. Comput. Graph. **18**(5), 695–706 (1994)

7. Andres, E., Jacob, M.A.: The discrete analytical hyperspheres. IEEE Trans. Vis. Comput. Graph. **3**(1), 75–86 (1997)

8. Andres, E., Roussillon, T.: Analytical description of digital circles. In: Debled-Rennesson, I., Domenjoud, E., Kerautret, B., Even, P. (eds.) DGCI 2011. LNCS, vol. 6607, pp. 235–246. Springer, Heidelberg (2011)

9. Toutant, J.-L., Andres, E., Largeteau-Skapin, G., Zrour, R.: Implicit digital surfaces in arbitrary dimensions. In: Barcucci, E., Frosini, A., Rinaldi, S. (eds.) DGCI 2014. LNCS, vol. 8668, pp. 332–343. Springer, Heidelberg (2014)

10. Toutant, J., Andres, E., Roussillon, T.: Digital circles, spheres and hyperspheres: from morphological models to analytical characterizations and topological properties. Discrete Appl. Math. **161**(16–17), 2662–2677 (2013)

11. Stelldinger, P., Köthe, U.: Towards a general sampling theory for shape preservation. Image Vis. Comput. **23**(2), 237–248 (2005)

12. Breton, R., Sivignon, I., Dupont, F., Andrès, É.: Towards an invertible euclidean reconstruction of a discrete object. In: Nyström, I., Sanniti di Baja, G., Svensson, S. (eds.) DGCI 2003. LNCS, vol. 2886, pp. 246–256. Springer, Heidelberg (2003)

13. Sivignon, I., Breton, R., Dupont, F., Andres, E.: Discrete analytical curve reconstruction without patches. Image Vis. Comput. **23**(2), 191–202 (2005)

14. Cohen-Or, D., Kaufman, A.E.: Fundamentals of surface voxelization. CVGIP **57**(6), 453–461 (1995)

15. Andres, E., Acharya, R., Sibata, C.: The supercover 3D polygon. In: Miguet, S., Montanvert, A., Ubéda, S. (eds.) Discrete Geometry for Computer Imagery. LNCS, vol. 1176, pp. 237–242. Springer, Heidelberg (1996)

16. Andrès, É.: Defining discrete objects for polygonalization: the standard model. In: Braquelaire, A., Lachaud, J.-O., Vialard, A. (eds.) DGCI 2002. LNCS, vol. 2301, pp. 313–325. Springer, Heidelberg (2002)

17. Andres, E.: Discrete linear objects in dimension n: the standard model. Graph. Models **65**(1–3), 92–111 (2003)

18. Duff, T.: Interval arithmetic recursive subdivision for implicit functions and constructive solid geometry. In: Proceedings of the 19th Annual Conference on Computer Graphics and Interactive Techniques, SIGGRAPH 1992, pp. 131–138 (1992)

On Functionality of Quadraginta Octants of Naive Sphere with Application to Circle Drawing

Ranita Biswas[✉] and Partha Bhowmick

Department of Computer Science and Engineering,
Indian Institute of Technology, Kharagpur, India
biswas.ranita@gmail.com, bhowmick@gmail.com

Abstract. Although the concept of *functional plane* for naive plane is studied and reported in the literature in great detail, no similar study is yet found for naive sphere. This article exposes the first study in this line, opening up further prospects of analyzing the topological properties of sphere in the discrete space. We show that each *quadraginta octant Q* of a naive sphere forms a bijection with its projected pixel set on a unique coordinate plane, which thereby serves as the functional plane of Q, and hence gives rise to merely *mono-jumps* during back projection. The other two coordinate planes serve as *para-functional* and *dia-functional* planes for Q, as the former is 'mono-jumping' but not bijective, whereas the latter holds neither of the two. Owing to this, the quadraginta octants form symmetry groups and subgroups with equivalent jump conditions. We also show a potential application in generating a special class of discrete 3D circles based on back projection and jump bridging by *Steiner voxels*. A circle in this class possesses 4-symmetry, uniqueness, and bounded distance from the underlying real sphere and real plane.

Keywords: Naive sphere · Quadraginta octants · Symmetry groups · Functional plane · Projective geometry

1 Introduction

Discretization models and their combinatorial structures have drawn a stronger attention of the research community over the last couple of decades [17,18]. Several series of works have been reported on characterization and modeling of different geometric objects like planes and hyperplanes, spheres and hyperspheres, polygons and polytopes, and the like, in the framework of digital geometry. The underlying concepts often vary one from the other while imposing additional criterion one over another. For example, a naive plane is discretized to satisfy the minimality in the number of constituent voxels along with the topological condition of separating the discrete space. However, the very condition of minimality is prohibitive to devising a proven mechanism for discretization of Euclidean primitives like lines, segments, triangles, or polygons as connected voxel sets on a naive plane. The 'naive model' of discrete plane is subsequently enhanced to 'graceful model'—first introduced in [8] and studied later in detail in [9–12].

© Springer International Publishing Switzerland 2016
N. Normand et al. (Eds.): DGCI 2016, LNCS 9647, pp. 256–267, 2016.
DOI: 10.1007/978-3-319-32360-2_20

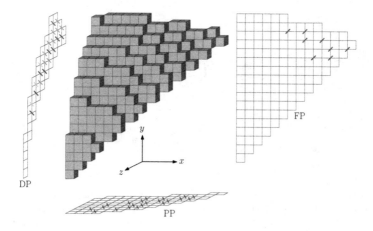

Fig. 1. 1st q-octant of naive sphere of radius 23 and its projections on FP, PP, and DP (dark green ticks = mono-jumps, other ticks = multi-jumps) (Color figure online).

1.1 Motivation

It is the *functional plane* that plays the leading role in characterization and construction of a discrete plane to its graceful model. As shown in [8], *jumps* are the root cause behind the failure of naive plane in construction of Euclidean primitives on its surface. It is worth mentioning here at this point that combinatorial configurations of jumps are given by the orientation of the functional plane. For the formal definitions of naive plane, graceful plane, jumps, and functional planes, we refer to [8,9].

The above concept, as a whole, is also relevant to discrete sphere and has the potential to address many theoretical issues in the context of primitive construction on a discrete spherical surface. No perceivable progress is however noticed in this line, which drives us to take up this work. To the best of our knowledge, this is the first work of its kind, which provides a new insight of analyzing the topological properties of discrete sphere in the integer space. More importantly, it indicates the immense possibility to make out symmetry groups and their topological characterization for various other 3D objects, a few of which evidently being hypersphere, ellipsoid, and hyper-ellipsoid.

1.2 Main Results

We summarize here the theoretical results obtained by us. The proofs and related details are discussed in the subsequent sections.

A naive sphere is made up of 48 basic symmetric parts, which are called *quadraginta octants*, or *q-octants* in short [3,6]. In this paper, we show how these 48 q-octants give rise to 3 groups when characterized by their respective *functional planes* (FP). Each group is further subdivided into two subgroups, each having a unique combination of its *para-functional plane* (PP) and *dia-functional*

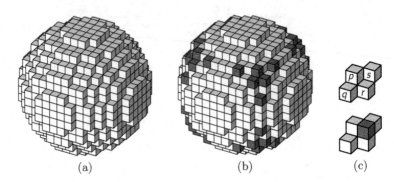

(a)   (b)   (c)

Fig. 2. (a) Naive sphere of radius 7. (b) Steiner voxels (dark green) corresponding to functional planes of the q-octants. (c) A jump (q, s) in (b) and its bridging by Steiner voxel (lying outside of the naive sphere, but may also lie inside) (Color figure online).

plane (DP) defined by *jump configurations*. We show how *mono-jumps* occur from FP and PP, and *multi-jumps* from DP, while taking back projection from them to the naive sphere. Figure 1 shows the 1st q-octant of a naive sphere, its projections on the three coordinate planes, and their respective jumps. Figure 2 illustrates how jumps in different q-octants are bridged by inclusion of some additional voxels in the naive sphere. By analogy to other geometric problems, we refer to these voxels as Steiner voxels.

The grouping of q-octants results to a functional gradation of the coordinate planes against the q-octant groups. Figure 3 shows an example. This, in turn, eventually leads to characterization and modeling of a special class of discrete 3D circles on the surface of a naive sphere, once we bridge the jump voxels by requisite Steiner voxels. We call these circles *ortho-coordinate circles*, as they are orthogonal to one of the three coordinate planes. Their construction is efficiently doable by a technique based on back projection from a coordinate plane based on its functionality w.r.t. the concerned q-octants and their group properties.

2   Preliminaries

In this section, we fix some basic notions and notations to be used in the sequel. For more details, we refer to [17]. We also go through the concepts from previous researches which deem useful in the context of our work.

2.1   Basic Notions and Notations

By discretization, we mean *rasterization* or *voxelation* of a real object (curve or surface), subject to certain topological constraints. The notion owes its origin to computer graphics and geometric modeling [7,14–16].

We define *x-distance*, *y-distance*, and *z-distance* between two (real or integer) points, $p(i, j, k)$ and $p'(i', j', k')$, as $d_x(p, p') = |i - i'|$, $d_y(p, p') = |j - j'|$, and $d_z(p, p') = |k - k'|$, respectively. Using these inter-point distances,

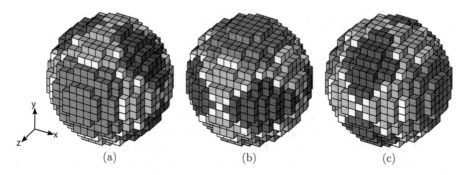

Fig. 3. Grouping of q-octants using functional gradation of coordinate planes. (a) Q-octants with xy-plane as FP are shown in green, as PP are shown in yellow, and as DP are shown in red. (b) Q-octants with zx-plane as FP are shown in green, as PP are shown in yellow, and as DP are shown in red. (c) Q-octants with yz-plane as FP are shown in green, as PP are shown in yellow, and as DP are shown in red. Voxels belonging to more than one group are shown in white in all three renditions (Color figure online).

we define the respective x-, y-, and z-distances between a point $p(i,j,k)$ and a surface S as follows. Let $d_x(p,S)$ be the x-distance between p and S. If there exists a point $p'(i',j',k')$ (the nearest, if there is more than one) in S such that $(j',k') = (j,k)$, then $d_x(p,S) = d_x(p,p')$; otherwise, $d_x(p,S) = \infty$. $d_y(p,S)$ and $d_z(p,S)$ are defined in a similar way. Between two points $p(i,j,k)$ and $p'(i',j',k')$, the *isothetic distance* is taken as the Minkowski norm [17], $d_\infty(p,p') = \max\{d_x(p,p'),d_y(p,p'),d_z(p,p')\}$; between a point $p(i,j,k)$ and a surface S, it is $d_\perp(p,S) = \min\{d_x(p,S), d_y(p,S), d_z(p,S)\}$.

A *voxel* is an integer point in 3D space, and equivalently, a 3-cell [17]. Two distinct voxels are said to be 0-*adjacent* if they share a vertex (0-cell), 1-*adjacent* if they share an edge (1-cell), and 2-*adjacent* if they share a face (2-cell). Thus, for $l \in \{0,1,2\}$, two distinct voxels $p(i,j,k)$ and $p'(i',j',k')$ are l-adjacent if $d_\infty(p,p') = 1$ and $d_x(p,p') + d_y(p,p') + d_z(p,p') \leqslant 3 - l$. Note that the 0-, 1-, and 2-neighborhood notations correspond respectively to the classical 26-, 18-, and 6-neighborhood notations [13,19].

For $l \in \{0,1,2\}$, an l-*path* in a 3D discrete object A (or the discrete space $\mathbb{Z}^3$) is a sequence of voxels from A such that every two consecutive voxels are l-adjacent. The object A is said to be l-*connected* if there is an l-path connecting any two points of A. An l-*component* is a maximal l-connected subset of A.

Let D be a subset of a discrete object A. If $A \setminus D$ is not l-connected, then the set D is l-*separating* in A. Let D be an l-separating discrete object in A such that $A \setminus D$ has exactly two l-components. A 3-cell $c \in D$ is said to be l-*simple* w.r.t. A if $D \setminus \{c\}$ is l-separating in A. An l-separating discrete object in A is l-*minimal* (or l-*irreducible*) if it does not contain any l-simple 3-cell w.r.t. A.

Given a discrete object $A \subseteq \mathbb{Z}^3$, we say that a coordinate plane, say, xy, is *functional* for A, if for every voxel $v = (x_0, y_0, z_0) \in A$ there is no other voxel in A with the same first two coordinates.

2.2   Naive Sphere and Quadraginta Octants

In our work, we consider sphere with integer radius and integer center. For simplicity and without loss of generality, we consider its center as $(0, 0, 0)$. We denote by S_r the real sphere of radius r, and its corresponding naive sphere by S_r. As shown in [3, 6], the voxels comprising S_r have isothetic distance less than $\frac{1}{2}$ from S_r, and they form a 1-connected, 2-minimal, and tunnel-free set, thereby conforming to the concepts proposed in [13]. It has nine planes of symmetry, which lead to $2^3 = 8$ *coordinate octants*, or *c-octants* in short. The three coordinate values of a c-octant can be ordered in $3! = 6$ ways, thereby further dividing the sphere into $8 \times 6 = 48$ *quadraginta octants*, or *q-octants* in short. For construction of S_r, we can generate only its first q-octant, namely $\mathsf{S}_r^{(1)}$, and reflect it about the planes of symmetry for obtaining S_r. As shown in [3, 6],

$$\mathsf{S}_r = \left\{ \begin{matrix} p \in \mathbb{Z}^3 : r^2 - \max(X) \leqslant s < r^2 + \max(X) \\ \wedge \left((s \neq r^2 + \max(X) - 1) \vee (\mathrm{mid}(X) \neq \max(X)) \right) \end{matrix} \right\} \qquad (1)$$

where $p = (i, j, k), s = i^2 + j^2 + k^2$, and $X = \{|i|, |j|, |k|\}$.

We follow the scheme proposed in [3, 6] for uniquely representing the c-octants and the q-octants, and give here a brief review. Each c-octant $\mathbb{C}_i$ is represented by a 3-tuple of signs of coordinate axes, namely $C_i := \left(c_i^{(1)}, c_i^{(2)}, c_i^{(3)} \right)$. For example, $C_1 = (+, +, +)$, $C_2 = (-, +, +)$, and so forth. On the contrary, the 3-tuple $Q_t := \left(q_t^{(1)}, q_t^{(2)}, q_t^{(3)} \right)$ for each q-octant $\mathbb{Q}_t$ represents the three signed coordinate axes. That is, each element $q_t^{(\cdot)}$ has two variables, namely ω and σ. The variable ω contains a literal (numeric form of the name of the coordinate axis) from $\{1, 2, 3\} := \{\mathbf{x}, \mathbf{y}, \mathbf{z}\}$, and the variable σ contains the sign of the corresponding coordinate. That is, $Q_1 = (+1, +2, +3)$, $Q_2 = (+2, +1, +3)$, $Q_3 = (+2, +3, +1)$, ..., $Q_{24} = (-1, +3, -2)$, ..., $Q_{48} = (-1, -3, -2)$. For example, Q_{24} has $\omega[q_{48}^{(1)}] = 1$, $\sigma[q_{48}^{(1)}] = \text{'}-\text{'}$, $\omega[q_{48}^{(2)}] = 3$, etc. The sequence of coordinates in Q_t tells us the increasing order of the absolute coordinate values of the integer points belonging to tth q-octant. We use this for grouping of q-octants.

3   Functional Gradation of Coordinate Planes

We first give a brief review on the graceful model of discrete plane. Let P be a 3D real plane, P and $\mathcal{P}$ be its corresponding naive and graceful planes, and let F be the functional plane. Let s and t be two voxels on P. Let s' and t' be the projections of s and t on F. Let $L(s', t')$ be the 2D digital straight segment (DSS) joining s' and t' on F. As there is an one-to-one correspondence between P and its projection pixels on F, we get a set of voxels on P corresponding to the pixels of $L(s', t')$. This set may not be connected due to the presence of one or more *jumps* [9][1]. A (mono-)jump is created by a pair of disconnected voxels—be it a plane or be it a sphere—as illustrated in Fig. 2(c). This problem is solved in $\mathcal{P}$ by inserting a *Steiner voxel* in between the two voxels forming a jump in P

---

[1] A 'jump' here is synonymous with 'mono-jump' in the context of our work.

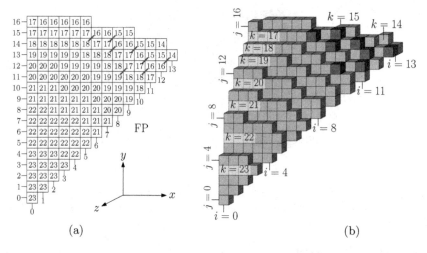

Fig. 4. (a) Jumps (dark green ticks) corresponding to FP of $S_r^{(1)}$ for $r = 23$. (b) Steiner voxels (dark green) bridging the jumps (Color figure online).

so that those two voxels become 0-connected in $\mathcal{P}$. A *tandem* is thus formed by the Steiner voxel and one of the jump voxels, which are 2-adjacent to each other (Fig. 2(c)). To ensure that $\mathcal{P}$ is a subset of the supercover $\mathbf{P}$ of P, each Steiner voxel is chosen only if it intersects P.

For a plane in general orientation, the functional plane (FP) is unique, and it is one of the coordinate planes. For a sphere, on the contrary, it is not so; rather, for each q-octant, the concept is analogous with plane. To explain this, we denote by $S_r^{(t)}$ the tth q-octant of S_r, where $t = 1, 2, \ldots, 48$, and define its FP as follows.

Definition 1 (FP). *The functional plane of $S_r^{(t)}$ is the coordinate plane on which its projection has a bijection with $S_r^{(t)}$.*

Each coordinate plane serves as the functional plane (FP) of 16 specific q-octants, as evident from the following lemma.

Lemma 1. *FP of $S_r^{(t)}$ is xy-, yz-, or zx-plane, depending on whether the value of $t \bmod 6$ belongs to $\{1,2\}$, $\{3,4\}$, or $\{5,0\}$, respectively.*

Proof. Follows from the construction of $S_r^{(t)}$ in conformance with Eq. 1. □

By Lemma 1, we get the FP for a q-octant and hence can apply the tandem configuration used for a graceful plane with the same FP. This owes to two facts: (i) $S_r^{(t)}$ has bijection with its FP projection; (ii) exactly one Steiner voxel can bridge a jump corresponding to FP (we prove it shortly), and such jump is a *mono-jump*. Jumps other than mono-jumps are called *multi-jumps*.

For a mono-jump, a Steiner voxel can be put either outside or inside of S_r, in between the voxels forming the mono-jump in order to bridge them, according to its intersection with S_r, ensuring its belongingness in the standard sphere [4].

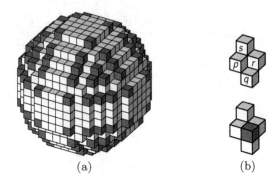

(a)   (b)

Fig. 5. (a) Steiner voxels (dark green) corresponding to PP of each q-octant for $r = 7$. (b) A cutout jump configuration showing how the Steiner voxel is positioned (Color figure online).

Figure 4 shows the projection of $S_r^{(1)}$ on its FP (xy-plane), the resultant mono-jumps, and the bridging Steiner voxels. See also Fig. 2, which shows how Steiner voxels are inserted for bridging all mono-jumps.

For the two coordinate planes other than the one forming FP for $S_r^{(t)}$, we do not get a bijection with its projection on either of them. Specifically, for these two non-functional planes, a *run of voxels* (2-path with two common coordinate values) gets projected to a single pixel. We categorize them as *para-functional plane* (PP) and *dia-functional plane* (DP), and define as follows.

Definition 2 (PP, DP). *PP of $S_r^{(t)}$ is the coordinate plane on which its projection corresponds to only mono-jumps but is not bijective with $S_r^{(t)}$. Its DP is the coordinate plane which is neither FP nor PP; in other words, the projection on DP is not bijective and corresponds to mono- or multi-jumps.*

We have the following theorem on mono-jumps.

Theorem 1 (FP, PP Jumps). *Jumps corresponding to FP and PP are always mono-jumps.*

Proof. W.l.o.g., consider the 1st q-octant of S_r. Let $(i, j, k), (i, j+1, k-d_1), (i+1, j, k-d_2) \in S_r^{(1)}$. By 2-minimal property of S_r, we get $d_1, d_2 \in \{0, 1\}$. Now, if $(i+1, j+1, k-d_3) \in S_r^{(1)}$, then $d_3 \in \{0, 1, 2\}$. Hence, corresponding to FP, a jump (q, s) arises with $q = (i, j, k) \in S_r^{(1)}$ and $s = (i+1, j+1, k-d_3) \in S_r^{(1)}$ if and only if d_3 attains its maximum value (i.e., 2). Thus, the jump is a mono-jump, as shown in the configuration in Fig. 2(c).

To prove the same for PP, observe that S_r is a collection of digital annuli, where, a z-value in $[-r, r]$ corresponds to a digital annulus bounded from inner and outer by two digital circles (as closed 0-paths) with real radii and $(0, 0, z)$ as their common center. By construction of S_r, the part of a (inner/outer) digital circle within $S_r^{(1)}$ always has a unique y-value for a given x-value. If a multi-jump occurs, then the concerned digital circle does not remain a closed 0-path. Therefore, a jump corresponding to PP is always a mono-jump.   □

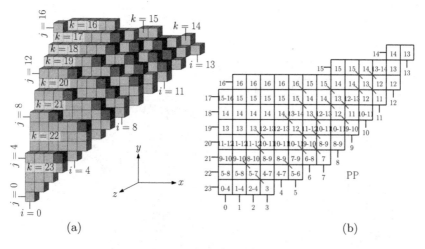

(a)   (b)

Fig. 6. Jumps corresponding to PP (zx-plane) of $\mathsf{S}_r^{(1)}$ with radius 23. (a) Steiner voxels (for clarity, outside ones are shown in dark green). (b) Projection of $\mathsf{S}_r^{(1)}$ on its PP; jumps shown in dark green ticks (Color figure online).

A jump configuration for PP is shown in Fig. 5(b), where p and q are two voxels from the outer digital circle of an annulus, and (q, s) forms the jump. Figure 6 shows the projection of $\mathsf{S}_r^{(1)}$ on its PP (zx-plane) and the requisite Steiner voxels bridging the jumps on $\mathsf{S}_r^{(1)}$ corresponding to PP. The mono-jump locations are marked using dark green ticks and multi-jumps using different colors (light green for two, yellow for three, and red for more requirement of Steiner voxels to fill up the jump). Figure 7 shows $\mathsf{S}_r^{(1)}$ and the Steiner voxels corresponding to DP.

A DP may contain coincident projection pixels for several runs of voxels from the q-octant. Also, to connect the runs whose projections are adjacent, we may need more than one Steiner voxel—a case of *multi-jump* (Fig. 7). We have the following corollary.

Corollary 1 (DP Jumps). *Jumps corresponding to DP are mono- or multi-jumps.*

4   Grouping of Quadraginta Octants

The following lemma explains the way of determining FP, PP, and DP for any q-octant of S_r.

Lemma 2 (Projection Planes). *FP, PP, and DP of $\mathsf{S}_r^{(t)}$ are determined by dropping from Q_t the coordinates $\omega[q_t^{(3)}]$, $\omega[q_t^{(2)}]$, and $\omega[q_t^{(1)}]$, respectively.*

Proof. As mentioned in Sect. 2.2, each q-octant $\mathsf{S}_r^{(t)}$ follows an ordering on the absolute values of the three coordinates captured in its 3-tuple, Q_t. For example,

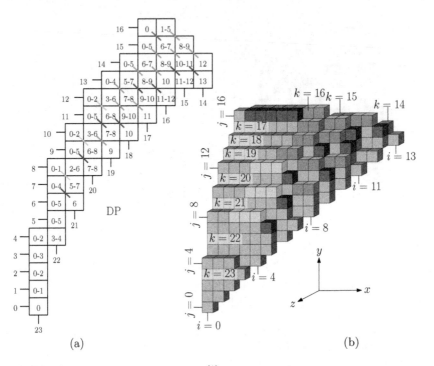

Fig. 7. Jumps corresponding to DP of $\mathsf{S}_r^{(1)}$ for radius 23. (a) Projection on DP (yz-plane), showing mono-jumps in dark green and multi-jumps in different colored ticks. (b) Steiner voxels, color-coded as per jump size (Color figure online).

$Q_1 = (+1, +2, +3)$, and so the ordering is $x \leqslant y \leqslant z$; on dropping z, we get xy-plane as the FP of $\mathsf{S}_r^{(1)}$. In general, for $t \in \{1, 2, \ldots, 48\}$, the coordinate plane obtained by dropping the coordinate of maximum absolute value in Q_t is the FP of $\mathsf{S}_r^{(t)}$. Similar characterizations are valid for PP and DP, whence the proof. $\square$

The functional gradation of coordinate planes leads to 3 groups covering all the 48 q-octants of naive sphere. Each group can be further subdivided into two subgroups, containing 8 q-octants each. As per the representation scheme (Sect. 2.2), the numeral set $N = \{1, 2, 3\}$ denotes the names of the three coordinates $(\mathbf{x}, \mathbf{y}, \mathbf{z})$. We use $\mathcal{G}_a$ to denote the group whose FP is defined by (the coordinates in) $N \setminus \{a\}$, $\forall a \in N$. We use $\mathcal{G}_{a:b}$ to denote the subgroup of $\mathcal{G}_a$ whose PP is defined by $N \setminus \{b\}$, $\forall b \in N \setminus \{a\}$. For example, $\mathcal{G}_3$ contains all the q-octants having xy-plane as FP, and its subgroup $\mathcal{G}_{3:1}$ contains the q-octants having yz-plane as PP. We have now the following proposition.

Proposition 1 (Grouping). *For each $a \in N$, the group of q-octants with their common FP defined by $N \setminus \{a\}$ is $\mathcal{G}_a = \{\mathsf{S}_r^{(t)} | \omega[q_t^{(3)}] = a\}$; and their subgroups in $\mathcal{G}_a$ with common PP defined by $N \setminus \{b\}$ is $\mathcal{G}_{a:b} = \{\mathsf{S}_r^{(t)} | (\omega[q_t^{(3)}], \omega[q_t^{(2)}]) = (a, b)\}$, where $b \in N \setminus \{a\}$.*

Proof. From Lemma 1, we get $\mathcal{G}_a = \{\mathsf{S}_r^{(t)} | t \bmod 6 \in \{2a-1, (2a) \bmod 6\}\}$, $\forall a \in N$. Now, by construction of S_r and referring to Lemma 2, the clause

"$t \bmod 6 \in \{2a - 1, (2a) \bmod 6\}$" in the above equation is equivalent with the clause "$\omega[q_t^{(3)}] = a$". This gives the proof of group formation.

The subgroups of $\mathcal{G}_a$ are $\{S_r^{(t)}|t \bmod 6 = 2a - 1\}$ and $\{S_r^{(t)}|t \bmod 6 = (2a) \bmod 6\}$, or, equivalently, they are $\{S_r^{(t)}|(\omega[q_t^{(3)}], \omega[q_t^{(2)}]) = (a, b_1)\}$ and $\{S_r^{(t)}|(\omega[q_t^{(3)}], \omega[q_t^{(2)}]) = (a, b_2)\}$, where $\{b_1, b_2\} = N \setminus \{a\}$. □

We refer back to Fig. 3 to visualize the distribution of q-octants by functional gradation of coordinate planes, which eventually leads to group and subgroup formation.

5   Circle Drawing—An Application

A limited research has been done on discretization of circles or curves in 3D space. Only in recent time, some progress is noticed, e.g., offset discretization scheme in $\mathbb{R}^3$ [1,2], discrete spherical paths and circles in $\mathbb{Z}^3$ [3–5], etc. In this section, we introduce a special class of 3D circle in $\mathbb{Z}^3$, defined as follows.

Definition 3. *A (naive) ortho-coordinate circle* $C_r^{\langle a,b,c \rangle}$ *is a discretization of the real circle* $C_r^{\langle a,b,c \rangle}$ *having radius r and lying on a real plane, with normal vector* $\langle a, b, c \rangle$*, that is orthogonal to one of the coordinate planes.*

Note that the circle $C_r^{\langle a,b,c \rangle}$ consists of voxels from the intersection of S_r and the naive plane with normal $\langle a, b, c \rangle$, with requisite Steiner voxels for ensuring connectivity. It is easily constructible using q-octant groups (for fixing the mono- and the multi-jumps in the 0-path defining $C_r^{\langle a,b,c \rangle}$), a line drawing algorithm [7] (for back projection), and the formulation of naive sphere (Eq. 1, for checking the belongingness of the voxels of $C_r^{\langle a,b,c \rangle}$ in S_r). Note that one of a, b, c is zero and the real plane is considered to pass through an integer point (w.l.o.g., $(0,0,0)$). We consider only integer values for r, a, b, c. The major steps are as follows.

1. Set $p = (0,0)$ and compute q from $\langle a, b, c \rangle$. Shoot a digital ray from p towards q, and produce up to s whose back projection on S_r includes a voxel with its coordinate value as zero which is also zero in $\langle a, b, c \rangle$ (e.g., $x = 0$ if $a = 0$).
2. For each pixel u in $\mathrm{DSS}(p, s)$, execute the following steps.
    (a) Use back projection from u to get a single voxel or a run of voxels on S_r.
    (b) If this voxel or voxel run is not connected with the last drawn voxel or voxel run, then identify the mono-jumps and the multi-jumps.
    (c) Compute and insert Steiner voxels for bridging the jumps. (Skip (b, c) if $u = p$.)
3. Use symmetry to construct the parts in other q-octants.

Figure 8 shows a demonstration. A circle $C_r^{\langle a,b,c \rangle}$ is unique for a given specification and conforms to 4-symmetry, since only the blue part needs to be generated by the algorithm, and the others just follow the symmetry. Due to the method of selection of Steiner voxels, each voxel of $C_r^{\langle a,b,c \rangle}$ belongs to the naive

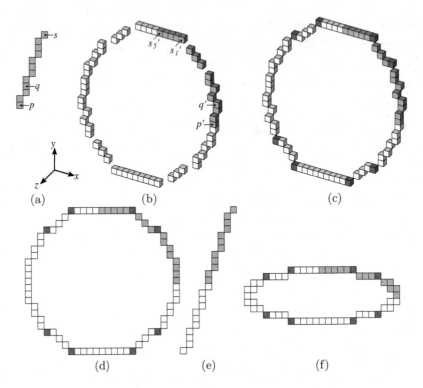

Fig. 8. Construction of an ortho-coordinate circle ($r = 12$ and plane normal $= \langle 0, -3, 1 \rangle$). (a) DSS on FP ($yz$-plane), which the plane ($-3y + z = 0$) of the circle is orthogonal to. (b) Disconnected circle on S_r by back projection from the DSS in (a). Parts drawn by symmetry are shown white. (c) $C_{12}^{\langle 0, -3, 1 \rangle}$ after incorporating requisite Steiner voxels. (d, e, f) Projections of $C_{12}^{\langle 0, -3, 1 \rangle}$ on xy-, yz-, and zx-planes (Color figure online).

plane with normal $\langle a, b, c \rangle$, and also to the standard sphere with radius r, thus giving an upper bound of isothetic distance $\frac{1}{2}$ from the real plane $\langle a, b, c \rangle$ and of 2 from the real sphere with radius r. A detailed study on isothetic distance bounds for plane and sphere can be seen in [4].

6   Concluding Note

The analysis and gradation of coordinate planes in view of their functionality is a novel proposition in this paper. Being 48-symmetric, a naive sphere can be divided into groups and subgroups based on this gradation. This grouping would have various applications, as shown by us for one such, in generating a special class of 3D circles lying on the sphere. In higher dimensions, the scope and challenge would be higher and better, as we foresee. Characterization of para-functional plane seems also interesting for objects like discrete planes and hyperplanes, which is yet to be studied.

References

1. Aveneau, L., Andres, E., Mora, F.: Expressing discrete geometry using the conformal model. In: AGACSE 2012, La Rochelle, France, July 2012
2. Aveneau, L., Fuchs, L., Andres, E.: Digital geometry from a geometric algebra perspective. In: Barcucci, E., Frosini, A., Rinaldi, S. (eds.) DGCI 2014. LNCS, vol. 8668, pp. 358–369. Springer, Heidelberg (2014)
3. Biswas, R., Bhowmick, P.: On finding spherical geodesic paths and circles in $\mathbb{Z}^3$. In: Barcucci, E., Frosini, A., Rinaldi, S. (eds.) DGCI 2014. LNCS, vol. 8668, pp. 396–409. Springer, Heidelberg (2014)
4. Biswas, R., Bhowmick, P.: On different topological classes of spherical geodesic paths and circles in $\mathbb{Z}^3$. Theoret. Comput. Sci. **605**, 146–163 (2015)
5. Biswas, R., Bhowmick, P., Brimkov, V.E.: On the connectivity and smoothness of discrete spherical circles. In: Barneva, R.P., et al. (eds.) IWCIA 2015. LNCS, vol. 9448, pp. 86–100. Springer, Heidelberg (2015)
6. Biswas, R., Bhowmick, P.: From prima quadraginta octant to lattice sphere through primitive integer operations. Theoret. Comput. Sci. (2015, in press). http://dx.doi.org/10.1016/j.tcs.2015.11.018
7. Bresenham, J.E.: Algorithm for for computer control of a digital plotter. IBM Syst. J. **4**(1), 25–30 (1965)
8. Brimkov, V.E., Barneva, R.P.: Graceful planes and thin tunnel-free meshes. In: Bertrand, G., Couprie, M., Perroton, L. (eds.) DGCI 1999. LNCS, vol. 1568, pp. 53–64. Springer, Heidelberg (1999)
9. Brimkov, V.E., Barneva, R.P.: Graceful planes and lines. Theoret. Comput. Sci. **283**(1), 151–170 (2002)
10. Brimkov, V.E., Barneva, R.P.: Connectivity of discrete planes. Theoret. Comput. Sci. **319**(1–3), 203–227 (2004)
11. Brimkov, V.E., Barneva, R.P.: Plane digitization and related combinatorial problems. Discrete Appl. Math. **147**(2–3), 169–186 (2005)
12. Brimkov, V.E., Coeurjolly, D., Klette, R.: Digital planarity–a review. Discrete Appl. Math. **155**(4), 468–495 (2007)
13. Cohen-Or, D., Kaufman, A.: Fundamentals of surface voxelization. Graph. Models Image Process. **57**(6), 453–461 (1995)
14. Cohen-Or, D., Kaufman, A.: 3D line voxelization and connectivity control. IEEE Comput. Graph. Appl. **17**(6), 80–87 (1997)
15. Gouraud, H.: Continuous shading of curved surfaces. IEEE Trans. Comput. **20**(6), 623–629 (1971)
16. Kaufman, A.: Efficient algorithms for 3D scan-conversion of parametric curves, surfaces, and volumes. In: SIGGRAPH 1987, pp. 171–179 (1987)
17. Klette, R., Rosenfeld, A.: Digital Geometry: Geometric Methods for Digital Picture Analysis. Morgan Kaufmann, San Francisco (2004)
18. Mukhopadhyay, J., Das, P.P., Chattopadhyay, S., Bhowmick, P., Chatterji, B.N.: Digital Geometry in Image Processing. Chapman and Hall/CRC, Boca Ration (2013)
19. Toutant, J.-L., Andres, E., Roussillon, T.: Digital circles, spheres and hyperspheres: from morphological models to analytical characterizations and topological properties. Discrete Appl. Math. **161**(16–17), 2662–2677 (2013)

Encoding Specific 3D Polyhedral Complexes Using 3D Binary Images

Rocio Gonzalez-Diaz, Maria-Jose Jimenez$^{(\boxtimes)}$, and Belen Medrano

Applied Mathematics Department I, School of Computer Engineering,
University of Seville, Seville, Spain
{rogodi,majiro,belenmg}@us.es
http://personal.us.es/

Abstract. We build upon the work developed in [4] in which we presented a method to "locally repair" the cubical complex $Q(I)$ associated to a 3D binary image I, to obtain a "well-composed" polyhedral complex $P(I)$, homotopy equivalent to $Q(I)$. There, we developed a new codification system for $P(I)$, called ExtendedCubeMap (ECM) representation, that encodes: (1) the (geometric) information of the cells of $P(I)$ (i.e., which cells are presented and where), under the form of a 3D grayscale image g_P; (2) the boundary face relations between the cells of $P(I)$, under the form of a set B_P of structuring elements.

In this paper, we simplify ECM representations, proving that geometric and topological information of cells can be encoded using just a 3D **binary** image, without the need of using colors or sets of structuring elements. We also outline a possible application in which well-composed polyhedral complexes can be useful.

Keywords: 3D binary image · Well-composedness · 3D cubical complex · 3D polyhedral complex

1 Introduction

Consider $\mathbb{Z}^3$ as the set of points with integer coordinates. in 3D space $\mathbb{R}^3$. A *3D binary digital image* is a set $I = (\mathbb{Z}^3, 26, 6, B)$ (or $I = (\mathbb{Z}^3, B)$, for short), where $B \subset \mathbb{Z}^3$ is the *foreground*, $B^c = \mathbb{Z}^3 \backslash B$ the *background*, and $(26, 6)$ is the adjacency relation for the foreground and background, respectively. A 3D binary image I is *well-composed* [9] if the boundary surface of its continuous analog is a 2D manifold. 3D well-composed images enjoy important topological and geometrical properties in such a way that several algorithms used in computer vision, computer graphics and image processing are simpler. Unfortunately natural and synthetic images are not a priori well-composed. There are

Authors partially supported by IMUS, Junta de Andalucia under grant FQM-369, Spanish Ministry under grants MTM2012-32706 and MTM2015-67072-P, and ACAT ESF Research Networking Programme.

© Springer International Publishing Switzerland 2016
N. Normand et al. (Eds.): DGCI 2016, LNCS 9647, pp. 268–281, 2016.
DOI: 10.1007/978-3-319-32360-2_21

several "repairing" methods for turning them into well-composed images (see, for example, [8, 10, 12, 13]).

In [11], the authors extended the notion of "digital well-composedness" to nD sets. In [1], they proved that the digital well-composedness implies the equivalence of connectivities of the level set components in nD. They proposed and proved a self-dual discrete (non-local) interpolation method, based on a sub-part of a quasi-linear algorithm that computes the morphological tree of shapes, whose result is always a digitally well-composed function. Besides, it is proven in [2] that the only local self-dual well-composed interpolation of a 2D image is obtained by the median operator.

In this paper, we first recall, in Sect. 2, a method to "locally repair" the cubical complex $Q(I)$ associated to a 3D binary image I, obtaining a "well-composed" polyhedral complex $P(I)$, homotopy equivalent to $Q(I)$, encoded using an ExtendedCubeMap (ECM) representation. A possible application of ECM representations is also sketched. In Sect. 3, we prove that the coordinates of the points in an ECM representation that encode the cells of $Q(I)$ contain all the information of the cells. This result is extended to $P(I)$ in Sect. 4. Therefore there is no need of using colors or sets of structuring elements to encode $P(I)$. The paper ends up with a section of conclusions and future work.

2   ECM Representations

The notion of well-composedness is extended in [4, 14] to 3D complete polyhedral complexes K embedded in $\mathbb{R}^3$. This way, K is *well-composed* if the boundary surface of K is a 2D manifold.

In [4], we presented a method to "locally repair" the 3D complete cubical complex $Q(I)$ (embedded in $\mathbb{R}^3$) associated to a voxel-based representation of a given image I, to obtain a well-composed polyhedral complex $P(I)$ homotopy equivalent to $Q(I)$. Our main motivation was that of (co)homology computations on the cell complex representing a 3D binary image I [3, 6]. We could take advantage of a well-composed-like representation since such computations could be performed only on the boundary surface of $P(I)$. A new codification system called *ExtendedCubeMap (ECM) representation*, for $P(I)$ were also developed in [4], encoding: (1) the (geometric) information of the cells of $P(I)$, under the form of a 3D grayscale image g_P; and (2) the boundary face relations between the cells of $P(I)$, under the form of a set B_P of structuring elements that can be stored as indexes in a look-up table. A naive demo implemented in Matlab for computing ECM representations can be downloaded in [5].

In this paper, we improve such codification system, showing that the geometric and topological information of the cells of $P(I)$ can be encoded using just a 3D binary image $J = (\mathbb{Z}^3, B)$. Geometric information of each cell σ can be obtained by examining the coordinates of the point p in B that encodes σ (no color is needed). Faces of such cell σ can be computed by examining the points in a neighborhood of p in B (no look-up table is needed). As far as we know, this is the first time that a family of polyhedral complexes more general than cubical ones are stored using binary images.

2.1   Towards Applications: 3D Printing

3D CAD modeling and recent 3D printing deal mainly with 3D meshes. Typically, the physical objects are modeled in standard STL format where the objects are discretized as finite meshes (encoded as a 3D polyhedral complexes). The interplay between abstract representations and the physical objects natural raise a question about the meshes that can be practically constructed (i.e. printed). Possible obstacles are complexes which boundary surfaces are not combinatorial manifolds. Example of such a configuration are two cubes touching in a vertex or in an edge. Tools that detect, and fix such configurations are a first step towards an automatic tool to check if an abstract representation is realizable as a physical object and/or to change it so that the physical object is more robust (taking into account that non-manifold parts will break immediately).

Although most of the meshes for 3D printing are simplicial, others are cubi-cal[1]. This way, a "manifoldization technique" could be as follows: (1) Start with any cubical mesh (i.e., cubical complex) modeling a physical object. (2) Make a "manifold version" of it by computing a well-composed polyhedral complex $P(I)$ homotopy equivalent to $Q(I)$, to obtain a more robust version of the physical object. (3) Encode $P(I)$ in a valid format for the 3D printing. The codification of $P(I)$ in a 3D binary format presented in this paper, could help in this last step.

3   3D Cubical Complexes

3D digital images are usually represented by **unit** closed cubes in $\mathbb{R}^3$ with square faces parallel to the coordinate planes, centered at points in $\mathbb{Z}^3$ (also called *voxels*). In this paper, **voxels are rescaled with factor** 4, so we consider size-4 cubes centered at points $4\mathbb{Z}^3$.

A set of voxels together with all their faces (*cells*) constitute a combinatorial structure called *cubical complex*, denoted by Q. The $0-faces$ of a given voxel c are its 8 corners (vertices), its $1-faces$ are its 12 edges, its $2-faces$ are its 6 square faces and, finally, its $3-face$ is the voxel itself.

Let Q be a cubical complex composed by a set of voxels (size-4 cubes) together with all their faces. Let $J = (\mathbb{Z}^3, B)$ be a 3D binary image. We say that J encodes Q if: (1) For any σ in Q, r_σ (the barycenter of σ) is in B. And (2), for any $p \in B$, p is the barycenter of a cell σ in Q. Then, a point $p \in 2\mathbb{Z}^3$ encodes:

- A vertex iff $p \in \mathcal{E}_0 = \{(4i + 2, 4j + 2, 4k + 2)\}_{i,j,k \in \mathbb{Z}}$.
- An edge iff $p \in \mathcal{E}_1 = \{(4i + 2, 4j + 2, 4k), (4i, 4j + 2, 4k + 2), (4i + 2, 4j, 4k + 2)\}_{i,j,k \in \mathbb{Z}}$.
- A square face iff $p \in \mathcal{E}_2 = \{(4i+2, 4j, 4k), (4i, 4j+2, 4k), (4i, 4j, 4k+2)\}_{i,j,k \in \mathbb{Z}}$.
- A voxel iff $p \in \mathcal{E}_3 = 4\mathbb{Z}^3$.

---

[1] http://www.shapeways.com/blog/archives/17972-shapeways-launches-svx-voxel-file-format-for-3d-printing.html.

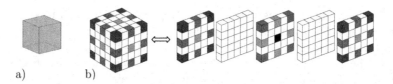

a)   b)

Fig. 1. (a) A voxel c. (b) Set of points encoding c. Color illustrates the type of coordinate of the point, which provides the dimension of the encoded cell: blue for $0-$cells (i.e. points in $\mathcal{E}_0$), red for $1-$cells (i.e. points in $\mathcal{E}_1$), green for $2-$cells (i.e. points in $\mathcal{E}_2$) and black for $3-$cells (i.e. points in $\mathcal{E}_3$) (Color figure online).

Now, given a point $p \in B$ encoding a cell $\sigma \in Q$, our aim is to find the points in B encoding the faces of σ. First, recall that: $N_6^\ell = \{(\pm\ell, 0, 0), (0, \pm\ell, 0),$ $(0, 0, \pm\ell)\}$; $N_{12}^\ell = \{(0, \pm\ell, \pm\ell), (\pm\ell, 0, \pm\ell), (\pm\ell, \pm\ell, 0)\}$; $N_8^\ell = \{(\pm\ell, \pm\ell, \pm\ell)\}$; and for any $N \subseteq \mathbb{R}^3$ and $p \in \mathbb{R}^3$, $N(p)$ denotes the set $\{p + q, q \in N\}$.

Proposition 1. *Let σ be an $\ell-$cell of Q encoded by $p = r_\sigma$. Then, the set of $k-$faces of σ, for $0 \le k < \ell$, is: (1) $N_6^2(p) \cap \mathcal{E}_k$ if $k = \ell - 1$; (2) $N_{12}^2(p) \cap \mathcal{E}_k$ if $k = \ell - 2$; (3) $N_8^2(p) \cap \mathcal{E}_k$ if $k = \ell - 3$.*

Proof. Let $\ell = 2$. First, $p = r_\sigma \in \mathcal{E}_2$ so, assume that $r_\sigma = (4i + 2, 4j, 4k)$ for some $i, j, k \in \mathbb{Z}^3$. Then, the barycenters of the $1-$faces of σ are $\{(4i + 2, 4j \pm 2, 4k), (4i + 2, 4j, 4k \pm 2)\}$. Second, $N_6^2(p) = \{(4i, 4j, 4k), (4i + 4, 4j, 4k), (4i + 2, 4j \pm 2, 4k), (4i + 2, 4j, 4k \pm 2)\}$. Finally, $N_6^2(p) \cap \mathcal{E}_1 = \{(4i + 2, 4j \pm 2, 4k), (4i + 2, 4j, 4k \pm 2)\}$. The other cases can be proven in a similar way. $\square$

See Fig. 1 in which a voxel c (on the right) is encoded by a set of points (on the left). Colors are used to distinguish dimension of cells encoded but, it follows from Proposition 1 that J can be stored as a binary image.

Remark 1. From now on, a cell in a 3D cubical complex (later, in the 3D polyhedral complex) will be sometimes identified with the point in $\mathbb{Z}^3$ encoding such cell.

4   Encoding Specific Polyhedral Complexes Using Binary Images

A 3D polyhedral complex K [7] is a combinatorial structure by which a space is decomposed into vertices ($0-$cells), edges ($1-$cells), polygons ($2-$cells) and polyhedra ($3-$cells) that are glued together by their boundaries (*faces*) such that the non-empty intersection of any two cells of K is also a cell of K. An $\ell-$cell $\sigma \in K$ is a face of an $\ell'-$cell $\sigma' \in K$ if σ lies in the boundary of σ' and $\ell \le \ell'$. A cell μ is *maximal* if it is not a face of any other cell $\sigma \in K$. A 3D polyhedral complex is *complete* if all its maximal cells have dimension 3. It is *well-composed* if its boundary surface is a 2D manifold. In [4] we developed a procedure to get as output a 3D complete well-composed polyhedral complex that is homotopy equivalent to the standard cubical complex representing an input 3D binary image. This type of 3D polyhedral complex will be called *well-composed polyhedral complex over a picture* and denoted by P. Besides, we say that a 3D binary image $J = (B, \mathbb{Z}^3)$ encodes P if B is the union of all the points encoding the polyhedra of P (together with all their faces). In this section, we show how to compute J and prove that J contains all the geometric and topological information of P.

We recall below the set S of 27 types of polyhedra used in [4] to construct P. For $v, w \in \mathcal{E}_0$, the edge with endpoints v and w (denoted by $e(v, w)$) is a size-4 edge. The size-2 cube centered at v with faces parallel to the coordinate planes is denoted by $c(v)$ being $s(v)$ a square face of $c(v)$. Finally, $k(v, w)$ will denote the pyramid with apex w and base the square face $s(v)$ whose barycenter lies on $e(v, w)$. Finally, $t(v, w)$ denotes a triangle face of $k(v, w)$.

Definition 1. *The set S consists of:*

(a) *The voxel c centered at a point in $4\mathbb{Z}^3$. See Fig. 1a.*

(b) *The size-2 cube $c(v)$ centered at a point $v \in 2\mathbb{Z}^3$. See Table 1b.1*

(c) *The pyramid $k(v, w)$, where $v, w \in \mathcal{E}_0$. See Table 1c.1*

(d) *The polyhedra $\{p_\ell(v, v_1, v_2, v_3)\}_{\ell=1,2,3,4}$, where $v, v_1, v_2, v_3 \in \mathcal{E}_0$ are distinct points forming a size-4 square s, being v adjacent to v_1 and v_2:*

- *$p_1(v, v_1, v_2, v_3)$ is determined by the the edges $e(v_1, v_3)$, $e(v_2, v_3)$, and the triangles $t(v, v_1)$, $t(v, v_2)$, whose barycenters lie inside s. See Table 1d.1*
- *$p_2(v, v_1, v_2, v_3)$ is determined by the edge $e(v_2, v_3)$, the triangles $t(v, v_2)$, $t(v_1, v_3)$, and the square face $s(\frac{v+v_1}{2})$, whose barycenters lie inside the square s. See Table 1e.1*
- *$p_3(v, v_1, v_2, v_3)$ is determined by the four triangles $t(v, v_1)$, $t(v, v_2)$, $t(v_3, v_1)$ and $t(v_3, v_2)$ whose barycenters lie inside s. See Table 1f.1*
- *$p_4(v, v_1, v_2, v_3)$ is determined by the square faces $s(\frac{v+v_1}{2})$, $s(\frac{v+v_3}{2})$, and the triangles $t(v, v_2)$, $t(v_2, v_3)$, whose barycenters lie inside s. See Table 1g.1*

(e) *The 22 hexahedra $\{h_\ell(v_1, \ldots v_8)\}_{\ell=1,\ldots,22}$, where $\{v_1, \ldots, v_8\} \subset \mathcal{E}_0$ are the vertices of a voxel c, are determined by a set of vertices $\{x_1, \ldots x_8\}$. Each vertex x_i is either v_i or the vertex w_i of $c(v_i)$, that lies inside c. Observe that $h_1(v_1, \ldots v_8)$ (when $x_\ell = v_\ell$ for $\ell = 1, \ldots, 8$) is the voxel c itself. Similarly, $h_{22}(v_1, \ldots v_8)$ (when $x_\ell = w_\ell$ for $\ell = 1, \ldots, 8$) is the size-2 cube $c(\frac{v_1+\cdots+v_8}{8})$ being $\frac{v_1+\cdots+v_8}{8} \in 4\mathbb{Z}^3$. All the other possible vertex combinations lead to the 20 hexahedra showed in Fig. 2.*

Although some of the 2$-$faces of the polyhedra $\{p_\ell(v, v_1, v_2, v_3)\}_{\ell=1,2,3,4}$, are not planar (hence they are not polygons), P is always a CW-complex.

Notice that all the coordinates used in the representation of a cubical complex were even coordinates. Now, odd coordinates will also be needed. We denote:

- $\mathcal{O}_0 = \{(2i + 1, 2j + 1, 2k + 1)\}_{i,j,k \in \mathbb{Z}}$
- $\mathcal{O}_1 = \{(2i, 2j + 1, 2k + 1), (2i + 1, 2j, 2k + 1), (2i + 1, 2j + 1, 2k)\}_{i,j,k \in \mathbb{Z}}$
- $\mathcal{O}_2 = \{(2i, 2j, 2k + 1), (2i, 2j + 1, 2k), (2i + 1, 2j, 2k)\}_{i,j,k \in \mathbb{Z}}.$

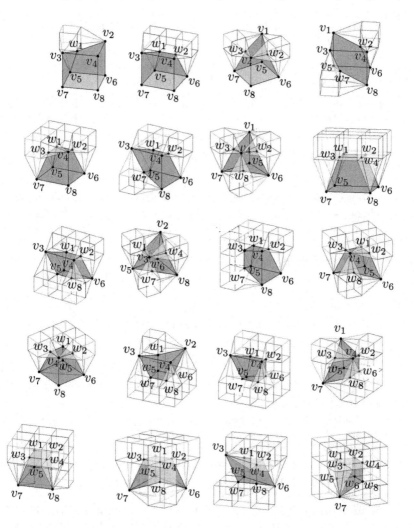

Fig. 2. List of 20 out of 22 types of hexahedra described in Definition 1e.

Now, from the ECM representation, (in fact, from the image g_P), we induce the coordinates of the points that codify the dimension, position and type of cell of any of the 27 different types of polyhedra in S and all their faces.

Proposition 2. *Given a 3D binary image $J = (B, \mathbb{Z}^3)$ encoding a well-composed polyhedral complex over a picture P, for a point $p \in B$ we have that:*

– *If $p \in \mathcal{E}_0$ then p is a vertex if $N_6^1(p) \cap B = \emptyset$; otherwise, it is the cube described in Definition 1b.*
– *If $p \in \mathcal{E}_1$ then p is an edge if $N_6^1(p) \cap B = \emptyset$; otherwise, it is either the pyramid described in Definition 1c or the cube described in Definition 1b.*
– *If $p \in \mathcal{E}_2$ then p is a square face if $N_6^1(p) \cap B = \emptyset$; otherwise, it is one of the polyhedra described in Definition 1d.*

- *If $p \in \mathcal{E}_3$ then p is one of the 22 hexahedra listed in Definition 1e.*
- *If $p \in \mathcal{O}_\ell$, $\ell = 0, 1, 2$ then p is an $\ell-$cell.*

Proof. Any vertex $v \in P$ (i.e., $v \in B$) satisfies that either $v \in \mathcal{E}_0$ or $v \in \mathcal{O}_0$. See the blue voxels in Fig. 1, second column of Table 1 and Fig. 3.

(a) A voxel c and its faces are encoded by their baricenters. This way, $c \in \mathcal{E}_3$ has square faces in $\mathcal{E}_2$, edges in $\mathcal{E}_1$ and vertices in $\mathcal{E}_0$. See Fig. 1b and Table 1a.1.

(b) A size-2 cube $c(v)$ centered at a point $v \in \mathcal{E}_0$ and its faces are encoded by their baricenters. This way, $c(v) \in \mathcal{E}_0$ has square faces in $\mathcal{O}_1$, edges in $\mathcal{O}_1$ and vertices in $\mathcal{O}_0$. See Table 1b.2.

(c) The 3$-$cell $k(v, w)$ is encoded by $p = r_{(e(v,w))} \in \mathcal{E}_1$. Without loss of generality, suppose that $p = (4i, 4j + 2, 4k + 2)$ for some $i, j, k \in \mathbb{Z}$. Then, the four triangle faces of $k(v, w)$ are $\{(4i, 4j+2\pm1, 4k+2), (4i, 4j+2, 4k+2\pm1)\} \subset \mathcal{O}_2$ (see green voxels in Table 1c.2). The four edges of $k(v, w)$ incident to w are $\{(4i, 4j + 2 \pm 1, 4k + 2 \pm 1)\} \subset \mathcal{O}_1$.

(d) The 3$-$cell $p_\ell(v, v_1, v_2, v_3)$, $\ell = 1, 2, 3, 4$, where $v, v_1, v_2, v_3 \in \mathcal{E}_0$, form a size$-4$ square face s, being v adjacent to v_1, v_2. Without loss of generality, suppose that $v = (i - 2, j - 2, k + 2)$, $v_1 = (i + 2, j - 2, k + 2)$, $v_2 = (i - 2, j + 2, k + 2)$, and $v_3 = (i + 2, j + 2, k + 2)$. This way, $p_\ell(v, v_1, v_2, v_3)$ is encoded by $r_s = (4i, 4j, 4k + 2) \in \mathcal{E}_2$. the two quadrangles of $p_\ell(v, v_1, v_2, v_3)$, $\ell = 1, 2, 3, 4$, are $\{(4i, 4j, 4k + 2 \pm 1)\} \subset \mathcal{O}_2$;
  - The two triangle faces of $p_1(v, v_1, v_2, v_3)$ are $\{(4i - 1, 4j, 4k + 2), (4i, 4j - 1, 4k+2)\} \subset \mathcal{O}_2$; the non-(size-4) edges incident to v_2 are $\{(4i-1, 4j, 4k + 2 \pm 1)\}$ and the ones incident to v_3 are $\{(4i, 4j - 1, 4k + 2 \pm 1)\} \subset \mathcal{O}_1$. See Table 1d.2.
  - The two triangle faces of $p_2(v, v_1, v_2, v_3)$ are $\{(4i\pm1, 4j, 4k+2)\} \subset \mathcal{O}_2$; the non-(size-4) edges incident to v_2 or v_3 are $\{(4i \pm 1, 4j, 4k + 2 \pm 1)\} \subset \mathcal{O}_1$. See Table 1e.2.
  - The four triangles of $p_3(v, v_1, v_2, v_3)$ are $\{(4i\pm1, 4j, 4k+2), (4i, 4j\pm1, 4k+2)\} \subset \mathcal{O}_2$; the eight edges incident to v_1 or v_2 are $\{(4i \pm 1, 4j, 4k + 2 \pm 1), (4i, 4j \pm 1, 4k + 2 \pm 1)\} \subset \mathcal{O}_1$. See Table 1f.2.
  - The two triangles of $p_4(v, v_1, v_2, v_3)$ are $\{(4i-1, 4j, 4k+2), (4i, 4j+1, 4k+2)\} \subset \mathcal{O}_2$; the four edges incident to v_2 are $\{(4i-1, 4j, 4k+2\pm1), (4i, 4j+1, 4k + 2 \pm 1)\} \subset \mathcal{O}_1$. See Table 1g.2.

(e) The 3$-$cell $h_\ell(v_1, \dots v_8)$, $\ell = 1, \dots, 22$, where vertices $v_1, \dots v_8 \in \mathcal{E}_0$ form a voxel c, is encoded by $p = \frac{v_1 + \dots + v_8}{2} \in \mathcal{E}_3$, (i.e., the barycenter of c). Let $p = (4i, 4j, 4k)$ for some $i, j, k \in \mathbb{Z}^3$. Then, the vertices v_ℓ, $\ell = 1, \dots, 8$, are $\{(4i\pm 2, 4j\pm2, 4k\pm2)\}$; and the vertices w_ℓ, $\ell = 1, \dots, 8$, are $\{(4i\pm1, 4j\pm1, 4k\pm1)\}$, such that, for example, if $v_\ell = (4i + 2, 4j - 2, 4k + 2)$ then $w_\ell = (4i + 1, 4j - 1, 4k + 1)$. The non-square quadrangle $\{x_{\ell_1}, x_{\ell_2}, x_{\ell_3}, x_{\ell_4}\}$ of $h_\ell(v_1, \dots v_8)$, is encoded by the barycenter of the square face $\{w_{\ell_1}, w_{\ell_2}, w_{\ell_3}, w_{\ell_4}\}$, which is in $\mathcal{O}_2$. Finally, a triangle $t(v_{\ell_i}, v_{\ell_j})$ is encoded by $r_{e(w_{\ell_i}, w_{\ell_j})} \in \mathcal{O}_1$. □

Now, given a cell $b \in B$, our aim is to directly compute the boundary faces of b without making use of any of the structuring elements listed in [4]. In that paper, boundary relations were obtained by running over a fixed set of 121 structuring elements (see Fig. 4) and looking for the ones that fitted around the point.

Table 1. First column: 7 out of 27 polyhedra used in the paper. Second column: their ECM representation. Third column: the proposed codification, where J is the 3D binary image encoding a polyhedral complex P.

Polyhedron	ECM rep.	Codification proposed in the paper
a.1) Voxel c	a.2)	a.3) **1** is at position $(i,j,k) \in 4\mathbb{Z}^3$ in J $$\begin{pmatrix}10101\\00000\\10101\\00000\\10101\end{pmatrix}\begin{pmatrix}00000\\00000\\00000\\00000\\00000\end{pmatrix}\begin{pmatrix}10101\\00000\\10101\\00000\\10101\end{pmatrix}\begin{pmatrix}00000\\00000\\00000\\00000\\00000\end{pmatrix}\begin{pmatrix}10101\\00000\\10101\\00000\\10101\end{pmatrix}$$
b.1) $c(v)$	b.2)	b.3) **1** is at position $(i,j,k) \in 2\mathbb{Z}^3$ in J $$\begin{pmatrix}111\\111\\111\end{pmatrix}\begin{pmatrix}111\\111\\111\end{pmatrix}\begin{pmatrix}111\\111\\111\end{pmatrix}$$
c.1) $k(v,w)$	c.2)	c.3) **1** is at position $(i,j,k) \in \mathcal{E}_1$ in J $$\begin{pmatrix}1100\\1100\\1100\end{pmatrix}\begin{pmatrix}1100\\1101\\1100\end{pmatrix}\begin{pmatrix}1100\\1100\\1100\end{pmatrix}$$
d.1) $p_1(\cdots)$	d.2)	d.3) **1** is at position $(x,y,z) \in \mathcal{E}_2$ in J $$\begin{pmatrix}0\ 0\\1\ 1\ 0\ 0\\1\ 1\ 0\ 0\\0\ 0\ 0\ 0\ 0\\0\ 0\ 0\ 0\ 0\end{pmatrix}\begin{pmatrix}0\ 1\\1\ 1\ 0\ 0\\1\ 1\ 0\ 1\\0\ 0\ 0\ 0\ 0\\1\ 0\ 1\ 0\ 1\end{pmatrix}\begin{pmatrix}0\ 0\\1\ 1\ 0\ 0\\1\ 1\ 0\ 0\\0\ 0\ 0\ 0\ 0\\0\ 0\ 0\ 0\ 0\end{pmatrix}$$
e.1) $p_2(\cdots)$	e.2)	e.3) **1** is at position $(x,y,z) \in \mathcal{E}_2$ in J $$\begin{pmatrix}1\ 1\ 1\\1\ 1\ 1\\0\ 0\ 0\ 0\ 0\\0\ 0\ 0\ 0\ 0\end{pmatrix}\begin{pmatrix}1\ 1\ 1\\1\ 1\ 1\\0\ 0\ 0\ 0\ 0\\1\ 0\ 1\ 0\ 1\end{pmatrix}\begin{pmatrix}1\ 1\ 1\\1\ 1\ 1\\0\ 0\ 0\ 0\ 0\\0\ 0\ 0\ 0\ 0\end{pmatrix}$$
f.1) $p_3(\cdots)$	f.2)	f.3) **1** is at position $(x,y,z) \in \mathcal{E}_2$ in J $$\begin{pmatrix}0\ 0\\1\ 1\ 0\ 0\\1\ 1\ 1\\0\ 0\ 1\ 1\\0\ 0\end{pmatrix}\begin{pmatrix}0\ 1\\1\ 1\ 0\ 0\\1\ 1\ 1\\0\ 0\ 1\ 1\\1\ 0\end{pmatrix}\begin{pmatrix}0\ 0\\1\ 1\ 0\ 0\\1\ 1\ 1\\0\ 0\ 1\ 1\\0\ 0\end{pmatrix}$$
g.1) $p_4(\cdots)$	g.2)	g.3) **1** is at position $(x,y,z) \in \mathcal{E}_2$ in J $$\begin{pmatrix}1\ 1\ 1\\1\ 1\ 1\\0\ 0\ 1\ 1\\0\ 0\end{pmatrix}\begin{pmatrix}1\ 1\ 1\\1\ 1\ 1\\0\ 0\ 1\ 1\\1\ 0\end{pmatrix}\begin{pmatrix}1\ 1\ 1\\1\ 1\ 1\\0\ 0\ 1\ 1\\0\ 0\end{pmatrix}$$

Fig. 3. Digital images encoding the 20 hexahedra showed in Fig. 2.

Proposition 3. *Let $p \in B$ be an ℓ−cell of P, $\ell = 1, \ldots, 3$. Then the $(\ell - 1)$−faces of p can be obtained by the following process. For each $q \in N_6^1(p)$:*

(1) if $q \in B \cap \mathcal{O}_{\ell-1}$, then q is an $(\ell - 1)$−face of p.
(2) if $q \notin B$, then
  (2.1) if $q' = p + 2(q - p) \in B \cap \mathcal{E}_{\ell-1}$, then q' is an $(\ell - 1)$−face of p;
  (2.2) else, if there exists $q'' \in B \cap N_8^1(q) \cap \mathcal{E}_{\ell-1}$ or $q'' \in B \cap N_{12}^1(q) \cap \mathcal{E}_{\ell-1}$, then q'' is an $(\ell - 1)$−face of p.

Proof. The cell $p \in B$ corresponds to a specific polyhedron or face of a polyhedron described in Definition 1, depending on the type of coordinates of p and on whether or not there are points of B around p (Proposition 2). For each of them, one of the structuring elements of Fig. 4 (modulo reflection and 90 degree rotation) provides its boundary faces. Hence, the proof consists in a case verification to set the equivalence between the known structuring element and the points satisfying either (1) and/or (2.1) and/or (2.2).

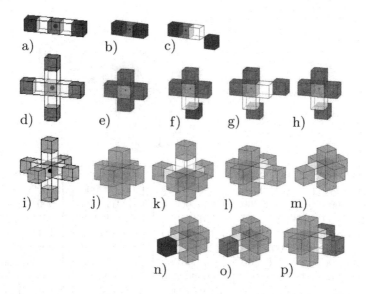

Fig. 4. Set of structuring elements (modulo reflections and 90 degree rotations) for computing boundary face relations used in [4]: 3 structuring elements for the boundary of a 1−cell (first row); 5 structuring elements for the boundary of a 2−cell (second row); 8 structuring elements for the boundary of a 3−cell (third and fourth rows).

- For $p \in \mathcal{E}_\ell$.
  - If $N_6^1(p) \cap B = \emptyset$, then one of the structuring elements in Fig. 4(a,d,i) fits around p [4] and the points in $N_6^2(p) \cap \mathcal{E}_{\ell-1}$ are the $(\ell - 1)$−faces of p. These points are obtained by applying (2.1).
  - Else,
    * If $p \in \mathcal{E}_0$, p is a size−2 cube, and hence, structuring element in Fig. 4j fits around p, whose points are obtained by step (1).
    * If $p \in \mathcal{E}_1$, then p is either a size−2 cube or a pyramid $k(v, w)$, so one of the structuring elements in Fig. 4(j,n) fits around p, whose points are obtained by applying (1).
    * If $p \in \mathcal{E}_2$, then $p = p_\ell(v, v_1, v_2, v_3)$, $\ell = 1, 2, 3, 4$, or p is a size-2 cube, so one of the structuring elements in Fig. 4(j,o,p) fits around p, whose points are obtained by applying (1).
    * If $p \in \mathcal{E}_3$, then p is one of $\{h_\ell(v_1, \ldots v_8)\}_{\ell=2,\ldots,22}$ and hence, one of the structuring elements in Fig. 4(j,k,l,m) fits around p, whose points are obtained by applying (1) and (2.1).
- For $p \in \mathcal{O}_\ell$:
  - If $p \in \mathcal{O}_0$, then p is a vertex of P and the proposition does not apply.
  - If $p \in \mathcal{O}_1$, then p is an edge of either a cube $c(v)$ or a pyramid $k(v, w)$. In the first case, the structuring element in Fig. 4b applies, that is obtained by (1). In the second case, the structuring element in Fig. 4c fits around p. Then, step (1) in the proposition provides one of the 1−faces and step (2.2) provides (unambiguously) the other one.

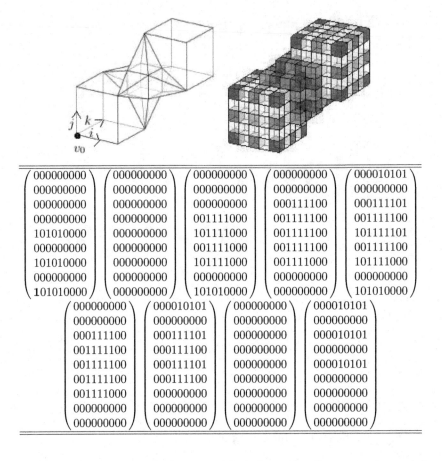

Fig. 5. On top, a naive example of a 3D polyhedral complex P_{ex} over a picture (where vertex v_0 has coordinates $(2,2,2)$) and its codification J (proposed in this paper) on the right. Color illustrates the dimension of the cell. Matrices on bottom: the binary image J seen as a set of binary matrices.

- If $p \in \mathcal{O}_2$, then p is one of the 2−cells on the boundary of the polyhedra in Definition 1(b,c,d):
  * If $p \in \{(2i+1, 4j+2, 4k+2), (4i+2, 2j+1, 4k+2), (4i+2, 4j+2, 2k+1)\}$, then p is a square face of a size−2 cube and, structuring element in Fig. 4e fits around p, whose points are obtained by applying (1).
  * If $p \in \{(4i+2, 4j, 2k+1), (4i, 4j+2, 2k+1), (2i+1, 4j+2, 4k), (2i+1, 4j, 4k+2), (4i+2, 2j+1, 4k), (4i, 2j+1, 4k+2)\}_{i,j,k \in \mathbb{Z}}$, then p is either a triangular face of a pyramid $k(v, w)$ or a square face of a size−2 cube. In the first case, structuring element in Fig. 4f fits around p and in the second case, the one in Fig. 4e. Both cases are obtained by applying (1).
  * If $p \in \{(4i, 4j, 2k+1), (2i+1, 4j, 4k), (4i, 2j+1, 4k)\}_{i,j,k \in \mathbb{Z}}$, then p is a quadrangular face of some polyhedron from Definition 1d. The

Table 2. Boundary-face representation of P_{ex} (see Fig. 5): E can be seen as a binary 22×54 matrix, F as a binary 54×48 matrix and C as a binary 48×15 matrix.

List $V = \{v_\ell\}$ of vertices, where $v_\ell = (i, j, k)$ represents a vertex in P_{ex}.
$\{(2, 2, 2), (6, 2, 2), (2, 6, 2), (6, 6, 2), (5, 5, 5),$
$(7, 5, 5), (5, 7, 5), (7, 7, 5), (2, 2, 6), (6, 2, 6),$
$(2, 6, 6), (10, 6, 6), (6, 10, 6), (10, 10, 6), (5, 5, 7),$
$(7, 5, 7), (5, 7, 7), (7, 7, 7), (6, 6, 10), (10, 6, 10),$
$(6, 10, 10), (10, 10, 10)\}$

List $E = \{e_\ell\}$ of edges, where $e_\ell = (i, j)$ represents an edge in P_{ex} with endpoints $v_i, v_j \in V$.
$\{(1, 2), (1, 3), (1, 9), (2, 4), (2, 10),$
$(3, 4), (3, 11), (4, 5), (4, 6), (4, 7),$
$(4, 8), (5, 6), (5, 7), (5, 10), (5, 11),$
$(5, 15), (6, 8), (6, 10), (6, 12), (6, 16),$
$(7, 8), (7, 11), (7, 13), (7, 17), (8, 12),$
$(8, 13), (8, 18), (9, 10), (9, 11), (10, 15)$
$(10, 16), (11, 15), (11, 17), (12, 14), (12, 16),$
$(12, 18), (12, 20), (13, 14), (13, 17), (13, 18),$
$(13, 21), (14, 22), (15, 16), (15, 17), (15, 19),$
$(16, 18), (16, 19), (17, 18), (17, 19), (18, 19),$
$(19, 20), (19, 21), (20, 22), (21, 22)\}$

List $F = \{f_\ell\}$ of faces, where $f_\ell = (i_1, \ldots, i_n)$ represents a 2-cell in P_{ex} bounded by edges $e_{i_1}, \ldots, e_{i_n} \in E$.
$\{(1, 2, 4, 6), (1, 3, 5, 28), (2, 3, 7, 29), (4, 5, 8, 14), (4, 5, 9, 18),$
$(6, 7, 8, 15), (6, 7, 10, 22), (14, 15, 28, 29), (28, 29, 30, 32), (8, 10, 13),$
$(10, 11, 21), (8, 9, 12), (9, 11, 17), (12, 14, 18), (14, 16, 30),$
$(18, 20, 31), (30, 31, 43), (15, 16, 32), (13, 15, 22), (22, 24, 33),$
$(32, 33, 44), (17, 19, 25), (19, 20, 35), (25, 27, 36), (35, 36, 46),$
$(21, 23, 26), (23, 24, 39), (26, 27, 40), (39, 40, 48), (46, 47, 50),$
$(48, 49, 50), (43, 45, 47), (44, 45, 49), (34, 37, 42, 53), (38, 41, 42, 54),$
$(51, 52, 53, 54), (25, 26, 34, 38), (34, 36, 38, 40), (35, 37, 47, 51), (36, 37, 50, 51),$
$(39, 41, 49, 52), (40, 41, 50, 52), (12, 13, 17, 21), (12, 16, 20, 43), (21, 24, 27, 48),$
$(43, 44, 46, 48), (17, 20, 27, 46), (13, 16, 24, 44)\}$

List $G = \{g_\ell\}$ of polyhedra, where $g_\ell = (i_1, \ldots, i_n) \in G$ represents a polyhedron in P_{ex} bounded by faces $f_{i_1}, \ldots, f_{i_n} \in F$.
$\{(1, 2, 3, 4, 6, 8), (34, 35, 36, 38, 40, 42), (43, 44, 45, 46, 47, 48),$
$(4, 5, 12, 14), (6, 7, 10, 19), (8, 9, 15, 18), (24, 28, 37, 38),$
$(25, 30, 39, 40), (29, 31, 41, 42), (10, 11, 12, 13, 43)$
$(14, 15, 16, 17, 44), (18, 19, 20, 21, 48), (22, 23, 24, 25, 47),$
$(26, 27, 28, 29, 45), (30, 31, 32, 33, 46)\}$

structuring elements that provide the faces in the different cases are those in Fig. 4(e,g,h). For the one in Fig. 4e, step (1) applies; in the other two cases, step (1) and (2.2) apply.

For each case, there are no other points $q \in \mathcal{O}_{\ell-1}$ or $q', q'' \in \mathcal{E}_{\ell-1}$ that could be returned by the proposed steps. This can be deduced from the type of coordinates of p and its neighbors.   □

In Fig. 5, a small example of a well-composed polyhedral complex over a picture is shown together with its codification under the form of a 3D binary digital image J. In fact, J can be given as a 3D binary matrix of size $9 \times 9 \times 9$, which are the nine $2D$ binary matrices shown on the bottom of Fig. 5. To give a naive comparison with our codification, we show in Table 2 the boundary-face representation of P. We can see that in our codification we just need to store a $9 \times 9 \times 9$ binary matrix plus the coordinates of vertex v_0. On the other hand, to store the boundary-face representation of P, we need to store the coordinates of all the vertices of P (list V) plus the three incidence matrices E, F and G of size 22×54, 54×48, 48×15, respectively.

5   Conclusions

Given a well-composed polyhedral complex P over a picture (see Sect. 4), we have proved that geometric and topological information of the cells of P can be encoded by just using a 3D binary image. This way, the type of 3D coordinates codifying each cell, together with the information encoded at some neighbor points provide both the geometry and the boundary faces of each cell without the need of using a set of structuring elements. More specifically, this means a more efficient treatment of the ECM representation developed in [4] to codify well-composed polyhedral complexes that are homotopy equivalent to the standard cubical complex representing a (non-necessarily well-composed) 3D binary image.

Acknowledgments. We would like to thank Pawel Dlotko for fruitful discussions about possible applications of ECM representations and to the reviewers for their valuable suggestions.

References

1. Boutry, N., Géraud, T., Najman, L.: How to make nD functions digitally well-composed in a self-dual way. In: Benediktsson, J.A., Chanussot, J., Najman, L., Talbot, H. (eds.) ISMM 2015. LNCS, vol. 9082, pp. 561–572. Springer, Heidelberg (2015)
2. Géraud, T., Carlinet, E., Crozet, S.: Self-duality and digital topology: links between the morphological tree of shapes and well-composed gray-level images. In: Benediktsson, J.A., Chanussot, J., Najman, L., Talbot, H. (eds.) ISMM 2015. LNCS, vol. 9082, pp. 573–584. Springer, Heidelberg (2015)

3. Gonzalez-Diaz, R., Jimenez, M.J., Medrano, B.: Cubical cohomology ring of 3D photographs. Int. J. Imaging Syst. Technol. **21**(1), 76–85 (2011)
4. Gonzalez-Diaz, R., Jimenez, M.-J., Medrano, B.: 3D well-composed polyhedral complexes. Discrete Appl. Math. **183**, 59–77 (2015)
5. Gonzalez-Diaz, R., Jimenez, M.-J., Medrano, B.: Demo: well-composed polyhedral complexes. In: Demo session of 18th IAPR International Conference on Discrete Geometry for Computer Imagery. http://grupo.us.es/cimagroup/DEMO_Matlab.zip, http://clem.dii.unisi.it/~dgci2014/demos/demo1.pdf
6. Gonzalez-Diaz, R., Lamar, J., Umble, R.: Computing cup products in Z2-cohomology of 3D polyhedral complexes. Found. Comput. Math. **14**(4), 721–744 (2014)
7. Kozlov, D.: Combinatorial Algebraic Topology. Algorithms and Computation in Maths, vol. 21. Springer, Heidelberg (2008)
8. Lachaud, J.O., Montanvert, A.: Continuous analogs of digital boundaries: a topological approach to iso-surfaces. Graph. Models **62**, 129–164 (2000)
9. Latecki, L.J.: 3D well-composed pictures. Graph. Models Image Process. **59**(3), 164–172 (1997)
10. Latecki, L.J.: Discrete Representation of Spatial Objects in Computer Vision. Springer Science+Business Media B.V., Dordrecht (1998)
11. Najman, L., Géraud, T.: Discrete set-valued continuity and interpolation. In: Hendriks, C.L.L., Borgefors, G., Strand, R. (eds.) ISMM 2013. LNCS, vol. 7883, pp. 37–48. Springer, Heidelberg (2013)
12. Siqueira, M., Latecki, L.J., Tustison, N., Gallier, J., Gee, J.: Topological repairing of 3D digital images. J. Math. Imaging Vis. **30**, 249–274 (2008)
13. Stelldinger, P., Latecki, L.J.: 3D object digitization: majority interpolation and marching cubes. In: Proceedings of the 18th IEEE International Conference on Pattern Recognition, pp. 1173–1176 (2006)
14. Stelldinger, P.: Image Digitization and Its Influence on Shape Properties in Finite Dimensions, p. 312. IOS Press, The Netherlands (2008)

Interactive Curvature Tensor Visualization
on Digital Surfaces

Hélène Perrier[1], Jérémy Levallois[1,2], David Coeurjolly[1(✉)],
Jean-Philippe Farrugia[1], Jean-Claude Iehl[1], and Jacques-Olivier Lachaud[2]

[1] Université de Lyon, CNRS LIRIS, UMR5205, 69622 Lyon, France
`david.coeurjolly@liris.cnrs.fr`
[2] Université de Savoie Mont Blanc, CNRS LAMA, UMR5127,
73776 Chambéry, France

Abstract. Interactive visualization is a very convenient tool to explore complex scientific data or to try different parameter settings for a given processing algorithm. In this article, we present a tool to efficiently analyze the curvature tensor on the boundary of potentially large and dynamic digital objects (mean and Gaussian curvatures, principal curvatures, principal directions and normal vector field). More precisely, we combine a fully parallel pipeline on GPU to extract an adaptive triangulated isosurface of the digital object, with a curvature tensor estimation at each surface point based on integral invariants. Integral invariants being parametrized by a given ball radius, our proposal allows to explore interactively different radii and thus select the appropriate scale at which the computation is performed and visualized.

Keywords: Isosurface visualization · Digital geometry · Curvature estimation · GPU

1 Introduction

Volumetric objects are being more and more popular in many applications ranging from object modeling and rendering in Computer Graphics to geometry processing in Medical Imaging or Material Sciences. When considering large volumetric data, interactive visualization of those objects (or isosurfaces) is a complex problem. Such issues become even more difficult when dynamic volumetric datasets are considered. Beside visualization, we are also interested in performing geometry processing on the digital object and to explore different parameter settings of the geometry processing tool. Here, we focus on curvature tensor estimation (mean/Gaussian curvature, principal curvatures directions...). Most curvature estimators require a parameter fixing the scale at which the computation is performed. For short, such parameter (integration radius, convolution kernel size...)

This work has been mainly funded by DIGITALSNOW ANR-11-BS02-009, KIDICO ANR-2010-BLAN-0205 and PRIMES LABEX ANR-11-LABX-0063/ANR-11-IDEX-0007 research grants.

© Springer International Publishing Switzerland 2016
N. Normand et al. (Eds.): DGCI 2016, LNCS 9647, pp. 282–294, 2016.
DOI: 10.1007/978-3-319-32360-2_22

specifies a scale for analyzing the object surface, and is naturally related to the amount allowed perturbations on input data. As a consequence, when using such estimators, exploring different values of this parameter is mandatory. However, this is usually an offline process, due to the amount of computations that need to be done.

Contributions. In this work, we propose a framework to perform interactive visualization of complex 3D digital structures combined with a dynamic curvature tensor estimation. We define a fully data parallel process on the GPU (Graphics Processor Unit) to both efficiently extract adaptive isosurface and compute per vertex curvature tensor using Integral Invariants estimators. This system allows us to visualize curvature tensor in real-time on large dynamic objects. Our approach combines a GPU implementation of pointerless octrees to represent the data, an adaptive viewpoint-dependent mesh extraction, and a GPU implementation of integral invariant curvature estimators.

Related Works. Extracting and visualizing isosurface on volumetric data has been widely investigated since the seminal Marching Cubes approach by LORENSEN and CLINE [11]. This approach being data parallel, GPU implementation of this method is very efficient [16]. However, such technique generates a lot of triangles which is not well suited to large digital data. Hence, adaptive approaches have been proposed in order to optimize the triangulated mesh resolution according to the object geometry or the viewpoint. In this topic, many solutions have been developed in Computer Graphics [6,9,10,14,15]. The method developed by LENGYEL et al. [6] suits best our needs. It combines an octree space partitioning with new Marching Cubes configurations to generate adaptive meshes on CPU from static or near static data. We propose here a full GPU pipeline inspired by this method, that maintains a view dependent triangulation. Our high framerate allows us to inspect dynamic 3D data in real-time.

Curvature estimation on discrete or digital surface has also been widely investigated. In [2], authors propose digital versions of Integral Invariant estimators [12,13] in order to estimate the complete curvature tensor (mean/Gaussian curvatures, principal curvatures, principal curvature directions, normal vector field) on digital surfaces. Such approaches are based on an integration principle using a ball kernel of a given radius. Additionally, authors have demonstrated that these estimators have multigrid convergence properties. In this article, we also propose an efficient GPU implementation of such estimators to visualize such curvature fields in real-time and to interactively change the value of the kernel radius.

In this paper, we first present the previous works on curvature tensor estimation. Then, we propose an efficient GPU approach to extract a triangulated isosurface from a digital object. Finally, we propose a fully-parallel GPU pipeline to compute curvature tensor in real-time and present our results.

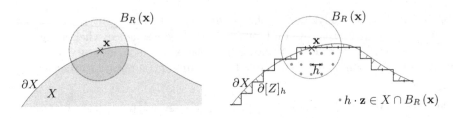

Fig. 1. Integral invariant computation (*left*) and notations (*right*) in dimension 2 [2].

2   Curvature Tensor Estimation

In our context, we consider digital shapes (any subset Z of $\mathbb{Z}^d$) and boundaries of digital shapes $Bd(Z)$. We denote by $\mathsf{G}_h(X)$ the Gauss digitization of a shape $X \subset \mathbb{R}^d$ in a d–dimensional grid with grid step h, *i.e.* $\mathsf{G}_h(X) := \{\mathbf{z} \in \mathbb{Z}^d, \ h \cdot \mathbf{z} \in X\}$. For such digitized set $Z := \mathsf{G}_h(X)$, $[Z]_h$ is a subset of $\mathbb{R}^d$ corresponding to the union of hypercubes centered at $h \cdot \mathbf{z}$ for $\mathbf{z} \in Z$ with edge length h. By doing so, both ∂X and $\partial[Z]_h$ are topological boundaries of objects lying in the same space (see Fig. 1-b). Note that the combinatorial digital boundary $Bd(Z)$ of Z made of cells in a cellular Cartesian complex (*pointels, linels, surfels, ...*), can be trivially embedded into $\mathbb{R}^d$ such that it coincides with $\partial[Z]_h$.

In [1], authors define a 2D digital curvature estimator $\hat{\kappa}^R$ and a 3D digital mean curvature estimator $\hat{H}^R$ based on the digital volume estimator $\widehat{\mathrm{Vol}}(Y, h) := h^d \mathrm{Card}(Y)$ (area estimator $\widehat{\mathrm{Area}}(Y, h)$ in dimension 2):

Definition 1. *Given the Gauss digitization $Z := \mathsf{G}_h(X)$ of a shape $X \subset \mathbb{R}^2$ (or $\mathbb{R}^3$ for the 3D mean curvature estimator), digital curvature estimators are defined for any point $\mathbf{x} \in \mathbb{R}^2$ (or $\mathbb{R}^3$) as:*

$$\forall 0 < h < R, \quad \hat{\kappa}^R(Z, \mathbf{x}, h) := \frac{3\pi}{2R} - \frac{3\widehat{\mathrm{Area}}(B_{R/h}(\mathbf{x}/h) \cap Z, h)}{R^3}, \tag{1}$$

$$\hat{H}^R(Z, \mathbf{x}, h) := \frac{8}{3R} - \frac{4\widehat{\mathrm{Vol}}(B_{R/h}(\mathbf{x}/h) \cap Z, h)}{\pi R^4}. \tag{2}$$

where $B_{R/h}(\mathbf{x}/h)$ is the ball of digital radius R/h centered on digital point $(\mathbf{x}/h) \in Z$.

Such curvature estimators have multigrid convergence properties [1]: when the digital object becomes finer and finer, *i.e.* when the digitization step h tends to zero, the estimated quantities on $\partial[\mathsf{G}_h(X)]_h$ converges (theoretically and experimentally) to the associated one on ∂X in $O\left(h^{\frac{1}{3}}\right)$ for convex shapes with at least C^3-boundary and bounded curvature (setting $R := kh^{\frac{1}{3}}$ for some $k \in \mathbb{R}$).

In [2], authors have also defined 3D digital principal curvature estimators $\hat{\kappa}_1^R$ and $\hat{\kappa}_2^R$ on $Z \subset \mathbb{Z}^3$ based on digital moments:

Definition 2. *Given the Gauss digitization* $Z := \mathsf{G}_h(X)$ *of a shape* $X \subset \mathbb{R}^3$, *3D digital principal curvature estimators are defined for any point* $\mathbf{x} \in \mathbb{R}^3$ *as:*

$$\forall 0 < h < R, \quad \hat{\kappa}_1^R(Z, \mathbf{x}, h) := \frac{6}{\pi R^6}(\hat{\lambda}_2 - 3\hat{\lambda}_1) + \frac{8}{5R}, \tag{3}$$

$$\hat{\kappa}_2^R(Z, \mathbf{x}, h) := \frac{6}{\pi R^6}(\hat{\lambda}_1 - 3\hat{\lambda}_2) + \frac{8}{5R}, \tag{4}$$

where $\hat{\lambda}_1$ *and* $\hat{\lambda}_2$ *are the two greatest eigenvalues of the covariance matrix of* $B_{R/h}(\mathbf{x}/h) \cap Z$.

The covariance matrix needs to compute digital moments of order 0, 1 and 2 (see Eq. 20 of [2] for more details). These estimators are proven convergent in $O\left(h^{\frac{1}{3}}\right)$ when setting the ball radius h in $R = kh^{\frac{1}{3}}$, where k is a constant related to the maximal curvature of the shape, on convex shapes with at least C^3-boundary and bounded curvature [2]. Additionally, eigenvectors associated to $\hat{\lambda}_1$ and $\hat{\lambda}_2$ of the covariance matrix are principal curvature direction estimators $\hat{\mathbf{w}}_1^R$ and $\hat{\mathbf{w}}_2^R$. The smallest eigenvector corresponds to the normal direction $\hat{\mathbf{n}}^R$ at $\mathbf{x}$. Convergence results can be found in [5].

It has been shown that the radius of the ball depends on the geometry of the underlying shape. In [7], a proposal was made for a parameter-free estimation of the radius of the ball by analyzing the shape *w.r.t.* the local shape geometry using maximal digital straight segments of the digital boundary. In [8], these estimators have been analyzed in scale-space (for a range of radii) for a given digital shape. This allows to detect features of the shape thanks to the behavior of estimators on singularities. As a consequence, for all these integral invariant based approaches, we need to consider different ball radius which could be time consuming when implemented on CPU. We propose here a fully parallel implementation on GPU allowing us to change the radius and thus update the estimated quantities in real-time.

3   Isosurface Extraction on GPU

In this section, we detail the adaptive isosurface extraction algorithm. The proposed approach uses an octree representation of the input object on which an adaptive Marching Cube builds the isosurface efficiently. Such hierarchical representation of the object allows us to handle large datasets and to locally adapt the level of details *w.r.t.* the geometry or camera position. We first present the octree representation and then the isosurface extraction.

3.1   Linear Octree Representation

Representing a hierarchical structure on GPU is usually challenging since such a data parallel component is unable to handle recursivity. Efficient spatial tree encoding can be achieved using pointerless structures such as linear quadtrees or octrees GARGANTINI [4]. This structure indexes each cell by a *Morton code*:

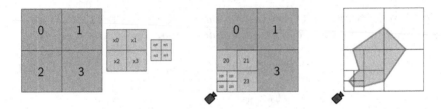

Fig. 2. Morton codes associated to cells of a linear quatree. Each morton code of a child cell is obtained by adding a suffix to its parent code *(left)*. The adaptive representation consists of quadtree cells whose depth is view point dependent *(middle)*. Finally, adaptive Marching Cubes is used to generate the triangulation *(right)*.

the code of children cells are defined by the code of the parent suffixed by two bits (in dimension 2, three bits in dimension 3) (see Fig. 2-*left*). A cell's code encodes its position *w.r.t.* its parent cell and its complete path to the tree root. Hence, the tree is fully represented as a linear vector of its leaves. Furthermore, cell operations such as subdivision, merging can be efficiently implemented using bitwise operations on the Morton code. In the following, we use the GPU friendly implementation proposed by DUPUY *et al.* [3].

3.2   Data Parallel and Adaptive Mesh Generation

Using this spatial data structure, a triangulated mesh can be constructed using Marching Cubes [11] (MC for short): the triangulation is generated from local triangle patches computed on local cell configurations. Such approach is fully parallel and easy to implement on the GPU. However, since adjacent cells may not have same depth in the octree, original LORENSEN and CLINE's rules need to be updated (see Fig. 2-*right*). Many authors have addressed this problem both for primal and dual meshes [6,9,10,14,15].

In the following, we use the algorithm proposed by LENGYEL *et al.* [6]. First, this approach forces the octree structure to make sure that the depth difference between any two adjacent cells is at most one. Then, LENGYEL *et al.* introduce the concept of transition cells. Those cells are defined to be inserted between two neighboring octree cells of different depth. With specific MC configurations for their triangulation, a crack free mesh can be extracted.

Similarly to original MC algorithm, this approach is well suited to a GPU implementation: given a set of cells (a vector of morton codes), each triangle patch can be extracted in parallel for both regular cells and transition cells.

3.3   Level of Details Criteria and Temporal Updates

As illustrated in Fig. 2-*middle*, we propose a viewpoint dependent criterion to decide if a cell needs to be refined: the closer we are to the camera, the finer the cells are. Such a criterion is also well suited to GPU since it can be evaluated independently on every cell. Figure 3 illustrates our level of details (LoD for short) criterion. In dimension 2, if α denotes the viewing angle, an object at a distance d from the camera has a projected size on screen of $2 \cdot d \cdot \tan(\alpha)$. (see Fig. 3-*left*). Our distance criterion is based on the ratio (*visibility ratio* in the following) between the cell diameter $l(c)$ (power of 2 depending on the depth), and its projected size. For a given cell c, split and merge decision are based on this visibility ratio:

- c is split if its children cells have a visibility ratio greater than constant k;
- c and its sibling cells are merged if their parent cell c' has a visibility ratio lower than k;
- otherwise, the cell c stays for the next frame.

Using such a criterion, split and merge decisions are computed in parallel from the morton codes of all cells.

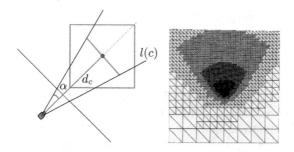

Fig. 3. Notations *(left)* and adaptive meshing in dimension 2 using the LoD distance and angular criterion *(right)*.

Once decisions have been made, a new set of cells is sent to the mesh generation step described above. Finally, before constructing the triangulation from remaining cells and transition cells, geometrical culling is performed in order to skip the triangle patch construction for cells that are not visible. Figure 4 illustrates the overall fully data parallel pipeline.

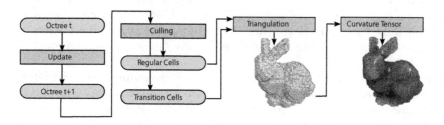

Fig. 4. GPU pipeline summary. Data buffers are represented in green and computations in red. Each computation retrieves data from a buffer and fills a new one (Color figure online).

4   Interactive Curvature Computation on GPU

We first present design principles for the GPU implementation and then present our approaches. The general idea is to perform an Integral Invariant computation on GPU at each vertex of the generated triangulated mesh. Since GPU have massively parallel architectures, we can do all those computations in parallel to obtain a very efficient implementation. Please note that when the LoD criterion is removed, each MC vertex is exactly centered at a surfel center. At different depth of the octree, MC vertices still correspond to surfel centers of subsampled versions of the input object. Hence, Integral Invariant framework defined in Sect. 2 is consistent with the triangulated surface obtained on GPU: triangles are used for visualization purposes but all computations are performed on the digital object $G_h(X)$ for estimators (1),(2),(3) and (4).

To implement the integration on $B_R(\mathbf{x}) \cap X$ (Fig. 1), several strategies have been evaluated.

4.1   Per Vertex Real-Time Computation on GPU

Naive Approach. A first simple solution consists in scanning all digital points, at a given resolution, lying inside the integration domain (Fig. 5-*left*) and then estimating the geometrical moment as the sum of the geometrical moments of each elementary cubes lying in the intersection (see Eqs. (1) to (4)). This is exactly similar to what is done on the CPU. On the GPU, we can exploit *mipmap* textures to obtain multi-resolution information. If the input binary object is stored in a 3D texture, GPU hardware constructs a multi-resolution pyramid (mipmap) such that 8 neighboring voxel intensities at level l are averaged to define the voxel value at level $l+1$. If the level 0 corresponds to the binary input object, at a given level l, a texture probe at a point (x, y, z) returns the fraction of $G_h(X)$ belonging to the cube of center (x, y, z) and edge length 2^l. As a consequence, we can approximate the volume of $B_R(\mathbf{x}) \cap X$ by considering mipmap values at a given resolution l (Fig. 5-*right*). In this case, errors only occurs for cells lying inside X (with density 1) not entirely covered by $B_R(\mathbf{x})$. Furthermore, the *mipmap* texture can be used to design, using a single texture probe, a fast inclusion test of a given cell c at level l into the shapes: we say that c is in X if its density (retrieved from the texture probe at level l) is greater than $1/2$. The idea here is to mimic a kind of adaptive Gauss digitization process.

Hierarchical Decomposition. Using the hierarchical nature of the mipmap texture, we could also consider hierarchical decompositions of $B_R(\mathbf{x})$. The idea is to decompose the ball into mipmap cells of different resolution in order to limit the number of texture access (important bottleneck on GPU hardwares) and to get better approximated quantities. On the GPU, computing a hierarchical decomposition of $B_R(\mathbf{x})$ at each point $\mathbf{x}$ is highly inefficient since the hardware optimizes the parallelism only when the micro-program described in the shader has a predictive execution flow. Hence, implementing the recursive algorithm (or its de-recursified version) requires a lot of branches (conditional *if* tests) in the flow.

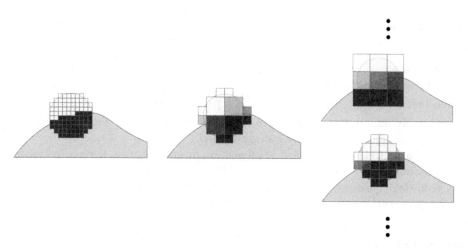

Fig. 5. Here are the three approaches we used to evaluate the integrand. (*left*) is the naive approach, where we sum all the digital points that are inside the integrand domain (*middle*) shows the hierarchical decomposition of the ball, thus limiting the number of texture probes to do. (*right*) illustrates the dynamic approach, we probe the textures at a higher resolution and we refine it at each frame until we reach the finer level.

We have thus considered a fast multi-resolution approximation as illustrated in Fig. 5-*middle*: for a given radius R, we precompute a hierarchical decomposition of B_R. Such decomposition is made of octree cells at different resolutions. At a given MC vertex $\mathbf{x}$, the algorithm becomes simple since we just scan the ball cells and estimate the area (or other moments) from the mipmap values associated to this cell. Note that these precomputed cells in the B_R decomposition may not be aligned with mipmap cells. However, we can use GPU which interpolates mipmap values if we probe at non-discrete positions. For a given radius R, the expected number of cells in B_R is in $O(\log R)$. Implementation details on the hierarchical octree representation can be found in the supplementary material (see Sect. 6).

Dynamic Refinement. In this last approach, we want to optimize the interactivity and the execution flow or parallelism on GPU. The integration is simply computed using a regular grid at different resolution l (Fig. 5-*right*). The shader code becomes trivial (simple triple loop) and a lot of texture fetches are involved, but interactivity can be easily obtained. Indeed, we consider the following multipass approach:

1. When the surface geometry has been computed, we set the current level l at an high level $l := l_{\max}$ of the mipmap pyramid.
2. We compute the integrals (and the curvature tensor) at the mipmap level l and send the estimated quantities to the visualization stage.
3. If the user changes the camera position or the computation parameters, we return to step 1.

4. If the current framerate is above a given threshold, we decrease l and return to step 2 if the final l_0 level has not been reached yet.

A consequence of the multi-pass approach is that when there is no interaction (camera settings, parameters), the GPU automatically refines the estimation. Even if step 2 is quite expensive, in $O\left(\left(\frac{R}{2^l}\right)^3\right)$, the interactive control in step 3 considerably improves the interactive exploration of the tensor with fast preliminary approximations which are quickly refined.

Compared to the hierarchical approach, no precomputation is required for a given radius R. As a consequence, we could even locally adapt the ball radius to the geometry (for instance following the octree depth of the current MC vertex). In the next section, we evaluate the performances of both approaches.

5   Experiments

5.1   Full Resolution Experiment

We first evaluate curvature estimations on a full-resolution geometry obtained by disabling the LoD criterion at the mesh generation step. Figure 6-top shows results of curvature tensor estimation (mean, Gaussian, first and second principal directions) on "OctaFlower" at level l_0 (considered as our ground truth in the following). Figure 6-bottom shows mean curvature estimation in real-time on a dynamic object. For this case, we simply evaluate the implicit expression at each vertex of a cell to decide if such cell is included or not into $B_R(\mathbf{x}) \cap X$ instead of updating the *mipmap texture* of densities at each time step. For all illustrations, we use directly normal vectors computed by algorithm discussed in Sect. 2.

Then, we compare the approximations made by computing curvature from level $l \geq l_0$ in Fig. 7. We can see that results using approximation seems to quickly converge to ground truth results. Table 1 shows numerical results of L_∞ error (maximal absolute difference for all vertices) and L_2 error (mean squared errors of all vertices), as well the number of *mipmap texture* fetches. We can

Fig. 6. *First row:* Mean and Gaussian curvature estimation, (zoom of) first and second principal directions estimation on "OctaFlower" with a digital domain of 130^3. *Second row:* Mean curvature computed in real-time (around 20 FPS) on a dynamic object.

Fig. 7. Illustration of mean curvature computation on "OctaFlower" (digital domain of 130^3) using mipmap approximation with different levels: l_3, l_2, l_1 and l_0 (*i.e.* no approximation).

also see that the number of texel fetch, required to compute the curvature, reduces drastically when computing approximations. However, at higher levels, approximation errors become more important. Those levels are thus never used in practice, they are presented here to illustrate how the refining process converges to the l_0 level.

We also compare them with the hierarchical algorithm (as discussed in Sect. 4.1). This method introduces a higher error when compared with l_0. This is due to the precomputation of the subdivision that no longer ensures to fetch data at the center of a mipmap cell. The GPU interpolation reduces the bias, but it does not remove it.

5.2 Fully Adaptive Evaluation

When dealing with large datasets, we cannot expect real-time curvature tensor estimation if we consider the full resolution geometry, due to the huge amount of data to process. By computing a dynamically refined approximate curvature tensor estimation joined with an adaptive triangulation (as discussed in Sect. 4.1), we manage to maintain a real-time framerate by giving control over the amount of data to process at each frame. In Fig. 8, we compare timings (in logarithmic scale) for a triangulation that is dynamically refined according to its distance to the camera. We measured those timings using the multiresolution regular and the hierarchical algorithm.

First we can note that the required time to compute the ground truth curvature for an object is usually as high as the time required to extract its geometry and to compute all of the above levels. Using approximations is thus mandatory. Another advantage with approximations is that it allows us to get a visualization of our object as soon as we run the application. This allows for real-time interactions with the object, required in order to change the visualized quantity, curvature radius, etc.

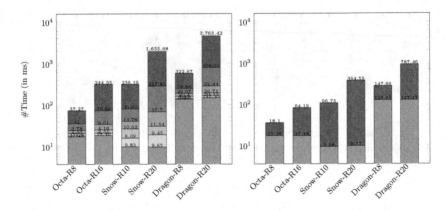

Fig. 8. Timings in milliseconds (in logscale) obtained while visualizing an adaptive triangulation on three objects – "OctaFlower" (digital domain of 130^3), "Snow microstructures" (233^3) and "XYZ-Dragon" (510^3) – by computing the curvature with a regular grid (*left*) and with hierarchical algorithm (*right*), with two different radii for each object. In *orange color:* time required to extract the triangulation. In *blue color:* time required to compute the curvature tensor at different levels: from l_4 (light blue) to l_0 (dark blue). Timings are given using a NVIDIA GeForce GTX 850 M GPU (Color figure online).

Table 1. Comparison of number of texture fetches, L_2 and L_∞ error obtained on "OctaFlower" with a digital domain of 130^3 when computing mean curvature with two radii: 8 and 16, with hierarchical algorithm (H) and $l \geq l_0$ approximation algorithms. The object is triangulated at full resolution with a regular grid and contains 282,396 vertices.

		H	l_4	l_3	l_2	l_1	l_0
Number of texture fetches	$R = 8$	1468		2	28	260	2104
	$R = 16$	5706	2	28	90	2120	17080
L_∞ error (*w.r.t.* l_0)	$R = 8$	0.051		0.306	0.085	0.047	0
	$R = 16$	0.053	0.146	0.041	0.017	0.005	0
L_2 error (*w.r.t.* l_0)	$R = 8$	3.51e-05		2.67e-04	7.57e-05	2.02e-05	0
	$R = 16$	4.43e-05	1.39e-04	4.16e-05	1.57e-05	2.07e-06	0

It is also visible in Fig. 8 that a hierarchical decomposition greatly reduces the curvature computation time, especially with big radii. However, due to the precomputation and the current hierarchical structure, this algorithm is biased and creates an error (presented in Table 1) that needs to be considered.

Figure 9 shows curvature computation and exploration in real-time on large datasets: "XYZ-Dragon" (with digital domain of 512^3) and "Snow microstructures" (233^3).

Fig. 9. *Left column:* Mean curvature, first and second principal directions and normal vector field estimation on "Snow microstructures" (233^3 and $R = 8$). *Right column:* Mean curvature, first and second principal directions and normal vector field estimation on "XYZ-Dragon" (510^3 and $R = 8$). Normal vectors are colored with a mapping of their component to RGB color space (Color figure online).

6   Conclusion and Discussion

In this article, we have proposed a fully data parallel framework on GPU hardware which combines an adaptive isosurface construction from digital data with a curvature tensor estimation at each vertex. Using this approach, we can explore in real-time different curvature measurements (mean, Gaussian, principal directions) with different ball radii on potentially large dynamic dataset. Our proposal relies on both a linear octree representation with Morton codes and an efficient integral computation on GPU. The source code and additional material (video, ...) are available on the project website (https://github.com/dcoeurjo/ICTV).

References

1. Coeurjolly, D., Lachaud, J.-O., Levallois, J.: Integral based curvature estimators in digital geometry. In: Gonzalez-Diaz, R., Jimenez, M.-J., Medrano, B. (eds.) DGCI 2013. LNCS, vol. 7749, pp. 215–227. Springer, Heidelberg (2013)
2. Coeurjolly, D., Lachaud, J.O., Levallois, J.: Multigrid convergent principal curvature estimators in digital geometry. Comput. Vis. Image Underst. **129**, 27–41 (2014)
3. Dupuy, J., Iehl, J.C., Poulin, P.: GPU Pro 5, chap. Quadtrees on the GPU. A K Peters/CRC Press. http://liris.cnrs.fr/publis/?id=6299
4. Gargantini, I.: An effective way to represent quadtrees. Commun. ACM **25**(12), 905–910 (1982)
5. Lachaud, J.O., Coeurjolly, D., Levallois, J.: Robust and convergent curvature and normal estimators with digital integral invariants. In: Modern Approaches to Discrete Curvature. Lecture Notes in Mathematics. Springer International Publishing (2016, forthcoming)
6. Lengyel, E.S., Owens, J.D.: Voxel-based terrain for real-time virtual simulations. University of California at Davis (2010)
7. Levallois, J., Coeurjolly, D., Lachaud, J.-O.: Parameter-free and multigrid convergent digital curvature estimators. In: Barucci, E., Frosini, A., Rinaldi, S. (eds.) DGCI 2014. LNCS, vol. 8668, pp. 162–175. Springer, Heidelberg (2014)
8. Levallois, J., Coeurjolly, D., Lachaud, J.O.: Scale-space feature extraction on digital surfaces. Comput. Graph. **51**, 177–189 (2015)
9. Lewiner, T., Mello, V., Peixoto, A., Pesco, S., Lopes, H.: Fast generation of pointerless octree duals. Comput. Graph. Forum **29**(5), 1661–1669 (2010). http://dx.doi.org/10.1111/j.1467-8659.2010.01775.x
10. Lobello, R.U., Dupont, F., Denis, F.: Out-of-core adaptive iso-surface extraction from binary volume data. Graph. Models **76**(6), 593–608 (2014). http://dx.doi.org/10.1016/j.gmod.2014.06.001
11. Lorensen, W.E., Cline, H.E.: Marching cubes: a high resolution 3d surface construction algorithm. ACM Comput. Graph. **21**(4), 163–169 (1987)
12. Pottmann, H., Wallner, J., Huang, Q., Yang, Y.: Integral invariants for robust geometry processing. Comput. Aided Geom. Des. **26**(1), 37–60 (2009)
13. Pottmann, H., Wallner, J., Yang, Y., Lai, Y., Hu, S.: Principal curvatures from the integral invariant viewpoint. Comput. Aided Geom. Des. **24**(8–9), 428–442 (2007)
14. Schaefer, S., Warren, J.: Dual marching cubes: primal contouring of dual grids. In: Proceedings of 12th Pacific Conference on Computer Graphics and Applications, pPG 2004, pp. 70–76. IEEE (2004)
15. Shu, R., Zhou, C., Kankanhalli, M.S.: Adaptive marching cubes. Visual Comput. **11**(4), 202–217 (1995)
16. Tatarchuk, N., Shopf, J., DeCoro, C.: Real-time isosurface extraction using the GPU programmable geometry pipeline. In: ACM SIGGRApPH 2007 Courses, pp. 122–137. ACM (2007)

Construction of Digital Ellipse by Recursive Integer Intervals

Papia Mahato$^{(\boxtimes)}$ and Partha Bhowmick

Department of Computer Science and Engineering, Indian Institute of Technology,
Kharagpur, India
papiamahatostar@gmail.com, bhowmick@gmail.com

Abstract. In this paper, we revisit the problem of ellipse construction in the integer plane. Our perspective is elementary number-theoretic analysis of a digital ellipse having an integer point as its center and two integer values specifying the lengths of its semi-major and semi-minor axes. We characterize a digital ellipse to derive certain recurrences on the integer intervals that contain the integer values of a specific square term corresponding to the integer points comprising the digital ellipse. This, in turn, helps in designing ellipse drawing algorithm on the integer plane. We propose two algorithms—one using floating-point-based distance computation, and another using purely integer operations. Some test results have also been presented to exhibit further research possibilities related to digital ellipse.

Keywords: Digital ellipse · Digital geometry · Integer intervals · Integer algorithm

1 Introduction

Ellipse is an important geometric primitive with a multitude of applications in different fields of science and engineering. Hence, like various other geometric primitives, its characterization and discretization is one of the necessary and interesting research topics in the subject of digital geometry. Its computational gamut in the discrete space brings in fundamental theoretical issues, which are related to several fields of computer science, and to name a few, these are computer graphics, computer vision, projective geometry, and image analysis.

Several methodologies have been proposed over the last fifty years for generation of digital ellipse, which may be seen in [3, 6, 8–13]. The adopted techniques are predominantly based on incremental raster approximation [8], double-step generation [11, 12], run slicing [13], and hybridization [5]. Most of these algorithms basically owe in concept to the one proposed in the very early stage of digitization [2].

Unlike other 2D geometric primitives like straight lines and circles, which have been continuously and extensively studied by the digital geometers, ellipse is possibly not yet studied up to its merit till date. This motivates us to focus on making out some novel characterization of ellipse in the discrete space—in particular, in $\mathbb{Z}^2$—with the objective of designing efficient algorithms as per the

© Springer International Publishing Switzerland 2016
N. Normand et al. (Eds.): DGCI 2016, LNCS 9647, pp. 295–308, 2016.
DOI: 10.1007/978-3-319-32360-2_23

requirement of different applications. We report in this paper our first set of results, which centers around some of the elementary number-theoretic properties of digital ellipse when it has an integer specification. It culminates to certain interesting recurrences on integer intervals that correspond to the integer points comprising a digital ellipse.

The paper is organized as follows. In Sect. 2, we explain the preliminary concepts and the theoretical framework adopted in our work. In Sect. 3, we derive some elementary number-theoretic properties of digital ellipse. These properties are used in Sect. 4 for designing efficient algorithms on construction of digital ellipse. Here we first propose an algorithm that uses a few floating-point operations, which we improve to a second algorithm that uses strictly integer operations. In Sect. 5, we present some test results and point out further research issues related to digital ellipse.

2   Preliminaries

We start with some definitions and metrics in 2D space, which are used in the sequel. We mostly follow the conventions from [7].

An *integer point* means a point with integer coordinates. A *pixel* is equivalent to a 2-*cell*, perceived as a unit square on the xy-plane, and hence uniquely identified by its center, as it is an integer point. Two pixels are said to be 0-*adjacent* if they share a vertex (0-*cell*) and 1-*adjacent* if they share an edge (1-*cell*). A 0-*connected* (digital) curve is a sequence of pixels such that every two consecutive pixels are 0- or 1-adjacent (i.e., in 8-neighborhood). For a 1-*connected* curve, every two consecutive pixels in the sequence have to be 1-adjacent (i.e., in 4-neighborhood). A curve is *open-ended* if it has two distinct endpoints, s and t. A curve is *closed* if it partitions the integer plane into an interior and an exterior. If removal of a pixel p from an open-ended curve does not break the connectedness between s and t, then p is said to be a *simple pixel*. For a closed curve, on the other hand, removal of a simple pixel does not give rise to connectedness between its interior and exterior. For further details on simple pixel, we refer to [7]. If a curve does not contain any *simple pixel*, then it is said to be *irreducible*. In an open-ended irreducible curve, s and t has one neighbor each and the rest have two neighbors each. For a closed irreducible curve, each constituent pixel has two neighbors each. Hence, for an open-ended irreducible k-connected curve, where $k \in \{0, 1\}$, there is always a unique k-path between any two of its pixels. And for a closed irreducible k-connected curve, there are exactly two k-paths between any two pixels.

Between two points $p(i, j)$ and $p'(i', j')$, we define x-*distance* as $d_x(p, p') = |i - i'|$, y-*distance* as $d_y(p, p') = |j - j'|$, and the isothetic distance (Minkowski norm [7]) as $d_\infty(p, p') = \max\{d_x(p, p'), d_y(p, p')\}$. Based on this, the isothetic distance of a point $p(i, j)$ from a 2D curve C (which is an ellipse in our work) is defined as $d_\perp(p, C) = \min\{d_x(p, C), d_y(p, C)\}$, where,

$$d_x(p, C) = \begin{cases} d_x(p, q) & \text{if there exists a point } q(x, j) \text{ on } C \\ \infty & \text{otherwise,} \end{cases}$$

$$d_y(p, C) = \begin{cases} d_y(p, q) & \text{if there exists a point } q(i, y) \text{ on } C \\ \infty & \text{otherwise.} \end{cases}$$

While measuring $d_x(p, q)$ and $d_y(p, q)$, if there lies more than one point q on C, then we consider the nearest. Figure 1 shows the basic idea.

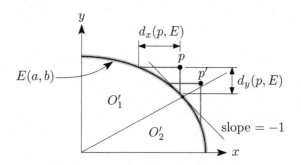

Fig. 1. Isothetic distance and its anomaly with pseudo-octants O_1' and O_2'. Both p and p' are in O_1', but their respective isothetic distances from $E(a, b)$ are $d_\perp(p, E) = d_y(p, E)$ and $d_\perp(p', E) = d_x(p', E)$, thus forming the anomaly. This anomaly gives rise to disconnectedness in the digital ellipse if p' is an integer point and $d_\perp(p', E) \leqslant \frac{1}{2}$.

We use $E(a, b)$ to denote a real ellipse, where a and b are the respective lengths of its semi-major and semi-minor axes. In our work, we consider a and b as positive integers. Further, an ellipse is considered in its *canonical form*, which means its center is $(0, 0)$ and its axes are parallel to the coordinate axes [13]. So, its equation is

$$\frac{x^2}{a^2} + \frac{y^2}{b^2} = 1 \qquad (1)$$

and the corresponding digital ellipse is defined as follows.

Definition 1. *A digital ellipse* $\mathsf{E}(a, b)$ *is the 0-connected irreducible sequence of integer points obtained by discretization of the real ellipse* $E(a, b)$ *such that each point in* $\mathsf{E}(a, b)$ *has an isothetic distance of at most* $\frac{1}{2}$ *from* $E(a, b)$.

2.1 Elliptic Octants

An ellipse, whether the real or the digital, possesses four-way symmetry about its major and minor axes. As a result, we get four symmetric arcs of an ellipse, which lie in four different quadrants. For digitization purpose, the real ellipse is considered, and its arc in the 1st quadrant—and so for each other quadrant—is further divided into two arcs by a radial line that passes through the point on the ellipse at which its slope is −1. The four quadrants are thus subdivided into eight *pseudo-octants*, in which the respective real arcs are digitized to generate the full digital ellipse. This can be seen in the existing literature; see, for example, [4,13], and the bibliographies therein.

The above idea works fine as far as a and b are comparable, but fails in case of a sufficiently high eccentricity—for example, when a is sufficiently large in value compared to b. A discussion on this can be seen in [13]. This happens due to the anomaly that for $a \gg b$, an integer point p may satisfy $d_y(p, E(a,b)) > d_x(p, E(a,b))$ although lying in the 1st pseudo-octant. An illustration is given in Fig. 1, and some examples of failure are shown in Fig. 2.

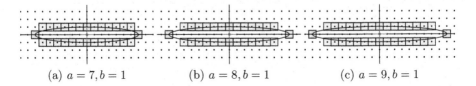

(a) $a = 7, b = 1$ (b) $a = 8, b = 1$ (c) $a = 9, b = 1$

Fig. 2. Disconnectedness arising due to anomaly of pseudo-octants for $a \geqslant 8, b = 1$.

To circumvent the aforesaid problem, we introduce the concept of *elliptic octant* w.r.t. $E(a,b)$, which we henceforth refer simply as *e-octant* for brevity. We use $\mathbb{O}_t$ to denote the tth ($1 \leqslant t \leqslant 8$) e-octant. We have the following proposition based on the notion of isothetic distance.

Proposition 1. *A point p lies in $\mathbb{O}_t$, where*

$$t(\bmod 4) \in \begin{cases} \{0,1\} \; if \; and \; only \; if \; d_y(p, E(a,b)) \leqslant d_x(p, E(a,b)) \\ \{2,3\} \; if \; and \; only \; if \; d_x(p, E(a,b)) \leqslant d_y(p, E(a,b)). \end{cases}$$

By Proposition 1, $\mathbb{O}_1$, $\mathbb{O}_4$, $\mathbb{O}_5$, and $\mathbb{O}_8$ become mutually symmetric with their isothetic distance as y-distance; and so also for $\mathbb{O}_2$, $\mathbb{O}_3$, $\mathbb{O}_6$, and $\mathbb{O}_7$ with their isothetic distance as x-distance. Owing to the above symmetry, it suffices to characterize the arc of a digital ellipse that lies in the first quadrant (i.e., $\mathbb{O}_1$ and $\mathbb{O}_2$). In $\mathbb{O}_1$, the y-distance serves as the isothetic distance; whereas in $\mathbb{O}_2$, it is the x-distance.

The above characterization by isothetic distance helps in generating a digital ellipse as a sequence of runs comprising its arcs in the eight e-octants. A *run* is defined as a maximum sequence of successive integer points with equal y- or equal x-coordinate. So, a digital ellipse in $\mathbb{O}_1$ consists of horizontal runs, and in $\mathbb{O}_2$ consists of vertical runs. This property is used during the construction of a digital ellipse. In $\mathbb{O}_1$, we start from the point $(0, b)$ and go on generating the subsequent points along the horizontal runs, as long as the isothetic distance of an integer point from $E(a,b)$ is its y-distance. In $\mathbb{O}_2$, we do the reverse—starting from the point $(a, 0)$, we generate the vertical runs, as long as the isothetic distance is x-distance. The other e-octants are easily constructed from the arcs obtained in $\mathbb{O}_1$ and $\mathbb{O}_2$ by using the 4-symmetry of $E(a,b)$.

3 Properties of Digital Ellipse

We denote by $E_t(a,b)$ the arc of the real ellipse $E(a,b)$ that lies in $\mathbb{O}_t$, and by $\mathsf{E}_t(a,b)$ that of the digital ellipse $\mathsf{E}(a,b)$ lying in $\mathbb{O}_t$. We have the following lemma on isothetic distance in an e-octant.

Lemma 1. *The isothetic distance of a point $p(i,j)$ from $E_t(a,b)$ is $d_\perp(p, E_t(a,b)) = d_y(p, E_t(a,b))$ if and only if $t(\mathrm{mod}\ 4) \in \{0,1\}$, or equivalently, $\left| ab|j| - b^2\sqrt{a^2 - i^2} \right| \leqslant \left| ab|i| - a^2\sqrt{b^2 - j^2} \right|$.*

Proof. By definition of isothetic distance and by Eq. 1, $d_y(p, E_t(a,b)) = \left| |j| - (b/a)\sqrt{a^2 - i^2} \right|$ and $d_x(p, E_t(a,b)) = \left| |i| - (a/b)\sqrt{b^2 - j^2} \right|$. So, by Proposition 1, $d_y(p, E_t(a,b)) \leqslant d_x(p, E_t(a,b))$ if and only if

$$t(\mathrm{mod}\ 4) \in \{0,1\}$$
$$\iff \left| |j| - (b/a)\sqrt{a^2 - i^2} \right| \leqslant \left| |i| - (a/b)\sqrt{b^2 - j^2} \right|$$
$$\iff \left| ab|j| - b^2\sqrt{a^2 - i^2} \right| \leqslant \left| ab|i| - a^2\sqrt{b^2 - j^2} \right|. \tag{2}$$

This completes the proof. □

We consider the 1st and the 2nd e-octants to obtain a characterization of digital ellipse. Owing to 4-symmetry, this will be applicable to other e-octants. Lemma 1 provides the way to decide whether a (real or integer) point p lies in $\mathbb{O}_1$ or in $\mathbb{O}_2$. But it does not tell whether p belongs to $\mathsf{E}_1(a,b)$ or to $\mathsf{E}_2(a,b)$. The following theorem states the necessary and sufficient condition to decide whether an integer point belongs to $\mathsf{E}_1(a,b)$.

Theorem 1. *An integer point $p(i,j)$ with $j > 0$ belongs to $\mathsf{E}_1(a,b)$ if and only if*

$$4b^2a^2 - a^2(2j+1)^2 \leqslant 4b^2i^2 \leqslant 4b^2a^2 - a^2(2j-1)^2 - 1. \tag{3}$$

Proof. A point $p(i,j)$ belongs to $\mathsf{E}_1(a,b)$ if and only if $d_y(p, E_1(a,b)) \leqslant \frac{1}{2}$ (by Proposition 1 and Definition 1), or equivalently, there exists a real point $q(i, j-\delta)$ on $E(a,b)$ such that $-\frac{1}{2} \leqslant \delta < \frac{1}{2}$. Note that $j > 0$ ensures that $q \in \mathbb{O}_1$. Further, to ensure that $\mathsf{E}_1(a,b)$ is irreducible (Definition 1), p is not considered to be in $\mathsf{E}_1(a,b)$ when $\delta = \frac{1}{2}$. Now, as mentioned in the proof of Lemma 1, $d_y(p, E_t(a,b)) = \left| |j| - (b/a)\sqrt{a^2 - i^2} \right|$, or, $d_y(p, E_1(a,b)) = (j - (b/a)\sqrt{a^2 - i^2})$. So, the last condition is true if and only if

$$-\frac{1}{2} \leqslant j - \frac{b}{a}\sqrt{a^2 - i^2} < \frac{1}{2}$$
$$\iff j - \frac{1}{2} < \frac{b}{a}\sqrt{a^2 - i^2} \leqslant j + \frac{1}{2}$$
$$\iff a(2j-1) < 2b\sqrt{a^2 - i^2} \leqslant a(2j+1)$$
$$\iff -4b^2a^2 + a^2(2j-1)^2 < -4b^2i^2 \leqslant -4b^2a^2 + a^2(2j+1)^2$$
$$\iff 4b^2a^2 - a^2(2j+1)^2 \leqslant 4b^2i^2 < 4b^2a^2 - a^2(2j-1)^2,$$

hence we get Eq. 3, as a, b, i, j are all integers. □

Theorem 1 provides all integer points of $E_1(a, b)$ with positive ordinate, i.e., $j > 0$ for a point $p(i, j) \in E_1(a, b)$. Now, Eq. 3 does not give a valid integer interval if $-a^2(2j + 1)^2 > -a^2(2j - 1)^2 - 1$, or, $-4a^2j > 4a^2j - 1$, or, $8a^2j < 1$, which implies $j = 0$, since a is a positive integer and j is a non-negative integer in $\mathbb{O}_1$. Consequently, we have the following corollary.

Corollary 1. *An integer point $p(i, j)$ belongs to $E_1(a, b)$ but does not satisfy Eq. 3 only if $j = 0$.*

The equation for points constituting $E_2(a, b)$ is given in the following corollary, which is also obtained in a similar way.

Corollary 2. *An integer point (i, j) with $i > 0$ belongs to $E_2(a, b)$ if and only if*

$$4b^2a^2 - b^2(2i + 1)^2 \leqslant 4a^2j^2 \leqslant 4b^2a^2 - b^2(2i - 1)^2 - 1. \tag{4}$$

Moreover, if $p(i, j)$ belongs to $E_2(a, b)$ but does not satisfy Eq. 4, then $i = 0$.

Now, to efficiently compute the integer points comprising the digital ellipse in the first e-octant, we need the following theorem.

Theorem 2. *An integer point (i, j) with $j > 0$ belongs to $E_1(a, b)$ if and only if $4b^2i^2$ lies in the interval $I_k = [u_k, v_k := u_k + l_k - 1]$, where $j = b - k$, $k \geqslant 0$, and u_k and l_k are given as follows.*

$$
\begin{aligned}
u_k &= \begin{cases} -4a^2b - a^2 & if \ k = 0 \\ u_{k-1} + l_{k-1} & otherwise \end{cases} \\
l_k &= \begin{cases} 8a^2b & if \ k = 0 \\ l_{k-1} - 8a^2 & otherwise \end{cases}
\end{aligned}
\tag{5}
$$

Proof. We get u_0 and l_0 corresponding to $k = 0$ by substituting $j = b$ in Eq. 3.

To get the recurrence of l_k for $k > 0$, observe that $l_k = 4b^2a^2 - a^2(2j - 1)^2 - 4b^2a^2 + a^2(2j + 1)^2 = 8a^2j = 8a^2(b - k)$, in accordance with Eq. 3. Hence, $l_k - l_{k-1} = 8a^2(b - k) - 8a^2(b - k + 1) = 8a^2$. To get the recurrence of u_k, we substitute $j = b - k$ in Eq. 3 to get $u_k = 4b^2a^2 - a^2(2(b - k) + 1)^2$, and substitute $j = b - k - 1$ to get $v_{k-1} = 4b^2a^2 - a^2(2(b - k - 1) - 1)^2 - 1 = u_k - 1$. Thus, $u_k - v_{k-1} = 1$, or, $u_k = v_{k-1} + 1 = u_{k-1} + l_{k-1}$. $\quad\square$

Since I_0 contains only perfect squares of the form $4b^2i^2$, can we reset $u_0 = 0$. However, to make the recurrence of u_k work for $k \geqslant 1$, we adhere to a negative value of u_0, and this does not affect the performance of the algorithms discussed in Sect. 4.

The recurrence relations for $E_2(a, b)$, obtained in a similar fashion, are put in the following corollary.

Corollary 3. *An integer point (i, j) with $i > 0$ belongs to $\mathsf{E}_2(a, b)$ if and only if $4a^2 j^2$ lies in the interval $I_k = [u_k, v_k := u_k + l_k - 1]$, where $i = a - k$, $k \geqslant 0$, and u_k and l_k are given as follows.*

$$
\begin{aligned}
u_k &= \begin{cases} -4b^2 a - b^2 & if \ k = 0 \\ u_{k-1} + l_{k-1} & otherwise \end{cases} \\
l_k &= \begin{cases} 8b^2 a & if \ k = 0 \\ l_{k-1} - 8b^2 & otherwise \end{cases}
\end{aligned}
\tag{6}
$$

We conclude this section with an important observation. Notice in the proof of Theorem 2 that $u_k = v_{k-1} + 1$, which holds for both $\mathbb{O}_1$ and $\mathbb{O}_2$. We put this in the following corollary.

Corollary 4. *For $k = 0, 1, 2, \ldots$, the intervals I_k are disjoint and contiguous.*

4 Algorithms for Ellipse Construction

The basic idea of digital ellipse construction by the proposed technique is demonstrated in Fig. 3. Let $p(i, j)$ be a point of $\mathsf{E}_1(a, b)$. Then, considering the clockwise traversal of $\mathsf{E}_1(a, b)$, the next point would be either $(i + 1, j)$ or $(i + 1, j - 1)$, the former being when it is the same run and the latter being for a change of run. The recurrences in Theorem 2 are used for this, as demonstrated in Fig. 3 for $\mathsf{E}_1(a, b)$ with $a = 6$ and $b = 4$. In the first e-octant, the first run starts at the point $(0, 4)$. We get $u_k = -612$ and $v_k = 539$ for the first run, and the points that satisfy the interval $[u_k, v_k]$ are $(0, 4)$, $(1, 4)$, and $(2, 4)$. After this, the value of j becomes 3 for the second run, and the values of u_k and v_k for $k = 1$ are computed recursively using Eq. 5. The process goes on till we are in $\mathbb{O}_1$. For $\mathbb{O}_2$, a similar process runs starting from $(6, 0)$, as shown in Fig. 3.

4.1 Algorithm Draw-Ellipse-Float

We have designed two algorithms for generation of a digital ellipse, which are based on the above-mentioned technique. We first discuss here Algorithm 1 that uses both

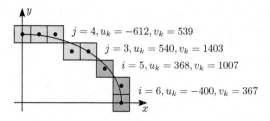

Fig. 3. An example showing generation of $\mathsf{E}_1(6, 4)$ (yellow cells) and $\mathsf{E}_2(6, 4)$ (green cells) by the proposed technique (Color figure online).

integer and floating-point operations, the latter type of operations being needed for determining the octant between $\mathbb{O}_1$ and $\mathbb{O}_2$ based on Lemma 1. Necessary variables are defined and initialized accordingly in Steps 1–2. The first **while** loop (Lines 4–5) and the second **while** loop (Lines 6–8) run in $\mathbb{O}_1$ and generate all integer points of $\mathsf{E}_1(a,b)$. In each iteration of the first **while** loop, the procedure `drawRunFloat` generates a horizontal run (i.e., a maximum-length sequence of successive points with same y-coordinate) of $\mathsf{E}_1(a,b)$. This procedure is again used later to generate the vertical runs of $\mathsf{E}_2(a,b)$ in the **while** loop for $\mathbb{O}_2$ (Lines 11–12). The value of t is set to 1 and to 2 while invoking it in $\mathbb{O}_1$ (Line 5) and in $\mathbb{O}_2$ (Line 12), respectively. The ellipse parameters (a,b) and the point coordinates (i,j) are passed in appropriate order to generate the respective horizontal and vertical

---

Algorithm 1. DRAW-ELLIPSE-FLOAT

Input: int a, b
Output: Digital ellipse with semi-major axis a and semi-minor axis b
1 int $i \leftarrow 0, j \leftarrow b, u \leftarrow -4a^2b - a^2, l \leftarrow 8a^2b, v \leftarrow 4a^2b - a^2 - 1, s \leftarrow 0$
2 float $d_y \leftarrow 0, d_x \leftarrow 0$
3 $\mathsf{E} \leftarrow \{\}$ ▷ output set of integer points
4 **while** $(d_y \leqslant d_x) \wedge (u \leqslant v)$ **do**
5 $\quad$ drawRunFloat$(i, j, a, b, 1, s, d_y, d_x, u, l, v, \mathsf{E})$ ▷ Lemma 1 & Theorem 1
6 **while** $(u > v) \wedge (i < a)$ **do**
7 $\quad$ include4SymPoints(i, j, E) ▷ Corollary 1
8 $\quad$ $i \leftarrow i + 1, s \leftarrow 4b^2i^2$
9 $i \leftarrow a, j \leftarrow 0, u \leftarrow -4b^2a - b^2, l \leftarrow 8b^2a, v \leftarrow 4b^2a - b^2 - 1, s \leftarrow 0$
10 $d_y \leftarrow 0, d_x \leftarrow 0$
11 **while** $(d_x \leqslant d_y) \wedge (u \leqslant v)$ **do**
12 $\quad$ drawRunFloat$(j, i, b, a, 2, s, d_y, d_x, u, l, v, \mathsf{E})$ ▷ Lemma 1 & Corollary 2
13 **while** $(u > v) \wedge (j < b)$ **do**
14 $\quad$ include4SymPoints(i, j, E) ▷ Corollary 2
15 $\quad$ $j \leftarrow j + 1, s \leftarrow 4a^2j^2$
16 **return** E

Procedure drawRunFloat $(i, j, a, b, t, s, d_y, d_x, u, l, v, \mathsf{E})$

1 **repeat**
2 $\quad$ **if** $t = 1$ **then** include4SymPoints(i, j, E)
3 $\quad$ **else** include4SymPoints(j, i, E) ▷ Proposition 1
4 $\quad$ $i \leftarrow i + 1, s \leftarrow 4b^2i^2$
5 **until** $s > v$;
6 $j \leftarrow j - 1$
7 **if** $t = 1$ **then** $d_y \leftarrow |abj - b^2\sqrt{a^2 - i^2}|, d_x \leftarrow |abi - a^2\sqrt{b^2 - j^2}|$
8 **else** $d_y \leftarrow |bai - a^2\sqrt{b^2 - j^2}|, d_x \leftarrow |baj - b^2\sqrt{a^2 - i^2}|$ ▷ Lemma 1
9 $u \leftarrow v + 1, l \leftarrow l - 8a^2, v \leftarrow u + l - 1$ ▷ Theorem 2, Corollary 3 & 4

Algorithm 2. DRAW-ELLIPSE-INT

Input: int a, b
Output: Digital ellipse with semi-major axis a and semi-minor axis b

1 int $i \leftarrow 0, j \leftarrow b, u \leftarrow -4a^2b - a^2, l \leftarrow 8a^2b, v \leftarrow 4a^2b - a^2 - 1, s \leftarrow 0, i_1, j_1$
2 $\mathsf{E} \leftarrow \{\}$
3 **while** $(u \leqslant s) \wedge (s \leqslant v)$ **do**
4 $\quad$ drawRunInt$(i, j, a, b, 1, s, u, l, v, i_1, j_1, \mathsf{E})$ $\triangleright$ Theorem 1

5 $i_1 \leftarrow i - 1, j_1 \leftarrow j + 1$
6 **while** $(u > v) \wedge (i < a)$ **do**
7 $\quad$ include4SymPoints(i, j, E) $\triangleright$ Corollary 1
8 $\quad$ $i_1 \leftarrow i, j_1 \leftarrow j, i \leftarrow i + 1, s \leftarrow 4b^2i^2$

9 $i \leftarrow a, j \leftarrow 0, u \leftarrow -4b^2a - b^2, l \leftarrow 8b^2a, v \leftarrow 4b^2a - b^2 - 1, s \leftarrow 0$
10 **while** $(u \leqslant s) \wedge (s \leqslant v) \wedge ((i \neq i_1) \vee (j \neq j_1))$ **do**
11 $\quad$ drawRunInt$(j, i, b, a, 2, s, u, l, v, i_1, j_1, \mathsf{E})$ $\triangleright$ Corollary 2

12 **while** $(u > v) \wedge (j < b)$ **do**
13 $\quad$ include4SymPoints(i, j, E) $\triangleright$ Corollary 2
14 $\quad$ $j \leftarrow j + 1, s \leftarrow 4a^2j^2$

15 **return** E

Procedure drawRunInt $(i, j, a, b, t, s, u, l, v, i_1, j_1, \mathsf{E})$

---

1 **repeat**
2 $\quad$ **if** $t = 1$ **then** include4SymPoints(i, j, E)
3 $\quad$ **else if** $(i = i_1) \wedge (j = j_1)$ **then return**
4 $\quad\quad$ **else** include4SymPoints(j, i, E) $\triangleright$ Proposition 1
5 $\quad$ $i \leftarrow i + 1, s \leftarrow 4b^2i^2$
6 **until** $s > v$;
7 $j \leftarrow j - 1$
8 $u \leftarrow v + 1, l \leftarrow l - 8a^2, v \leftarrow u + l - 1$ $\triangleright$ Theorem 2, Corollary 3 & 4

---

runs for $\mathsf{E}_1(a, b)$ and $\mathsf{E}_2(a, b)$. For $\mathbb{O}_2$, the parameters are properly initialized in Lines 9–10.

Inside the procedure **drawRunFloat**, there is a **repeat-until** loop (Lines 1–5) that computes the points of the $k (\geqslant 0)$th horizontal (if $t = 1$) or vertical (if $t = 2$) run using the integer interval $[u, v]$. For every s generated in succession and lying in $[u, v]$, the procedure **include4SymPoints** includes the corresponding point $p(i, j)$ and its three symmetric points $\{(i, -j), (-i, -j), (-i, j)\}$ in E. The parameters (i, j) are passed in **include4SymPoints** depending on the value of t. After generation of all the points of the kth run, all the related parameters are updated in Lines 6–8 of **drawRunFloat** to generate the points of the next run.

As discussed in Sect. 3 (Corollary 1), integer points of the form $(i, j = 0)$ and belonging to $\mathsf{E}_1(a, b)$ cannot be tracked by the integer intervals used in the first **while** loop (Lines 4–5). This happens particularly when a is significantly large compared to b. In order to track such points, the second **while** loop (Lines 6–8) is used with necessary conditional checks. A similar **while** loop (Lines 13–15) is also used in $\mathbb{O}_2$ to ensure the completeness of $\mathsf{E}_2(a,b)$.

4.2   Algorithm Draw-Ellipse-Int

Algorithm 1 is not free from floating-point operations and hence susceptible to computational pitfalls. So, as an improvement, we design Algorithm 2 where we use only integer operations. This algorithm, contrary to the previous, does not resort to computation of isothetic distance. Instead, it uses two extra integer variables, i_1 and j_1, meant to store the coordinates of the last point in $\mathsf{E}_1(a,b)$.

It is worth mentioning here now a few details concerning x- and y-distances of a (real or integer) point $p(i,j)$ from $E(a,b)$. Let, w.l.o.g., p lie in $\mathbb{O}_1$. Then, by Proposition 1, $d_y(p, E_1(a,b)) \leqslant d_x(p, E_1(a,b))$. Now, $p \in \mathsf{E}_1(a,b)$ if and only if $d_y(p, E_1(a,b)) \leqslant \frac{1}{2}$, which, however, does not give any idea about the value of $d_x(p, E_1(a,b))$. Interestingly, it may also happen that $d_x(p, E_1(a,b)) \leqslant \frac{1}{2}$ for the same integer point p. In such situation, the point p would be included only in $\mathsf{E}_1(a,b)$ by Algorithm 1. However, if neither of i and j is zero, then p would satisfy the integer intervals corresponding to both $\mathsf{E}_1(a,b)$ and $\mathsf{E}_2(a,b)$ (Eqs. 3 and 4).

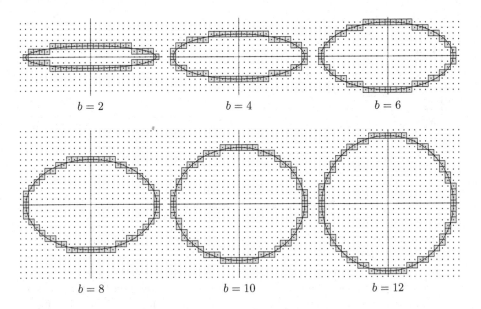

Fig. 4. Digital ellipses with $a = 12$ and increasing values of b.

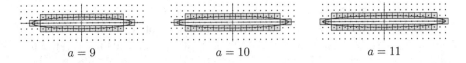

Fig. 5. Digital ellipses with $b = 1$ and increasing values of a.

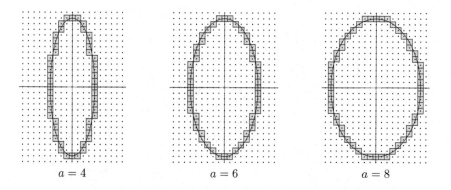

Fig. 6. Digital ellipses with $b = 12$ and increasing values of a.

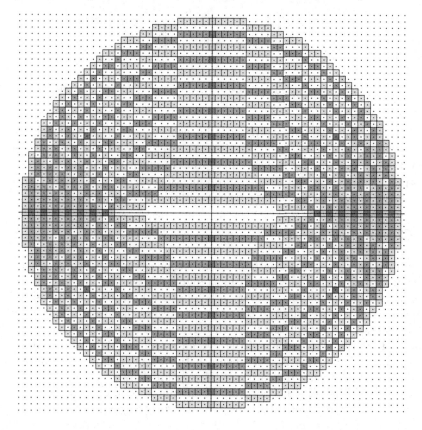

Fig. 7. A set of digital ellipses: $\{\mathsf{E}(a,b) : (a,b) = (16,2), (17,4), (18,6), \ldots, (30,30)\}$.

As a result, this point p would be generated twice—first in $\mathbb{O}_1$ and then in $\mathbb{O}_2$. In fact, there may occur many such points in succession—around the junction of $\mathbb{O}_1$ and $\mathbb{O}_2$—which would result to overlap between the two digital arcs $\mathsf{E}_1(a,b)$

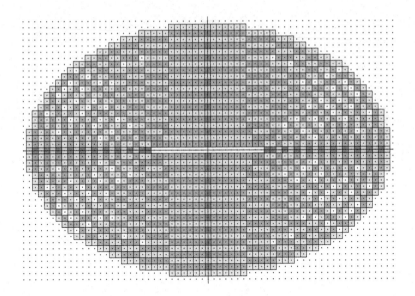

Fig. 8. Another set: $\{\mathsf{E}(a,b) : (a,b) = (11,1), (12,2), (13,3), \ldots, (30,20)\}$.

and $\mathsf{E}_2(a,b)$. This overlap is prevented by using i_1 and j_1 in Algorithm 2, as mentioned above. The necessary conditional check is put in Line 10. Updating the values of i_1 and j_1 are done appropriately when the algorithm executes for $\mathbb{O}_1$. Notice that, except for the above-mentioned points, the **while** loops of Algorithm 2 work in the same way as those of Algorithm 1.

5   Concluding Notes

Algorithms 1 and 2 both run with an optimal time complexity, which is linear in the number of integer points comprising a digital ellipse, i.e., $\Theta(a+b)$. This owes to the fact that the number of every type of operations (comparison, addition, multiplication, etc.) required to report each integer point of a digital ellipse is upper-bounded by a small constant. To derive tight asymptotic bounds on the number of operations in either of the two algorithms, a thorough analysis has to be done. Nature of distribution of runs with change in major- and minor-axis lengths can also be studied as a prospective future work.

We present here some instances of digital ellipse produced by the proposed technique. These are identical irrespective of the algorithm used.

Figure 4 displays some digital ellipses with a fixed semi-major axis $(a = 12)$ and increasing values of b. The corresponding real ellipses are also shown; it helps us to understand the pattern of points forming their digitization. Figures 5 and 6 display two different sets of results with a fixed semi-minor axis—one for small $b(=1)$ and another with usual $b(=12)$—and with increasing values of a. All these results show the 4-symmetry of digital ellipse along with the properties of their composition as 0-connected irreducible sequences.

Figure 7 contains a set of 15 digital ellipses whose semi-major axis increases from 16 to 30 units at unit step, and semi-minor axis from 2 to 30 at double step, finally reaching a digital circle of radius 30. Notice for this set that all the fifteen digital ellipses are disjoint, as there is no common pixel between any two. Some 'gaps' are left between two consecutive digital ellipses, which indicates an interesting similitude with concentric integer circles, since the latter class also exhibits the aforesaid disjointness property, as recently shown in [1]. Characterization of such an ellipse class would be an interesting research issue in the context of covering the integer plane by coaxial ellipses.

In Fig. 8, we have shown another set containing 20 digital ellipses, whose semi-major axis increases from 11 to 30, and semi-minor axis from 1 to 20— both at unit steps. The gaps formed here are less in number, since there is overlap between two consecutive ellipses. This set, together with the previous set, indicates that overlapping ellipses can reduce the gaps, although not completely. What follows as a natural question is to find a minimal or a minimum set of coaxial digital ellipses with major- and minor-axis lengths specified by two rational numbers, so as to cover the interior of an ellipse of given integer specification. Characterization of this set and designing efficient algorithms for digital ellipse construction with non-integer (rational, for definiteness) specification are two important follow-up research problems in the context of our work.

References

1. Bera, S., Bhowmick, P., Stelldinger, P., Bhattacharya, B.: On covering a digital disc with concentric circles in $\mathbb{Z}^2$. Theoret. Comput. Sci. **506**, 1–16 (2013)
2. Bresenham, J.E.: Algorithm for computer control of a digital plotter. IBM Syst. J. **4**(1), 25–30 (1965)
3. Fellner, D.W., Helmberg, C.: Robust rendering of general ellipses and elliptical arcs. ACM Trans. Graph. **12**(3), 251–276 (1993)
4. Foley, J.D., Dam, A.V., Feiner, S.K., Hughes, J.F.: Computer Graphics: Principles and Practice. Addison-Wesley, Reading (1993)
5. Haiwen, F., Lianqiang, N.: A hybrid generating algorithm for fast ellipses drawing. In: International Conference on Computer Science and Information Processing (CSIP), pp. 1022–1025 (2012)
6. Kappel, M.: An ellipse-drawing algorithm for raster displays. In: Earnshaw, R.A. (ed.) Fundamental Algorithms for Computer Graphics. NATO ASI Series, vol. 17, pp. 257–280. Springer, Heidelberg (1991)
7. Klette, R., Rosenfeld, A.: Digital Geometry: Geometric Methods for Digital Picture Analysis. Morgan Kaufmann, San Francisco (2004)
8. McIlroy, M.D.: Getting raster ellipses right. ACM Trans. Graph. **11**(3), 259–275 (1992)
9. Pitteway, M.L.V.: Algorithm for drawing ellipses or hyperbolae with a digital plotter. Comput. J. **10**, 282–289 (1967)
10. Van Aken, J., Novak, M.: Curve-drawing algorithms for raster displays. ACM Trans. Graph. **4**(2), 147–169 (1985)
11. Van Aken, J.: An efficient ellipse-drawing algorithm. IEEE Comput. Graphics Appl. **4**(9), 24–35 (1984)

12. Wu, X., Rokne, J.: Double-step generation of ellipses. IEEE Comput. Graphics Appl. **9**(3), 56–69 (1989)
13. Yao, C., Rokne, J.G.: Run-length slice algorithms for the scan-conversion of ellipses. Comput. Graph. **22**(4), 463–477 (1998)

Morphological Analysis

Parallelization Strategy for Elementary Morphological Operators on Graphs

Imane Youkana[1,2]([⊠]), Jean Cousty[1], Rachida Saouli[1,2], and Mohamed Akil[1]

[1] Laboratoire d'Informatique Gaspard-Monge, ESIEE Paris,
Université Paris-Est, Paris, France
{imane.youkana,jean.cousty,rachida.saouli,mohamed.akil}@esiee.fr
[2] Département d'Informatique, Université de Biskra, Biskra, Algeria

Abstract. This article focuses on the graph-based mathematical morphology operators presented in *[J. Cousty et al., "Morphological filtering on graphs", CVIU 2013]*. These operators depend on a size parameter that specifies the number of iterations of elementary dilations/erosions. Thus, the associated running times increase with the size parameter. In this article, we present distance maps that allow us to recover (by thresholding) all considered dilations and erosions. The algorithms based on distance maps allow the operators to be computed with a single linear-time iteration, without any dependence to the size parameter. Then, we investigate a parallelization strategy to compute these distance maps. The idea is to build iteratively the successive level-sets of the distance maps, each level set being traversed in parallel. Under some reasonable assumptions about the graph and sets to be dilated, our parallel algorithm runs in $O(n/p + K \log_2 p)$ where n, p, and K are the size of the graph, the number of available processors, and the number of distinct level-sets of the distance map, respectively.

1 Introduction

Mathematical morphology provides a set of filtering and segmenting tools that are very useful in applications to image analysis. There is a growing interest for considering digital objects not only composed of points but also composed of elements lying between them and carrying structural information about how the points are glued together. The simplest of these representations are the graphs. The domain of an image is considered as a graph (which can be planar or not) whose vertex set is made of the pixels and whose edge set is given by an adjacency relation on these pixels. Note that this adjacency relation can be either spatially invariant or spatially variant leading to operators that are either spatially invariant or spatially variant. Graphs are also useful to process other kinds of discrete structures defined for instance on 3-dimensional meshes. In this context, it becomes relevant to consider morphological transformations acting on the subsets of vertices, the subsets of edges and the subgraphs of a graphs and not only those acting on the set of all subsets of pixels.

Mathematical morphology on graphs was pioneered by Vincent [12] who proposes operators relying on a dilatation (and its adjunct erosion) that act

© Springer International Publishing Switzerland 2016
N. Normand et al. (Eds.): DGCI 2016, LNCS 9647, pp. 311–322, 2016.
DOI: 10.1007/978-3-319-32360-2_24

on the vertices of a graph. More recently, [3,6] introduce basic dilatations and erosions that map a set of vertices to a set of edges and a set of edges to a set of vertices. It was shown in [3] that these operators can be combined in order to obtain operators acting on the subsets of edges, on the subsets of vertices and on the subgraphs of a given graph. In particular, interesting openings and closings (and then the associated alternate sequential filters) are obtained by iteration of the basic operators. The number of iterations constitutes a filtering parameter related to the size of the features to be preserved or removed. Therefore, based on the straightforward definition, the time-complexity of the associated algorithm increases with the size parameter. More precisely, for a parameter value of λ the algorithm runs in $O(\lambda.n)$ time, where n is the size of the underlying graph. In this article, our main contributions are twofold: we first propose to use distance maps in order to avoid the dependence to the parameter λ when computing the results of the operators of [3]; and then we propose a parallelization strategy leading to fast computation, in particular, for multicore/multithread architectures.

After presenting background notions about morphology and graphs in Sect. 2, we investigate in Sect. 3 some distance maps that lead to characterizations of the dilations and erosions presented in [3]. Since we are interested in operators that map sets of edges to sets of vertices and sets of vertices to sets of edges, we introduce edge-vertex and vertex-edge distance maps. Given a set of edges (resp. vertices), the edge-vertex (resp. vertex-edge) distance map provides for each vertex (resp. edge) a geodesic distance to the closest edge (resp. vertex) of the input set. In order to computed these distance maps, we adapt classical linear-time algorithm for distance maps in unweighted graphs. These algorithms derive from breadth first search. Whatever the size parameter, any dilation, erosion, opening and closing of [3] can be obtained by thresholding these distance maps. Therefore, the time complexity of the associated algorithms is linear with respect to the size of the graph, without any dependence to the size parameter.

In Sect. 4, we propose a parallel algorithm to compute the proposed distance maps, hence the morphological operators of [3]. Parallel and/or separable algorithms for morphological operators and distance maps on images have been widely studied [1,2,5,7–11]. Based on the regular structure of the space, such computations use a static partitioning of the image into rows, columns or blocks processed in parallel. In order to handle the non-regular structure of a graph, our parallelization strategy is based on dynamic partitioning which depends on the input set and which is iteratively computed during the execution. The time complexity of our parallel algorithm is analyzed. In particular it depends of the complexity of two auxiliary functions to manage the dynamic partitions. These functions are presented in Sect. 5. Under some reasonable assumptions about the graph and set under consideration, our algorithm runs in $O(n/p + K \log_2 p)$ time, where n, p, and K are the size of the underlying graph, the number of available processors and the number of distinct level sets of the distance map, respectively. In the considered practical cases, this complexity is dominated by the $O(n/p)$ term.

2   Background Notions for Morphology on Graphs

A *(undirected) graph* is a pair $X = (X^\bullet, X^\times)$ where $X^\bullet$ is a set and $X^\times$ is composed of unordered pairs of distinct elements in $X^\bullet$, *i.e.*, $X^\times$ is a subset of $\{\{x, y\} \subseteq X^\bullet \mid x \neq y\}$. Each element of $X^\bullet$ is called a *vertex or a point (of X)*, and each element of $X^\times$ is called an *edge (of X)*.

Important Remark. Hereafter, the workspace is a graph $\mathbb{G} = (\mathbb{G}^\bullet, \mathbb{G}^\times)$ and we consider the sets $\mathcal{G}^\bullet$, $\mathcal{G}^\times$ and $\mathcal{G}$ of respectively all subsets of $\mathbb{G}^\bullet$, all subsets of $\mathbb{G}^\times$ and all subgraphs of $\mathbb{G}$.

Mathematical morphology on graphs, as introduced in [3], relies on four basic operators. The operators $\delta^\bullet$ and $\epsilon^\bullet$ are defined from $\mathcal{G}^\times$ to $\mathcal{G}^\bullet$ by:

$$\delta^\bullet(X^\times) = \{x \in \mathbb{G}^\bullet \mid \exists \{x, y\} \in X^\times\}, \text{ for any } X^\times \subseteq \mathbb{G}^\times; \text{ and} \tag{1}$$

$$\epsilon^\bullet(X^\times) = \{x \in \mathbb{G}^\bullet \mid \forall \{x, y\} \in \mathbb{G}^\times, \{x, y\} \in X^\times\}, \text{ for any } X^\times \subseteq \mathbb{G}^\times. \tag{2}$$

The operators $\epsilon^\times$, and $\delta^\times$ are defined from $\mathcal{G}^\bullet$ to $\mathcal{G}^\times$ by:

$$\epsilon^\times(X^\bullet) = \{\{x, y\} \in \mathbb{G}^\times \mid x \in X^\bullet \text{ and } y \in X^\bullet\}, \quad \text{for any } X^\bullet \subseteq \mathbb{G}^\bullet; \text{ and} \tag{3}$$

$$\delta^\times(X^\bullet) = \{\{x, y\} \in \mathbb{G}^\times \mid x \in X^\bullet \text{ or } y \in X^\bullet\}, \quad \text{for any } X^\bullet \subseteq \mathbb{G}^\bullet. \tag{4}$$

In order to obtain efficient filters (opening, closings, and associated alternate sequential filters), which are parametrized by a integer value related to a notion of size of the features to be preserved or removed, one needs to consider iterated versions of the basic building blocks presented above. Let α be an operator acting on $\mathcal{G}^\bullet$ or on $\mathcal{G}^\times$ and let i be a non negative integer. The operator α^i is defined by the identity when $i = 0$ and by $\alpha \circ \alpha^{i-1}$ otherwise.

Since the operators defined above map the elements of $\mathcal{G}^\bullet$ (*i.e.*, subsets of vertices) to those of $\mathcal{G}^\times$ (*i.e.*, subsets of edges) or the elements of $\mathcal{G}^\times$ to those of $\mathcal{G}^\bullet$, they cannot be directly iterated. However, any composition of an operator acting from $\mathcal{G}^\bullet$ to $\mathcal{G}^\times$ (resp. from $\mathcal{G}^\times$ to $\mathcal{G}^\bullet$) with an operator from $\mathcal{G}^\times$ to $\mathcal{G}^\bullet$ (resp.from $\mathcal{G}^\bullet$ to $\mathcal{G}^\times$) leads to an operator on $\mathcal{G}^\bullet$ (resp. on $\mathcal{G}^\times$). Then, such composition can be iterated and eventually followed again by an operator from $\mathcal{G}^\bullet$ to $\mathcal{G}^\times$ (resp. from $\mathcal{G}^\times$ to $\mathcal{G}^\bullet$). Therefore, to define iterated operators on graphs, two cases can be distinguish depending whether a final composition with an operator from $\mathcal{G}^\bullet$ to $\mathcal{G}^\times$ (resp. from $\mathcal{G}^\times$ to $\mathcal{G}^\bullet$) is considered or not.

Definition 1 (Iterated Dilations/Erosions). *Let λ be a nonnegative integer.*

Case 1 (Even Values of λ). *If λ is even, the operators $\delta_{\lambda/2}$ and $\epsilon_{\lambda/2}$ are defined on $\mathcal{G}^\bullet$ by $\delta_{\lambda/2} = (\delta^\bullet \circ \delta^\times)^{\lambda/2}$ and $\epsilon_{\lambda/2} = (\epsilon^\bullet \circ \epsilon^\times)^{\lambda/2}$; the operators $\Delta_{\lambda/2}$ and $\varepsilon_{\lambda/2}$ are defined on $\mathcal{G}^\times$ by $\Delta_{\lambda/2} = (\delta^\times \circ \delta^\bullet)^{\lambda/2}$ and $\varepsilon_{\lambda/2} = (\epsilon^\times \circ \epsilon^\bullet)^{\lambda/2}$.*

Case 2 (Odd Values of λ). *If λ is odd, the operators $\delta_{\lambda/2}$ and $\epsilon_{\lambda/2}$ are defined from $\mathcal{G}^\bullet$ to $\mathcal{G}^\times$ by $\delta_{\lambda/2} = \delta^\times \circ (\delta^\bullet \circ \delta^\times)^{(\lambda-1)/2}$ and $\epsilon_{\lambda/2} = \epsilon^\times \circ (\epsilon^\bullet \circ \epsilon^\times)^{(\lambda-1)/2}$; the operators $\Delta_{\lambda/2}$ and $\varepsilon_{\lambda/2}$ are defined from $\mathcal{G}^\times$ to $\mathcal{G}^\bullet$ by $\Delta_{\lambda/2} = \delta^\bullet \circ (\delta^\times \circ \delta^\bullet)^{(\lambda-1)/2}$ and $\varepsilon_{\lambda/2} = \epsilon^\bullet \circ (\epsilon^\times \circ \epsilon^\bullet)^{(\lambda-1)/2}$.*

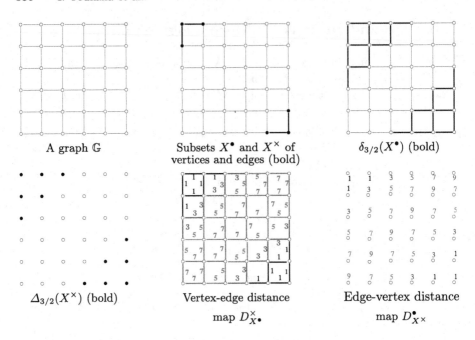

Fig. 1. Illustration of some morphological operators on graphs and of vertex-edge and edge-vertex distance maps.

Illustrations of the operators $\delta_{\lambda/2}$ and $\Delta_{\lambda/2}$ are provided in Fig. 1 for $\lambda = 3$.

The operators $\delta^\bullet, \delta^\times, \delta_{\lambda/2}$ and $\Delta_{\lambda/2}$ are all morphological dilations and the operators $\epsilon^\times, \epsilon^\bullet, \epsilon_{\lambda/2}$ and $\varepsilon_{\lambda/2}$ are their adjunct erosions. Thus, any composition of one of these dilations with its adjunct erosion leads to a morphological filter which is either an opening or a closing depending on the composition order. In particular, when the integer parameter λ is even (resp. odd), the compositions of $\delta_{\lambda/2} \circ \epsilon_{\lambda/2}$ and $\epsilon_{\lambda/2} \circ \delta_{\lambda/2}$ (resp. $\Delta_{\lambda/2} \circ \epsilon_{\lambda/2}$ and $\varepsilon_{\lambda/2} \circ \delta_{\lambda/2}$) filters on $G^\bullet$ and the compositions of $\Delta_{\lambda/2} \circ \varepsilon_{\lambda/2}$ and $\varepsilon_{\lambda/2} \circ \Delta_{\lambda/2}$ (resp. $\delta_{\lambda/2} \circ \epsilon_{\lambda/2}$ and $\epsilon_{\lambda/2} \circ \Delta_{\lambda/2}$) filters on $G^\times$. The simultaneous application of these compositions on the vertices and on the edges of any element in $\mathcal{G}$ (i.e., on any subgraph of $\mathbb{G}$) leads to a subgraph of $\mathbb{G}$, hence morphological filtering on subgraphs.

When the size parameter λ is an even integer, the operators $\delta_{\lambda/2}$ and $\epsilon_{\lambda/2}$ correspond to the dilation and erosion proposed in [12]. It is known [12] that the result of these operators can be obtained by thresholding a (geodesic) distance map instead of iterating the basic dilation or erosion.

Let x and y be two vertices in $\mathbb{G}^\bullet$. A *(vertex-vertex) path from x to y* is a sequence $(x_0, u_0, \ldots, u_{\ell-1}, x_\ell)$ such that $x_0 = x$, $x_\ell = y$, and, for any i in $\{0, \ldots, \ell-1\}$, we have $u_i = \{x_i, x_{i+1}\}$. The *length of a path* $\pi = (x_0, u_0, \ldots, u_{\ell-1}, x_\ell)$ is the number of its elements minus one, i.e., the integer value 2ℓ. A *shortest path from x to y* is a path of minimal length from x to y. We denote by $L(x, y)$ the length of a shortest path from x to y. The *(vertex-vertex)* *distance map* $D_{X^\bullet}$ *to a set* $X^\bullet \subseteq \mathbb{G}^\bullet$ *is the map from* $\mathbb{G}^\bullet$ *to the set of integers such that:*

$$D_{X^\bullet}(x) = \min\{L(x,y) \mid y \in X^\bullet\}, \text{ for any } x \in X^\bullet. \tag{5}$$

Then, when λ is even, the following relation characterizes the dilation $\delta_{\lambda/2}$:

$$\delta_{\lambda/2}(X^\bullet) = \{x \in \mathbb{G}^\bullet \mid D_{X^\bullet}^\bullet(x) \le \lambda\}, \text{ for any } X^\bullet \in \mathcal{G}^\bullet. \tag{6}$$

Based on Eq. 6, to obtain the dilation of a set of vertices, one needs to compute a distance map and to threshold it. An advantage, compared to the computation based on the iterative definition, is to avoid the dependence to the parameter λ in the algorithm time-complexity. More precisely, it is known that the distance map and thresholding computations (see *e.g.* Algorithm 1 for distance map) can be done in linear-time with respect to the size of the graph $\mathbb{G}$. In particular, Algorithm 1 is a variation on breadth-first search, which is a linear-time algorithm with respect to the size of $\mathbb{G}$. Observe that at line 8 of Algorithm 1, the distance value given to y is equal to the one of its predecessor x plus two. This is indeed correct with respect to the above definition of the length of a path, for which, *e.g.*, the distance between two neighbors is equal to two.

It can be deduced from the duality properties stated in [3] that $(\delta_{\lambda/2}, \epsilon_{\lambda/2})$ and $(\Delta_{\lambda/2}, \varepsilon_{\lambda/2})$ are pairs of dual operators, meaning that one operator in the pair can be easily computed from the other due to complementation operations. Hence, in order to provide efficient algorithms for these operators, we only need to focus on the operators $\delta_{\lambda/2}$ and $\Delta_{\lambda/2}$ and deduce the others by duality. For instance, the erosion $\epsilon_{\lambda/2}(X^\bullet)$ can be obtained with the same algorithm as $\delta_{\lambda/2}(X^\bullet)$ provided a complementation on both the input and output of the dilation algorithm. It is also straightforward to obtain a similar linear-time algorithms, based on an *edge-edge distance map*, for the edge dilation $\Delta_{\lambda/2}$ and erosion $\varepsilon_{\lambda/2}$ when λ is even.

Algorithm 1. Sequential vertex-vertex distance map.

Data: a connected graph $\mathbb{G} = (\mathbb{G}^\bullet, \mathbb{G}^\times)$, a subset $X^\bullet$ of $\mathbb{G}^\bullet$.
Result: the distance map $D_{X^\bullet}^\bullet$ to the set $X^\bullet$.

1   $\mathcal{Q} :=$ an empty queue with FIFO property;
2   **foreach** *vertex* x *in* $\mathbb{G}^\bullet$ **do**
3   **if** $x \in X^\bullet$ **then** $\mathcal{Q}.push(x)$; $D_{X^\bullet}^\times(u) := 0$;
4   **else** $D_{X^\bullet}^\times(x) := \infty$;

5   **while** $\mathcal{Q}.isNotEmpty()$ **do**
6      $x := \mathcal{Q}.pop()$;
7   **foreach** *vertex* y *adjacent to* x *in* $\mathbb{G}$ **do** // *i.e., when* $\{x,y\} \in \mathbb{G}^\times$
8   **if** $D_{X^\bullet}^\bullet(y) = \infty$ **then** $\mathcal{Q}.push(y)$; $D_{X^\bullet}^\bullet(y) := D_{X^\bullet}^\bullet(x) + 2$;

The next section presents an approach based on distance maps to obtain linear time algorithms for $\Delta_{\lambda/2}$ and $\delta_{\lambda/2}$ when λ is odd. Then, Sect. 4 presents a parallelization strategy leading to efficient parallel algorithms for all morphological operators on graphs presented in [3].

3 Vertex-Edge and Edge-Vertex Distance Maps

When considering an odd value of λ, an important difference with the even case is that the results and arguments of the dilations $\delta_{\lambda/2}$ and $\Delta_{\lambda/2}$ are not homogeneous: one of them is a set of edges whereas the other one is a set of vertices. In order to deal with this inhomogeneity, we introduce the edge-vertex and vertex-edge distance maps. Given a set of edges (resp. vertices), the edge-vertex (resp. vertex-edge) distance map provides for each vertex (resp. edge) of $\mathbb{G}$ a distance to the closest edge (resp. vertex) of the input set. These distance maps allow us to characterize (by thresholding) the dilations $\delta_{\lambda/2}$ and $\Delta_{\lambda/2}$ when λ is odd. Finally, Algorithm 1 is adapted to compute these distance maps.

The distance maps considered in this section rely on the lengths of shortest paths from vertices to edges. A *(vertex-edge) path from a vertex x of $\mathbb{G}$ to an edge u of $\mathbb{G}$* is a sequence $(x_0, u_0, \ldots, x_\ell, u_\ell)$ such that $u_\ell = u$, $x_\ell \in u_\ell$, and $(x_0, u_0, \ldots, x_\ell)$ is a vertex-vertex path from x to x_ℓ. The *length of a path* $(x_0, u_0, \ldots, x_l, u_\ell)$ is the number of its elements minus one, *i.e.*, the integer value $2\ell + 1$. A *shortest path from a vertex x of $\mathbb{G}$ to an edge u of $\mathbb{G}$* is a path of minimal length from x to u. We denote by $L(x, u)$ the length of a shortest path from x to u. Finally, given a subset $X^\bullet$ of vertices of $\mathbb{G}$, we define the *vertex-edge distance map to $X^\bullet$* as the map $D_{X^\bullet}^\times$ from $\mathbb{G}^\times$ to the set of integers such that:

$$D_{X^\bullet}^\times(u) = \min\{L(x, u) \mid x \in X^\bullet\}, \text{for any } u \in X^\times. \tag{7}$$

Dualy, given a subset $X^\times$ of edges, the *edge-vertex distance map to the set $X^\times$* is the map $D_{X^\times}^\bullet$ from $\mathbb{G}^\bullet$ to the set of integers such that:

$$D_{X^\times}^\bullet(x) = \min\{L(x, u) \mid u \in X^\times\}, \text{ for any } x \in X^\bullet. \tag{8}$$

Edge-vertex and vertex-edge distance maps are illustrated in Fig. 1.

The next property states that the dilations $\delta_{\lambda/2}$ and $\Delta_{\lambda/2}$ can also be characterized with distance maps when λ is odd.

Property 2. *Let λ be any odd positive integer. The following relations hold true:*

$$\delta_{\lambda/2}(X^\bullet) = \{u \in \mathbb{G}^\times \mid D_{X^\bullet}^\times(u) \leq \lambda\}, \text{ for any } X^\bullet \in \mathcal{G}^\bullet; \text{ and}$$
$$\Delta_{\lambda/2}(X^\times) = \{x \in \mathbb{G}^\bullet \mid D_{X^\times}^\bullet(x) \leq \lambda\}, \text{ for any } X^\times \in \mathcal{G}^\times.$$

Algorithms 2 and 3 presented below compute these distance maps in linear time with respect to size $|\mathbb{G}^\bullet| + |\mathbb{G}^\times|$ of $\mathbb{G}$.

Algorithm 2. Vertex-edge distance map.

Data: A connected graph $(\mathbb{G}^\bullet, \mathbb{G}^\times)$, A subset $X^\bullet$ of $\mathbb{G}^\bullet$.
Result: The vertex-edge distance map $D_{X^\bullet}^\times$ to the set $X^\bullet$.

1 $\mathcal{Q} :=$ an empty queue with FIFO property;
2 **foreach** *edge* $u = \{x, y\}$ *in* $\mathbb{G}^\times$ **do**
3   if $x \in X^\bullet$ *or* $y \in X^\bullet$ **then** $\mathcal{Q}$.push(u); $D_{X^\bullet}^\times(u) := 1$;
4   else $D_{X^\bullet}^\times(u) := \infty$;

5 **while** $\mathcal{Q}.isNotEmpty()$ **do**
6     $u := \mathcal{Q}$.pop() ;
7   **foreach** *edge* v *in* $\mathbb{G}^\times$ *adjacent to* u **do** // *i.e., when we have* $v \cap u \neq \emptyset$
8   if $D_{X^\bullet}^\times(v) = \infty$ **then** $\mathcal{Q}$.push(v); $D_{X^\bullet}^\times(v) := D_{X^\bullet}^\times(u) + 2$;

Algorithm 3. Edge-vertex distance map.

Data: A connected graph $(\mathbb{G}^\bullet, \mathbb{G}^\times)$, a subset $X^\times$ of $\mathbb{G}^\times$.
Result: The edge-vertex distance map $D_{X^\times}^\bullet$ to the set $X^\times$.

1 $\mathcal{Q} :=$ an empty queue with FIFO property;
2 **foreach** *vertex* x *in* $\mathbb{G}^\bullet$ **do** $D_{X^\times}^\bullet(x) := \infty$;
3 **foreach** *edge* $u = \{x, y\}$ *in* $\mathbb{G}^\times$ **do**
4   if $D_{X^\times}^\bullet(x) = \infty$ **then** $\mathcal{Q}$.push(x); $D_{X^\times}^\bullet(x) := 1$;
5   if $D_{X^\times}^\bullet(y) = \infty$ **then** $\mathcal{Q}$.push(y); $D_{X^\times}^\bullet(y) := 1$;

6 **while** $\mathcal{Q}.isNotEmpty()$ **do**
7     $x := \mathcal{Q}$.pop() ;
8   **foreach** *vertex* y *in* $G^\bullet$ *adjacent to* x **do** // *i.e., when* $\{x, y\} \in \mathbb{G}^\times$
9   if $D_{X^\times}^\bullet(y) = \infty$ **then** $\mathcal{Q}$.push(y); $D_{X^\times}^\bullet(y) := D_{X^\times}^\bullet(x) + 2$;

---

4   Parallel Algorithm for Distance Maps on Graphs

Contrary to the parallel computation of distance maps on an image, which is often based on a static partitioning of the image into rows, columns or blocks processed in parallel, our parallelization strategy on graphs is based on dynamic partitioning. The partition depends on the input set and is iteratively computed during the execution. More precisely, our strategy iteratively considers the successive level-sets of the distance maps, each level set being partitioned and then traversed in parallel. In this section, our parallel algorithm is presented and its complexity is analyzed assuming that partitioning can be done efficiently. Efficient parallel management of partitions is the topic of the next section.

For the sake of simplicity, we only describe the case of vertex-vertex distance maps, but our strategy can also be adapted to edge-edge, vertex-edge and edge-vertex distance maps computations.

Let us first present our algorithm from a high level point of view. To this end, we recall the notion of a level set. Given an integer λ and a (distance) map D from $\mathbb{G}^\bullet$ in the set of integers, the λ-*level set* of D is the set of all elements of value λ for D (*i.e.*, the set $\{x \in \mathbb{G}^\bullet \mid D(x) = \lambda\}$). Given a subset $X^\bullet$ of $\mathbb{G}^\bullet$, after an initialization step where an integer variable λ is set to 0 and where the elements of $X^\bullet$ are inserted in a variable set E (hence E is the ($\lambda = 0$)-level-set of $D_{X^\bullet}^\bullet$), our algorithm can be sketched as follows:

1. Partition E (*i.e.*, the λ-level set of $D_{X^\bullet}^\bullet$) into p balanced subsets $E_1, \ldots, E_p$.
2. Assign each of the p subsets $E_1, \ldots, E_p$ to one of the p processors
3. Let, in parallel, each processor insert the non already traversed neighbors of the elements in its assigned subset E_i into a private variable set S_i and set the distance map value of the elements in S_i to $\lambda + 2$.
4. Merge the private sets $\{S_i \mid i \in \{1, \ldots, p\}\}$ and store the result in E so that E becomes the ($\lambda + 2$)-level set of $D_{X^\bullet}^\bullet$.
5. Increment λ and repeat steps 1–4 until E becomes empty.

In Step 3, in order to concurrently check if a vertex has been already traversed, we need to equip each vertex with a synchronization Boolean variable that is handled with an atomic *test-and-set* instruction. The test-and-set instruction sets a given variable to true and returns its old value as a single atomic (*i.e.*, non-interruptible) instruction.

Algorithm 4 provides the precise description of our parallel approach. It uses two auxiliary functions called *Partition* and *Union*. In the next section, we provide algorithms for these two functions. The efficiency of Algorithm 4 depends on these functions. As we will see, the function *Partition* runs in linear time with respect to n/p and the function *Union* runs in $O(n/(Kp) + log_2 p)$ amortized time, where n, p and K are the size of the graph, the number of processors, and the number of level-sets of the produced distance map. Furthermore, any class of the produced partition contains either n/p or $n/p + 1$ elements.

Finally, in order to state the time complexity of Algorithm 4, we need to make two assumptions about the graph and the set of vertices under consideration.

The *degree of a vertex x of* $\mathbb{G}$ is the number of edges that contain x (*i.e.*, the cardinality of the set $\{y \in \mathbb{G}^\bullet \mid \{x, y\} \in \mathbb{G}^\times\}$). Let β be any positive integers. We say that $\mathbb{G}$ is β-*balanced* if the degrees of any two vertices of $\mathbb{G}$ differ by at most β. Let $X^\bullet$ be a subset of $\mathbb{G}^\bullet$. We say that $X^\bullet$ is β-*balanced* if every nonempty level-set of $D_{X^\bullet}^\bullet$ contains at least β elements.

Note that when $X^\bullet$ is p-*balanced*, the distance map $D_{X^\bullet}^\bullet$ has at most $|\mathbb{G}^\bullet|/p$ nonempty level-sets, then the while loop at line 7 is executed at most $|\mathbb{G}^\bullet|/p$ times. Furthermore, if a given level set E contains n vertices, any of the $\{E_i \mid i \in \{1, \ldots, p\}\}$ obtained at line 8 contains at most $n/p + 1$ vertices, which allows us to deduce that the loop line 11 runs in $O(|\mathbb{G}^\bullet|/p)$ time since the level-sets of $D_{X^\bullet}^\bullet$ partition $\mathbb{G}^\bullet$. As any of the $\{E_i \mid i \in \{1, \ldots, p\}\}$ contains at most $n/p+1$ vertices, when $\mathbb{G}$ is β-balanced, we can bound the number of edges that contain an element in E_i by $m/p + d_{min} + \beta n/p + \beta$, where m is the total number of edges that contain an element in E_i and where d_{min} is the minimal degree of a vertex of $\mathbb{G}$. Thus, we also have $|S_i| \leq m/p + d_{min} + \beta n/p + \beta$, where S_i is the set obtained after the execution of foreach loop line 11. Hence, since the level sets of $D_{X^\bullet}^\bullet$ partition $\mathbb{G}^\bullet$, it can be shown that the insertion operation on S_i at line 14 is executed at most $(3|\mathbb{G}^\times| + 2\beta|\mathbb{G}^\bullet|)/p$ times by each of the p processors during the

---

Algorithm 4. Parallel vertex-vertex distance map.

Data: A connected graph $(\mathbb{G}^\bullet, \mathbb{G}^\times)$, a subset $X^\bullet$ of $\mathbb{G}^\bullet$, the number p of processors.

Result: The distance map $D_{X^\bullet}^\bullet$ to the set $X^\bullet$.

1 $E := \emptyset$; $\lambda := 0$;
2 Set to False all elements of a shared Boolean array *Traversed* of size $|\mathbb{G}^\bullet|$
3 $(E_1, \ldots, E_p) :=$ Partition$(X^\bullet, p)$;
4 **foreach** *processor* i *in* $\{1, \ldots, p\}$ **do in parallel**
5   **foreach** *vertex* $x \in E_i$ **do** $D_{X^\bullet}^\bullet(x) := \lambda$; $Traversed[x] := True$; ;
6 $E := $Union$(p, E_1, \ldots, E_p)$;
7 **while** $E \neq \emptyset$ **do**
8      $(E_1, \ldots, E_p) :=$ Partition(E, p);
9   **foreach** *processor* i *in* $\{1, \ldots, p\}$ **do in parallel**
10        $S_i := \emptyset$;
11   **foreach** x *in* E_i **do**
12   **foreach** *vertex* y *adjacent to* x *in* $\mathbb{G}$ **do** // *i.e., when* $\{x, y\} \in \mathbb{G}^\times$
13   **if** *test-and-set*$(Traversed[y]) = False$ **then**
14                 $S_i := S_i \cup \{y\}$;
15                 $D_{X^\bullet}^\bullet(y) := \lambda + 2$;

16     $E := $Union$(p, S_1, \ldots, S_p)$;

---

overall execution and the continuation condition of the loop at line 12 must be tested less than $3|\mathbb{G}^\times|/p + 2(\beta+1)|\mathbb{G}^\bullet|/p$ times. Hence, using an array of linked lists to represent the graph $\mathbb{G}$ and using simple arrays for all sets, we deduce that the time complexity of the main part (lines 9 to 15) of Algorithm 4 is linear with respect to $(|\mathbb{G}^\times| + |\mathbb{G}^\bullet|)/p$. Considering also the auxiliary functions *Union* and *Partition*, the overall time complexity of Algorithm 4 can be established.

Theorem 3. *Algorithm 4 outputs a map $D_{X^\bullet}^\bullet$ which is the vertex-vertex distance map to the set $X^\bullet$. Let p be the number of available processors. Let us assume that β is a constant integer such that $\mathbb{G}$ is β-balanced and that $X^\bullet$ is p-balanced. Then, Algorithm 4 runs in $O((|\mathbb{G}^\bullet| + |G^\times|)/p + K \log_2 p)$ time, where K is the number of nonempty level-sets of $D_{X^\bullet}^\bullet$.*

Under the assumption of Theorem 3, the distance map D_X contains at most $|\mathbb{G}^\bullet|/p$ nonempty level-sets. Thus the time complexity of Algorithm 4 is less than $O((|\mathbb{G}^\bullet| + |G^\times|)/p + |\mathbb{G}^\bullet|(\log_2 p)/p)$.

The assumptions in Theorem 3 hold, in general, true when the graph-based morphological operators of [3] are applied to image processing. In particular, when we consider a 2-dimensional image equipped with the 4- or 8- adjacency relation, the degrees of any two vertices are the same (except on the image borders), and the number of distinct level-sets is of the order of $\sqrt{|G^\bullet|}$, meaning that in average, each level set contains $\sqrt{|G^\bullet|}$ vertices. Furthermore, in practice, we generally have $K. \log_2 p \leq (|\mathbb{G}^\bullet| + |\mathbb{G}^\times|)/p$. Thus, roughly speaking, we can say that the time-complexity is, in general, dominated by $(|\mathbb{G}^\bullet| + |G^\times|)/p$.

5   Parallel Partition and Disjoint Union Algorithms

In this section, we present efficient parallel algorithms for the partition and union function used in Algorithm 4 and we analyze their time-complexity.

The parallel partition algorithm (see Algorithm 5) consists of computing in parallel, with p processors, a *balanced partition* $\{E_1, \ldots, E_p\}$ of a set E. The partition is balanced in the sense that the k-first sets of the partition contain $|E|/p + 1$ elements whereas the following ones contain $|E|/p$ elements, where k is the remainder in the integer division of $|E|$ by p. The elements of E, stored in an array, are moved to arrays previously allocated for the subsets $E_1, \ldots, E_p$ in the order of their indices: the first set receives the first elements of the array E and so on (see Fig. 2). Thus, each processor computes the index of the first and of the last element that must be copied (lines 2 to 7) before actually copying the elements of E located between the computed indices (line 8). The computation of the first and of the last indices can be done in constant time and the copying step is done in linear time with respect to $|E|/p$ (each processor moves at most $|E|/p + 1$ elements).

Algorithm 5. Partition.

Data: An array E of $n = |E|$ elements, the number p of processors.
Result: A balance partition $(E_1, \ldots, E_p)$ of E.

1 **foreach** *processor i in* $\{1, \ldots, p\}$ **do in parallel**
2 | **if** $i \leq (n \mod p)$ **then**
3 | | $start[i] := (i - 1) * (n/p + 1)$;
4 | | $end[i] := start[i] + n/p$;
5 | **else**
6 | | $start[i] := (n \mod p) * (n/p + 1) + (i - 1 - (n \mod p)) * (n/p)$;
7 | | $end[i] := start[i] + n/p$ - 1;
8 | **foreach** j_i *in* $\{start[i], \ldots, end[i]\}$ **do** $E_i[j_i - start[i]] := E[j_i]$;

Our parallel *Union* algorithm (see Algorithm 6) computes the union of p disjoint sets $\{S_1, \ldots, S_p\}$ with p processors. The elements of each set are stored in an array and each processor i copies the elements of the array S_i in the array E. The elements of S_i are stored in the resulting array E from the index $start[i]$, where $start[i]$ is the sum of the cardinalities of the sets $S_1, \ldots, S_{i-1}$ (see Fig. 3). Thus, our algorithm first computes the values $start[i]$ for any i in $\{1, \ldots, p\}$ (line 1) before actually copying the elements into E (line 3). Given the cardinalities $|S_1|, \ldots, |S_p|$, computing the values $start[i]$ for any i in $\{1, \ldots, p\}$ is known as the prefix-sum problem. It can be solved in parallel with p processors with

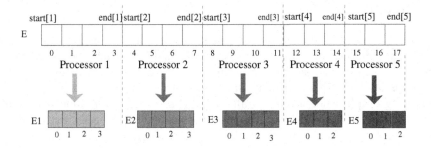

Fig. 2. Illustration of the *Partition* algorithm with $p = 5$ processors.

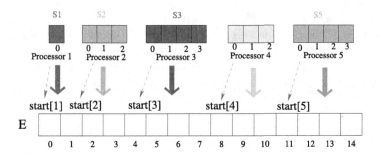

Fig. 3. Illustration of the *Union* algorithm with $p = 5$ processors.

a $O(\log_2 p)$ running-time algorithm [4]. Then, each processor i copies (line 3) the elements of S_i into E at the correct position. Let us consider the amortized-time complexity of this operation for a sequence of calls to *Union* as used in Algorithm 4, under the assumptions of Theorem 3. Let K be the number of distinct level sets of $D_{X\bullet}^{\bullet}$. There is one call to *Union* for each level-set of the distance map $D_X^{\bullet}$. Thus, there are K calls to *Union*. We have seen in Sect. 4 that there are at most $(3|\mathbb{G}^{\times}| + 2\beta|\mathbb{G}^{\bullet}|)/p$ insertions in S_i. Any element inserted in S_i is considered exactly once at line 3 of Algorithm 6. Thus, the amortized time-complexity of line 3 is $O((|\mathbb{G}^{\bullet}| + |\mathbb{G}^{\times}|)/(Kp))$ and the one of Algorithm 6 is $O((|\mathbb{G}^{\bullet}| + |\mathbb{G}^{\times}|)/(Kp) + \log_2 p)$.

Algorithm 6. Union.

Data: A series $S_1, \ldots, S_p$ of p sets, and the number p of processors.
Result: An array E whose elements constitutes the union of $\{S_1, \ldots, S_p\}$.
1 start = ParallelPrefixSum($|S_1|, \ldots, |S_p|$);
2 **foreach** *processor* i *in* $\{1, \ldots, p\}$ **do in parallel**
3 $\quad\lfloor$ **foreach** j_i *in* $\{0, \ldots, |S_i| - 1\}$ **do** $E[start[i] + j_i] := S_i[j_i]$;

---

6   Conclusion

In this article efficient sequential and parallel algorithms for the (binary) graph-based mathematical morphology operators defined in [3] have been proposed. These algorithms are based on distance maps computation in unweighted graphs. The sequential algorithms run in linear time with respect to the size of the underlying graph, whereas the parallel algorithms run (under some reasonable assumptions) in $O(n/p + K \log_2 p)$ time, where n, p, and K are the size of the underlying graph, the number of available processors, and the number of distinct level-sets of the distance map, respectively.

From a computational point of view, future work will include experimental studies of the execution times, variations on our parallel algorithms with improved load balancing, as well as algorithms for the so-called "grayscale case" in order to filter functions as well as binary sets. On the methodological point of view, the use of distance maps in unweighted graphs opens the door towards the investigation of morphological operators on graphs embedded in metric spaces (or more generally on weighted graphs) where the result of an operator depends on the "length" of the edges according to the metric.

References

1. Chia, T.L., Wang, K.B., Chen, Z., Lou, D.C.: Parallel distance transforms on a linear array architecture. IPL **82**(2), 73–81 (2002)
2. Coeurjolly, D.: 2D subquadratic separable distance transformation for path-based norms. In: Barcucci, E., Frosini, A., Rinaldi, S. (eds.) DGCI 2014. LNCS, vol. 8668, pp. 75–87. Springer, Heidelberg (2014)
3. Cousty, J., Najman, L., Dias, F., Serra, J.: Morphological filtering on graphs. CVIU **117**(4), 370–385 (2013)
4. Ladner, R.E., Fischer, M.J.: Parallel prefix computation. JACM **27**(4), 831–838 (1980)
5. Man, D., Uda, K., Ueyama, H., Ito, Y., Nakano, K.: Implementations of parallel computation of Euclidean distance map in multicore processors and GPUs. In: ICNC, pp. 120–127 (2010)
6. Meyer, F., Angulo, J.: Micro-viscous morphological operators. In: ISMM, pp. 165–176 (2007)
7. Pham, T.Q.: Parallel implementation of geodesic distance transform with application in superpixel segmentation. In: DICTA, pp. 1–8. IEEE (2013)
8. Saito, T., Toriwaki, J.I.: New algorithms for euclidean distance transformation of an n-dimensional digitized picture with applications. Pattern Recogn. **27**(11), 1551–1565 (1994)
9. Shyu, S.J., Chou, T., Chia, T.L.: Distance transformation in parallel. In: Proceedings of Workshop Combinatorial Mathematics and Computation Theory, pp. 298–304 (2006)
10. Soille, P., Breen, E.J., Jones, R.: Recursive implementation of erosions and dilations along discrete lines at arbitrary angles. PAMI **18**(5), 562–567 (1996)
11. Svolos, A.I., Konstantopoulos, C.G., Kaklamanis, C.: Efficient binary morphological algorithms on a massively parallel processor. In: IPDPS, p. 281 (2000)
12. Vincent, L.: Graphs and mathematical morphology. Sig. Proc. **16**(4), 365–388 (1989)

Digitization of Partitions and Tessellations

Jean Serra[1]([✉]) and B. Ravi Kiran[2]

[1] A3SI-ESIEE LIGM, Université Paris-Est, Noisy-le-Grand, France
jean.serra@esiee.fr
[2] Center de Robotique, MINES ParisTech, PSL-Research University, Paris, France
ravi.kiran@mines-paritech.fr

Abstract. We study hierarchies of partitions in a topological space where the interiors of the classes and their frontiers are simultaneously represented. In both continuous and discrete cases our approach rests on tessellations whose classes are $\mathcal{R}$-open sets. In the discrete case, the passage from partitions to tessellations is expressed by Alexandrov topology and yields double resolutions. A new topology is proposed to remove the ambiguities of the diagonal configurations. It leads to the triangular grid in $\mathbb{Z}^2$ and the centered cubic grid in $\mathbb{Z}^3$, which are the only translation invariant grids which preserve connectivity and permit the use of saliency functions.

Keywords: Tessellations · Regular sets · Hierarchies · Khalimsky topology · Simplicial complexes · 3-D grids

1 Introduction

When a set is partitioned, the frontiers between classes are not materialized and the operations which need these frontiers may not be easy. Does the frontier between two classes belong to one of them, to both, or to none? How to combine interiors and frontiers in a unique representation? This is the matter of the present study.

When one goes up in a hierarchy of partitions, the frontier elements which disappear are only those separating classes, e.g. the two medians in Fig. 1b, and not the barbs of Fig. 1c. How to differentiate the first ones from second? An answer is given by the notion of tessellation (Sect. 3).

Connectivity also may set trouble. How to merge the squares of Fig. 1a? Must we firstly link the "2", or rather the "8"? In 2D, the classical digital approach focuses on the boundaries, which are usually supposed to be Jordan curves [13,22]. One also constructs "well composed set", whose boundaries have no diagonal configurations [14]. We believe that the classes themselves, and not their boundaries, provide a more convenient input to the present problem. We will see that Jordan curves are totally useless in what follows, though the particular structure of the classes ($\mathcal{R}$-open sets) is crucial.

Image processing rests either on Euclidean background, or on the digital one, but the problem of the frontiers in a partition is set upstream this distinction,

© Springer International Publishing Switzerland 2016
N. Normand et al. (Eds.): DGCI 2016, LNCS 9647, pp. 323–334, 2016.
DOI: 10.1007/978-3-319-32360-2_25

2	2	2	8	8	8
2	2	2	8	8	8
2	2	2	8	8	8
8	8	8	2	2	2
8	8	8	2	2	2
8	8	8	2	2	2

a   b   c

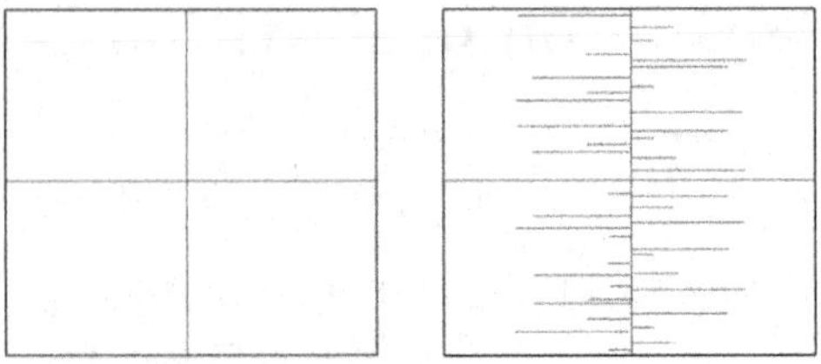

Fig. 1. (a) In a hierarchy, must we firstly connect the "2", or the "8"? Do the frontiers in Figure (c) play the same role as in Figure (b)?

and its solution will apply to both continuous and discrete models. In the second case Khalimsky topology is a convenient starting point, for it identifies $\mathbb{Z}^n$ with Euclidean n-cubes [1,17]. It involves simplicial complexes which lead to double resolution techniques [3,4,9] (Sect. 5). But it does not preserve connectivity (see Fig. 1a), unlike the topological variant developed in Sect. 6.

2 Lattice of the Regular Open Sets (Reminder)

Let E be a set equipped with a (non necessarily separated) topology, and $\mathcal{G}$ be the class of its open sets. $\mathcal{G}$ is a complete lattice, but not complemented. Now, the notion of a complement is essential, and this orients us toward the family $\mathcal{R} = \mathcal{R}(E)$ of the *regular open sets* of E. An open set B is said to be regular, or $\mathcal{R}$-open, when it does not change when one takes its adherence $\overline{B}$, and then one takes the interior of the latter, i.e. $B = (\overline{B})^\circ$. It is the case of Fig. 1b for example, but not of Fig. 1c[1]. The main result about $\mathcal{R}$ is the following theorem [6,7,21]:

Theorem 1. $\mathcal{R}$ *is a complete lattice for the inclusion ordering, where the supremum and the infimum are given by*

$$\vee\, B_i = (\overline{\cup B_i})^\circ; \quad \wedge\, B_i = (\cap B_i)^\circ. \tag{1}$$

Lattice $\mathcal{R}$ is completely distributive and with unique complement

$$compl\, B = (\complement B)^\circ. \tag{2}$$

---

[1] In mathematical morphology, the three major bibliographic sources come from G. Matheron, Ch. Ronse, and H. Heijmans. The notion of an $\mathcal{R}$-open (closed) set recurrently appears in Matheron works, from 1969 until 1996. In [16], p.156–157, he builds a σ-algebra of regular closed sets and proves a series of characteristic properties; in [15] he associates a random set to every pair $(A^\circ, \overline{A}), A \in \mathcal{P}(E)$; in [23] he interprets the regularization operator as a strong morphological filter, and gives a middle element between $\overline{(\overline{A})^\circ}$ and $(\overline{A^\circ})^\circ$. For his part, Ch. Ronse shows in [21] that any complete boolean lattice is isomorphic to a lattice of regular open sets, and indicates as a watermark that the associated topology is Alexandrov.

Here the symbol $\complement$ designates the set complement operator. For example, in the two Fig. 4a and b, the complement of each class is the supremum, in $\mathcal{R}$, of the three other ones.

3   Tessellations

A "tessellation" is a partition of a topological space where both interiors and their boundaries are classes [25]:

Definition 1. *We define a tessellation τ of a topological space E as any family $\{B_i, i \in I\}$ of disjoint open sets called "classes":*

$$\tau = \{B_i, i \in I\} \ with \ i \neq j \Rightarrow B_i \cap B_j = \varnothing \tag{3}$$

such that the union of all B_i and of all boundaries $Fr(B_i, B_j) = \overline{B}_i \cap \overline{B}_j$ covers the space E:

$$E = \cup\{B_i, i \in I\} \cup \{\overline{B}_i \cap \overline{B}_j, \ i,j \in J, \ i \neq j\}, \tag{4}$$

We will designate by N (N for "net") the set $\cup\{\overline{B}_i \cap \overline{B}_j, i, j \in E, i \neq j\}$ of all boundaries between classes.

3.1   Tessellations and $\mathcal{R}$ open sets

In what follows, $S = (\overline{B})^{\circ}$ designates the $\mathcal{R}$-open transform of the open set $B \in \mathcal{G}(E)$. The operation $B \to S = (\overline{B})^{\circ}$ is an algebraic closing on the set $\mathcal{G}$ of the open sets of E, and the image of $\mathcal{G}$ is $\mathcal{R}$. For example, if we take Fig. 1c for B, we obtain Fig. 1b for transform S. This closing means that S is the smallest $\mathcal{R}$-open set that contains B. Indeed, if another $\mathcal{R}$-open S' contains B, then $S' = (\overline{S'})^{\circ} \supseteq (\overline{B})^{\circ} = S$. Note that

$$S = (\overline{B})^{\circ} \ \Rightarrow \ S \supseteq B \ and \ \overline{S} = \overline{B}. \qquad B \in \mathcal{G} \tag{5}$$

Besides, the set difference $\Delta = S \backslash B$ between S and B coincides with the difference between the boundaries $Fr(B)$ and $Fr(S)$:

$$\Delta = S \backslash B \ \Leftrightarrow \ Fr(S) = Fr(B) \backslash \Delta, \tag{6}$$

and the frontier of B contains that of S.

The open sets of a tessellation cannot admit lacunae or fissures of empty interior, which makes the tessellation impossible. Thus the convenient class to consider are that of the $\mathcal{R}$-open sets. Indeed,

Theorem 2. *All classes B_i of any tessellation $\tau = \{B_i, i \in I\}$ are necessarily $\mathcal{R}$-open.*

Proof. We prove the theorem in the negative, by showing that if $B_i \neq (\overline{B_i})^\circ = S_i$, then the B_i do not generate a tessellation. We firstly observe that the B_i are disjoint iff the corresponding $\mathcal{R}$-open sets S_i are disjoint. We have

$$B_i \cap B_j = \emptyset \Leftrightarrow B_i \cap \overline{B_j} = \emptyset \Rightarrow B_i \cap (\overline{B_j})^\circ = \emptyset = B_i \cap S_j \Rightarrow S_i \cap S_j = \emptyset$$

as well as the inverse implication, since $S_i \supseteq B_i, \forall i$. Besides, we have, from Rel. (5), $\cup\{\overline{B_i} \cap \overline{B_j}, i, j \in, i \neq j\} = \cup\{\overline{S_i} \cap \overline{S_j}, i, j \in, i \neq j\}$ so that the S_i form a tessellation of E. If there exists one B_i at least which is strictly included in S_i then $z \in S_i \setminus B_i$ does not belong to any B_i or to any boundary $\overline{B_i} \cap \overline{B_j}$ since S_i is open. Therefore the B_i do not cover set E, which achieves the proof.

All in all, just as a partition of E is a family in $\mathcal{P}(E)$ whose each element is the complement of the union of all the others, a tessellation of E is a family in $\mathcal{R}(E)$ whose each element is the complement of the supremum of all the others, both complement and supremum being taken in $\mathcal{R}$.   □

The algebraic closing $B \to S = (\overline{B})^\circ$ fills up the fine fissure, isthmuses and the point lacunae of B. However, some separations, too narrow, may not be filled up. Take for example, B composed of two disjoint open squares of $\mathbb{R}^2$ whose adherence share one vertex only, then the adherence $\overline{B_1 \cup B_2}$ joins the two squares, but the interior $(\overline{B_1 \cup B_2})^\circ$ separates them again.

In the literature, an example of tessellation is given by the clefts in the perfect fusion graphs of J. Cousty and G. Bertrand [2]. Theorem 2 reminds us of Jordan's one, though it is true in any topological space, and it does not focus on the frontiers, but on the classes. In $\mathbb{R}^2$ every Jordan curve induces a tessellation [9], but a tessellation into two open classes, even connected, can have a frontier which is not a Jordan curve. Figure 4b depicts a digital contour which separates $\mathcal{R}$-open sets and which is thick. Note also that connectivity is not involved in the above theorem, whereas it is essential in Jordan's one.

3.2   Structure of the Tessellations

We now introduce a minimal tessellation τ_0 of the space E . The classes $\{s_i\}$ of τ_0 are called "the leaves", and are supposed in locally finite number. These leaves are indivisible $\mathcal{R}$-open sets, i.e. each class of a larger tessellation contains one leaf at least and is disjoint from those that it does not contain. The family $\mathcal{T}$ of all tessellations of E is obviously ordered by the following relation:

$$\tau \leq \tau' \quad \Leftrightarrow \quad S(x) \subseteq S'(x) \quad x \in \tau_0 \; \tau, \tau' \in \mathcal{T} \tag{7}$$

where $S(x)$ (resp. $S'(x)$) is the class of τ (resp. τ') at point x. This ordering provides $\mathcal{T}$ with the structure of a complete lattice. More precisely, consider a set P of labels and the family $\{\tau_p, p \in P\}$:

Proposition 1. *The set $\mathcal{T}$ of all tessellations $\tau \geq \tau_0$ of E forms a complete lattice for the ordering of Rel. (7). Its universal elements are τ_0 and E. The infimum of family $\{\tau_p, p \in P, \tau_p \geq \tau_0\}$ is the tessellation whose class at point x is the infimum, in $\mathcal{R}$, of the classes of the τ_p at point x, and the supremum is the smallest tessellation whose classes are suprema of the classes of the τ_p in $\mathcal{R}$.*

Proof. Let x and y be two classes of τ_0, and $S_p(x)$ (resp. $S_p(y)$) be the class of τ_p at point x (resp. y). We have to prove that $\{[\cap S_p(x)]^\circ, x \in \tau_0\}$ is a tessellation. If so, then this tessellation will be the greatest lower bound of the τ_p, since on cannot find at point x an $\mathcal{R}$-open class greater than $[\cap S_p(x)]^\circ$ and which is included in all $S_p(x)$.

The point $y \in S_p(x)$ iff $x \in S_p(y)$; therefore, if $y \in [\cap S_p(x)]$ then $[\cap S_p(x)] = [\cap S_p(y)]$ and the two $\mathcal{R}$-infima $[\cap S_p(x)]^\circ$ and $[\cap S_p(y)]^\circ$ are equal. If for some p the two sets $S_p(x)$ and $S_p(y)$ are disjoint, then *a fortiori* the $\mathcal{R}$-infima $[\cap S_p(x)]^\circ$ and $[\cap S_p(y)]^\circ$ are also disjoint. Finally, as $\cup\{x, x \in \tau_0\} \subseteq \cup\{[\cap S_p(x)]^\circ, x \in \tau_0\}$, and as $\cup\{x, x \in \tau_0\} = E$, the family $\{[\cap S_p(x)]^\circ, x \in \tau_0\}$ turns out to be a tessellation of E, and $\mathcal{T}$ is a complete inf-semi Lattice. But in addition $\mathcal{T}$ admits a greatest element, namely E itself, so that $\mathcal{T}$ is a complete lattice, which achieves the proof.

3.3 Hierarchies of Tessellations

The tessellations met in image processing often form hierarchies, i.e. totally ordered closed sequences starting from the leaves and ending at E itself, considered as an R-open class.

One cannot bring the classes of a tessellation τ of the hierarchy down to the union of their leaves, for the portions of frontiers between adjacent leaves (i.e. those whose intersection of the adherences is not empty) would neither belong to classes, nor to the background net N which separates the classes. We must find out another law of composition. Consider a partition of the totality of the leaves into J sub-sets $B_1, ..B_j, ..B_J$, then

Proposition 2.   *The unique tessellation whose each class includes exactly one B_j, $1 \leq j \leq J$ has the $\mathcal{R}$-open sets $S_j = (\overline{B_j})^\circ$, $1 \leq j \leq J$ as classes.*

Proof. Theorem 2 shows that we must replace each B_j with its $\mathcal{R}$-open version S_j for finding again a tessellation. Moreover, there is no tessellation $\{S'_j\}$ which keeps disjoint the family $\{S_j\}$ and such that $S'_j \supset S_j \; \forall j$. Such a strict inclusion would mean that one could find an open set containing $x \in Fr(S_j) \cap S'_j$, such that S'_j is not disjoint from S'_i, which is impossible.   □

The supremum and the infimum of a family of classes are those involved in the lattice $\mathcal{R}$ of the $\mathcal{R}$-open sets. Consequently, in a hierarchy of tessellations, every point of the net N is absorbed by a class, sooner or later.

4   Connected Classes and Saliency Function

Connectivity does not intervene in the paradigm of the hierarchies of tessellations. However the leaves are often connected, and the question of preserving the
connectivity of the classes arises. Now this requirement is not always possible,
neither in Euclidean topologies, nor in digital ones. Even when the leaves $\{s_i\}$
are connected, the $\mathcal{R}$-open sets $S_j = (\overline{B_j})^{\circ}$ of Proposition 2 may be not connected. One can also think of the example of the two open squares of $\mathbb{R}^2$ whose
adherences have only one point in common. They are connected, regular, their
adherence is connected, but their union, though regular, is again not connected.

To overcome this issue, we have to more specify the topological space E,
which is now supposed to be locally arcwise connected, i.e. where the arcwise-
connected components of an open set are open. Furthermore, we say that a pair
of two regular sets S_i and S_j of E is strongly when the sets are adjacent(i.e. the
intersection of their adherences is not empty) and when one can find a point x
in their common boundary which is included in a small neighborhood $\delta(x)$ itself
included in the union of their adherences, i.e.

$$S_i \cap S_j = \emptyset \text{ and } \exists x \in \overline{S_i} \cap \overline{S_j} \mid x \in \delta(x) \subseteq \overline{S_i} \cup \overline{S_j}. \tag{8}$$

Hierarchies of tessellations require this strong adjacency to transmit the connectivity from the leaves to their clusters into classes:

Theorem 3. *let E be a locally arcwise connected set, and τ_0 be a tessellation of
E into $\mathcal{R}$-open connected leaves $\{s_i, i \in I\}$ where every adjacent pair is strongly
adjacency (8). The $\mathcal{R}$-supremum of any finite sequence $\{s_k, 1 \leq k \leq n\}$ of leaves
is then connected.*

Proof. Let $S = \vee\{s_k, 1 \leq k \leq n\}$ be the $\mathcal{R}$-supremum of the family. From Proposition 2, S is the union of the $\{s_k\}$ and of the boundaries between pairs$\{s_k, s_l\}$.
Consider two strongly adjacent leaves $\{s_k, s_l\}$. By strong adjacency, there exists
a point x of the boundary $\overline{s_k} \cap \overline{s_l}$ which has a neighborhood $\delta(x) \subseteq \overline{s_k} \cup \overline{s_l}$. As E is
locally arcwise connected, there exists also a connected neighborhood $\delta' \subseteq \delta(x)$
which thus contains points of s_k and of s_l. Therefore the union $s_k \cup s_l \cup Fr(s_k, s_l)$
is arcwise connected, and so is S, which achieves the proof.   □

In particular, the usual spaces $\mathbb{R}^n$ and $\mathbb{Z}^n$ are locally arcwise connected. Note
that a same leave may appear several times in the sequence $\{s_p, 1 \leq p \leq n\}$.
When a hierarchy H with n levels $\tau_p, 1 \leq p \leq n$ of tessellations satisfies the
conditions of Theorem 3 then one can allocate a numerical value to each point
of the net N_0 of the frontiers between the leaves of H. This function on N_0, s
say, indicates the level when the frontier element adjacent between two classes
vanishes, and the classes merge.

This is nothing but the classical saliency function [19], here generalized, and
without assuming that the boundaries are Jordan curves (in the case of $\mathbb{R}^2$).
When the conditions of Theorem 3 are not satisfied, as in Fig. 1a, the saliency
function may be undefined at some crucial points. Finally, when the classes
are possibly not connected, the saliency function may no longer summarizes
exhaustively the hierarchy, as shown in Fig. 2.

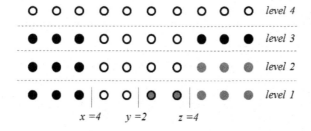

Fig. 2. The space E is the 10 points of $\mathbb{Z}$. The saliency function $\{x, y, z\}$ does not detect the non connected class (in black) which appears at level 3 of this hierarchy.

5   Tessellations of $\mathbb{Z}^n$ and Khalimsky spaces

The notion of a tessellation rests on $\mathcal{R}$-open sets, and to apply it to $\mathbb{Z}^n$, we first need to provide a topology to this space.

5.1   Reminder on Khalimsky Topology

E. Khalimsky topology clearly shows the analogies between the tessellations of $\mathbb{R}^n$ and those of $\mathbb{Z}^n$. It was published in Russian in the sixties, but it is better known by more recent papers in English [8]. It is just sketched here, but the reader will find a more detailed presentation in papers by E. Melin [18], L. Mazo et al. [17], and the lecture notes of Ch. Kiselman [11].

In 1937, P.S. Alexandrov introduced topological spaces E with smallest neighborhood (in short sn-topology), where it is assumed that the class of the open sets is closed under intersection [20]. Khalimsky topology belongs to this category. In $\mathbb{R}$, it associates the open interval $]m - \frac{1}{2}, m' + \frac{1}{2}[$ with every pair $m \leq m'$ of odd integers, and the closed interval $[n - \frac{1}{2}, n' + \frac{1}{2}]$ with every pair $n \leq n'$ of even integers. The unions of open (resp. closed) intervals generate a non separated sn-topology. When $m = m'$ and $n = n'$ then $\mathbb{R}$ is partitioned into unit intervals, and is thus connected. The product topology of n Khalimsky lines generates the topology in $\mathbb{R}^n$. When all coordinates of their centers are odd (resp. even), then the n-cubes are open (resp.closed), and the others n-cubes are said to be mixed.

In $\mathbb{R}^2$ the Kovalevsky cells provide an equivalent approach, which allow us to display these topologies [12]. This structure is akin to the simplicial complexes [1], to fusion grids [2] and to the planar graphs.

According to Theorem 2, in a Khalimsky space the tessellations are $\mathcal{R}$-open sets, which simplifies the basic elements. In $\mathbb{Z}^2$ for example, the isolated points and the pending edges are removed. The background net N is made of loops contouring the classes and where open edges and points alternate (see Fig. 3).

5.2   Khalimsky digital tessellations

The use of Khalimsky topology in digital image processing is classical [17,18]. In this section we indicate its links with the $\mathcal{R}$-open sets and tessellations,

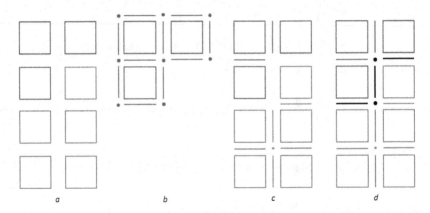

Fig. 3. *Simple resolution:* (a) two classes of a partition of $\mathbb{Z}^2$. *Double resolution:* (b) Khaminsky closure of the first class of Fig. a; (c) regular open versions of the two classes; (d) corresponding tessellation (in black, the net between classes).

and we show that this topology does not solve the question of the diagonal configurations.

Khalimsky topology on $\mathbb{Z}$ is obtained by identifying each unit interval of $\mathbb{R}$ with the corresponding integer. The open intervals are obtained in the same manner from $]m - \frac{1}{2}, m' + \frac{1}{2}[$ where m and m' are odd and $m \leq m'$. In this topology $\mathbb{Z}$ is connected. The extension to $\mathbb{Z}^n$ is made as previously with $\mathbb{R}^n$.

Interpret the points of a set $X \subseteq \mathbb{Z}^2$ as points of odd coordinates in a Khalimsky plane $\mathbb{K}^2$ which contains twice more points by line and twice more lines [3,4,9]. In Kovalevsky representation, the initial points of $X \subseteq \mathbb{Z}^2$ become squares and the additional points become segments and points, as depicted in Fig. 3b. The adherence $X \rightarrow \overline{X}$ is the union of the adherences of its basic elements, and the interior $\overline{X} \rightarrow (\overline{X})^\circ$ is the complement of the adherence of the complement set. Figure 3b depicts the step $X \rightarrow \overline{X}$ for one class of X, and Fig. 3c $\overline{X} \rightarrow (\overline{X})^\circ$ depicts the construction of the regularized version $(\overline{X})^\circ$. The two classes of tessellation $(\overline{X})^\circ$ are separated by the net in bold of Fig. 3d. In the double resolution plane $\mathbb{K}^2$, all points of the background have odd coordinates. This example illustrates the following property:

Proposition 3. *Let $\mathbb{K}^n$ be the Khalimsky space of n dimensions and $\mathbb{Z}^n$ be the sub-space formed by the points of $\mathbb{K}^n$ whose all coordinates are odd. Every partition π of $\mathbb{Z}^n$ induces in $\mathbb{K}^n$ a unique tessellation τ whose each class contains one class of π, and the correspondence between π and τ is biunivocal.*

Proof. When $\mathbb{Z}^n$ is embedded in $\mathbb{K}^n$, each class B_i of π becomes open, since all its coordinates are odd. Replace all B_i by their regularized versions $S_i = (\overline{B_i})^\circ$, and let z be a point of $\mathbb{K}^n$ which does not belongs to any class, i.e. $z \in [\cup S_i]^c$. Point z, which has one even coordinate at least, is the center of a unit cube which meets two classes S_i at least, which amounts to say that the $\overline{S_i}$ cover the space. Thus the family $\{S_i\}$ forms a tessellation of $\mathbb{K}^n$. Conversely, let τ be a tessellation of $\mathbb{K}^n$. Those points whose all coordinates are odd lie necessarily in some classes, as they are open, therefore the inverse passage to $\mathbb{Z}^n$ forms a partition.

This proposition theoretically justifies the classical rule of the double sampling [3,4,10]. The fine mesh displays the net of the contours, which are topologically closed in $\mathbb{R}^n$ as in $\mathbb{K}^n$, and which envelop the connected open sets. The odd points of $\mathbb{K}^n$, i.e. the unit open cubes play exactly the same role as the leaves in a hierarchy. But the topology of $\mathbb{K}^n$ does not remove the ambiguity of the diagonals, as shown in Fig. 1a. We must introduce another topology.

6   Tessellations of $\mathbb{Z}^2$ and $\mathbb{Z}^3$ by Voronoi polyhedra

Let $X \subseteq \mathbb{R}^n$ be a locally finite set of *centers*. One can always associate with each center $x \in X$ the so-called Voronoi polyhedron $Q(x)$ of all points $y \in \mathbb{R}^n$ closer to x than to any other center. $Q(x)$ is convex and open, hence regular, so that the set $\{Q(x), x \in X\}$ of all Voronoi polyhedra generates a tessellation of $\mathbb{R}^n$. In particular in $\mathbb{R}$, when the centers are the points m of odd integer abscissae m, the corresponding Voronoi polyhedron is $]m-1, m+1[$. We find again Khalimsky topology.

Let $\mathcal{H}^3$ represent the truncated octahedron. Let us come back to $\mathbb{R}^n$ and impose the following two conditions to the Voronoi polyhedra:

1. *they must be identical, up to a translation (i.e. regular grid);*
2. *the adherences of two adjacent polyhedra always have a common face of n-1 dimension.*

Concerning the first constraint, the mineralogist E.S. Fedorov proved that there are only two solutions in $\mathbb{R}^2$, the square and the hexagon, and five in $\mathbb{R}^3$, the cube, the hexagonal prism, the truncated octahedron, and the two elongated and rhombic dodecahedra [5].

The second condition imposes strong adjacencies, thus preserves connectivity, according to Theorem 3. This condition reduces Fedorov possibilities to the only hexagon in the plane, and only truncated octahedron in the space. In the 2-D case the centers describe the triangular grid, in the 3-D case the centered cubic grid [24] (see Fig. 5a). If we are not interested in preserving connectivity, this second condition becomes cumbersome.

6.1   Hexagonal Tessellation of $\mathbb{Z}^2$

In spite of its advantages the hexagonal grid is not often used (though it recently re-appeared about the simplicial complexes for digital watersheds [3]). Consider in $\mathbb{R}^2$ three axes of coordinates at 120°, and the origin O. Take for centers all points of the plane whose coordinates are odd on each of the three axes. The associated Voronoi polygons are open hexagons. The other open sets are obtained by unions of these hexagons plus the edges adjacent between them. They result in a sn-topology where the triple points are closed. In Fig. 4b the

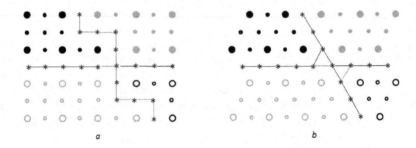

Fig. 4. Square versus hexagonal tessellation. The big discs are the elementary open sets; the unions of big and small discs form the regular open sets the small discs represent the segments and the points (square grid); and the contours with segments and asterisks delineate the background net. In the hexagonal case the quadruple point vanishes (thus its inherent ambiguity), but the background is no longer threadlike (small triangles).

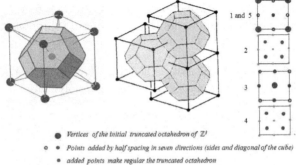

● Vertices of the initial truncated octahedron of $\mathbb{Z}^3$
○ ● Points added by half spacing in seven directions (sides and diagonal of the cube)
● added points make regular the truncated octahedron

Fig. 5. *Left*: Truncated octahedron and centered cubic grid. *Center*: Partition the space. *Right*: Top-down sections of the truncated octahedron. The big points are the open truncated octahedric cells of $\mathbb{Z}^3$, the small points and the rings are the points added to make $\mathbb{H}^3$. The regularization is obtained by union of big and small voxels.

previous squares and hexagons are replaced by the unique symbol of big discs, which indicate the pixels of $X \subseteq \mathbb{Z}^2$, and the small points are added to give the $(\overline{X})°$. The asterisks indicate the net of the frontiers of the tessellation in the double resolution. By comparison with Khalimsky tessellation in square grid (Fig. 4a) two major differences appear. On the one hand, the frontiers are no longer Jordan arcs (clusters of pixels in the hexagonal grid),and on the other hand the ambiguous diagonal were removed by suppression of the quadruples points. If Fig. 4b is interpreted as the leaves of a hierarchy, a unique value is allocated to each crossing point, namely that of the level when North-East and South-West classes merge.

6.2   Tessellation of $\mathbb{Z}^3$ by Truncated Octahedra

The Voronoi polyhedra of the centered cubic grid are the truncated octahedra depicted in Fig. 5. They partition $\mathbb{R}^3$ in open polyhedra, square and hexagonal faces, triple edges and quadruple vertices. These elements generate a sn-topology. The regularization fills up the internal 1-D or 2-D fissures of zero thickness, and the background net is a connected union of faces and edges which completely envelops the classes.

In the digital version, the double resolution rule for inducing a tessellation is indicated in Fig. 5. One starts from three horizontal planes of the cubic grid containing the vertices of the unit cube (n° 1 and 5 in the Fig.), and the center (n° 3). The planes n° 2 and 4 are added for generating a centered cubic grid twice finer. In the three directions of the cube and the four ones of the main diagonals alternate points of $\mathbb{Z}^3$ with those added for forming the double resolution space.

Again the tessellation reduces the cells to the two types of the (open) truncated octahedra, and the (closed) square or hexagonal faces, i.e. something that can be described in terms of graphs. If the first elements are displayed by points and the second by asterisks, like in Fig. 4 in 2-D, and if we connect the asterisks which share an edge, then each class turns out to be a cluster of points completely surrounded by a net of asterisks. In practice, the centered cubic grid can easily be emulated by shifting horizontally the even planes by the vector $(1, 1, 0)$.

7   Conclusion

In this paper, we characterized the partitions whose interiors and boundaries of the classes are jointly represented. They are the tessellations, and they require special types of classes (namely to be $\mathcal{R}$-open sets). No conditions like to be Jordan curves (in 2D) hold on the boundaries, which can have a certain thickness. The theory applies to both Euclidean and digital spaces, which gives more soundness to the numerical techniques. It was shown in detail how the notions of tessellation, Alexandrov topology, and double resolution interfere. For $\mathbb{Z}^3$ we proposed a variant of Khalimsky topology which suppresses the ambiguities of the diagonal configurations, it is based on the centerd-cubic grid.

Acknowledgment. The authors acknowledge G. Bertrand J. Cousty and T. Geraud for their useful comments, and the two reviewers for their pertinent remarks.

References

1. Bertrand, G.: Completions and simplicial complexes. In: Domenjoud, E., Kerautret, B., Even, P., Debled-Rennesson, I. (eds.) DGCI 2011. LNCS, vol. 6607, pp. 129–140. Springer, Heidelberg (2011)
2. Cousty, J., Bertrand, G.: Uniqueness of the perfect fusion grid on z^d. JMIV **34**(3), 291–306 (2009)
3. Cousty, J., Bertrand, G., Couprie, M., Najman, L.: Collapses and watersheds in pseudomanifolds of arbitrary dimension. JMIV **50**(3), 261–285 (2014)

4. Cousty, J., Najman, L., Serra, J.: Some morphological operators in graph spaces. In: Wilkinson, M.H.F., Roerdink, J.B.T.M. (eds.) ISMM 2009. LNCS, vol. 5720, pp. 149–160. Springer, Heidelberg (2009)

5. Fedorov, E.S.: A Course in Crystallography (In Russian). R. K. Rikker, Saint-Petersburg (1901)

6. Matheron, G.: Les treillis compacts. Technical report, Ecole des Mines, Paris (1996)

7. Heijmans, M.: Morphological Image Operators. Academic Press, Boston (1994)

8. Khalimsky, E.: Topological structures in computer science. Int. J. Stochast. Anal. **1**(1), 25–40 (1987)

9. Kiran, B.R., Serra, J.: Fusion of ground truths and hierarchies of segmentations. Pattern Recogn. Lett. **47**, 63–71 (2014)

10. Kiran, B.R.: Energetic-Lattice based optimization. Ph.D. thesis, Université Paris-Est (2014)

11. Kiselman, C.O.: Digital geometry and mathematical morphology. Lecture Notes, Uppsala University, Departement of Mathematics (2002)

12. Kovalevsky, V.A.: Finite topology as applied to image analysis. Comput. Vis. Graph. Image Process. **46**(2), 141–161 (1989)

13. Kronheimer, E.: The topology of digital images. Topol. Its Appl. **46**(3), 279–303 (1992)

14. Latecki, L., Eckhardt, U., Rosenfeld, A.: Well-composed sets. Comput. Vis. Image Underst. **61**(1), 70–83 (1995)

15. Matheron, G.: Random Sets and Integral Geometry. Wiley, New York (1975)

16. Matheron, G.: Eléments pour une Théorie des Milieux Poreux. Masson, Paris (1967)

17. Mazo, L., Passat, N., Couprie, M., Ronse, C.: Paths, homotopy and reduction in digital images. Acta Applicandae Math. **113**(2), 167–193 (2011)

18. Melin, E.: Digital surfaces and boundaries in khalimsky spaces. JMIV **28**(2), 169–177 (2007)

19. Najman, L., Schmitt, M.: Geodesic saliency of watershed contours and hierarchical segmentation. IEEE Trans. PAMI **18**(12), 1163–1173 (1996)

20. Alexandroff, P.: Diskrete raüme rec. math. [mat. sbornik] (1937)

21. Ronse, C.: Regular open or closed sets. Working Document WD59, Philips Research Lab., Brussels (1990)

22. Rosenfeld, A.: Digital topology. Am. Math. Mon. **86**, 621–630 (1979)

23. Serra, J.: Image Analysis and Mathematical Morphology. Theoretical Advances, vol. 2. Academic Press, London (1988)

24. Serra, J.: Cube, cube-octahedron or rhombododecahedron as bases for 3-d shape descriptions. In: Arcelli, C., et al. (eds.) Advances in Visual Form Analysis, pp. 502–519 (1997)

25. Serra, J.: Image Analysis and Mathematical Morphology. Academic Press Inc., Orlando (1983)

Geometric Transforms

nD Quasi-Affine Transformations

Marie-Andrée Jacob-Da Col and Loïc Mazo[(✉)]

ICube-UMR 7357, 300 Bd Sébastien Brant - CS10413,
67412 Illkirch Cedex, France
{dacolm,mazo}@unistra.fr

Abstract. A Quasi-Affine Transformation (QAT) is a transformation on $\mathbb{Z}^n$ which corresponds to the composition of a rational affine transformation and an integer part function. The aim of this paper is twofold. Firstly, it brings new insight into the periodic structures involved by a QAT. Secondly, some new results in nD are presented specifically about the behavior under iteration of a QAT.

1 Introduction

Affine transformations (scaling, rotating, etc.) are widely used in image analysis and processing, for instance to register images. Nevertheless, two issues have to be considered when one uses such a transformation in the digital world. Indeed digital images only embed a finite number of pixels (or voxels) and it is well known that for instance continuous rotations when applied on $\mathbb{Z}^n$ are generally not bijective [11,12], nor topology preserving [10] (*e.g.* holes can appear in a simply connected object). Moreover, very few affine transformations can be exactly calculated in computers. For instance, an 1/8th of a turn involves an irrational coefficient that cannot be represented by a floating number. In order to study the properties of the "affine" transformations embedded in computers, the first author of the present paper has proposed in [4] to model them by Quasi-Affine Transformations (QAT) which are actual affine transformations on $\mathbb{Q}^n$ followed by a floor function. We stress that this model slightly differs from the model used for instance by Nouvel *et al.* or Ngo *et al.* who consider Euclidean affine transformations (in $\mathbb{R}^n$) followed by some rounding function.

At this point, we need to introduce a few notations. Points of $\mathbb{R}^n$, mixed up with vectors, are denoted with a bold font and for any $c \in \mathbb{R}$, $\boldsymbol{c}$ denotes the vector whose coordinates are all equal to c (in particular, $\boldsymbol{0}$ is the null vector). Binary relations between vectors (*e.g.* $\boldsymbol{v} \leq \boldsymbol{w}$, or $\boldsymbol{v} \leq \boldsymbol{0}$), are to be understood coordinatewise. The floor function is denoted by $\lfloor \cdot \rfloor$ and, ω being a positive integer, we write $\lfloor x \rfloor_\omega$ for $\lfloor \frac{1}{\omega} x \rfloor$. If $g \colon \mathbb{Z}^n \to \mathbb{Z}^n$ is an affine transformation of the n-dimensional $\mathbb{Z}$-module, we note g_0 the linear part of g and v its constant part: $g = g_0 + v$. The *Quasi-Affine Transformation* (QAT) associated with g and ω is the transformation $\lfloor g \rfloor_\omega \colon \mathbb{Z}^n \to \mathbb{Z}^n$ defined by $\lfloor g \rfloor_\omega(\boldsymbol{x}) = \lfloor \frac{1}{\omega} g(\boldsymbol{x}) \rfloor$.

We can now explain how a QAT $\lfloor g \rfloor_\omega$ can be derived from an Euclidean affine transformation t. To fix ideas, let t be the rotation of $\pi/4$ radians such that $t(\boldsymbol{0}) = (1/3, 1/7)$ and $\boldsymbol{u}$ be the Euclidean vector with coordinates $(\pi, 7/3)$.

N. Normand et al. (Eds.): DGCI 2016, LNCS 9647, pp. 337–348, 2016.
DOI: 10.1007/978-3-319-32360-2_26

The first way to obtain a QAT from the Euclidean rotation t is to approximate the real coefficients of (the matrix of) t by rationals whose greatest common denominator is ω leading to a rational affine transformation $t' : \mathbb{Q}^n \to \mathbb{Q}^n$. Then the application of t' on the integer vectors $\lfloor x \rfloor$, $x \in \mathbb{R}^n$, followed by the floor function leads to the QAT $\lfloor g \rfloor_\omega : \mathbb{Z}^n \to \mathbb{Z}^n$ where $g = \omega t'$ is an affine transformation on $\mathbb{Z}^n$. For instance, one can set $\omega = 1000$ and $g(x) = g_0(x) + v$ where g_0 is defined by its matrix $\left(\begin{smallmatrix} 707 & -708 \\ 707 & 707 \end{smallmatrix} \right)$ and v is the vector $\left(\begin{smallmatrix} 333 \\ 142 \end{smallmatrix} \right)$. Then, $t(u)$ is computed as

$$\widehat{t(u)} = \lfloor g \rfloor_\omega (\lfloor u \rfloor) = \left\lfloor \tfrac{1}{1000} \left(\left(\begin{smallmatrix} 707 & -708 \\ 707 & 707 \end{smallmatrix} \right) \left(\begin{smallmatrix} 3 \\ 2 \end{smallmatrix} \right) + \left(\begin{smallmatrix} 333 \\ 142 \end{smallmatrix} \right) \right) \right\rfloor .$$

We have proposed in [1] another way to discretize an affine transformation in order to introduce the multigrid convergence scheme in the framework of QATs. Firstly, each real number is 'projected' on the integer line by a scaling operation followed by a rounding as in $\pi \to \lfloor \omega \pi \rfloor = 3141$ taking $\omega = 1000$ as the scaling factor[1]. Then, any multiplication must be followed by an integer division. Indeed, consider for instance the calculus $\sqrt{2} \times \pi + 6/7$ with the precision $\omega = 1000$. This calculation can be done by $\lfloor 1414 \times 3141 \rfloor_{1000} + 857 = 4441 + 857 = 5298$. Then, a "back-projection" on the real line gives the final result 5.298. In this setting, the Euclidean transformation t corresponds to the integer affine transformation $\lfloor g \rfloor_\omega : x \mapsto \lfloor g_0(x) \rfloor_\omega + v$, that is to the QAT $\lfloor g_0 + \omega v \rfloor_\omega$ and $t(u)$ is computed as

$$\widehat{t(u)} = \frac{1}{\omega} \left(\lfloor g \rfloor_\omega (\lfloor \omega u \rfloor) \right) = \tfrac{1}{1000} \left\lfloor \tfrac{1}{1000} \left(\begin{smallmatrix} 707 & -708 \\ 707 & 707 \end{smallmatrix} \right) \left(\begin{smallmatrix} 3141 \\ 2333 \end{smallmatrix} \right) + \left(\begin{smallmatrix} 333 \\ 142 \end{smallmatrix} \right) \right\rfloor .$$

However, here it doesn't matter what the discretization scheme is. Indeed, the present paper is only interested in the properties of the QATs by themselves. More precisely, it is devoted to the understanding of the periodic structure created by a QAT, which basically expresses the lack of bijectivity of the transformation, and on the behavior under iteration, which brings out the loss of precision. We present new results about these two topics in the general case and also give a new perspective on the links with the lattice group theory. The article is organized as follows. In Sect. 2, we study tilings generated by QATs: a tile is the set of the inverse images of a given point. Indeed, as explained above, a QAT derived from a bijective affine transformation in the Euclidean world will generally not be bijective and fibers can be empty, contain one or several points. Then, if we only focus on the difference between the affine transformation and the quasi-affine transformation, tiles yield a lattice structure in the discrete space. In 2D and 3D, explicit formulas have been proposed to compute a minimal basis of the QAT periodic structure [3,6]. In the n-dimensional case, an upper bound on the number of distinct tiles is given in [5]. We now give a closed formula for the period in any dimension. In Sect. 3, we study the behavior under iterations of a contracting QAT. If g is a contracting affine transformation of $\mathbb{R}^n$ then g has a unique fixed point and for each $x \in \mathbb{R}^n$ the sequence $g^n(X)$

---

[1] One can use two distinct scale factors for the digitization of the space and the quantification of the transformation.

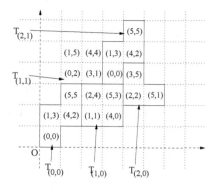

Fig. 1. Example of tiles and remainders. A point of $\mathbb{Z}^2$ is represented by a unit square whose bottom-left corner corresponds to the represented point. For each point in a tile we provide its corresponding remainder. Tiles $T_{(2,1)}$ and $T_{(0,0)}$ are arithmetically equivalent. The tiles $T_{(1,0)}$ and $T_{(1,1)}$ have the same shapes but they are not equivalent.

tends toward this fixed point. But the corresponding QAT has not necessarily a unique fixed point. The behavior under iteration of the 2D contracting QATs has been studied in [4, 8, 9]. In this paper, we study the nD case and prove that a QAT has a unique fixed point if and only if there are no cycle among the points of norm 1.

2   Periodicity of the Tiles

In order to obtain efficient computations for the affine transformation of discrete images, we can use some properties of QATs. In [7] the 2D case has been treated and extended to 3D in [3]. More precisely, we have seen in these papers that the periodicity properties of the tilings associated to QATs improve considerably the computation of the discrete affine transformations. This periodicity has been studied in the n-dimensional case in [2, 5]. Nevertheless, no closed formula for the tile period was given in these papers. We begin by recalling some definitions about QATs and their tiles.

A nonsingular QAT, that is a QAT derived from a bijective affine transformation, is generally not bijective and for a given QAT $\lfloor g \rfloor_\omega$ the preimages $\lfloor g \rfloor_\omega^{-1}(\boldsymbol{y})$ can have either none, one or several elements. As these preimages define a tessellation of the space $\mathbb{Z}^n$, we call them *tiles*. In the sequel we assume a QAT $\lfloor g \rfloor_\omega = \lfloor \frac{1}{\omega}(g_0 + v) \rfloor$ and we note $T_{\boldsymbol{y}}$ the tile of $\boldsymbol{y}$ for this QAT. The *remainder* of a tile $T_{\boldsymbol{y}}$, noted $r_{\boldsymbol{y}}$, is the set of the remainders modulo ω of the vectors $g(\boldsymbol{x})$, $\boldsymbol{x} \in T_{\boldsymbol{y}}$, that is

$$r_{\boldsymbol{y}} = \{g(\boldsymbol{x}) - \omega\,\boldsymbol{y} \mid x \in T_{\boldsymbol{y}}\}.$$

Observe that the sets $T_{\boldsymbol{y}}$ and $r_{\boldsymbol{y}}$ have the same cardinal since g is nonsingular. Two tiles $T_{\boldsymbol{y}}$ and $T_{\boldsymbol{y}'}$ are said to be *(arithmetically) equivalent* when their remainders are the same. We then write $T_y \equiv T_{y'}$. Two vectors y and y' of $\mathbb{Z}^n$ are *equivalent modulo* $\lfloor g \rfloor_\omega$ – we write $y \equiv y'$ – if the tiles T_y and $T_{y'}$ are equivalent. Eventually, we write $p \wedge q$, resp. $p \vee q$, for the gcd, resp. the lcm, of the integers p and q. Two equivalent tiles of a QAT $\lfloor g \rfloor_\omega$ share the same shape but the converse is false. Figure 1 exhibits some tiles of the QAT defined by the integer matrix $\begin{pmatrix} 3 & 1 \\ -1 & 3 \end{pmatrix}$, the vector $v = 0$ and the integer $\omega = 6$.

The next proposition and its corollary describes the periodic structures induced by the equivalence modulo $\lfloor g \rfloor_\omega$ and the corresponding tile equivalence.

Proposition 1. *Let $\lfloor g \rfloor_\omega$ be a non singular QAT. The equivalence modulo $\lfloor g \rfloor_\omega$ has a translational symmetry whose lattice group is $G = g_0(\frac{1}{\omega} \mathbb{Z}^n) \cap \mathbb{Z}^n$. In particular, two non empty tiles T_y and $T_{y'}$ are equivalent iff $y - y' \in \frac{1}{\omega} g_0(\mathbb{Z}^n)$. Moreover, the multi-valued function defined by $q \mapsto T_{\lfloor q \rfloor}$ induces a surjection from the quotient space $(\frac{1}{\omega} g(\mathbb{Z}^n) \cup \mathbb{Z}^n)/G$ onto the tile classe set $\{T_y\}_{y \in \mathbb{Z}^n}/ \equiv.$*

Proof. Let y, y' in $\mathbb{Z}^n$ such that $y' - y = \frac{1}{\omega} g_0(z)$ for some $z \in \mathbb{Z}^n$. We shall prove that the tiles T_y and $T_{y'}$ have the same remainders. Let $x \in T_y$, if such an integer vector exists. The remainder of $g(x)$ modulo ω is $r(x) = g(x) - \omega\, y$. We observe that $\frac{1}{\omega} g(x + z) - y' = \frac{1}{\omega} g(x) + \frac{1}{\omega} g_0(z) - y' = \frac{1}{\omega} g(x) - y = \frac{1}{\omega} r(x)$ by definition of z and $r(x)$. Thus, the integer vector $x' = x + z$ lies in $T_{y'}$ and the remainder of $g(x')$ modulo ω is equal to the remainder of $g(x)$ modulo ω. Thereby, we proved that either $r_y = \emptyset$ or $\emptyset \neq r_y \subseteq r_{y'}$. It is plain that, by the same reasoning, we can also prove $r_{y'} = \emptyset$ or $\emptyset \neq r_{y'} \subseteq r_y$. Hence, either $T_y = T_{y'} = \emptyset$ or $r_y = r_{y'}$, that is, in both cases, $T_y \equiv T_{y'}$.

Conversely, if two non empty tiles T_y and $T_{y'}$ are equivalent, there exist two vectors $x \in T_y$ and $x' \in T_{y'}$ such that $g(x) \equiv g(x') \pmod{\omega}$. Thus $g_0(x' - x) \equiv 0 \pmod{\omega}$ so $z = \frac{1}{\omega} g_0(x' - x) \in \mathbb{Z}^n$. Then, $\lfloor g(x') \rfloor_\omega = \lfloor \frac{1}{\omega}(g(x) + g_0(x' - x)) \rfloor = \lfloor \frac{1}{\omega} g(x) + z \rfloor = \lfloor \frac{1}{\omega} g(x) \rfloor + z = y + z$ where $z \in \frac{1}{\omega} g_0(\mathbb{Z}^n)$. We cannot assert that for two empty tiles T_y and $T_{y'}$ there exists $z \in \frac{1}{\omega} g_0(\mathbb{Z}^n)$ such that $y' = y + z$. Nevertheless, this is not necessary to obtain the result since the tiles cannot all be empty.

Incidentally, we have shown that, for any x, x' in $\mathbb{Z}^n$, if $g(x) \equiv g(x') \pmod{\omega}$, then $\lfloor g(x') \rfloor_\omega = \lfloor g(x) \rfloor_\omega + z$ where $z \in \frac{1}{\omega} g_0(\mathbb{Z}^n)$ which, from the first part of the proof, implies that $T_{\lfloor g(x') \rfloor_\omega} \equiv T_{\lfloor g(x) \rfloor_\omega}.$ $\square$

Note that, from a practical point of view, the last assertion of Proposition 1 means that one can prove that two tiles are equivalent by just exhibiting two integer points in these tiles whose images under $\frac{1}{\omega} g$ have the same fractional part.

Corollary 1. *Let $\lfloor g \rfloor_\omega$ be a non singular QAT. The tiles of $\lfloor g \rfloor_\omega$ have a translational symmetry whose lattice group is $g_0^{-1}(\omega \mathbb{Z}^n) \cap \mathbb{Z}^n$.*

Proof. Let $y, z \in \mathbb{Z}^n$ and t the translation of vector $g_0^{-1}(\omega z)$. Assume that the tiles T_y and T_{y+z} are equivalent. Then $r_y = r_{y+z}$ and

$$T_{y+z} = \{g^{-1}(\omega\,(y+z)+r) \mid r \in r_{y+z}\}$$
$$= \{g^{-1}(\omega\,(y+z)+r) \mid r \in r_y\}$$
$$= \{g^{-1}(\omega\,y+r)+g_0^{-1}(\omega z) \mid r \in r_y\}$$
$$= t(T_y).$$

In particular, $g_0^{-1}(\omega z)$ is an integer vector.

Conversely, let $y, u \in \mathbb{Z}^n$ and let t be the translation of vector u. Assume that there exists $z \in \mathbb{Z}^n$ such that $u = g_0^{-1}(\omega z)$. Since $z = \frac{1}{\omega}g_0(u)$, we derive from Proposition 1 that $T_{y+z} \equiv T_y$. Then, from the first part of the proof, $t(T_y) = T_{y+z}$. Thus, $t(T_y)$ is a tile and this tile is equivalent to T_y. □

In [5], the authors give upper and lower bounds for the number of tile equivalent classes. Thanks to the Smith normal form of integer matrices, we give hereafter the exact number of classes. We also provide the cardinal of any (maximal) union of representative tiles. Firstly, we give two lemmas that describe the role of unimodular transformations on the lattices associated to a QAT.

Lemma 1 ([5]). *Let u be an unimodular transformation. The QAT $\lfloor g \rfloor_\omega$ and $\lfloor g \circ u \rfloor_\omega$ have the same tile remainders. In particular, they share the same tile equivalence and the same empty tiles.*

Proof. We note T_y^u, resp. r_y^u, the tile of y, resp. the remainder of T_y^u, for the QAT $\lfloor g \circ u \rfloor_\omega$ while T_y and r_y are the tile of y, resp. the remainder of T_y, for the QAT $\lfloor g \rfloor_\omega$. We have $x \in T_y^u \iff u(x) \in T_y$, thus $T_y = u(T_y^u)$. Furthermore,

$$r_y^u = \{g \circ u(x) - \omega\,y \mid x \in T_y^u\} = \{g(z) - \omega\,y \mid z \in T_y\} = r_y\ . \qquad \square$$

Lemma 2. *Let u be an unimodular transformation. The cardinals of the fundamental domains of the equivalences modulo $\lfloor g \rfloor_\omega$ and modulo $\lfloor u \circ g \rfloor_\omega$ are equal.*

Proof. The lattice groups of the equivalences modulo $\lfloor g \rfloor_\omega$ and modulo $\lfloor u \circ g \rfloor_\omega$ are $L_1 = \mathbb{Z}^n \cap \frac{1}{\omega}g_0(\mathbb{Z}^n)$ and $L_2 = \mathbb{Z}^n \cap \frac{1}{\omega}u \circ g_0(\mathbb{Z}^n)$. Note that $L_2 = u(L_1)$ since $u(\mathbb{Z}^n) = \mathbb{Z}^n$ and u is bijective (u is unimodular). As $|\det(u)| = 1$, we get that the volumes of the parallelepipeds wrapping L_1 and L_2 are equal. Then, from Picks's theorem, we derive that the numbers of integer vectors in these domains are identical. □

The following lemma describes the lattice groups of QATs whose matrix is diagonal.

Lemma 3. *Let $\lfloor g \rfloor_\omega$ be a nonsingular QAT whose matrix $(d_{i,j})$ is diagonal. Then,*

- *the lattice group of the equivalence modulo $\lfloor g \rfloor_\omega$ is $\prod_{i=1}^n \left(\frac{d_{i,i}}{\omega \wedge d_{i,i}}\right)\mathbb{Z}$;*
- *the lattice group of the non empty tiles is $\prod_{i=1}^n \left(\frac{\omega}{\omega \wedge d_{i,i}}\right)\mathbb{Z}$.*

Proof

- From Proposition 1, the lattice group of the equivalence modulo $\lfloor g \rfloor_\omega$ is $\mathbb{Z}^n \cap g_0\left(\frac{1}{\omega}\mathbb{Z}^n\right) = \frac{1}{\omega}(\omega\,\mathbb{Z}^n \cap g_0(\mathbb{Z}^n))$ and $g_0(\mathbb{Z}^n) = \prod_{i=1}^n (d_{i,i}\,\mathbb{Z})$ for g_0 is diagonal. Then, the relation $\omega \vee d_{i,i} = (\omega\,d_{i,i})/(\omega \wedge d_{i,i})$ yields the first part of the result.
- From Corollary 1, the lattice group of the non empty tiles is $g_0^{-1}(\omega\,\mathbb{Z}^n) \cap \mathbb{Z}^n = \prod_{i=1}^n \frac{1}{d_{i,i}}(\omega \vee d_{i,i})\mathbb{Z} = \prod_{i=1}^n \frac{\omega}{\omega \wedge d_{i,i}}\mathbb{Z}$. □

Thanks to the Smith normal form of the affine transformations of $\mathbb{Z}^n$, we now derive from the three preceding lemmas the cardinal of the fundamental domains of the lattice structures involved in a QAT.

Theorem 1. *Let $\lfloor g \rfloor_\omega$ be a nonsingular QAT and D be a fundamental domain of the equivalence modulo $\lfloor g \rfloor_\omega$. Noting $s = (s_{i,j})$ the Smith normal form of the matrix of g_0 and δ its determinant, one has*

$$\#D = \frac{\delta}{\prod_{i=1}^n \omega \wedge s_{i,i}} \quad and \quad \sum_{y \in D} \#T_y = \frac{\omega^n}{\prod_{i=1}^n \omega \wedge s_{i,i}}.$$

Proof

- From Proposition 1, the cardinal of D only depends on g_0. Let $s = u \circ g_0 \circ v$ be the Smith normal form of g_0. From Lemmas 1 and 2, we derive that $\#D = \#D'$ where D' is the fundamental domain of the equivalence modulo $\lfloor s \rfloor_\omega$. Then Lemma 3 gives the first result.
- From Corollary 1, $\sum_{y \in D} \#T_y$ is equal to the cardinal of the fundamental domain of $g_0^{-1}(\omega\,\mathbb{Z}^n) \cap \mathbb{Z}^n$. Then, thanks to the properties of unimodular transformations, we have

$$\begin{aligned} g_0^{-1}(\omega\,\mathbb{Z}^n) \cap \mathbb{Z}^n &= \left(v \circ s^{-1} \circ u\right)(\omega\,\mathbb{Z}^n) \cap \mathbb{Z}^n \\ &= \left(v \circ s^{-1}\right)(\omega\,\mathbb{Z}^n) \cap \mathbb{Z}^n \\ &= v\left(s^{-1}(\omega\,\mathbb{Z}^n) \cap v^{-1}(\mathbb{Z}^n)\right) \\ &= v\left(s^{-1}(\omega\,\mathbb{Z}^n) \cap \mathbb{Z}^n\right). \end{aligned}$$

Hence, the lattice of the non empty tiles of g is the image by an unimodular transformation of the non empty tile lattice of s. We conclude straightforwardly by invoking the second part of Lemma 3. □

Example: Figure 2 illustrates a QAT in $\mathbb{Z}^2$ (g_0: $\left(\begin{smallmatrix} 12 & -11 \\ 18 & 36 \end{smallmatrix}\right)$, $v = \mathbf{0}$, $\omega = 84$) whose fundamental domain contains 15 tiles (the tiles with same color are equivalent). In this example, for all $i, j \in \mathbb{N}$, $T_{(i+5,j)} \equiv T_{(i,j)}$ and $T_{(i+2,j-3)} \equiv T_{(i,j)}$. The fundamental domain of this QAT is the set $\{T_{(i,j)} \mid i = 0, 1, 2, 3, 4,\ j = 0, 1, 2\}$. The Smith normal form of the QAT is $\left(\begin{smallmatrix} 1 & 0 \\ 0 & 630 \end{smallmatrix}\right)$. From Theorem 1, we derive that there is no empty tile ($\#D = 15$) and there are 168 integer points in $\bigcup_{0 \leq i \leq 4, 0 \leq j \leq 2} T_{(i,j)}$.

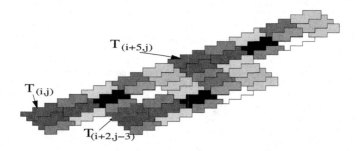

Fig. 2. Periodicity of thes tiles of a 2D QAT (see text)

In [3] we used the periodicity to improve the transformation of a 3D image by a linear transformation. In order to use this periodicity we need to determine a basis of the equivalence modulo $\lfloor g \rfloor_\omega$ lattice group such that, for any $m \leq n$, $(\boldsymbol{y}_i)_{i=1}^m$ is a basis of $\mathbb{Z}^m \times \{\boldsymbol{0}\}^{n-m}$.

Proposition 2. *Let $\lfloor g \rfloor_\omega$ be a non singular QAT. There exists a basis $(\boldsymbol{y}_i)_{i=1}^n$ of the equivalence modulo $\lfloor g \rfloor_\omega$ lattice group such that, for any $m \leq n$, $(\boldsymbol{y}_i)_{i=1}^m$ is a basis of $\mathbb{Z}^m$ (more specifically, $\mathbb{Z}^m \times \{\boldsymbol{0}\}^{n-m}$ that we identify with $\mathbb{Z}^m$). Each vector $\boldsymbol{y}_i$, $1 \leq i \leq n$, is defined by $\boldsymbol{y}_i = h_i(\boldsymbol{\lambda}_i)$ where h_i is the restriction to $\mathbb{Z}^i$ of the Hermite normal form of g_0 and $\boldsymbol{\lambda}_i = (\lambda_{i,1}, \dots, \lambda_{i,i})$ is the solution of $h_i(\boldsymbol{\lambda}_i) = \boldsymbol{0} \pmod{\omega}$ such that $\lambda_{i,i}$ is minimal.*

Proof. The modulo $\lfloor g \rfloor_\omega$ equivalence lattice group is $g_0 \left(\frac{1}{\omega} \mathbb{Z}^n \right) \cap \mathbb{Z}^n$. Let h be the Hermite normal form of g_0. As $h^{-1} \circ g_0$ is unimodular, the group $g_0(\mathbb{Z}^n)$ is generated by the column vectors of h. Let $\boldsymbol{c}_j = (h_{i,j})_{i=1}^n$ be the j-th column vector of h. Note that, since h is triangular, $\boldsymbol{c}_j \in \mathbb{Z}^j \times \{\boldsymbol{0}\}^{n-j}$. Then finding the basis $(\boldsymbol{y}_i)$ amounts to find the integer tuples $(\lambda_{i,1}, \dots, \lambda_{i,i})$ such that $\boldsymbol{y}_i = \sum_{j=1}^i \lambda_{i,j} \boldsymbol{c}_j \in \omega \mathbb{Z}^n$ and $\lambda_{i,i}$ is minimal. $\qquad\qquad\square$

Proposition 2 shows that we can compute the basis $(\boldsymbol{y}_i)_{i=1}^n$ by solving triangular systems of i linear equations in the modules $(\mathbb{Z}/\omega\mathbb{Z})^i$, $1 \leq i \leq n$, which can be done by Gaussian elimination in polynomial time.

3 Behavior Under Iteration of a Quasi-Linear Transformation

In this section we consider affine transformations g whose vector $\boldsymbol{v}$ is such that $\boldsymbol{0} \leq \boldsymbol{v} < \boldsymbol{\omega}$. Then, the associated QAT can be seen as the composition of a linear transformation with some rounding operator. For this reason, we say that such a QAT is a *quasi linear transformation* (QLT). Since we are interested in fixed points, or more generally in cycles, we restrict our study to non expansive transformations for the infinite norm, which is defined on a linear transformation f whose matrix is $A = (a_{i,j})$ by $\|f\|_\infty = \|A\|_\infty = \max_i \left(\sum_j |a_{i,j}| \right)$. Thus we

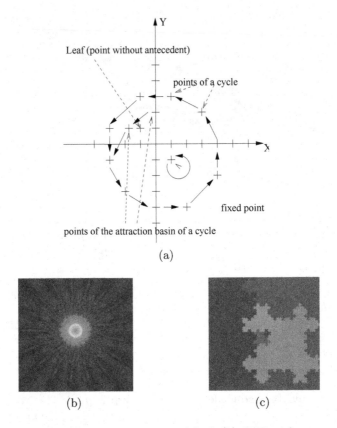

Fig. 3. (a) Example of cycle, fixed point and leaf. (b) QLT with many cycles. The points of a cycle and those that reach this cycle have the same colours. (c) QLT with a unique fixed point. The colour of a point is determined by the number of iterations necessary to reach the fixed point (Color figure online).

have for any vector $\boldsymbol{x}$, $\|f(\boldsymbol{x})\|_\infty \leq \|f\|_\infty \times \|\boldsymbol{x}\|_\infty$. It is well known that if f is a contracting linear transformation of $\mathbb{R}^n$ then f has the origin as unique fixed point and for each $\boldsymbol{x} \in \mathbb{R}^n$ the sequence $f^n(\boldsymbol{x})$ tends toward this fixed point. But a QLT built from a non expansive linear transformation has not necessarily a unique fixed point. Consider the sequence $\boldsymbol{y}_n = (\lfloor g \rfloor_\omega(\boldsymbol{y}_{n-1}))_{n \geq 0} = (\lfloor g \rfloor_\omega{}^n(\boldsymbol{y}_0))_{n \geq 0}$ with $\boldsymbol{y}_0 \in \mathbb{Z}^n$. Depending on $\boldsymbol{y}_0$ we can obtain a *cycle* ($\boldsymbol{y}_p = \boldsymbol{y}_q$ for some integer pair $p < q$), a *fixed point* ($\boldsymbol{y}_n = \boldsymbol{y}_{n+1}$ for some $n \in \mathbb{N}$), or *leaves* (points without antecedent). Figures 3a, b and c illustrate this behavior. In Fig. 3b the QLT has many cycles. It is defined by $\boldsymbol{y} = \frac{1}{4495}\left(\begin{smallmatrix} 4187 & -1622 \\ 1622 & 4187 \end{smallmatrix}\right)\boldsymbol{x}$. In Fig. 3c the QLT which is defined by $\boldsymbol{y} = \frac{1}{3}\left(\begin{smallmatrix} -1 & 1 \\ -1 & -1 \end{smallmatrix}\right)\boldsymbol{x}$ has a unique fixed point.

The first results concern the localisation of the cycles. Firstly, the following lemma gives a bound on the difference between a QLT and the corresponding linear transformation.

Lemma 4. *Let $\lfloor g \rfloor_\omega$ be a QLT, $d = \lfloor g \rfloor_\omega - \frac{1}{\omega} g_0$ and $s = \max(\frac{\|v\|_\infty}{\omega}, 1 - \frac{\|v\|_\infty}{\omega})$. Then, for any integer vector $\boldsymbol{x}$, $\|d(\boldsymbol{x})\|_\infty \leq s$, the inequality being strict if $s > \frac{1}{2}$.*

Proof. Consider $\boldsymbol{x} \in \mathbb{Z}^n$, $\boldsymbol{y} = \lfloor g \rfloor_\omega (\boldsymbol{x}) \in \mathbb{Z}^n$, $\boldsymbol{y}' = \frac{1}{\omega} g_0(\boldsymbol{x}) \in \mathbb{Q}^n$ and $d(\boldsymbol{x}) = \boldsymbol{y} - \boldsymbol{y}'$. We have

$$
\begin{aligned}
d(\boldsymbol{x}) &= \left\lfloor \frac{1}{\omega} (g_0(\boldsymbol{x}) + \boldsymbol{v}) \right\rfloor - \frac{1}{\omega} g_0(\boldsymbol{x}) \\
&= \frac{1}{\omega} (g_0(\boldsymbol{x}) + \boldsymbol{v}) - \boldsymbol{r} - \frac{1}{\omega} g_0(\boldsymbol{x}) \quad \text{with } 0 \le \boldsymbol{r} < 1 \\
&= \frac{1}{\omega} \boldsymbol{v} - \boldsymbol{r} \quad \text{with } 0 \le \boldsymbol{r} < 1
\end{aligned}
$$

It follows that $\frac{1}{\omega} \boldsymbol{v} - 1 < d(\boldsymbol{x}) \le \frac{1}{\omega} \boldsymbol{v}$. As we assume in the current section that $0 \le \boldsymbol{v} < \omega$, we get $\| d(\boldsymbol{x}) \|_\infty \le s$, the inequality being strict if $s = 1 - \frac{1}{\omega} \| \boldsymbol{v} \|_\infty$, that is to say if $\| \boldsymbol{v} \|_\infty < \frac{\omega}{2}$. $\qquad \square$

As a consequence of Lemma 4, we show that a QLT corresponding to a non expansive linear transformation is itself non expansive from the origin.

Corollary 2. *If* $\lfloor g \rfloor_\omega$ *is a QLT and* $\left\| \frac{1}{\omega} g_0 \right\|_\infty \le 1$, *then* $\| \lfloor g \rfloor_\omega (\boldsymbol{x}) \|_\infty \le \| \boldsymbol{x} \|_\infty$ *for any* $\boldsymbol{x} \in \mathbb{Z}^n$.

Proof. Let $\lfloor g \rfloor_\omega$ be a QLT such that $\left\| \frac{1}{\omega} g_0 \right\|_\infty \le 1$. With the notations of Lemma 4, we have:

$$
\begin{aligned}
\| \lfloor g \rfloor_\omega (\boldsymbol{x}) \|_\infty &= \left\| \frac{1}{\omega} g_0(\boldsymbol{x}) + d(\boldsymbol{x}) \right\|_\infty \\
&\le \left\| \frac{1}{\omega} g_0(\boldsymbol{x}) \right\|_\infty + \| d(\boldsymbol{x}) \|_\infty \\
&\le \frac{1}{\omega} \| g_0 \|_\infty \| \boldsymbol{x} \|_\infty + \| d(\boldsymbol{x}) \|_\infty \\
&\le \| \boldsymbol{x} \|_\infty + \| d(\boldsymbol{x}) \|_\infty \\
&< \| \boldsymbol{x} \|_\infty + 1 \quad \text{(from Lemma 4)} \\
&\le \| \boldsymbol{x} \|_\infty \quad \text{for } \| \lfloor g \rfloor_\omega (\boldsymbol{x}) \|_\infty \text{ and } \| \boldsymbol{x} \|_\infty \text{ are integers.} \qquad \square
\end{aligned}
$$

Remark 1. From Corollary 2, we derive that, if $\lfloor g \rfloor_\omega$ is a QLT, $\frac{1}{\omega} g_0$ is non expansive and $\boldsymbol{x}$ belongs to a cycle, then $\| \lfloor g \rfloor_\omega (\boldsymbol{x}) \|_\infty = \| \boldsymbol{x} \|_\infty$.

Thanks to Lemma 4, we can also prove that the cycles of non expansive QLTs are 'not too far' from the origin. Thereby, we extend to the general case a result that was obtained in [4] for the 2D space.

Theorem 2. *Consider a QLT* $\lfloor g \rfloor_\omega$ *such that* $\frac{1}{\omega} g_0$ *is contracting and note* $s = \max(\frac{\| \boldsymbol{v} \|_\infty}{\omega}, 1 - \frac{\| \boldsymbol{v} \|_\infty}{\omega})$. *If* $\boldsymbol{x}$ *belongs to a cycle then* $\| \boldsymbol{x} \|_\infty \le \frac{s}{1 - \| g_0 \|_\infty / \omega}$, *the inequality being strict if* $s > \frac{1}{2}$.

Proof. Let $g_1 = (1/\omega)\, g_0$. From Lemma 4, inductively we get

$$\lfloor g \rfloor_\omega^{\,k}(\boldsymbol{x}) = g_1^k(\boldsymbol{x}) + g_1^{k-1}(d_1(\boldsymbol{x})) + g_1^{k-2}(d_2(\boldsymbol{x})) + \cdots + g_1(d_{k-1}(\boldsymbol{x})) + d_k(\boldsymbol{x})$$

with $\|d_i(\boldsymbol{x})\|_\infty \le s$ for $i = 1, 2, \ldots, k$ each inequality being strict if $s > \frac{1}{2}$. It follows that

$$\left\| \lfloor g \rfloor_\omega^{\,k}(\boldsymbol{x}) \right\|_\infty \le \left\| g_1^k(\boldsymbol{x}) \right\|_\infty + \sum_{i=1}^k \left\| g_1^{k-i}(d_i(\boldsymbol{x})) \right\|_\infty$$

$$\le \|g_1\|_\infty^k \|\boldsymbol{x}\|_\infty + \sum_{i=1}^k \|g_1\|_\infty^{k-i} \|d_i(\boldsymbol{x})\|_\infty$$

$$\le \|g_1\|_\infty^k \|\boldsymbol{x}\|_\infty + s \sum_{i=1}^k \|g_1\|_\infty^{k-i}$$

$$\le \|g_1\|_\infty^k \|\boldsymbol{x}\|_\infty + s \frac{1 - \|g_1\|_\infty^k}{1 - \|g_1\|_\infty} \ .$$

If $\boldsymbol{x}$ belongs to a cycle, it exits $k \in \mathbb{N}$ such that $\lfloor g \rfloor_\omega^{\,k}(\boldsymbol{x}) = \boldsymbol{x}$ and so

$$\|\boldsymbol{x}\|_\infty (1 - \|g_1\|_\infty^k) \le s \frac{1 - \|g_1\|_\infty^k}{1 - \|g_1\|_\infty} \ .$$

Finally, $\|\boldsymbol{x}\|_\infty \le s/(1 - \|g_1\|_\infty)$, the inequality being strict if $s > \frac{1}{2}$.   □

As noted in Remark 1, the points of a cycle of a non expansive QLT share the same infinite norm. Then, the following lemma shows that, for contracting QLTs with null constant part, those points cannot lie in the first hyperoctant.

Lemma 5. *Let $\lfloor g \rfloor_\omega$ a QLT such that $\left\| \frac{1}{\omega} g_0 \right\|_\infty < 1$ and $\boldsymbol{v} = \boldsymbol{0}$. Consider $\boldsymbol{y} = \lfloor g \rfloor_\omega(\boldsymbol{x})$ where $\boldsymbol{x} \in \mathbb{Z}^n$. If $\|\boldsymbol{y}\|_\infty = \|\boldsymbol{x}\|_\infty$, then the coordinates of $\boldsymbol{y}$ with maximal absolute value are negative.*

Proof. From the hypotheses, we have

$$\left\| \tfrac{1}{\omega} g_0(\boldsymbol{x}) \right\|_\infty \le \left\| \tfrac{1}{\omega} g_0 \right\|_\infty \|\boldsymbol{x}\|_\infty < \|\boldsymbol{y}\|_\infty \ .$$

Thus, any positive coordinate of $\frac{1}{\omega} g_0(\boldsymbol{x})$ is less than $\|\boldsymbol{y}\|_\infty$ which prove the result.   □

Corollary 3 applies Theorem 2 to two common particular cases: $\|\boldsymbol{v}\|_\infty = \omega/2$ (rounding half up) and $\boldsymbol{v} = \boldsymbol{0}$ (rounding down). In this corollary and in the following theorem we denote by S_{n-1} the unit sphere $\{\boldsymbol{x} \mid \|\boldsymbol{x}\|_\infty = 1\}$.

Corollary 3. *Let $\lfloor g \rfloor_\omega$ be a QLT.*

- *If $\|\boldsymbol{v}\|_\infty = \frac{\omega}{2}$ and $\|g_0\|_\infty < \frac{3}{4}\omega$ then any cycle is included in the unit sphere S_{n-1}.*

– *If $v = 0$ and $\|g_0\|_\infty < \frac{1}{2}\omega$, then any cycle is included in the unit cube $\{-1,0\}^n$.*

Proof. The first assertion is a direct consequence of Theorem 2. The second assertion is a consequence of Lemma 5 and Theorem 2.   □

Eventually with the next theorem we show that not only the cycles of a non expansive QLT are not too far from the origin but moreover one of these cycles (if such a cycle exists) lies in the unit sphere S_{n-1}. Thereby, we solve a conjecture that was stated in [4] for the 2D case.

Theorem 3. *Any QLT $\lfloor g \rfloor_\omega$ derived from a non expansive linear transformation $(1/\omega)\, g_0$ has a cycle in the unit sphere S_{n-1} or has no cycle.*

In order to prove the theorem we need the following technical lemma.

Lemma 6. *Let $\lfloor g \rfloor_\omega$ be a QLT such that $\left\|\frac{1}{\omega} g_0\right\|_\infty \le 1$. Let $x \in \mathbb{Z}^n \setminus \{0\}$, $p \in S_{n-1}$ be such that $\|x - p\|_\infty$ is minimal. If $y = \lfloor g \rfloor_\omega(x)$ has the same norm as x, then $q = \lfloor g \rfloor_\omega(p)$ is in S_{n-1} and $\|y - q\|_\infty$ is minimal.*

Proof. As $p \in S_{n-1}$, Corollary 2 induces that $q \in S_{n-1} \cup \{0\}$. Moreover, again from Corollary 2, we derive that:

$$
\begin{aligned}
\|y - q\|_\infty &= \| \lfloor g(x) \rfloor_\omega - \lfloor g(p) \rfloor_\omega \|_\infty \\
&= \| \lfloor g_0(x - p) + g(p) \rfloor_\omega - \lfloor g(p) \rfloor_\omega \|_\infty \\
&= \| \lfloor g_0(x - p) + u \rfloor_\omega \|_\infty \quad \text{where } u = g(p) - \omega \lfloor g(p) \rfloor_\omega \\
&\le \|x - p\|_\infty \quad \text{for } x \mapsto g_0(x) + u \text{ is a non expansive QLT.}
\end{aligned}
$$

Since $\|x - p\|_\infty$ is minimal and $p \in S_{n-1}$, one has $\|x - p\|_\infty = \|x\|_\infty - 1 = \|y\|_\infty - 1$. Then,

$$
\|y - q\|_\infty < \|y\|_\infty .
$$

We conclude that $q \neq 0$ so that $q \in S_{n-1}$ and $\|y - q\|_\infty$ is minimal.   □

Proof (Theorem 3). Consider x a point of a cycle and a point p defined as in Lemma 6 and consider the sequences $x_i = \lfloor g \rfloor_\omega^i(x)$ and $p_i = \lfloor g \rfloor_\omega^i(p)$. For each $i \in \mathbb{N}$ the points x_i and p_i verify the condition of Lemma 6. The sequence p_i is then an infinite sequence of points in S_{n-1}. But it exits a finite number of points in S_{n-1} so there is a cycle among these points.   □

Remark 2. If $v = 0$ and $\|g_0\|_\infty < \omega$, from Lemma 5 and Theorem 3, we derive that there exists a cycle in the unit cube $\{-1,0\}^n$.

4   Conclusion

In this paper we completed and improved some theoretical results abouts QATs. We gave a closed formula for the period of the tiles generated by QATs and proved a conjecture on the localization of their cycles. We plan to use these results in a multigrid and multiprecision framework that we are developing in a joint work between Poitiers University and Strasbourg University [1]. The goal of this project is to build a formal theory in which the consequences of the change of universe between affine and quasi-affine transforms, that is between real numbers and integers, is precisely described so that one could choose a calculus precision and an image resolution to ensure a set of properties up to a tolerance threshold. For instance, at any precision and resolution, almost all quasi-affine rotations have non-singular tiles. Nevertheless, it would be desirable firstly to prove that though the size in pixels of the tiles grows with the precision, their 'true' sizes, taking into account the space resolution, tend toward zero and secondly to bound the convergence speed. Alike, we have to ensure for example that, despite the increasing complexity of cycles of quasi-affine rotations as the precision goes to infinity, they converge, together with fixed points, toward the center of the rotation when the pair precision - resolution tends toward infinity.

References

1. Andres, E., Da Col, M., Fuchs, L., Largeteau-Skapin, G., Magaud, N., Mazo, L., Zrour, R.: Les omega-aqa: Représentation discrète des applications affines. In: Passat, N. (ed.) GéoDis (2014)
2. Blot, V., Coeurjolly, D.: Quasi-affine transformation in higher dimension. In: Brlek, S., Reutenauer, C., Provençal, X. (eds.) DGCI 2009. LNCS, vol. 5810, pp. 493–504. Springer, Heidelberg (2009)
3. Coeurjolly, D., Blot, V., Jacob-Da Col, M.-A.: Quasi-affine transformation in 3-D: theory and algorithms. In: Wiederhold, P., Barneva, R.P. (eds.) IWCIA 2009. LNCS, vol. 5852, pp. 68–81. Springer, Heidelberg (2009)
4. Jacob, M.A.: Applications quasi-affines. Ph.D. thesis, Université Louis Pasteur, Strasbourg(1993)
5. Jacob-Da Col, M.A., Tellier, P.: Quasi-linear transformations and discrete tilings. Theor. Comput. Sci. **410**(2123), 2126–2134 (2009)
6. Jacob-Da Col, M.A.: Applications quasi-affines et pavages du plan discret. Theor. Comput. Sci. **259**, 245–269 (2001). Elsiever Science. https://dpt-info.u-strasbg.fr/~jacob
7. Jacob, M.-A., Reveilles, J.-P.: Gaussian numeration systems. In: 5ème colloque DGCI, pp. 37–47 (1995). available at jacob https://dpt-info.u-strasbg.fr/~jacob
8. Nehlig, P.: Applications Affines Discrètes et Antialiassage. Ph.D. thesis, Université Louis Pasteur, Strasbourg (1992)
9. Nehlig, P.: Applications quasi-affines: pavages par images réciproques. Theor. Comput. Sci. **156**, 1–38 (1995). Elsiever Science
10. Ngo, P., Passat, N., Kenmochi, Y., Talbot, H.: Topology-preserving rigid transformation of 2d digital images. IEEE Trans. Image Process. **23**(2), 885–897 (2014)
11. Nouvel, B., Rémila, E.: Configurations induced by discrete rotations: periodicity and quasi-periodicity properties. Discrete Appl. Math. **147**(2), 325–343 (2005)
12. Nouvel, B., Rémila, É.: Characterization of bijective discretized rotations. In: Klette, R., Žunić, J. (eds.) IWCIA 2004. LNCS, vol. 3322, pp. 248–259. Springer, Heidelberg (2004)

Minimal Paths by Sum of Distance Transforms

Robin Strand$^{(\boxtimes)}$

Centre for Image Analysis, Uppsala University, Uppsala, Sweden
Robin.Strand@it.uu.se

Abstract. Minimal path extraction is a frequently used tool in image processing with applications in for example centerline extraction and edge tracking. This paper presents results and methods for (i) extracting minimal paths and for (ii) utilizing local direction in the minimal path computation. Minimal path extraction is based on sum of distance transforms resulting in a stable method, without need for local decisions, which is needed for methods based on backtracking. Local direction utilization is discrete derivative based concepts such as the structure tensor, the Hessian and the Beltrami framework. The combination of minimal path extraction and local direction utilization gives a strong framework for minimal path extraction.

1   Introduction

Many image processing and analysis applications use distance transforms [1,3,23,25,30,31]. By taking the pixel intensity values into consideration, the homogeneous regions can be extracted [5,6,25,26]. Not only pixel intensity values, but also other high-level features, based on for example pointwise texture features or local direction, can be utilized. Minimal path is a frequently used tool for, e.g., centerline or boundary extraction. One approach for minimal path extraction is based on backtracking the distance transform values from a given point to the closest seed point by steepest descent [15,33]. A limitation with this approach is that it is based on local decisions, which, in general, does not give unique solutions. This manuscript presents an alternative approach based on sum of distance transforms for finding minimal paths. This approach has been used in mathematical morphology [27,28], level-set based distance computation [13] and for a specific family of geodesic distance functions [10]. Here, we show that this approach is valid for a family of distance functions, including for example the fuzzy distance [6,25,27] and the geodesic distance [6,32]. In Sect. 4, the approach presented in Sect. 3 is applied to minimal path formulations that utilize prior knowledge on the prefered direction at each point.

2   Notation

Given some cost function $C : \pi_{p,q} \mapsto \mathbb{R}$, where p, q are points in a digital image and $\pi_{p,q}$ is a path between p and q, the minimal path between two points p and q is $\arg\min_{\pi \in \Pi_{p,q}} C(\pi)$, where $\Pi_{p,q}$ is the set of all paths between p and q. The cost of the minimal cost path between p and q is $C_{\mathrm{opt}}(p, q) = \min_{\pi \in \Pi_{p,q}} C(\pi)$.

© Springer International Publishing Switzerland 2016
N. Normand et al. (Eds.): DGCI 2016, LNCS 9647, pp. 349–358, 2016.
DOI: 10.1007/978-3-319-32360-2_27

Definition 1. *A path* $\pi = \langle p = p_0, p_1, \ldots, p_n = q \rangle$ *from* p *to* q *is a minimal path if there is no other path between* p *and* q *with lower cost/distance.*

A function d such that, for all x, y, z in its domain is a metric if

1. $d(x, x) = 0$ (identity)
2. $d(x, y) > 0$ for all $x \neq y$ (positivity)
3. $d(x, y) = d(y, x)$ (symmetry)
4. $d(x, z) \leq d(x, y) + d(y, z)$ (triangle inequality)

A distance transform from a single source point p is denoted DT_p and

$$DT_p(q) = d(p, q).$$

Intensity-weighted distance functions can be used to model some intuitive, expected distance in the image, such that the obtained minimal paths are optimal on dark or bright regions in the image, or at strong edges in the image and higher-level features can be used to extract minimal paths on a centerline or boundary of an object. Distance transforms utilizing higher-order features can for example be obtained by pre-filtering the image by, e.g., an edge detector or a "vesselness" filter [19], before using the distance functions below. Other approaches, where the prior information is incorporated in the distance function, will be given in Sect. 4.

3   Minimal Paths as Minima in the Sum of Distance Transforms

Minimal cost paths can be computed by an assumption that the minimal cost path between two points p and q is in the set of minima in $DT_{pq}(x) = DT_p(x) + DT_q(x)$ [10,13,27,28].

The results in this section are valid for distance functions such that all metricity properties fulfilled, except that the strict positivity is replaced by non-negativity, and the following property holds.

Property 1. For any minimal path $\langle p = p_0, p_1, \ldots, p_n = q \rangle$, it holds that

$$d(p, q) = d(p, p_i) + d(p_i, q) \text{ for all } i : 0 \leq i \leq n.$$

Examples of distance functions that fulfill Property 1 are

– *The fuzzy distance* [6,25,27]. The fuzzy distance is the minimal cost path when the cost between two adjacent points is

$$c(p, q) = \frac{f(p) + f(q)}{2} \cdot \|p - q\|,$$

where $f(p)$ is the intensity value at pixel p.

– *The geodesic distance* [6,32]. The geodesic distance is the minimal cost path
when the cost between two adjacent points is

$$c(p, q) = \omega \, |f(p) - f(q)| + \|p - q\|.$$

The parameter ω affects trade-off between the fuzzy membership values and
the Euclidean distance.

In Theorem 1, we prove that the minimal path between two points is in the
set of minimum of sum distance transforms from the points. For the proof we will
use Lemmas 1 and 2, both related to the property that line segments between
two given points contain points of the shortest path between the points [12,20].

Lemma 1. *Let* $M = \{x : d(p, q) = d(p, x) + d(x, q)\}$ *and* $R = \{x : DT_{pq}(x) = \min_x DT_{pq}(x)\}$. *Then* $M = R$.

Proof. This is a direct consequence of the triangle inequality: For any point r,
$d(p, q) \leq d(p, r) + d(r, q)$. If $r \notin M$, then $DT_{pq}(r) = DT_p(r) + DT_q(r) = d(p, r) + d(r, q) > d(p, q) = \min_x DT_{pq}(x)$. If $r \in M$, then $DT_{pq}(r) = DT_p(r) + DT_q(r) = d(p, r) + d(r, q) = d(p, q) = \min_x DT_{pq}(x)$. □

Lemma 2. *If* $x \in M = \{x : d(p, q) = d(p, x) + d(x, q)\}$, *then* x *is in a minimal
path between two points* p *and* q.

Proof. If $d(p, q) = d(p, x) + d(x, q)$, then the path $\pi = \langle p = p_0, p_1, \ldots, p_m = x \rangle \cdot \langle x = q_0, q_1, \ldots, q_n = q \rangle$ is minimal, since its cost is $d(p, x) + d(x, q) = d(p, q)$,
which is optimal by the triangular inequality. □

Theorem 1. *Given a (pseudo)-metric such that Property 1 is fulfilled, the points
in minimal paths between* p *and* q *are contained in the set*

$$R = \left\{ x : DT_{pq}(x) = \min_x DT_{pq}(x) \right\} \tag{1}$$

where $DT_{pq}(x) = DT_p(x) + DT_q(x)$.

Proof. The theorem follows directly from Lemmas 1 and 2. □

Theorem 1 can be applied by Algorithm 1. Examples of realizations of
Algorithm 1 are shown in Fig. 1.

Algorithm 1. Shortest route algorithm [10].

Input: A grey-scale digital image $\mathcal{D}$, two point p, q.
Output: A set containing a minimal cost path between p and q.
Compute the distance image DT_p from source point p
Compute the distance image DT_q from source point q
Compute the route distance $DT_{pq}(x) = DT_p(x) + DT_q(x)$
Mark points such that $DT_{pq}(x) = \min_x DT_{pq}(x)$

---

4   Minimal Paths with Relative Position Prior

A common approach for extracting vessels and tubular structures is to use a so-called vesselness filter [4,7], which is based on the Hessian matrix, i.e., second derivatives of the intensity values in the image data.

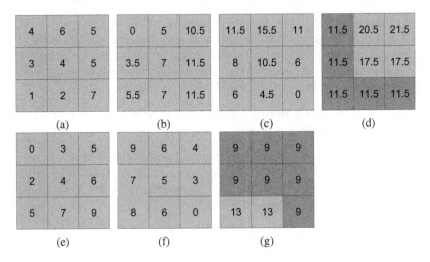

(a)   (b)   (c)   (d)

(e)   (f)   (g)

Fig. 1. Sets containing the minimal paths (in dark grey) between the upper left point to the lower right point obtained by Algorithm 1. (a): Input image with intensity values. (b–d): Fuzzy distance, (e–g): Geodesic distance with $\omega = 1$.

4.1   Minimal Paths by Prefiltering the Input Image

Given a window, or smoothing, function W, the structure tensor S is obtained by integrating the outer product of the gradient $\nabla f = [f_x f_y]$ with itself over a region given by a window function:

$$S = \nabla f \nabla f^T * W = \begin{bmatrix} f_x f_x & f_x f_y \\ f_y f_x & f_y f_y \end{bmatrix} * W,$$

where $*$ is convolution and W is some window function or a Gaussian. Note that the derivatives should be derived using an appropriate discrete operator [9,34]. The structure tensor, which captures information on the predominant directions of the gradient vector field in a specified neighborhood of a point, and its variations, has been widely used for enhancing local structure and direction [2,14,18]. Let λ_1 be the largest eigenvalue of S and λ_2 the smallest eigenvalue. Then

- The derivative is close to zero in all directions in constant areas ($|\lambda_1| \approx |\lambda_2| \approx 0$),
- there is a dominant direction with a strong gradient magnitude at edges ($|\lambda_1| \gg |\lambda_2| \approx 0$) and

– the derivative is strong also in the orthogonal directon at corners, so $|\lambda_1| \approx |\lambda_2| \gg 0$.

The structure tensor is based on the first derivative and is therefore not ideal for tubular structures since both f_x and f_y are close to zero at the center of a tubular structure. For tubular structures, the second derivative has properties that can be utilized since we can expect low intensity variation along the tubular structure and a local max/min in the perpendicular direction, i.e., a high second derivative.

To enhance tubular structures, the Hessian

$$H = \nabla^2 f = \begin{bmatrix} f_{xx} & f_{xy} \\ f_{yx} & f_{yy} \end{bmatrix},$$

which is based on second derivatives, is a good choice. Again, let λ_1 be the largest eigenvalue and λ_2 the smallest eigenvalue. Now,

– $\lambda_1 \approx \lambda_2 \approx 0$ in constant areas,
– $|\lambda_1| \gg |\lambda_2| \approx 0$ at tubular structures and
– $\lambda_1 \approx \lambda_2$ and $|\lambda_2| \gg 0$ for blob-like structures.

Based on this, different measures of "vesselness" and "blobness" have been proposed [8, 16, 17].

4.2   Minimal Paths by Prior Knowledge on Prefered Directions

The local "direction" in the image can be extracted in many ways, for example by the Hessian or structure tensor as in Sect. 4.1 and the prefered local direction can be based on measures on local features corresponding to, e.g., vessels, blobs, edges or corners. The resulting preferred local directions can be represented by a vector field, which, in turn can be used to guide the path cost function. In this section, we will explain how the local direction can be used for guiding the minimal path extraction by including information based on the gradient in the definition of the grey-weighted distance function.

Minimal Paths Guided by the Gradient

Metrics can be defined on the topographic representation of the image intensity function f, or rather, on the graph of f defined as $\varphi(x, y) \rightarrow (x, y, f(x, y))$. If the metric is a (pseudo-) Riemannian metric expressed as a metric tensor g on the graph of f, then the metric can be pulled back to the image plane by the Beltrami framework, [11]. The pullback metric is given by $J^T g J$, where J is the Jacobian of φ, i.e.,

$$J = \begin{bmatrix} 1 & 0 \\ 0 & 1 \\ f_x & f_y \end{bmatrix}.$$

Following the Beltrami framework, the distance (pullback metric) between two points p and $q = p + u$ is given by

$$c(p, q) = \sqrt{u^T B u},\tag{2}$$

where

$$B = J^T g J = \begin{bmatrix} 1 + f_x^2 & f_x f_y \\ f_y f_x & 1 + f_x^2 \end{bmatrix} = \begin{bmatrix} 1 & 0 \\ 0 & 1 \end{bmatrix} + \nabla f \nabla f^T.$$

If we set

$$g = \begin{bmatrix} 1 & 0 & 0 \\ 0 & 1 & 0 \\ 0 & 0 & 1 \end{bmatrix},$$

we get the geodesic path length on the surface. Similar to the geodesic distance, a topographic representation of the image is considered, where the intensity values correspond to the height.

Also similar to the geodesic distance, we can set a parameter that gives the trade-off between the intensity range and the spatial distance between x and y;

$$g = \begin{bmatrix} 1 & 0 & 0 \\ 0 & 1 & 0 \\ 0 & 0 & \alpha^2 \end{bmatrix},$$

corresponds to scaling of f with α, leading to the following following effect on B:

$$B = \begin{bmatrix} 1 & 0 \\ 0 & 1 \end{bmatrix} + \alpha^2 \nabla f \nabla f^T,$$

This idea has also been generalized by using the structure tensor instead of the outer product of gradients, i.e., exchanging B for

$$S = \begin{bmatrix} 1 & 0 \\ 0 & 1 \end{bmatrix} + \alpha^2 \nabla f \nabla f^T * G_\sigma,$$

where G_σ is a Gaussian function with standard deviation σ [18]. This is a generalization since the standard Beltrami pullback metric is achieved by setting $\sigma = 0$ and the generalization extends the Beltrami framework to penalize both image gradients (edges) and ridges (lines) by the structure tensor [18].

Note that this general framework allows other (pseudo-) metrics to be used and pulled back to the image plane. For example,

$$g = \begin{bmatrix} 0 & 0 & 0 \\ 0 & 0 & 0 \\ 0 & 0 & 1 \end{bmatrix}$$

gives a pseudo-metric that does not take spatial distance in the image plane into account. This idea has been used for edge detection in color images [24].

4.3   General Metric Tensor Minimal Path-Cost

By combining path costs given by the intensity values in the image, as given in Eq. 2, with the cost given by the local direction, a more general path cost framework is obtained by using a path cost functions on the form

$$c(p, p + u) = E\left(c_{\text{intensity}}(p, p + u), c_{\text{tensor}}(p, p + u)\right),$$

where E is an appropriate combination of a grey-weighted path cost function $c_{\text{intensity}}$ and a tensor-based path cost function c_{tensor} that takes local direction into account.

The tensor-based path cost function c_{tensor} is defined by a metric tensor $M(p)$ i.e.,

$$c_{\text{tensor}}(p, p + u) = \sqrt{u^T M(p) u}, \tag{3}$$

where $M(p)$ typically is restricted to symmetric and positive definite matrices. The positive definiteness is sometimes relaxed to non-negativity and sometimes $M(p)$ is restricted to diagonal matrices [22].

Example 1. Circular paths, for example, annual rings in log ends, can be extracted by utilizing the polar geometry of the paths [21,29]. By matrix diagonalization, we can define a tensor field where eigenvalues and eigenvectors are given by P and D:

$$M(p) = PDP^{-1}, \tag{4}$$

where

$$P = \begin{bmatrix} -\sin\theta & \cos\theta \\ \cos\theta & \sin\theta \end{bmatrix} \text{ and } D = \begin{bmatrix} 1/r^2 & 0 \\ 0 & 1 \end{bmatrix} \tag{5}$$

and r and θ are polar coordinate representation of p.

A path cost function based on multiplicative weighting of image intensity differences and spatial distance in polar coordinates can be defined by

$$c(p, p + v) = \frac{f(p) + f(p + v)}{2} \cdot \sqrt{v^T M(p) v},$$

with $M(p)$ as given in Eq. 4. An example of distance transforms and a minimal path using this path-cost function is shown in Fig. 2. By applying a standard Taylor expansion of $(p - (p + v))'PDP^{-1}(p - (p + v))$, P and D are as in Eq. 5 and

$$p = \begin{bmatrix} r\cos\theta \\ r\sin\theta \end{bmatrix} \text{ and } p + v = \begin{bmatrix} (r + \Delta r)\cos(\theta + \Delta\theta) \\ (r + \Delta r)\sin(\theta + \Delta\theta) \end{bmatrix},$$

we get

$$c(p, p + v) \approx \frac{f(p) + f(p + v)}{2} \cdot \sqrt{\Delta\theta^2 + \Delta r^2 + R_2(\Delta r, \Delta\theta)},$$

where $R_2(\Delta r, \Delta\theta)$ is the remainder term (a weighted sum of monomials of degrees higher than 2). This leads to a special case of the path cost function used in [21] for extracting circular paths:

$$c(p, p+v) = \frac{f(p) + f(p+v)}{2} \cdot \sqrt{(\omega_r \cdot \Delta r(p,v))^2 + (\omega_\theta \cdot \Delta\theta(p,v))^2},$$

where $\Delta r(p,v) = |r_p - r_{p+v}|$ and $\Delta\theta(p,v) = \min\left(|\theta_p - \theta_{p+v}|, 2\pi - |\theta_p - \theta_{p+v}|\right)$.

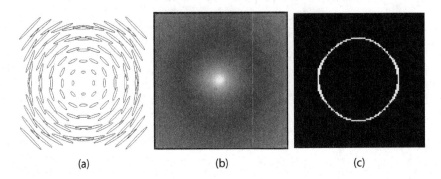

(a)   (b)   (c)

Fig. 2. Distance transform in polar coordinates on a constant intensity input image by the general metric tensor and minimal path extraction. (a): Visualization of the tensor field. (b): The sum of distance transforms from two source points on equal distance from the center of the image using polar weights. (c): The set containing the minimal path by Algorithm 1.

5   Conclusion

Results on minimal path extraction for path-based distance functions have been presented together with results on path cost functions that utilizes local direction. This combination gives a general and practically useful framework for minimal path extraction with many potential applications based on for example, centerline extraction and edge tracking.

References

1. Borgefors, G.: On digital distance transforms in three dimensions. Comput. Vis. Image Underst. **64**(3), 368–376 (1996)
2. Budde, M.D., Frank, J.A.: Examining brain microstructure using structure tensor analysis of histological sections. NeuroImage **63**(1), 1–10 (2012)
3. Danielsson, P.E.: Euclidean distance mapping. Comput. Graph. Image Process. **14**, 227–248 (1980)
4. Dufour, A., Tankyevych, O., Naegel, B., Talbot, H., Ronse, C., Baruthio, J., Dokládal, P., Passat, N.: Filtering and segmentation of 3D angiographic data: advances based on mathematical morphology. Med. Image Anal. **17**(2), 147–164 (2013)
5. Falcão, A.X., Stolfi, J., de Alencar Lotufo, R., Lotufo, R.A.: The image foresting transform: theory, algorithms, and applications. IEEE Trans. Pattern Anal. Mach. Intell. **26**(1), 19–29 (2004)

6. Fouard, C., Gedda, M.: An objective comparison between gray weighted distance transforms and weighted distance transforms on curved spaces. In: Kuba, A., Nyúl, L.G., Palágyi, K. (eds.) DGCI 2006. LNCS, vol. 4245, pp. 259–270. Springer, Heidelberg (2006)
7. Frangi, A., Niessen, W., Hoogeveen, R., Van Walsum, T., Viergever, M.: Model-based quantitation of 3-d magnetic resonance angiographic images. IEEE Trans. Med. Imaging **18**(10), 946–956 (1999)
8. Frangi, A.F., Niessen, W.J., Vincken, K.L., Viergever, M.A.: Multiscale vessel enhancement filtering. In: Wells, W.M., Colchester, A.C.F., Delp, S.L. (eds.) MICCAI 1998. LNCS, vol. 1496, pp. 130–137. Springer, Heidelberg (1998)
9. Grady, L.J., Polimeni, J.R.: Discrete Calculus, Applied Analysis on Graphs for Computational Science. Springer, London (2010)
10. Ikonen, L., Toivanen, P.: Shortest routes on varying height surfaces using grey-level distance transforms. Image Vis. Comput. **23**(2), 133–141 (2005)
11. Iwaniec, T., Martin, G.: The beltrami equation. Mem. Am. Math. Soc. **191**, 893 (2008)
12. Kim, C.E., Rosenfeld, A.: Digital straight lines and convexity of digital regions. IEEE Trans. Pattern Anal. Mach. Intell. **4**(2), 149–153 (1982)
13. Kimmel, R., Amir, A., Bruckstein, A.M.: Finding shortest paths on surfaces using level sets propagation. IEEE Trans. Pattern Anal. Mach. Intell. **17**(6), 635–640 (1995)
14. Knutsson, H.: Representing local structure using tensors. In: The 6th Scandinavian Conference on Image Analysis, pp. 244–251, Oulu, Finland, June 1989
15. Lengyel, J., Reichert, M., Donald, B.R., Greenberg, D.P.: Real-time robot motion planning using rasterizing computer graphics hardware, vol. 24. ACM (1990)
16. Lindeberg, T.: Edge detection and ridge detection with automatic scale selection. In: IEEE Proceedings of the Computer Vision and Pattern Recognition, CVPR 1996, pp. 465–470 (1996)
17. Lorenz, C., Carlsen, I.C., Buzug, T., Fassnacht, C., Weese, J.: Multi-scale line segmentation with automatic estimation of width, contrast and tangential direction in 2D and 3D medical images. In: Troccaz, J., Mösges, R., Grimson, W.E.L. (eds.) CVRMed-MRCAS 1997, CVRMed 1997, and MRCAS 1997. LNCS, vol. 1205, pp. 233–242. Springer, Heidelberg (1997)
18. Malm, P., Brun, A.: Closing curves with riemannian dilation: application to segmentation in automated cervical cancer screening. In: Bebis, G., Boyle, R., Parvin, B., Koracin, D., Kuno, Y., Wang, J., Wang, J.-X., Wang, J., Pajarola, R., Lindstrom, P., Hinkenjann, A., Encarnação, M.L., Silva, C.T., Coming, D. (eds.) ISVC 2009, Part I. LNCS, vol. 5875, pp. 337–346. Springer, Heidelberg (2009)
19. Manniesing, R., Viergever, M.A., Niessen, W.J.: Vessel enhancing diffusion: a scale space representation of vessel structures. Med. Image Anal. **10**(6), 815–825 (2006)
20. Nagy, B., Orosz, A.: Simple digital objects on Z^2. In: 7th International Conference on Applied Informatics, ICAI 2007, Eger, Hungary, vol. I, pp. 139–146 (2007)
21. Norell, K., Lindblad, J., Svensson, S.: Grey weighted polar distance transform for outlining circular and approximately circular objects. In: 14th International Conference on Image Analysis and Processing, pp. 647–652 (2007)
22. Ramanan, D., Baker, S.: Local distance functions: a taxonomy, new algorithms, and an evaluation. IEEE Trans. Pattern Anal. Mach. Intell. **33**(4), 794–806 (2011)
23. Rosenfeld, A., Pfaltz, J.L.: Distance functions on digital pictures. Pattern Recogn. **1**, 33–61 (1968)

24. Rousseau, S., Helbert, D., Carre, P., Blanc-Talon, J.: Metric tensor for multicomponent edge detection. In: 17th IEEE International Conference on Image Processing. pp. 1953–1956 (2010)
25. Saha, P.K., Wehrli, F.W., Gomberg, B.R.: Fuzzy distance transform: theory, algorithms, and applications. Comput. Vis. Image Underst. **86**, 171–190 (2002)
26. Sethian, J.A.: Level Set Methods and Fast Marching Methods. Cambridge University Press, Cambridge (1999)
27. Soille, P.: Generalized geodesy via geodesic time. Pattern Recogn. Lett. **15**(12), 1235–1240 (1994)
28. Soille, P.: Morphological image analysis: principles and applications. In: ter Haar Romeny, B.M. (ed.) Geometry-Driven Diffusion in Computer Vision, pp. 229–254. Springer, Heidelberg (2013)
29. Strand, R., Norell, K.: The polar distance transform by fast-marching. In: 19th International Conference on Pattern Recognition, pp. 1–4 (2008)
30. Strand, R., Ciesielski, K.C., Malmberg, F., Saha, P.K.: The minimum barrier distance. Comput. Vis. Image Underst. **117**(4), 429–437 (2013). special issue on Discrete Geometry for Computer Imagery
31. Strand, R., Nagy, B., Borgefors, G.: Digital distance functions on three-dimensional grids. Theor. Comput. Sci. **15**(412), 1350–1363 (2011)
32. Toivanen, P.: New geodesic distance transforms for grey-scale images. Pattern Recogn. Lett. **17**, 437–450 (1996)
33. Verbeek, P., Dorst, L., Verwer, B.J., Groen, F.C.: Collision avoidance and path finding through constrained distance transformation in robot state space. In: Proceedings of Intelligent Autonomous Systems, pp. 634–641. Elsevier, Amsterdam (1986)
34. Zenzo, S.D.: A note on the gradient of a multi-image. Comput. Vis. Graph. Image Process. **33**(1), 116–125 (1986)

Bijective Rigid Motions
of the 2D Cartesian Grid

Kacper Pluta[1,2(✉)], Pascal Romon[2], Yukiko Kenmochi[1], and Nicolas Passat[3]

[1] Université Paris-Est, LIGM, CNRS-ESIEE, Paris, France
{kacper.pluta,yukiko.kenmochi}@esiee.fr
[2] Université Paris-Est, LAMA, Paris, France
pascal.romon@u-pem.fr
[3] Université de Reims Champagne-Ardenne, CReSTIC, Reims, France
nicolas.passat@univ-reims.fr

Abstract. Rigid motions are fundamental operations in image processing. While they are bijective and isometric in $\mathbb{R}^2$, they lose these properties when digitized in $\mathbb{Z}^2$. To investigate these defects, we first extend a combinatorial model of the local behavior of rigid motions on $\mathbb{Z}^2$, initially proposed by Nouvel and Rémila for rotations on $\mathbb{Z}^2$. This allows us to study bijective rigid motions on $\mathbb{Z}^2$, and to propose two algorithms for verifying whether a given rigid motion restricted to a given finite subset of $\mathbb{Z}^2$ is bijective.

1 Introduction

Rigid motions (i.e., rotations, translations and their compositions) defined on $\mathbb{Z}^2$ are simple yet crucial operations in many image processing applications involving 2D data. One way to design rigid motions on $\mathbb{Z}^2$ is to combine continuous rigid motions defined on $\mathbb{R}^2$ with a digitization operator that maps the results back into $\mathbb{Z}^2$. However, the digitized rigid motion, though uniformly "close" to its continuous origin, often no longer satisfies the same properties. In particular, bijectivity is lost in general. In this context, it is useful to understand the combinatorial, geometrical and topological alterations associated with digitized rigid motions. More precisely, we observe the impact of rigid motions on the structure of $\mathbb{Z}^2$ at a local scale. Few efforts were already devoted to such topic, in particular for digitized rotations. Especially, pioneering works by Nouvel and Rémila [6] led to an approach for characterizing bijective digitized rotations [7], and more generally studying non-bijective ones [8].

Our contribution is threefold. We first show the usefulness of a combinatorial model of the local behavior of rigid motions on $\mathbb{Z}^2$, proposed initially for rotations on $\mathbb{Z}^2$ [6,8], called neighborhood motion maps. By using this model, we characterize *bijective* rigid motions on $\mathbb{Z}^2$, similarly to the characterization of bijective digitized rotations [7]. As such characterization is made locally, we also show that the local bijectivity of rigid motions on $\mathbb{Z}^2$, i.e. bijectivity of rigid motions of finite sets on $\mathbb{Z}^2$, can be verified by using neighborhood motion maps.

© Springer International Publishing Switzerland 2016
N. Normand et al. (Eds.): DGCI 2016, LNCS 9647, pp. 359–371, 2016.
DOI: 10.1007/978-3-319-32360-2_28

This article is organized as follows. In Sect. 2 we recall basic definitions and generalize to digitized rigid motions on $\mathbb{Z}^2$ the combinatorial model proposed previously by Nouvel and Rémila for rotations on $\mathbb{Z}^2$ [6,8]. Section 3 provides a characterization of bijective rigid motions on $\mathbb{Z}^2$; this characterization is an extension of the one proposed in [7]. In Sect. 4 we provide new algorithms for the verification whether a given rigid motion is bijective or not when restricted to a finite subset of $\mathbb{Z}^2$. Finally, in Sect. 5 we conclude this article and provide some perspectives.

2   Basic Notions

2.1   Digitized Rigid Motions

Rigid motions on $\mathbb{R}^2$ are bijective isometric maps defined as

$$\left| \begin{array}{l} \mathcal{U} : \mathbb{R}^2 \to \mathbb{R}^2 \\ \quad \mathbf{x} \; \mapsto \mathbf{R}\mathbf{x} + \mathbf{t} \end{array} \right. \tag{1}$$

where $\mathbf{t} = (t_x, t_y)^t \in \mathbb{R}^2$ is a translation vector and $\mathbf{R}$ is a rotation matrix with $\theta \in [0, 2\pi)$ its rotation angle. This leads to the representation of rigid motions by a triple of parameters $(\theta, t_x, t_y) \in [0, 2\pi) \times \mathbb{R}^2$.

According to Eq. (1), we generally have $\mathcal{U}(\mathbb{Z}^2) \not\subseteq \mathbb{Z}^2$. As a consequence, in order to define digitized rigid motions as maps from $\mathbb{Z}^2$ to $\mathbb{Z}^2$, we commonly apply rigid motions on $\mathbb{Z}^2$ as a part of $\mathbb{R}^2$, and then combine the real results with a digitization operator

$$\left| \begin{array}{l} \mathcal{D} : \mathbb{R}^2 \quad \to \mathbb{Z}^2 \\ \quad (x, y) \mapsto \left(\lfloor x + \tfrac{1}{2} \rfloor, \lfloor y + \tfrac{1}{2} \rfloor \right) \end{array} \right. \tag{2}$$

where $\lfloor z \rfloor$ denotes the largest integer not greater than z. Digitized rigid motions are then defined as $U = \mathcal{D} \circ \mathcal{U}_{|\mathbb{Z}^2}$. Due to the behavior of $\mathcal{D}$ that maps $\mathbb{R}^2$ onto $\mathbb{Z}^2$, digitized rigid motions are, most of the time, non-bijective. This leads us to define a notion of point status with respect to digitized rigid motions [5].

Definition 1. *Let* $\mathbf{y} \in \mathbb{Z}^2$ *be an integer point. The set of preimages of* $\mathbf{y}$ *with respect to* U *is defined as* $S_U(\mathbf{y}) = \{\mathbf{x} \in \mathbb{Z}^2 \mid U(\mathbf{x}) = \mathbf{y}\}$, *and* $\mathbf{y}$ *is referred to as a* s-point, *where* $s = |S_U(\mathbf{y})|$ *is called the* status *of* $\mathbf{y}$.

Remark 1. *In* $\mathbb{Z}^2$, $|S_U(\mathbf{y})| \in \{0, 1, 2\}$ *and only points* $\mathbf{p}$ *and* $\mathbf{q}$ *such that* $|\mathbf{p} - \mathbf{q}| = 1$ *can be preimages of a* 2-point [3].

The non-injective and non-surjective behaviors of a digitized rigid motion result in the existence of 2- and 0-points.

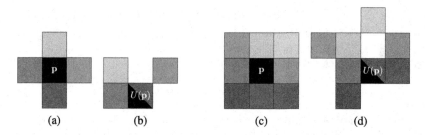

Fig. 1. Visual representation of neighborhood motion maps $\mathcal{G}_r^U(\mathbf{p})$ with colors: reference neighborhoods $\mathcal{N}_1(\mathbf{p})$ (a) and $\mathcal{N}_2(\mathbf{p})$ (c), and examples of $\mathcal{G}_1^U(\mathbf{p})$ (b) and $\mathcal{G}_2^U(\mathbf{p})$ (d) (Color figure online).

2.2 Neighborhood Motion Map

Let us consider the notion of neighborhood in $\mathbb{Z}^2$.

Definition 2. *The* neighborhood *of* $\mathbf{p} \in \mathbb{Z}^2$ *(of squared radius* $r \in \mathbb{R}_+$*), denoted* $\mathcal{N}_r(\mathbf{p})$, *is defined as* $\mathcal{N}_r(\mathbf{p}) = \{\mathbf{p} + \mathbf{d} \in \mathbb{Z}^2 \mid \|\mathbf{d}\|^2 \leq r\}$.

Remark 2. $\mathcal{N}_1$ *and* $\mathcal{N}_2$ correspond to the 4- and 8-neighborhoods, widely used in digital geometry [4].

In order to track local alterations of a neighborhood during rigid motions, we introduce the notion of a *neighborhood motion map*.

Definition 3. *Given a digitized rigid motion* U*, the* neighborhood motion map *of* $\mathbf{p} \in \mathbb{Z}^2$ *for* $r \in \mathbb{R}_+$ *is the function* $\mathcal{G}_r^U(\mathbf{p}) : \mathcal{N}_r(\mathbf{0}) \to \mathcal{N}_{r'}(\mathbf{0})$ *(with* $r' \geq r$*) defined as* $\mathcal{G}_r^U(\mathbf{p}) : \mathbf{d} \mapsto U(\mathbf{p} + \mathbf{d}) - U(\mathbf{p})$.

In other words, $\mathcal{G}_r^U(\mathbf{p})$ associates to each relative position of an integer point $\mathbf{p} + \mathbf{d}$ in the neighborhood of $\mathbf{p}$, the relative position of the image $U(\mathbf{p} + \mathbf{d})$ in the neighborhood of $U(\mathbf{p})$. The squared radius r' of $\mathcal{N}_{r'}(U(\mathbf{p}))$ is slightly larger than r. For instance, we have $r' = 2$ (resp. 5) for $r = 1$ (resp. 2). Figure 1 presents a visual presentation of $\mathcal{G}_r^U(\mathbf{p})$, where $r = 1, 2$. Note that a similar notion for digitized rotations was previously proposed by Nouvel and Rémila [6].

2.3 Remainder Range Partitioning

Digitized rigid motions $U = \mathcal{D} \circ \mathcal{U}$ are piecewise constant, which is a consequence of the nature of $\mathcal{D}$. In other words, the neighborhood motion map $\mathcal{G}_r^U(\mathbf{p})$ evolves non-continuously according to the parameters of $\mathcal{U}$ that underlies U. Our purpose is now to express how $\mathcal{G}_r^U(\mathbf{p})$ evolves.

Let us consider an integer point $\mathbf{p} + \mathbf{d}$ in the neighborhood $\mathcal{N}_r(\mathbf{p})$ of $\mathbf{p}$. From Eq. (1), we have

$$\mathcal{U}(\mathbf{p} + \mathbf{d}) = \mathbf{R}\mathbf{d} + \mathcal{U}(\mathbf{p}). \tag{3}$$

We know that $\mathcal{U}(\mathbf{p})$ lies in the unit square centered at the integer point $U(\mathbf{p})$, which implies that there exists a value $\rho(\mathbf{p}) = \mathcal{U}(\mathbf{p}) - U(\mathbf{p}) \in [-\frac{1}{2}, \frac{1}{2})^2$. The coordinates of $\rho(\mathbf{p})$, called the *remainder of* $\mathbf{p}$ *under* $\mathcal{U}$, are the fractional parts of the coordinates of $\mathcal{U}(\mathbf{p})$; and ρ is called the *remainder map under* U. As $\rho(\mathbf{p}) \in [-\frac{1}{2}, \frac{1}{2})^2$, this range, denoted by $\mathcal{P} = [-\frac{1}{2}, \frac{1}{2})^2$, is called the *remainder range*. Using ρ, we can rewrite Eq. (3) by

$$\mathcal{U}(\mathbf{p} + \mathbf{d}) = \mathbf{R}\mathbf{d} + \rho(\mathbf{p}) + U(\mathbf{p}).$$

Without loss of generality, we can consider that $U(\mathbf{p})$ is the origin of a local coordinate frame of the image space, i.e. $\mathcal{U}(\mathbf{p}) \in \mathcal{P}$. In such local coordinate frame, the former equation rewrites as

$$\mathcal{U}(\mathbf{p} + \mathbf{d}) = \mathbf{R}\mathbf{d} + \rho(\mathbf{p}). \tag{4}$$

Still under this assumption, studying the non-continuous evolution of the neighborhood motion map $\mathcal{G}_r^U(\mathbf{p})$ is equivalent to studying the behavior of $U(\mathbf{p} + \mathbf{d}) = \mathcal{D} \circ \mathcal{U}(\mathbf{p} + \mathbf{d})$ for $\mathbf{d} \in \mathcal{N}_r(\mathbf{0})$ and $\mathbf{p} \in \mathbb{Z}^2$, with respect to the rotation parameter θ defining $\mathbf{R}$ and the translation parameters embedded in $\rho(\mathbf{p})$, that deterministically depend on (t_x, t_y, θ). The discontinuities of $U(\mathbf{p} + \mathbf{d})$ occur when $\mathcal{U}(\mathbf{p} + \mathbf{d})$ is on the boundary of a digitization cell. Setting $\rho(\mathbf{p}) = (x, y)^t \in \mathcal{P}$ and $\mathbf{d} = (u, v)^t \in \mathcal{N}_r(\mathbf{0})$, this is formulated by one of the following two formulas

$$x + u \cos\theta - v \sin\theta = k_x + 1/2 \tag{5}$$
$$y + u \sin\theta + v \cos\theta = k_y + 1/2 \tag{6}$$

where $k_x, k_y \in \mathbb{Z}$. For a given $\mathbf{d} = (u, v)^t$ and k_x (resp. k_y), Eq. (5) (resp. (6)) defines a vertical (resp. horizontal) line in the remainder range $\mathcal{P}$, called a vertical (resp. horizontal) critical line. These critical lines with different $\mathbf{d}$, k_x and k_y, divide the remainder range $\mathcal{P}$ into rectangular regions called *frames*. Note that for $r = 1$ (resp. $r = 2$) there are 4 (resp. 8) vertical and horizontal critical lines, respectively. As long as coordinates of $\rho(\mathbf{p})$ belong to a same frame, the associated neighborhood motion map $\mathcal{G}_r^U(\mathbf{p})$ remains constant.

Proposition 4. *For any* $\mathbf{p}, \mathbf{q} \in \mathbb{Z}^2$, $\mathcal{G}_r^U(\mathbf{p}) = \mathcal{G}_r^U(\mathbf{q})$ *iff* $\rho(\mathbf{p})$ *and* $\rho(\mathbf{q})$ *are in the same frame.*

A similar proposition was already shown in [6] for the case $r = 1$ and digitized rotations. The above result is then an extension for general cases, such that $r \geq 1$ and digitized rigid motions. An example of the remainder range partitioning is presented in Fig. 2.

Remark 3. Equations (5) and (6) of critical lines are similar to those for digitized rotations, since the translation part is embedded only in $\rho(\mathbf{p}) = (x, y)^t$, as seen in Eq. (4).

2.4   Non-surjective and Non-injective Frames

Some frames correspond to neighborhood motion maps that exhibit 0- or 2-points, implying non-surjectivity or non-injectivity [7].

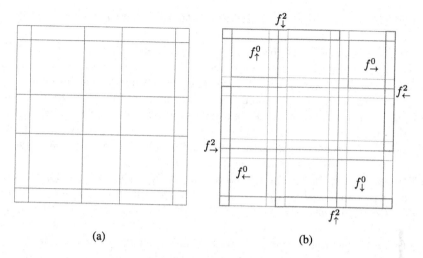

$$f_\downarrow^2$$
$$f_\uparrow^0$$
$$f_\to^0$$
$$f_\leftarrow^2$$
$$f_\to^2$$
$$f_\leftarrow^0$$
$$f_\downarrow^0$$
$$f_\uparrow^2$$

(a)   (b)

Fig. 2. Examples of remainder range partitioning for: $r = 1$ (a), and $r = 2$ (b). Non-injective zones f_*^2 and non-surjective zones f_*^0 are illustrated by red and brown rectangles, respectively (Color figure online).

Lemma 5. $U(\mathbf{p}) + \mathbf{d}_*$ *is a* 0-*point if and only if* $\rho(\mathbf{p})$ *is in one of the zones* f_*^0 *(union of frames themselves) defined as follows:*

$$f_\uparrow^0 = (1/2 - \cos\theta, \sin\theta - 1/2) \times (3/2 - \cos\theta - \sin\theta, 1/2),$$
$$f_\to^0 = (3/2 - \cos\theta - \sin\theta, 1/2) \times (1/2 - \sin\theta, \cos\theta - 1/2),$$
$$f_\downarrow^0 = (1/2 - \sin\theta, \cos\theta - 1/2) \times (-1/2, \cos\theta + \sin\theta - 3/2),$$
$$f_\leftarrow^0 = (-1/2, \cos\theta + \sin\theta - 3/2) \times (1/2 - \cos\theta, \sin\theta - 1/2),$$

where $* \in \{\uparrow, \to, \downarrow, \leftarrow\}$ *and* $\mathbf{d}_\uparrow = (0,1)^t, \mathbf{d}_\to = (1,0)^t, \mathbf{d}_\downarrow = (0,-1)^t,$ $\mathbf{d}_\leftarrow = (-1,0)^t.$

Lemma 6. $U(\mathbf{p})$ *is a* 2-*point whose preimages are* $\mathbf{p}$ *and* $\mathbf{p} + \mathbf{d}_*$ *if and only if* $\rho(\mathbf{p})$ *is in one of the zones* f_*^2 *defined as follows:*

$$f_\uparrow^2 = (\sin\theta - 1/2, 1/2) \times (-1/2, 1/2 - \cos\theta),$$
$$f_\to^2 = (-1/2, 1/2 - \cos\theta) \times (-1/2, 1/2 - \sin\theta),$$
$$f_\downarrow^2 = (-1/2, 1/2 - \sin\theta) \times (\cos\theta - 1/2, 1/2),$$
$$f_\leftarrow^2 = (\cos\theta - 1/2, 1/2) \times (\sin\theta - 1/2, 1/2).$$

We can characterize the non-surjectivity and non-injectivity of a digitized rigid motion by the presence of $\rho(\mathbf{p})$ in these specific zones. Both types of the zones are presented in Fig. 2.

3   Globally Bijective Digitized Rigid Motions

A digitized rigid motion is bijective if and only if there is no $\rho(\mathbf{p})$, for all $\mathbf{p} \in \mathbb{Z}^2$, in non-surjective and non-injective zones of $\mathscr{P}$. In this section, we characterize bijective rigid motions on $\mathbb{Z}^2$ while investigating such local conditions.

Let us start with the rotational part of the motion. We know from [7] that rotations with any angle of irrational sine or cosine are non-bijective; indeed such rotations have a dense image by ρ (there exists $\mathbf{p} \in \mathbb{Z}^2$ such that $\rho(\mathbf{p})$ lies in a non-surjective or/and non-injective zone of $\mathscr{P}$). This result is also applied to U, whatever translation part is added.

Therefore, we focus on rigid motions for which both cosine and sine of the angle θ are rational. Such angles are called *Pythagorean angles* [7] and are defined by *primitive Pythagorean triples* $(a, b, c) = (p^2 - q^2, 2pq, p^2 + q^2)$ with $p, q \in \mathbb{Z}, p > q$ and $p - q$ is odd, such that (a, b, c) are pairwise coprime, and cosine and sine of such angles are $\frac{a}{c}$ and $\frac{b}{c}$, respectively. The image of $\mathbb{Z}^2$ by ρ when U is a digitized rational rotation corresponds to a cyclic group $\mathcal{G}$ on the remainder range $\mathscr{P}$, which is generated by $\boldsymbol{\psi} = \left(\frac{p}{c}, \frac{q}{c}\right)^t$ and $\boldsymbol{\omega} = \left(-\frac{q}{c}, \frac{p}{c}\right)^t$ and whose order is equal to $c = p^2 + q^2$ [7]. When U contains a translation part, the image of ρ in $\mathscr{P}$, which we will denote by $\mathcal{G}'$, is obtained by translating $\mathcal{G}$ (modulo $\mathbb{Z}^2$) and $|\mathcal{G}'|$ is equal to the order of $\mathcal{G}$, its underlying group. Note also that a digitized rational rotation is bijective (the intersection of $\mathcal{G}$ with non-injective and non-surjective regions is empty) iff its angle comes from a twin Pythagorean triple—a primitive Pythagorean triple with the additional condition $p = q + 1$—see Nouvel and Rémila [7] and, more recently, Roussillon and Cœurjolly [9].

Our question is then whether a digitized rigid motion can be bijective, even when the corresponding rotation is not. In order to answer this question, we use the following equivalence property: digitized rational rotations are bijective if they are surjective *or* injective [7]. Indeed, this allows us to focus only on non-surjective zones.

Proposition 7. *A digitized rigid motion whose rotational part is given by a non-twin Pythagorean primitive triple is always non-surjective.*

Proof. We show that no translation factor can prevent the existence of an element of $\mathcal{G}'$ in a non-surjective zone. We consider the length of a side of f_*^0, given by $L_1 = \frac{2q(p-q)}{c}$, and the side of the bounding box of a fundamental square in $\mathcal{G}$, given by $L_2 = \frac{p+q}{c}$. Note that any non-surjective zone f_*^0 also forms a square. As $p > q + 1$, $L_2 < L_1$, and thus $\mathcal{G}' \cap f_*^0 \neq \varnothing$ (see Fig. 3(a)). $\square$

If, on the contrary, the rotational part is given by a twin Pythagorean triple, i.e. is bijective, the rigid motion is also bijective, under the following condition.

Proposition 8. *A digitized rigid motion is bijective if and only if it is composed of a rotation by an angle defined by a twin Pythagorean triple and a translation* $\mathbf{t} = \mathbf{t}' + \mathbb{Z}\boldsymbol{\psi} + \mathbb{Z}\boldsymbol{\omega}$, *where* $\mathbf{t}' \in \left(-\frac{1}{2c}, \frac{1}{2c}\right)^2$.

Proof. Let us first consider the case $\mathbf{t} = \mathbf{0}$. Since $L_2 > L_1$, there exists a fundamental square in $\mathcal{G}$, i.e. whose vertices are $(n\boldsymbol{\omega}+m\boldsymbol{\psi})$, $((n+1)\boldsymbol{\omega}+m\boldsymbol{\psi})$, $((n+1)\boldsymbol{\omega}+(m+1)\boldsymbol{\psi})$, $(n\boldsymbol{\psi}+(m+1)\boldsymbol{\psi})$, where $n, m \in \mathbb{Z}$, and the vertices lie outside of $f_{\downarrow}^0$, at N_∞ distance $1/2c$ (see Fig. 3(b)). Now let us consider the case $\mathbf{t} \neq \mathbf{0}$. The above four vertices are the elements of $\mathcal{G}$ closest to $f_{\downarrow}^0$, therefore if $N_\infty(\{\mathbf{t}\}) < 1/2c$, where $\{.\}$ stands for the fractional part function, then $\mathcal{G}' \cap f_{\downarrow}^0 = \varnothing$. Moreover, if $N_\infty(\{\mathbf{t}\})$ is slightly above $1/2c$, then it is plain that some point of $\mathcal{G}'$ will enter the frame $f_{\downarrow}^0$. But $\mathcal{G}$ is periodic with periods $\boldsymbol{\omega}$ and $\boldsymbol{\psi}$, so that the set of admissible vectors $\mathbf{t}$ has the same periods. Then we see that the admissible vectors form a square (i.e. a N_∞ ball of radius $1/2c$) modulo $\mathbb{Z}\boldsymbol{\psi}+\mathbb{Z}\boldsymbol{\omega}$ (see Fig. 3(c)). $\square$

4   Locally Bijective Digitized Rigid Motions

As seen above, the bijective digitized rigid motions, though numerous, are not dense in the set of all digitized rigid motions. We may thus generally expect defects such as 2-points. However, in practical applications, the bijectivity of a given U on the whole $\mathbb{Z}^2$ is not the main issue; rather, one usually works on a finite subset of the plane (e.g., a rectangular digital image). The relevant question is then: "given a finite subset $S \subset \mathbb{Z}^2$, is U restricted to S bijective?". Actually, the notion of bijectivity in this question can be replaced by the notion of injectivity since the surjectivity is trivial, due to the definition of U that maps S to $U(S)$.

The basic idea for such local bijectivity verification is quite natural. Because of its quasi-isometric property, a digitized rigid motion U can send at most two 4-neighbors to a same point (Remark 1). Thus, the lack of injectivity is a purely local matter, suitably handled by the neighborhood motion maps via the remainder map. Indeed, in accordance with Lemma 6, U is non-injective, with respect to S iff there exists $\mathbf{p} \in S$ such that $\rho(\mathbf{p})$ lies in the union $\mathcal{F} = f_{\downarrow}^2 \cup f_{\uparrow}^2 \cup f_{\leftarrow}^2 \cup f_{\rightarrow}^2$ of all non-injective zones. We propose two algorithms making use of the remainder map information, as an alternative to a brute force verification.

The first—*forward*—algorithm, verifies for each point $\mathbf{p} \in S$ the inclusion of $\rho(\mathbf{p})$ in one of the non-injective regions of $\mathcal{F}$. The second—*backward*—algorithm first finds all points $\mathbf{w}$ in $\mathcal{G}' \cap \mathcal{F}$, called the *non-injective remainder set*, and then verifies if their preimages $\rho^{-1}(\mathbf{w})$ are in S.

Both algorithms apply to rational motions, i.e., with a Pythagorean angle given by a primitive Pythagorean triple $(a, b, c) = (p^2 - q^2, 2pq, p^2 + q^2)$ and a rational translation vector $\mathbf{t} = (t_x, t_y)^t$. We capture essentially the behavior for all angles and translation vectors, since rational motions are dense. Methods for angle approximation by Pythagorean triples up to a given precision may be found in [1]. These assumptions guarantee the exact computations of the algorithms, which are based on integer numbers.

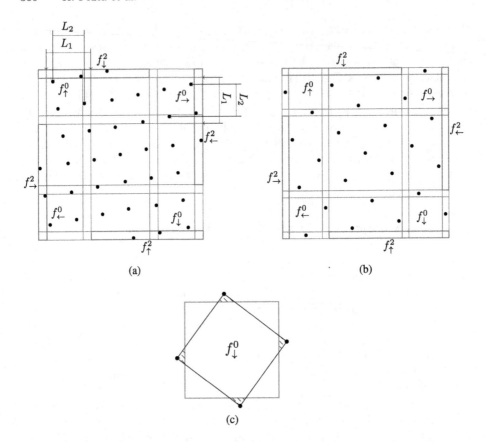

(a)   (b)

(c)

Fig. 3. Examples of remainder range partitioning together with $\mathcal{G}$ obtained for rotations by Pythagorean angles. (a) The non-bijective digitized rotation defined by the primitive Pythagorean triple $(12, 35, 37)$ and (b) the bijective digitized rotation defined by the twin Pythagorean triple $(7, 24, 25)$. The non-surjective and non-injective zones are illustrated by brown and red rectangles, respectively. (c) A fundamental square in $\mathcal{G}$ whose vertices are $(n\boldsymbol{\omega} + m\boldsymbol{\psi})$, $((n+1)\boldsymbol{\omega} + m\boldsymbol{\psi})$, $((n+1)\boldsymbol{\omega} + (m+1)\boldsymbol{\psi})$, $(n\boldsymbol{\psi} + (m+1)\boldsymbol{\psi})$, represented by black circles, and $f_\downarrow^0$ in brown. The union of the areas filled with a line pattern form a square (i.e. a $\mathcal{N}_\infty$ ball of radius $\frac{1}{2c}$) of the admissible translation vectors modulo $\mathbb{Z}\boldsymbol{\psi} + \mathbb{Z}\boldsymbol{\omega}$ (Color figure online).

4.1   Forward Algorithm

The strategy consists of checking whether the remainder map $\rho(\mathbf{p})$ of each $\mathbf{p} \in S$ belongs to one of non-injective zones f_*^2 defined in Lemma 6; if this is the case, we check additionally whether $\mathbf{p} + \mathbf{d}_* \in \mathcal{N}_1(\mathbf{p})$ belongs to S; otherwise, there is no 2-point with $\mathbf{p}$ under $U_{|S}$.

This leads to the forward algorithm, which returns the set B of all pairs of points having the same image. We can then conclude that $U_{|S}$ is bijective iff $B = \varnothing$; in other words, U is injective on $S \backslash B$. The break statement on line

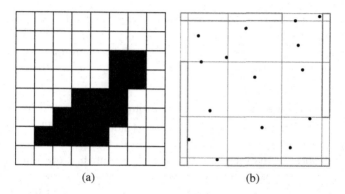

(a)   (b)

Fig. 4. (a) An initial finite set $S \subset \mathbb{Z}^2$, colored in black. (b) The remainder map image of S, i.e. $\rho(\mathbf{p})$ for all $\mathbf{p} \in S$, under U— given by parameters $\arccos \frac{6}{37}, 0, \frac{2}{25}$. Since no point $\rho(\mathbf{p})$ lies within the non-injective zone $\mathcal{F}$, we have a *visual* proof that U restricted to S is injective (Color figure online).

comes from the fact that there is no 3-point. Also from Remark 1, we restrict the internal loop to the set $\{\rightarrow, \downarrow\}$.

The main advantage of the forward algorithm lies in its simplicity. In particular, we can directly check which neighbor $\mathbf{p} + \mathbf{d}_*$ of $\mathbf{p}$ constitutes a 2-point with $\mathbf{p}$. Because rational rigid motions are exactly represented by integers, it can be verified without numerical error and in a constant time if $\rho(\mathbf{p}) \in \mathcal{F}$. The time complexity of this algorithm is $\mathcal{O}(|S|)$. Figure 4 illustrates the forward algorithm.

Remark 4. The forward algorithm can be used with non-rational rotations, at the cost of a numerical error.

4.2   Backward Algorithm

In this section, we consider a rectangular finite set S as the input; this setting is not abnormal as we can find a bounding box for any finite set. The strategy

---

Forward algorithm. A point-wise injectivity verification of $U_{|S}$.

Data: A finite set $S \subset \mathbb{Z}^2$; a digitized rigid motion U.
Result: The subset $B \subseteq S$ whose points are not injective under U.

1 $B \leftarrow \varnothing$
2 **foreach** $\mathbf{p} \in S$ **do**
3   **foreach** $* \in \{\rightarrow, \downarrow\}$ **do**
4   **if** $\mathbf{p} + \mathbf{d}_* \in S$ **and** $\rho(\mathbf{p}) \in f_*^2$ **then**
5              $B \leftarrow B \cup \{\{\mathbf{p}, \mathbf{p} + \mathbf{d}_*\}\}$
6              $S \leftarrow S \setminus \{\mathbf{p}, \mathbf{p} + \mathbf{d}_*\}$
7   **break**
8   **return** B

of the proposed backward algorithm consists of: Step 1: for a given U, i.e. a Pythagorean triple and a rational translation vector, listing all the points $\mathbf{w}$ in the non-injective remainder set; Step 2: finding all their preimages $\rho^{-1}(\mathbf{w})$; Step 3: finding among them those in S.

Step 1. As explained in Sect. 3, the cyclic group $\mathcal{G}$ is generated by $\psi = \left(\frac{p}{c}, \frac{q}{c}\right)^t$ and $\omega = \left(-\frac{q}{c}, \frac{p}{c}\right)^t$, and $\mathcal{G}'$ is its translation (modulo $\mathbb{Z}^2$). Therefore, all the points in $\mathcal{G}'$ can be expressed as $\mathbb{Z}\psi + \mathbb{Z}\omega + \{\mathbf{t}\}$. To find these points of $\mathcal{G}'$ in the non-injective zones, let us focus on $f_\downarrow^2$ given in Lemma 6. Note that the similar discussion is valid for other non-injective zones given by Lemma 6. The set of remainder points $\mathbb{Z}\psi + \mathbb{Z}\omega + \{\mathbf{t}\}$ lying in $f_\downarrow^2$ is then formulated by the following four linear inequalities, and we define the *non-injective remainder index set* $C_\downarrow$ such that

$$C_\downarrow = \left\{ (i,j) \in \mathbb{Z}^2 \;\middle|\; \begin{array}{c} -\frac{1}{2} < \frac{p}{c}i - \frac{q}{c}j + \{t_x\} < \frac{1}{2} - \frac{2pq}{p^2+q^2}, \\ \frac{p^2-q^2}{p^2+q^2} - \frac{1}{2} < \frac{q}{c}i + \frac{p}{c}j + \{t_y\} < \frac{1}{2} \end{array} \right\}. \tag{7}$$

Solving the system of inequalities in (7) consists of finding all pairs $(i,j) \in \mathbb{Z}^2$ inside the given rectangle. This is carried out by mapping $\mathbb{Z}\psi + \mathbb{Z}\omega + \{\mathbf{t}\}$ to $\mathbb{Z}^2$ using a similarity; denoting by $\hat{f}_\downarrow^2$ the image of $f_\downarrow^2$ under this transformation (Fig. 5).

To list all the integer points in $(i,j) \in C_\downarrow$, we first find the upper and lower corners of the rectangular region $\hat{f}_\downarrow^2$ given by Eq. (7): $\left(\frac{p-3q}{2}, \frac{p-q}{2}\right)$ and $\left(\frac{q-p}{2}, \frac{p+q}{2}\right)$. We then find all the horizontal lines $j = k$ where $k \in \mathbb{Z} \cap (\frac{p-q}{2}, \frac{p+q}{2})$. For each line $j = k$, we obtain the two intersections with the boundary of $\hat{f}_\downarrow^2$ as the maximal and minimal integers for i.

The complexity of this step depends on the number of integer lines crossing $\hat{f}_\downarrow^2$, which is q, and thus it is $\mathcal{O}(q)$. This completes Step 1.

Step 2. We seek for the set of all preimages of $i\psi + j\omega + \{\mathbf{t}\}$ for each $(i,j) \in C_\downarrow$, or equivalently, preimages of $i\psi + j\omega$ by the translationless remainder map. (The fact that this point is in $f_\downarrow^2$ plays no role in this step.) This is a Diophantine system (modulo $\mathbb{Z}^2$) and the set of preimages of a point $i\psi + j\omega + \{\mathbf{t}\}$ is given by a sublattice of $\mathbb{Z}^2$:

$$\mathbb{T}(i,j) = p\frac{\mu - v}{2} \begin{pmatrix} i \\ j \end{pmatrix} + \mathbb{Z}\begin{pmatrix} a \\ -b \end{pmatrix} + \mathbb{Z}\begin{pmatrix} c\sigma \\ c\tau \end{pmatrix} \tag{8}$$

where μ, v and σ, τ are the Bézout coefficients satisfying $\mu p^2 + v q^2 = 1$ and $\sigma a + \tau b = 1$, respectively.

To find these Bézout coefficients, we use the extended Euclidean Algorithm. The time complexity of finding μ and v (resp. σ and τ) is $\mathcal{O}(\log q)$ (resp. $\mathcal{O}(\log \min(a, b))$) [2]. As the Bézout coefficients are computed once for all $(i,j) \in C_\downarrow$, the time complexity of Step 2 is $\mathcal{O}(\log q) + \mathcal{O}(\log \min(a, b)) = \mathcal{O}(\log \min(a, b))$.

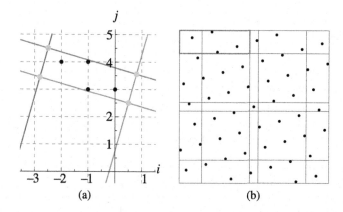

Fig. 5. (a) Geometrical interpretation of the system of linear inequalities in Eq. (7), in the (i,j)-plane for $(p,q) = (7,2)$. The region surrounded by the four lines is $\hat{f}_\downarrow^2$, and the integer points within are marked by black circles. (b) The remainder range, $\mathcal{G}'$ and $f_\downarrow^2$ illustrated by a red square, which corresponds to $\hat{f}_\downarrow^2$ in (a) (Color figure online).

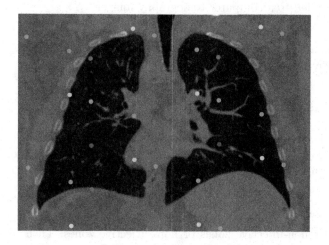

Fig. 6. Non-injective pixels obtained for a computer tomography image of a human chest by the backward algorithm (preimages of points in $\mathcal{G}' \cap f_\downarrow^2$). Each non-injective pixel is marked by a circle. Pixels of the same color are positioned periodically and are preimages of the same point in $\mathcal{G}' \cap f_\downarrow^2$. The result was obtained for $\theta = \arcsin \frac{591}{19409} \approx 1.7°$ and $\mathbf{t} = \mathbf{0}$. Note that from the point of view of backward algorithm the content of the image does not matter (Color figure online).

Step 3. We now consider the union of lattices $\mathbb{T}(i,j)$ for all couples (i,j) in $C_\downarrow$ obtained in Step 1. To find their intersection with S, we apply to each an algorithm similar to Step 1, with an affine transformation mapping the basis $\left(\begin{smallmatrix} a \\ -b \end{smallmatrix}\right), \left(\begin{smallmatrix} c\sigma \\ c\tau \end{smallmatrix}\right)$ to $\left(\begin{smallmatrix} 1 \\ 0 \end{smallmatrix}\right), \left(\begin{smallmatrix} 0 \\ 1 \end{smallmatrix}\right)$ and $p\frac{\mu-v}{2}\left(\begin{smallmatrix} i \\ j \end{smallmatrix}\right)$ to $\left(\begin{smallmatrix} 0 \\ 0 \end{smallmatrix}\right)$. Thus, a rectangular S maps to a quadrangular $\hat{S}$ after such an affine transformation, and we find the set of integer points in $\hat{S}$. Note that the involved transformation is the same for all the lattices, up to a translation.

The complexity of listing all the preimages is given by $|C_\downarrow|$ times the number of horizontal lines $j = k, k \in \mathbb{Z}$, passing $\hat{S}$, denoted by K. The cardinality of $C_\downarrow$ is related to the area of $f_\downarrow^2$ given by $\frac{2q^2(p-q)^2}{(p^2+q^2)^2}$ which cannot be larger than $\frac{3}{2} - \sqrt{2}$. As $|\mathcal{G}'| = c$ and $|C_\downarrow| = |\mathcal{G}' \cap f_\downarrow^2|$, $|C_\downarrow| \leq (\frac{3}{2} - \sqrt{2})c$. On the other hand, K is bounded by d_S/c, where d_S stands for the diagonal of S. As the complexity of d_S is given by $\mathcal{O}(\sqrt{|S|})$, the final complexity of Step 3 is $\mathcal{O}(\sqrt{|S|})$.

Remark 5. A possible refinement consists of ruling out false positives at border points $\mathbf{p}$ of S, by checking whether $\mathbf{p} + \mathbf{d}_*$ belongs to S, where $\mathbf{d}_*$ is given by the above procedure (thus avoiding the case when $\mathbf{p}$ and $\mathbf{p} + \mathbf{d}_*$ are mapped to the same point but $\mathbf{p} + \mathbf{d}_*$ is not in S). This can be achieved during Step 3.

All the steps together allow us to state that the backward algorithm, whose time complexity is $\mathcal{O}(q + \log \min(a, b) + \sqrt{|S|})$, identifies non-injective points in finite rectangular sets. Figure 6 presents an experimental result of the backward algorithm applied to a computer tomography image.

Remark 6. Even though the backward algorithm works with rectangles, one can approximate any set S by a union of rectangles and run the backward algorithm on each of them.

5   Conclusion

In this article, we have extended the neighborhood motion maps—which was previously proposed by Nouvel and Rémila [7] for digitized rotations—and we have shown that they are useful to characterize the bijectivity of rigid motions on $\mathbb{Z}^2$.

We first proved some necessary and sufficient conditions of bijective rigid motions on $\mathbb{Z}^2$; i.e., rigid motions such that no point $\mathbf{p} \in \mathbb{Z}^2$ has the image $\rho(\mathbf{p})$ in either non-injective zones or non-surjective zones. Then, from a more practical point of view, we focused on finite sets of $\mathbb{Z}^2$ rather than the whole $\mathbb{Z}^2$. In particular, we proposed two efficient algorithms for verifying whether a given digitized rigid motion is bijective when restricted to a finite set S. The forward algorithm consists of checking if points of S have preimages in non-injective zones. On the other hand, we used the reverse strategy to propose the backward algorithm consisting of the identification of points in $\mathcal{G}' \cap \mathcal{F}$ and their preimages in S. The complexities of the forward and backward algorithms are $\mathcal{O}(|S|)$ and $\mathcal{O}(q + \log \min(a, b) + \sqrt{|S|})$, respectively.

One of our perspectives is to extend the proposed framework to 3D digitized rigid motions.

Acknowledgments. The authors express their thanks to Laurent Najman of ESIEE Paris for his remarks concerning the backward algorithm, and to Mariusz Jędrzejczyk of Norbert Barlicki Memorial Teaching Hospital No. 1, Department of Radiology and Diagnostic Imaging, for providing the computer tomography image of a human chest.

The research leading to these results has received funding from the Programme d'Investissements d'Avenir (LabEx Bézout, ANR-10-LABX-58).

References

1. Anglin, W.S.: Using Pythagorean triangles to approximate angles. Am. Math. Mon. **95**(6), 540–541 (1988)
2. Hunter, D.J.: Essentials of Discrete Mathematics, 2nd edn. Jones and Bartlett Learning, Sudbury (2010)
3. Jacob, M.A., Andres, E.: On discrete rotations. In: Fifth International Workshop on Discrete Geometry for Computer Imagery, pp. 161–174. Université de Clermont-Ferrand I (1995)
4. Klette, R., Rosenfeld, A.: Digital Geometry: Geometric Methods for Digital Picture Analysis. Elsevier, Amsterdam (2004)
5. Ngo, P., Kenmochi, Y., Passat, N., Talbot, H.: Topology-preserving conditions for 2D digital images under rigid transformations. J. Math. Imaging Vis. **49**(2), 418–433 (2014)
6. Nouvel, B., Rémila, É.: On colorations induced by discrete rotations. In: Nyström, I., Sanniti di Baja, G., Svensson, S. (eds.) DGCI 2003. LNCS, vol. 2886, pp. 174–183. Springer, Heidelberg (2003)
7. Nouvel, B., Rémila, É.: Characterization of bijective discretized rotations. In: Klette, R., Žunić, J. (eds.) IWCIA 2004. LNCS, vol. 3322, pp. 248–259. Springer, Heidelberg (2004)
8. Nouvel, B., Rémila, É.: Configurations induced by discrete rotations: periodicity and quasi-periodicity properties. Discrete Appl. Math. **147**(2–3), 325–343 (2005)
9. Roussillon, T., Cœurjolly, D.: Characterization of bijective discretized rotations by Gaussian integers. Research report, LIRIS UMR CNRS 5205, January 2016. https://hal.archives-ouvertes.fr/hal-01259826

On Weighted Distances on the Khalimsky Grid

Gergely Kovács[1], Benedek Nagy[2,3($\boxtimes$)], and Béla Vizvári[3]

[1] Edutus College, Tatabánya, Hungary
kovacs.gergely@edutus.hu
[2] Faculty of Informatics, University of Debrecen, Debrecen, Hungary
nbenedek.inf@gmail.com
[3] Eastern Mediterranean University, Mersin-10, Famagusta, North Cyprus, Turkey
bela.vizvari@emu.edu.tr

Abstract. In this paper we introduce the weighted distances on the Khalimsky grid. There are two types of natural neighborhood relations and one semi-neighborhood is also defined. Based on them two types of weighted distances are defined. We give formulae for computing the weighted distance of any point-pair on the Khalimsky grid in both cases.

1 Introduction

Both in theory and practice of image processing and computer graphics, digital geometry is of high importance. In 2D pixels, in 3D voxels play the role of units. Digital geometry differs from the usual Euclidean geometry: neighbor relation is understood and play important role among the units (*e.g.*, pixels). In digital geometry (since the domain is discrete), digital, path-based distances are considered. The history of digital geometry has been started by the seminal paper [19], in which the cityblock (Manhattan) and the chessboard distances were defined on the square grid. Since these distances are very rough approximations of the Euclidean distance, various techniques are developed. By neighborhood sequences one may vary the type of used neighborhood along a path [3,4,10,13,24]. In weighted distances the minimal weighted paths are counted in which steps to different type of neighbors may count with different weights [1,20]. It is well known that based on a larger "neighborhood", *i.e.*, allowing direct steps to a larger set of pixels, better approximation of the Euclidean distance can be obtained [1,2,23]. There are also mixed solutions [17] to obtain errorless approximations on a 'perimeter' of a square. The field has developed involving other, non-traditional grids, *e.g.*, hexagonal and triangular grids in 2D [5,9,11] and various 3D grids [16,22]. Using various grids appropriate coordinate systems have large advantages [6,12,15] allowing relatively simple formulations of theoretical results which is also very useful for practical purposes.

Digital geometry has also some very interesting properties by comparing it to the usual, Euclidean geometry. One of these properties is known as a 'topological paradox': considering the two diagonals of a chessboard, one of them consist of only black, the other consists of only white pixels: these two lines cross each other without having an intersection pixel. There are various solutions to overcome this

N. Normand et al. (Eds.): DGCI 2016, LNCS 9647, pp. 372–384, 2016.
DOI: 10.1007/978-3-319-32360-2_29

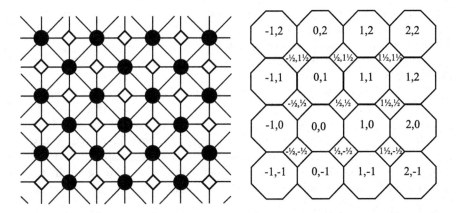

Fig. 1. The usual representation of the Khalimsky grid: neighbor points are connected by edges (left); and a part of the grid T(8,8,4): pixels with their Cartesian coordinates (right).

paradox, *e.g.*, to use a non-traditional grid, *e.g.*, the hexagonal grid in this case; or to use a topological coordinate system addressing not only the pixels of the grid [8,14]; or to use a restricted neighborhood, see, *e.g.*, [7,21]. In Khalimsky grid the 4-neighbor and the 8-neighbor relations are used for the points of the grid, alternately, depending on their positions (see Fig. 1, left, also).

Motivated by these facts, in this paper we are considering the Khalimsky grid (to avoid some topological paradoxes) and we define weighted distances on this grid. We also show some interesting phenomena, and future directions of research.

In the next section, we describe the Khalimsky grid with an appropriate coordinate system, then in Sect. 3 weighted distances on this grid are defined. In Sect. 4, we present results based on two types of natural neighborhood relations, while in Sect. 5 three types of neighborhood relations are used including a semi-neighborhood relation.

2   Description of T(8,8,4)

Figure 1 (left) shows a usual representation of the Khalimsky grid. As one can observe, the grid (the digital plane) is not homogeneous, and there are various ways to choose pixels. In our paper, we use the basic form of the grid, considering each vertex of the grid as a pixel. Consequently, our grid is T(8,8,4) if dual notation is used, see, *e.g.*, [18]. We have also rotated the grid with $\pi/4$ and in this way a convenient coordinate system can be assigned to the pixels. In this representation octagons represent the points for which 8-neighborhood is used and squares (diamonds) the points for which 4-neighborhood is used (see Fig. 1, right).

Each pixel of the T(8,8,4) grid is an octagon or a square. An octagon has 8 neighbors with common sides (4 octagons and 4 squares). A square has 4 neighbors with common sides (4 octagons).

Each pixel (octagon or square) of the grid can be described by a unique coordinate pair.

There is an octagon with coordinates $(0,0)$. The axes are horizontal or vertical. At every time when we step from an octagon to another one crossing their common side, the step is done parallel to one of the axes. If the step is in the direction of an axis, then the respective coordinate value is increased by 1, while in case the step is in opposite direction to an axis, the respective coordinate value is decreased by 1. In this way every octagon gets a unique coordinate pair with integer values. The octagons $p(p(1), p(2))$ and $q(q(1), q(2))$ are neighbors, if $|p(1) - q(1)| + |p(2) - q(2)| = 1$.

When we step from an octagon to a square crossing their common side, the step is done diagonally. If the step is in the half-direction of an axis, then the respective coordinate value is increased by 0.5, while in case the step is in opposite direction, the respective coordinate value is decreased by 0.5. In this way every square gets a unique coordinate pair with fraction values (*i.e.* each value has the form of an integer plus a half). The octagon p and the square q are neighbors, if $|p(1) - q(1)| + |p(2) - q(2)| = 1$.

There are no squares with common side. But every square p has 4 semi-neighbor square, for example q, for which $|p(1) - q(1)| + |p(2) - q(2)| = 1$. In this case, square p and square q have two common octagon neighbors, and the common side of the octagon neighbors is situated between the two squares as an edge connecting them.

3 Definition of Weighted Distances

Let $\alpha, \beta \in \mathbb{R}^+$ be positive weights. The simplest weighted distances allow to step from an octagon to a cityblock neighbor octagon by changing only 1 coordinate value by ± 1 with weight α. Similarly it allows to step from an octagon to a diagonal neighbor square (and vice versa) by changing both the coordinate values by ± 0.5 with weight β.

Let the weight of a step from a square to a semi-neighbor square be γ. This step changes only 1 coordinate value by ± 1 (We use this kind of step only at the second part of this article, in Sect. 5 about extended weighted distances.).

We can define the weighted distance of any two points (octagons or squares) of the grid. Let p and q be two points of the grid. A finite point sequence of points of the form $p = p_0, p_1, \ldots, p_m = q$, where p_{i-1}, p_i are neighbors (or semi-neighbors) for $1 \leq i \leq m$, is called a path from p to q. A path can be seen as consecutive steps to neighbors (or semi-neighbors). Then the cost of the path is the sum of the weights of its steps.

Finally, let the weighted distance $d(p, q; \alpha, \beta)$ of p and q by the weights α, β be the cost of the minimal (basic) weighted paths between p and q. Similarly, the weighted distance $d(p, q; \alpha, \beta, \gamma)$ is the cost of the minimal extended weighted paths between p and q using the weights α, β, γ.

A technical definition is used through the paper. The difference $w_{p,q} = (w(1), w(2))$ of two points p and q is defined by $w(i) = |q(i) - p(i)|$.

4   Minimal Weighted Paths

Firstly we use only the steps between neighbors (steps with weight γ is forbidden in this basic case).

There are various paths with various sums of weights that can be found between any two points. When the weights α and β are known the optimal search (the Dijkstra algorithm) can be used. However depending on the actual ratios and values of the weights α, β one can compute a minimum weighted paths. Using a combinatorial approach, we give methods for these computations for each possible case.

The relative weights give the separation of the possible cases.

4.1   Case $\alpha \leq \beta$

If p and q are octagons, then the use of β-steps is not efficient. A minimal path can be constructed only by α-steps. In every step the absolute value of a coordinate difference is decreasing by 1. Thus

$$d(p, q; \alpha, \beta) = \alpha \cdot (w(1) + w(2)). \tag{1}$$

For example the weighted distance of the octagons $(0,0)$ and $(3,1)$ is 4α.

If p and q are an octagon and a square, then the minimal path contains the minimal path between the octagon and another octagon, which is the closest octagon neighbor of the square. Moreover the minimal path contains one β-step between the above mentioned octagon and the square. Thus

$$d(p, q; \alpha, \beta) = \alpha \cdot (\lfloor w(1) \rfloor + \lfloor w(2) \rfloor) + \beta, \tag{2}$$

where $w(i)$ is a fraction value and $\lfloor w(i) \rfloor$ is the next lowest integer value of $w(i)$ (floor function). For example the weighted distance of the octagon $(0,0)$ and the square $(3.5,1.5)$ is $4\alpha + \beta$.

If p and q are squares, then the minimal path uses only two β-steps, the first and the last steps. Between these steps the minimal path contains the minimal path between the closest octagon neighbors of the squares using only α-steps. Thus

$$d(p, q; \alpha, \beta) = \alpha \cdot (\lfloor |w(1) - 0.5| \rfloor + \lfloor |w(2) - 0.5| \rfloor) + 2\beta, \tag{3}$$

where $w(i)$ is a non-negative integer, thus $|w(i) - 0.5|$ is a non-negative fraction. For example, the distance of the squares $(0.5,0.5)$ and $(3.5,1.5)$ is $2\alpha + 2\beta$. But $d((0.5, 0.5), (3.5, 0.5); \alpha, \beta) = 2\alpha + 2\beta$, too. Also, $d((0.5, 1.5), (3.5, 0.5); \alpha, \beta) = 2\alpha + 2\beta$.

Let us assume, that $2\alpha < \beta$, and let us determine the set of points, which distance from the octagon $(0,0)$ is less or equal to 2α. Then the distance of the 4 octagon-neighbors of the original octagon is α, *i.e.* these are in the set: $(1,0)$, $(0,1)$, $(-1,0)$, $(0,-1)$. The distances of every octagon-neighbors of these 4 neighbors are 2α, *i.e.* these are in the set, too: $(2,0)$, $(0,2)$, $(-2,0)$, $(0,-2)$, $(1,1)$, $(-1,1)$, $(-1,-1)$, $(1,-1)$. The distances of all other octagons are more than 2α, and the distances of the square-neighbors are more than 2α, too. It means that this set contains 13 octagon-elements, and it has 4 holes. For example, the square-neighbor $(0.5, 0.5)$ is not in the set, but all of its neighbors $((0,0), (1,0), (0,1),$ and $(1,1))$ are elements of the set.

4.2 Case $\beta \le \alpha \le 2\beta$

Two α-steps, a vertical and after that another horizontal, can be substituted by two consecutive diagonal β-steps, because $2\beta \le 2\alpha$. We can do this change $\min\{w(i)\}$ times. Thus if p and q are octagons, then

$$d(p,q;\alpha,\beta) = 2\beta \cdot \min\{w(i)\} + \alpha \cdot (\max\{w(i)\} - \min\{w(i)\}). \qquad (4)$$

For example the weighted distance of the octagons $(0,0)$ and $(3,1)$ is $2\beta + 2\alpha$.

Similarly, if p and q are an octagon and a square, then the minimal path contains the minimal path between two octagons and a final step between an octagon and a square. Thus

$$d(p,q;\alpha,\beta) = 2\beta \cdot \min\{\lfloor w(i) \rfloor\} + \alpha \cdot (\max\{\lfloor w(i) \rfloor\} - \min\{\lfloor w(i) \rfloor\}) + \beta =$$
$$= 2\beta \cdot \min\{w(i)\} + \alpha \cdot (\max\{w(i)\} - \min\{w(i)\}). \qquad (5)$$

For example $d((0,0), (3.5, 1.5); \alpha, \beta) = 3\beta + 2\alpha$.

Finally, if p and q are squares, then the minimal path contains the minimal path between two octagons and the starting and the final step between an octagon and a square. Thus

$$d(p,q;\alpha,\beta) = 2\beta \cdot \min\{(\lfloor |w(i) - 0.5| \rfloor\} + \alpha \cdot (\max\{(\lfloor |w(i) - 0.5| \rfloor\} - \min\{(\lfloor |w(i) - 0.5| \rfloor\}) + 2\beta \qquad (6)$$

For example the weighted distance of the squares $(0.5,0.5)$ and $(3.5,1.5)$ is $2\alpha + 2\beta$. But if the coordinates of the squares are $(0.5,0.5)$ and $(3.5,0.5)$, then $w(1) = 3$ and $w(2) = 0$. In this case the weighted distance is equal to $2\alpha + 2\beta$, too. This is not equal to $2\beta \cdot \min\{w(i)\} + \alpha \cdot (\max\{w(i)\} - \min\{w(i)\})$, which value is 3α.

4.3 Case $2\beta \le \alpha$

Every α-step can be substituted by two consecutive β-steps. Moreover on a diagonal way, a vertical and after that another horizontal α-step can be substituted by two consecutive parallel diagonal β-steps. It means that in every two steps the maximum of the absolute value of the coordinate differences is decreasing by 1. Thus if p and q are octagons, then

$$d(p,q;\alpha,\beta) = 2\beta \cdot \max\{w(i)\}.$$

It means that for example $d((0,0),(3,1);\alpha,\beta) = 6\beta$.

If p and q are an octagon and a square, then the minimal path contains the minimal path between two octagons and a final step between an octagon and a square. Thus

$$d(p,q;\alpha,\beta) = 2\beta \cdot \max\{\lfloor w(i) \rfloor\} + \beta = 2\beta \cdot \max\{w(i)\}.$$

For example $d((0,0),(3.5,1.5);\alpha,\beta) = 7\beta$.

If p and q are squares, then the minimal path contains the minimal path between two octagons and the starting and the final step between an octagon and a square (Of course if the two squares are semi-neighbors, the minimal path contains only two steps: the first and the final step.). Thus

$$d(p,q;\alpha,\beta) = 2\beta \cdot \max\{\lfloor |w(i) - 0.5| \rfloor\} + 2\beta = 2\beta \cdot \max\{w(i)\},$$

because $\max\{w(i)\} \geq 1$, and that case $\max\{\lfloor |w(i) - 0.5| \rfloor\} = \max\{w(i)\} - 1$. For example the weighted distance of the squares (0.5,0.5) and (3.5,1.5) is 6β.

5   Minimum Weighted Paths by Using the Semi-neighbor Step

In this case the γ-weighted step is not forbidden and thus, extended weighted distances are obtained. The relative weights give the separation of the possible cases again.

5.1   Case $\alpha \leq \beta$

Case $\alpha \leq \beta \leq \gamma$. If p and q are octagons, then similarly to the basic case, when $\alpha \leq \beta$, the minimal path can be constructed only by α-steps. In every α-step the absolute value of a coordinate difference is decreasing by 1. Every γ-step decreases the absolute value of a coordinate difference only by 1, too. Thus, the use of γ-steps is not efficient, because $\alpha \leq \gamma$. It means that $d(p,q;\alpha,\beta,\gamma)$ is computed by Eq. 1, i.e., it is $\alpha \cdot (w(1) + w(2))$.

Similarly, if p and q are an octagon and a square, this solution does not use γ-steps, either. Thus Eq. 2 produces $d(p,q;\alpha,\beta,\gamma) = \alpha \cdot (\lfloor w(1) \rfloor + \lfloor w(2) \rfloor) + \beta$.

If p and q are squares, then the solution of the basic case contains two β-steps. If these two steps are the same in coordinates (for example (+0.5, +0.5), then we could only change them by two γ-steps, but it is not efficient.

If $\min\{w(i)\} \geq 1$, then these two β-steps are the same in coordinates, i.e. the solution in this case is the original, see Eq. 3. For example the weighted distance of the squares (0.5,0.5) and (3.5,1.5) is $2\alpha + 2\beta$.

But if $\min\{w(i)\} = 0$, then one coordinate of the two β-steps is different (for example (+0.5, +0.5) and (+0.5, -0.5)). In this case we can change it only by one γ-step. For example if $\gamma \leq 2\beta$; $\min\{w(i)\} = 0$ and $\max\{w(i)\} = 1$, then the minimal path contains only one γ-step (without two β-steps).

If $\min\{w(i)\} = 0$, we have two alternative paths:

(1) $\max\{w(i)\}$ γ-steps, the total weight of it is $\gamma \cdot \max\{w(i)\}$,
(2) the first and the last steps are β-steps between squares and octagons; and between these steps there are $\max\{w(i)\} - 1$ α-steps, so the total weight of it is $2\beta + \alpha \cdot (\max\{w(i)\} - 1)$.

If $\max\{w(i)\}=1$, then (1) is the minimal path with weight γ, because $\gamma \leq 2\beta$. Otherwise path (1) is the minimal one, if

$$\max\{w(i)\} \leq \frac{2\beta - \alpha}{\gamma - \alpha}.$$

Thus if $\min\{w(i)\} = 0$, then

$$d(p,q;\alpha,\beta,\gamma) = \begin{cases} \gamma \cdot \max\{w(i)\}, & \text{if } \max\{w(i)\} \leq \frac{2\beta-\alpha}{\gamma-\alpha}; \\ \alpha \cdot (\max\{w(i)\} - 1) + 2\beta, & \text{if } \max\{w(i)\} \geq \frac{2\beta-\alpha}{\gamma-\alpha}. \end{cases}$$

For example, the weighted distance of the squares (0.5,0.5) and (3.5,0.5) is 3γ or $2\alpha + 2\beta$.

Case $\alpha \leq \gamma \leq \beta$. Similarly to the previous case the use of γ-steps instead of α-steps is not efficient, because $\alpha \leq \gamma$. It means that the minimal path between two octagons, and between an octagon and a square is the same as in the previous case.

If p and q are squares, then the solution of the original case contains two β-steps. It means that minimal path can contain only γ-steps, because $\gamma \leq \beta$.

In this case we have two alternative paths:

(1) The original path, see Eq. 3, $\alpha \cdot (\lfloor |w(1) - 0.5| \rfloor + \lfloor |w(2) - 0.5| \rfloor) + 2\beta$.
(2) The path containing only γ-steps, the total weight of it is $\gamma(w(1) + w(2))$. The minimal path is the better one.

Case $\gamma \leq \alpha \leq \beta$. If p and q are squares, then the minimal path can be constructed only by γ-steps. In every step the absolute value of a coordinate difference is decreasing by 1. Thus

$$d(p,q;\alpha,\beta,\gamma) = \gamma \cdot (w(1) + w(2)).$$

For example, the weighted distance of the squares (0.5,0.5) and (3.5,0.5) is 3γ.

Similarly, if p and q are an octagon and a square, then the minimal path contains one β-step and only γ-steps. This solution does not use α-steps, either. Thus

$$d(p,q;\alpha,\beta,\gamma) = \gamma \cdot (\lfloor w(1) \rfloor + \lfloor w(2) \rfloor) + \beta.$$

For example $d((0,0),(3.5,1.5);\alpha,\beta,\gamma) = 4\gamma + \beta$.

If p and q are octagons, then the two alternative path are following:

(1) The original minimal path, the total weight of it is $\alpha \cdot (w(1) + w(2))$.
(2) The path containing two β-steps (the first and the last) and only γ-steps, the total weight of it is $\gamma \cdot (\lfloor |w(1) - 0.5| \rfloor + \lfloor |w(2) - 0.5| \rfloor) + 2\beta$.
    The minimal path is the better one.

5.2 Case $\beta \leq \alpha \leq 2\beta$

Case $\beta \leq \alpha \leq 2\beta \leq \gamma$. Every γ-step can be substituted by two consecutive β-steps. If $2\beta \leq \gamma$, then the use of γ-steps is not efficient. It means that the computing the distances is the same as we did it above in the basic case.

Case $\beta \leq \alpha \leq \gamma \leq 2\beta$. If at least one of p and q is an octagon then neither the β-steps, nor the α steps can be substituted by γ steps in economic way. It means, that still the formulae (4) and (5) give the distances.

If both p and q are squares and $\min\{w(i)\} \geq 1$, then the solution of the basic case contains only β-steps with same coordinates. It means that the use of γ-steps is not efficient, the solution is given by the original Eq. 6.

But if $\min\{w(i)\} = 0$, then one coordinate of the two β-steps is different (for example $(+0.5, +0.5)$ and $(+0.5, -0.5)$). In this case we can change it only by one γ-step.

If $\min\{w(i)\} = 0$, we have two alternative paths:

(1) $\max\{w(i)\}$ γ-steps, the total weight of it is $\gamma \cdot \max\{w(i)\}$,
(2) the first and the last steps are β-steps between squares and octagons; and between these steps there are $\max\{w(i)\} - 1$ α-steps, so the total weight of it is $2\beta + \alpha \cdot (\max\{w(i)\} - 1)$.

If $\max\{w(i)\}=1$, then (1) is the minimal path with weight γ, because $\gamma \leq 2\beta$. Otherwise path (1) is the minimal one, if

$$\max\{w(i)\} \leq \frac{2\beta - \alpha}{\gamma - \alpha}.$$

Thus if $\min\{w(i)\} = 0$, then

$$d(p, q; \alpha, \beta, \gamma) = \begin{cases} \gamma \cdot \max\{w(i)\}, & \text{if } \max\{w(i)\} \leq \frac{2\beta-\alpha}{\gamma-\alpha}; \\ \alpha \cdot (\max\{w(i)\} - 1) + 2\beta, & \text{if } \max\{w(i)\} \geq \frac{2\beta-\alpha}{\gamma-\alpha}. \end{cases}$$

For example the weighted distance of the squares (0.5,0.5) and (3.5,0.5) is 3γ or $2\alpha + 2\beta$.

Case $\beta \le \gamma \le \alpha \le 2\beta$. If p and q are squares, then the use of α-steps is not efficient, because $\gamma \le \alpha$. If two β-steps are the same in coordinates (for example $(+0.5, +0.5)$, then we could only change them by two γ-steps, but it is not efficient. But if one coordinate of two consecutive β-steps is different (for example $(+0.5, +0.5)$ and $(+0.5, -0.5)$), then we can change it by one γ-step. It means that:

$$d(p, q; \alpha, \beta, \gamma) = 2\beta \cdot \min\{w(i)\} + \gamma \cdot (\max\{w(i)\} - \min\{w(i)\}).$$

For example the weighted distance of the squares $(0.5, 0.5)$ and $(3.5, 1.5)$ is $2\beta + 2\gamma$.

Similarly, if p and q are a square and an octagon, then the minimal path contains the minimal path between two squares and a final step between a square and an octagon. Thus,

$$d(p, q; \alpha, \beta, \gamma) = 2\beta \cdot \min\{\lfloor w(i) \rfloor\} + \gamma \cdot (\max\{\lfloor w(i) \rfloor\} - \min\{\lfloor w(i) \rfloor\}) + \beta =$$
$$= 2\beta \cdot \min\{w(i)\} + \gamma \cdot (\max\{w(i)\} - \min\{w(i)\}).$$

For example, $d((0, 0), (3.5, 1.5); \alpha, \beta, \gamma) = 3\beta + 2\gamma$.

If p and q are octagons, and $\min\{w(i)\} \ge 1$, then the solution of the basic case contains β-steps. In this case we are able to change α-steps to γ-steps (first and final steps need to be β-step). Thus in this case:

$$d(p, q; \alpha, \beta, \gamma) = 2\beta \cdot \min\{w(i)\} + \gamma \cdot (\max\{w(i)\} - \min\{w(i)\}).$$

For example the weighted distance of the octagons $(0, 0)$ and $(3, 1)$ is $2\beta + 2\gamma$.

In any other cases if $\min\{w(i)\} = 0$, we have two alternative paths:

(1) $\max\{w(i)\}$ α-steps, the total weight of it is $\alpha \cdot \max\{w(i)\}$,
(2) the first and the last steps are β-steps between octagons and squares; and between these steps there are $\max\{w(i)\} - 1$ γ-steps, so the total weight of it is $2\beta + \gamma \cdot (\max\{w(i)\} - 1)$.

If $\max\{w(i)\} = 1$, then (1) is the minimal path with weight α, because $\alpha \le 2\beta$. Otherwise path (1) is the minimal one, if

$$\max\{w(i)\} \le \frac{2\beta - \gamma}{\alpha - \gamma}.$$

Thus if $\min\{w(i)\} = 0$, then

$$d(p, q; \alpha, \beta, \gamma) = \begin{cases} \alpha \cdot \max\{w(i)\}, & \text{if } \max\{w(i)\} \le \frac{2\beta - \gamma}{\alpha - \gamma}; \\ \gamma \cdot (\max\{w(i)\} - 1) + 2\beta, & \text{if } \max\{w(i)\} \ge \frac{2\beta - \gamma}{\alpha - \gamma}. \end{cases}$$

For example the weighted distance of the octagons $(0, 0)$ and $(3, 0)$ is 3α or $2\gamma + 2\beta$.

Case $\gamma \leq \beta \leq \alpha \leq 2\beta$. If p and q are squares, then the minimal path can be constructed only by γ-steps. Thus

$$d(p, q; \alpha, \beta, \gamma) = \gamma \cdot (w(1) + w(2)).$$

For example, the weighted distance of the squares $(0.5, 0.5)$ and $(3.5, 1.5)$ is 4γ.

Similarly, if p and q are a square and an octagon, then the minimal path contains the minimal γ-path between two squares and a final β-step between a square and an octagon. Thus,

$$d(p, q; \alpha, \beta, \gamma) = \gamma \cdot (\lfloor w(1) \rfloor + \lfloor w(2) \rfloor) + \beta.$$

For example, $d((0, 0), (3.5, 1.5); \alpha, \beta, \gamma) = 4\gamma + \beta$.

If p and q are octagons, and $\min\{w(i)\} \geq 1$, then the solution contains two β-steps (first and last steps), and γ-steps between them. In this case we are able to change all steps to α-steps, but it is not efficient, because all β-steps and all γ-steps decrease the absolute value of the coordinate difference by 1 (It is not true in case of $\min\{w(i)\} = 0$, when the two β-steps are not the same in coordinates.). Thus in this case:

$$d(p, q; \alpha, \beta, \gamma) = \gamma \cdot (\lfloor |w(1) - 0.5| \rfloor + \lfloor |w(2) - 0.5| \rfloor) + 2\beta.$$

For example the weighted distance of the octagons $(0, 0)$ and $(3, 1)$ is $2\gamma + 2\beta$.

In any other cases if $\min\{w(i)\} = 0$, we have two alternative paths as in the previous case (All is similar.). Thus if $\min\{w(i)\} = 0$, then

$$d(p, q; \alpha, \beta, \gamma) = \begin{cases} \alpha \cdot \max\{w(i)\}, & \text{if } \max\{w(i)\} \leq \frac{2\beta - \gamma}{\alpha - \gamma}; \\ \gamma \cdot (\max\{w(i)\} - 1) + 2\beta, & \text{if } \max\{w(i)\} \geq \frac{2\beta - \gamma}{\alpha - \gamma}. \end{cases}$$

5.3 Case $2\beta \leq \alpha$

Case $2\beta \leq \alpha$ and $2\beta \leq \gamma$. If $2\beta \leq \gamma$, then using of γ-steps is not efficient. It means that the computing the distances is the same as we did it in the basic case.

Case $\beta \leq \gamma \leq 2\beta \leq \alpha$. If p and q are squares, then similarly to the basic case, when $2\beta \leq \alpha$, the minimal path can be constructed only by 2β-steps (α-steps are not efficient at all when $\alpha > 2\beta$). Every γ-step decreases the sum of the absolute value of coordinate differences by 1. It means, that if two consecutive β-steps decrease the sum of the absolute value of the coordinate differences by 1, too, then the use of γ-steps is efficient. But if two consecutive β-steps are parallel diagonal steps, then we are able only to change them for two γ-steps and it is not efficient. It means that

$$d(p, q; \alpha, \beta, \gamma) = 2\beta \cdot \min\{w(i)\} + \gamma \cdot (\max\{w(i)\} - \min\{w(i)\}).$$

For example the weighted distance of the octagons $(0, 0)$ and $(3, 1)$ is $2\beta + 2\gamma$.

If p and q are a square and an octagon, then the minimal path contains the minimal path between two squares and a final step between a square and an octagon (the use of α-steps is not efficient). Thus

$$d(p, q; \alpha, \beta, \gamma) = 2\beta \cdot \min\{\lfloor w(i) \rfloor\} + \gamma \cdot (\max\{\lfloor w(i) \rfloor\} - \min\{\lfloor w(i) \rfloor\}) + \beta =$$
$$= 2\beta \cdot \min\{w(i)\} + \gamma \cdot (\max\{w(i)\} - \min\{w(i)\}).$$

For example $d((0,0),(3.5,1.5); \alpha, \beta, \gamma) = 3\beta + 2\gamma$.

Finally, if p and q are octagons, then the minimal path contains the minimal path between two squares and the starting and the final step between an octagon and a square. Thus

$$d(p, q; \alpha, \beta, \gamma) =$$

$$2\beta \cdot \min\{\lfloor |w(i) - 0.5| \rfloor\} + \gamma \cdot (\max\{\lfloor |w(i) - 0.5| \rfloor\} - \min\{\lfloor |w(i) - 0.5| \rfloor\}) + 2\beta.$$

For example the weighted distance of the squares (0.5,0.5) and (3.5,1.5) is $2\gamma + 2\beta$.

Case $\gamma \leq \beta$ and $2\beta \leq \alpha$. If p and q are squares, then the minimal path can be constructed only by γ-steps. Thus

$$d(p, q; \alpha, \beta, \gamma) = \gamma \cdot (w(1) + w(2)).$$

For example, the weighted distance of the squares (0.5,0.5) and (3.5,1.5) is 4γ.

Similarly, if p and q are a square and an octagon, then the minimal path contains the minimal γ-path between two squares and a final β-step between a square and an octagon. Thus,

$$d(p, q; \alpha, \beta, \gamma) = \gamma \cdot (\lfloor w(1) \rfloor + \lfloor w(2) \rfloor) + \beta.$$

For example, $d((0,0),(3.5,1.5); \alpha, \beta, \gamma) = 4\gamma + \beta$.

If p and q are octagons, and $\min\{w(i)\} \geq 1$, then the solution contains two β-steps (first and last steps), and γ-steps between them. Thus in this case:

$$d(p, q; \alpha, \beta, \gamma) = \gamma \cdot (\lfloor |w(1) - 0.5| \rfloor + \lfloor |w(2) - 0.5| \rfloor) + 2\beta.$$

For example the weighted distance of the octagons (0,0) and (3,1) is $2\gamma + 2\beta$.

6   Summary and Further Thoughts

Minimal weighted paths can obviously computed by Dijkstra algorithm. However, based on the regular structure of the grid and by the help of an appropriate coordinate system direct formulae are provided to compute them. There are various cases depending on the relation of the used weights. A summary of the results is shown in Table 1 for various cases. Note that each of the distance functions detailed in this manuscripts are metric. It is a task of a future study to compare these new distance functions to other, well-known digital distances. Digital disks are usually defined based on digital distances to analyse various properties of

Table 1. Value of $d(p, q; \alpha, \beta, \gamma)$ (or $d(p, q; \alpha, \beta)$ in Basic case shown as 'B')

Cases	Between		
	Two octagons	An octagon and a square	Two squares
$\alpha \leq \beta$ and ('B' or $\alpha \leq \gamma$)	$\alpha(w(1) + w(2))$	$\alpha(\lfloor w(1) \rfloor + \lfloor w(2) \rfloor) + \beta$	Subcases
$\beta \leq \alpha \leq 2\beta$ and ('B' or $\alpha \leq \gamma$)	$2\beta \min\{w(i)\} + \alpha(\max\{w(i)\} - \min\{w(i)\})$		Subcases
$\beta \leq \gamma$ and $\gamma \leq \alpha$ and $\gamma \leq 2\beta$	Subcases	$2\beta \min\{w(i)\} + \gamma(\max\{w(i)\} - \min\{w(i)\})$	
$2\beta \leq \alpha$ and ('B' or $2\beta \leq \gamma$)	$2\beta \max\{w(i)\}$		
$\gamma \leq \beta$ and $\gamma \leq \alpha$	Subcases	$\beta + \gamma(\lfloor w(1) \rfloor + \lfloor w(2) \rfloor)$	$\gamma(w(1) + w(2))$

these distances. It is noted (see, *e.g.*, Subsect. 4.2) that in some cases digital disks based on weighted distances on the grid T(8,8,4) have holes, *i.e.*, they are not digitally convex (in any sense). It is a future work to describe and analyze these digital disks, develop algorithms for distance transform etc. Measuring the approximation quality of the Euclidean distance is also a usual task with digital distances: based on that, optimal weights can also be computed. We believe that the nice topological properties of the Khalimsky grid can effectively be used in various applications using the weighted distances.

References

1. Borgefors, G.: Distance transformations in digital images. Comput. Vis. Graph. Image Process. **34**(3), 344–371 (1986)
2. Borgefors, G.: Another comment on "a note on 'distance transformations in digital images'". CVGIP Image Underst. **54**(2), 301–306 (1991)
3. Das, P.P., Chakrabarti, P.P., Chatterji, B.N.: Generalized distances in digital geometry. Inf. Sci. **42**, 51–67 (1987)
4. Das, P.P., Chakrabarti, P.P., Chatterji, B.N.: Distance functions in digital geometry. Inf. Sci. **42**, 113–136 (1987)
5. Deutsch, E.S.: Thinning algorithms on rectangular, hexagonal and triangular arrays. Commun. ACM **15**, 827–837 (1972)
6. Her, I.: Geometric transformations on the hexagonal grid. IEEE Trans. Image Process. **4**(9), 1213–1222 (1995)
7. Khalimsky, E.D., Kopperman, R., Meyer, P.R.: Computer graphics and connected topologies on finite ordered sets. Topol. Appl. **36**, 1–17 (1990)
8. Kovalevsky, V.: Algorithms in digital geometry based on cellular topology. In: Klette, R., Žunić, J. (eds.) IWCIA 2004. LNCS, vol. 3322, pp. 366–393. Springer, Heidelberg (2004)
9. Luczak, E., Rosenfeld, A.: Distance on a hexagonal grid. IEEE Trans. Comput. **C–25**(5), 532–533 (1976)
10. Nagy, B.: Distance functions based on neighbourhood sequences. Publicationes Math. **63**(3), 483–493 (2003)

11. Nagy, B.: Characterization of digital circles in triangular grid. Pattern Recogn. Lett. **25**(11), 1231–1242 (2004)

12. Nagy, B.: Generalized triangular grids in digital geometry. Acta Math. Acad. Paedagogicae Nyíregyháziensis **20**, 63–78 (2004)

13. Nagy, B.: Distances with generalized neighbourhood sequences in nD and ∞D. Discrete Appl. Math. **156**, 2344–2351 (2008)

14. Nagy, B.: Cellular topology and topological coordinate systems on the hexagonal and on the triangular grids. Ann. Math. Artif. Intell. **75**(1-2), 117–134 (2015)

15. Nagy, B., Strand, R.: A connection between $\mathbb{Z}^n$ and generalized triangular grids. In: Peters, J., Klosowski, J., Arns, L., Chun, Y.K., Rhyne, T.-M., Monroe, L. (eds.) ISVC 2008, Part II. LNCS, vol. 5359, pp. 1157–1166. Springer, Heidelberg (2008)

16. Nagy, B., Strand, R.: Non-traditional grids embedded in $\mathbb{Z}^n$. Int. J. Shape Model. IJSM (World Scientific) **14**(2), 209–228 (2008)

17. Nagy, B., Strand, R., Normand, N.: A weight sequence distance function. In: Hendriks, C.L.L., Borgefors, G., Strand, R. (eds.) ISMM 2013. LNCS, vol. 7883, pp. 292–301. Springer, Heidelberg (2013)

18. Radványi, A.G.: On the rectangular grid representation of general CNN networks. Int. J. Circ. Theor. Appl. **30**, 181–193 (2002)

19. Rosenfeld, A., Pfaltz, J.L.: Distance functions on digital pictures. Pattern Recogn. **1**, 33–61 (1968)

20. Sintorn, I.-M., Borgefors, G.: Weighted distance transforms in rectangular grids. In: ICIAP, pp. 322–326 (2001)

21. Slapal, J.: Convenient adjacencies for structuring the digital plane. Ann. Math. Artif. Intell. **75**(1-2), 69–88 (2015)

22. Strand, R., Nagy, B., Borgefors, G.: Digital distance functions on three-dimensional grids. Theor. Comput. Sci. **412**, 1350–1363 (2011)

23. Svensson, S., Borgefors, G.: Distance transforms in 3D using four different weights. Pattern Recogn. Lett. **23**(12), 1407–1418 (2002)

24. Yamashita, M., Honda, N.: Distance functions defined by variable neighborhood sequences. Pattern Recogn. **17**(5), 509–513 (1984)

Digital Disks by Weighted Distances in the Triangular Grid

Benedek Nagy[1,2]([⊠]) and Hamid Mir-Mohammad-Sadeghi[2]

[1] Department of Computer Science, Faculty of Informatics,
University of Debrecen, Debrecen, Hungary
nbenedek.inf@gmail.com
[2] Department of Mathematics, Faculty of Arts and Sciences,
Eastern Mediterranean University, Mersin-10, Famagusta, North Cyprus, Turkey
ha.sadeghi@gmail.com

Abstract. Weighted (or with other name, chamfer) distances on the triangular grid was introduced recently based on the three possible neighborhood. In this paper, digital disks are defined and analyzed based on these weighted distances: geometric and combinatorial properties are studied. Approximation of the Euclidean circles and distances are shown. The obtained disks are usually dodecagons, enneagons (nonagons) or hexagons. They are proven to be digitally convex.

Keywords: Digital distances · Chamfer distances · Digital disks · Approximation of the Euclidean distance · Non-traditional grids

1 Introduction

Distance functions play important roles in several fields including theory in mathematics and geometry, and also, applications, e.g., engineering and various computing disciplines. The most usual distance function is the Euclidean distance and that is the base of Euclidean geometry. However, in image processing and computer graphics discrete space is preferred. In these spaces there are several phenomena which do not occur in the Euclidean plane, and vice versa. For instance, there is a point (moreover there are infinitely many distinct points) between any two distinct points of the Euclidean space. Opposite to this fact, there are neighbor points (pixels) in digital spaces, e.g., in regular grids. The points of a discrete grid having distance r from a given point of the grid (e.g., the Origin) do not form a circle (in the usual sense), but usually they form a small finite set that is not connected in any sense. Therefore, digital distances are of high importance; they are used in various applications instead of the Euclidean distance [7].

There are three regular tessellations of the plane: the square, the hexagonal and the triangular grids. The points of these regular grids can be addressed by integer coordinate values. In the square grid, the Cartesian coordinate system is used. The pixels of the hexagonal grid can be addressed with two integers [8],

© Springer International Publishing Switzerland 2016
N. Normand et al. (Eds.): DGCI 2016, LNCS 9647, pp. 385–397, 2016.
DOI: 10.1007/978-3-319-32360-2_30

or with a more elegant solution, with three coordinate values whose sum is zero reflecting the symmetry of the grid [6,9]. Similarly, in the triangular grid three coordinate values can effectively be used which are not independent [11,12,22]; and in this way easily captured the three types of neighborhood ([3], see also Fig. 1) in a mathematical way.

Digital distances are path-based and they are defined by connecting pixels/points by paths through neighbor pixels/points. The cityblock and the chessboard distances, the first two digital distances, are based on the number of steps connecting the points where 4-neighbor or 8-neighbor pixels are considered in each step on the square grid, respectively. Since they are very rough approximations of the Euclidean distance, the theory of digital distances are developed in various ways. In neighborhood sequences, the allowed steps may vary in a path [2,15]. In weighted distances various neighbors have various weights, and the sum of the weights is used in the path. These distances are also called chamfer distances [21,23]. These approaches could also be mixed, see, e.g., weight sequences in [20]. Another way to lower the rotational dependency of the digital distances is based on non-traditional grids. Both the hexagonal and the triangular grids have better symmetric properties than the square grid has: they have more symmetry axes and smaller rotations transform the grid to itself. Digital disks can be obtained by digitizing Euclidean circles [7,14], they are the digitized circles and disks. Here, in this paper, we use the term digital disks for objects defined by digital distances. One of the goodness measures of digital distances, that is used in various applications, gives how good is the approximation of the Euclidean distance by them [1]. It can be done by measuring the compactness ratio of these polygons, i.e., digital disks obtained by digital distances. It is known that the (Euclidean) circles/disks are the most compact objects in the plane, their ratio, i.e., the perimeter square over the area, is $4\pi \approx 12.566$, the smallest among all objects'. By measuring this value for the digital disks, the approximation of the Euclidean distance is measured. The most compactness of the circles can also be used to define another type of digital disks: the most compact grid objects try to inherit this characteristic property of the Euclidean circles/disks; they are characterized in [17,18,24] on various grids. Digital disks (spheres) are analyzed in [10] in nD rectangular grids based on weighted distances. The theory of distances based on neighborhood sequences on the triangular grid is also well developed [11,12]. Digital circles/disks and their types are analyzed in [13] while the approximation of the Euclidean circles/distance is done in [19] using the dual grid notation. The weighted distances have been recently investigated on the triangular grid [16]. Here digital disks, and approximation of the Euclidean distance, are provided for various cases depending on the used weights.

2   Preliminaries

In this section the description of the triangular grid and the definition of weighted distances are given.

The triangular grid is a regular tessellation of the plane with same size regular triangles. Actually, it is not a lattice, since there are grid vectors that do not transform the grid to itself. This is due to the fact that there are two types of orientations of the triangles. The grid is described by three coordinate axes x, y, and z (see Fig. 1, right). In this paper we refer for the triangle pixels, as points, and usually, we will use their center, i.e., the dual, hexagonal grid notation. Each point of grid is described by a unique coordinate triplet using only integer values. However, the three values are not independent: the sum of coordinate values can be 0 (even point) or 1 (odd point). Each gridpoint is either even or odd, and these subsets of triangles have two different shapes $\triangle$ and ∇, respectively. The vector through the mid-point of the edge to the opposite corner point is parallel/antiparallel with one of the axes (see also Fig. 1, right). Further we refer to the set of points of the triangular grid by $\mathbb{Z}_*^3$ ($\mathbb{Z}_*^3 = \{(x, y, z)| x + y + z \in \{0, 1\}\}$).

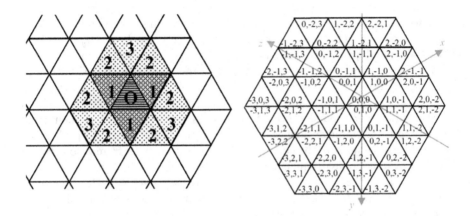

Fig. 1. Type of neighbors (left); coordinate system for the triangular grid (right).

There are three different types of neighborhood on the triangular grid. (In the rectangular grid there are only two types of basic neighborhood.) Two distinct points (triangles) are 1-neighbors iff they have a side in common; they are strict 2-neighbors if there is a triangle which is 1-neighbor of both of them; further, the two points are strict 3-neighbors iff they share a corner, but they are not 1- and 2-neighbors. Two pixels are 2-neighbor if they are strict 2-neighbors or 1-neighbors, and two pixels are 3-neighbors if they are strict 3-neighbors or 2-neighbors. Formally: Let $p = (p(1), p(2), p(3))$ and $q = (q(1), q(2), q(3))$ be two points of $\mathbb{Z}_*^3$, they are m-neighbors ($m = 1, 2, 3$), if the following two conditions hold:

1. $|p(i) - q(i)| \leq 1$ for every $i \in \{1, 2, 3\}$,
2. $\sum_{i=1}^{3} |p(i) - q(i)| \leq m$.

Equality in the second equation defines the strict m-neighborhood relation. In the following we describe some notations which are needed further.

By α-movement we denote a step from a point to one of its 1-neighbor; by β-movement a step to a strict 2-neighbor, and similarly by γ-movement a step to a strict 3-neighbor point. The strict 3-neighbor points of the Origin o is represented by $o_x = (-1, 1, 1)$, $o_y = (1, -1, 1)$ and $o_z = (1, 1, -1)$. A *lane* is the set of points with one of the coordinate value fixed. We say that two lanes are parallel if the same coordinate is fixed, e.g., the lane $\{p(p(1), 0, p(3))\}$ is parallel to $\{p(p(1), 3, p(3))\}$; both of them are perpendicular to axis y. Let $L_{x,-y}$ denote the half-lane $\{p = (p(1), p(2), 0) \mid p(2) \leq 0\}$, that is actually a half lane perpendicular to axis z (between the positive part of axis x and the negative part of axis y) starting from o. Let $L_{y,-x} = \{p = (p(1), p(2), 0) \mid p(1) \leq 0\}$ and similarly, for other four directions the half-lanes can be defined as $L_{x,-z} = \{p = (p(1), 0, p(3)) \mid p(3) \leq 0\}$, $L_{z,-x} = \{p = (p(1), 0, p(3)) \mid p(1) \leq 0\}$, $L_{y,-z} = \{p = (0, p(2), p(3)) \mid p(1) \leq 0\}$ and $L_{z,-y} = \{p = (0, p(2), p(3)) \mid p(2) \leq 0\}$. We can arrange these half-lanes into a cyclic list in positive direction (that is counterclockwise direction), and we can denote by $\triangleright L_{i,-j}$ the next element of $L_{i,-j}$ $(i, j \in \{x, y, z\}, i \neq j)$ in the cyclic list $(L_{x,-y}, L_{z,-y}, L_{z,-x}, L_{y,-x}, L_{y,-z}, L_{x,-z})$. Further, let p and q be two points in a common lane, then let L_{pq} be the set of points on this common lane between p to q. If p and q has a same parity, then let $L_{pq}^* \subseteq L_{pq}$ be the subset having only points with the same parity as p (and q).

Let $p = (p(1), p(2), p(3))$ and $q = (q(1), q(2), q(3))$ be two 3-neighbor points, then the number of their coordinate differences gives the order of their strict neighborhood (as we have defined above). Weighted distances are path based distances, thus we need paths connecting the points. A path from p to q can be defined by a finite point sequence $p = p_0, ..., p_n = q$ in which the points p_{i-1} and p_i are 3-neighbors for every $1 \leq i \leq n$. Thus this path can also be seen as sequence of n steps, such that in each step we move to a 3-neighbor point of the previous one. The weight of the path is equal to $\alpha n_1 + \beta n_2 + \gamma n_3$, where n_i is the number of steps to strict i-neighbors in the path ($n = n_1 + n_2 + n_3$). There are several different paths from p to q with various weights. The weighted distance $d(p, q; \alpha, \beta, \gamma)$ of p and q with weights α, β, γ is the sum of weights of a/the minimal weighted path between p and q. In this paper, the natural condition of the weights, that is $0 < \alpha \leq \beta \leq \gamma$, used. There are various cases regarding the relative ratio of the weights (see [16]). Let the sidelength of the triangle pixels be 1 unit (in the Euclidean plane $\mathbb{R}^2$). By using the dual grid notation, i.e., the center of each triangle pixel instead of the pixel itself, a movement to a

- 1-neighbor means a step by length $\frac{\sqrt{3}}{3} \approx 0.57735$,
- to a strict 2-neighbor means a step by length 1, and
- to a strict 3-neighbor means a step by length $\frac{2\sqrt{3}}{3} \approx 1.1547$.

Comparing it to the square grid a movement to a cityblock neighbor has length 1, while to a diagonal (that is a chessboard neighbor that is not a cityblock) neighbor has length $\sqrt{2}$. One may observe that in the triangular grid, the step to a strict 3-neighbor has length twice than the length of a step to a 1-neighbor. In this paper, we do not allow paths which contain so large difference between some consecutive steps in a path. For this reason, we are working here all the possible cases introduced in [16], but those for which $\gamma + \alpha < 2\beta$ and thus, steps to 1-neighbors and strict 3-neighbors could alternate on some paths with minimal weights. In this way, at the considered cases, the shortest paths may contain steps to 2-neighbors (including 1-neighbors), or only steps to strict 2- and 3-neighbors. Since we work in the dual representation of the triangular grid, the elements of $\mathbb{Z}_*^3$ will refer for the center points of the triangle pixels. In this way, we can easily define convex hull of any (finite) set D of grid points: let $\bar{D} \subset \mathbb{R}^2$ be the convex polygon with smallest area such that each point of D is included in $\bar{D}$. Formally:

$$\bar{D} = \left\{ \sum_{j=1}^{n} \lambda_j p_j \;\middle|\; \sum_{j=1}^{n} \lambda_j, \lambda_j \geq 0 \text{ and } j \in \bar{D} \right\}.$$

Further, the digital set $D \subset \mathbb{Z}_*^3$ is (digitally) *H-convex* if $D = \bar{D} \cap \mathbb{Z}_*^3$ [4,7].

3 Digital Disks

In this section digital disks are described as convex hulls of the point sets reached with paths having weights at most a given value, called, radius. This way, the usage of dual representation of the triangular grid, i.e., the vertices of the hexagonal grid, allows us to compute standard geometric measures, such as the usual perimeter and area of these objects. Thus let us consider the centers of the pixels instead of them. Formally, let

$$D(o, r; \alpha, \beta, \gamma) = \{p \mid d(o, p; \alpha, \beta, \gamma) \leq r\};$$

it is the digital disk defined by weights α, β, γ (with origin o) and radius r.

The weights α, β and γ can have various relations defining various cases. In different cases different paths become optimal, i.e., with minimum weight, and thus, the formula for the distance depends on the considered case [16]. The formulae for computing the weighted distance may depend not only on the weights, but on the type (parity) of the points.

Now, let $\bar{D}(o, r; \alpha, \beta, \gamma)$ denote the convex hull of $D(o, r; \alpha, \beta, \gamma)$ in $\mathbb{R}^2$. Since various formulae are used for computing the distance d, convex hulls can be obtained in various shapes. Actually, the following conditions are fulfilled by the convex hulls of digital disks:

- the weighted distance of the origin and the corners of the convex hull cannot be greater than our fixed radius r.
- those points of $D(o, r; \alpha, \beta, \gamma)$ are chosen to be corners that can make a maximum area of our convex hull with respect to the given radius r.

The relative values of the weights define various cases. In the following subsections we describe the shapes of the possible objects $\bar{D}(o, r; \alpha, \beta, \gamma)$, and characterize them, by their side lengths l, perimeters P, areas A and their isoperimetric ratios κ. The isoperimetric ratio is defined as $\kappa = \frac{P^2}{A}$ which will be used to approximate the Euclidean circle.

Lemma 1. *Let $p \in L_{i,-j}$ and $q \in \triangleright L_{i,-j}$ with $d(o, p; \alpha, \beta, \gamma) = d(o, q; \alpha, \beta, \gamma) = n\alpha$ (for some $n \in \mathbb{N}$, $n \geq 2$) having a shortest path between o and p (and between o and q, respectively) containing $n\alpha$−movements, then there is a lane containing both p and q and that lane does not contain the point o. Moreover p and q have the same parity.*

The proof of Lemma 1 goes in a combinatorial way by noticing that p and q have coordinates 0, $-\lfloor \frac{n}{2} \rfloor$ and $\lceil \frac{n}{2} \rceil$. By a similar technique one can also prove:

Lemma 2. *Let $p \in L_{i,-j}$ and $q \in \triangleright L_{i,-j}$ such that one of the following conditions is fulfilled:*

(a) *$d(o, p; \alpha, \beta, \gamma) = d(o, q; \alpha, \beta, \gamma) = n\beta$ (for some $n \in \mathbb{N}$, $n \geq 2$), and there is a shortest path between o and p containing n β-movements,*

(b) *$d(o, p; \alpha, \beta, \gamma) = d(o, q; \alpha, \beta, \gamma) = n\beta + \alpha$ (for some $n \in \mathbb{N}, n \geq 2$), and there is a shortest path between o and p containing an α-movement and n β-movements,*

(c) *$d(o, p; \alpha, \beta, \gamma) = d(o, q; \alpha, \beta, \gamma) = n\beta$ such that there is a shortest path between o and p containing n β-movements, and there are p', q' strict 3-neighbor points of p and q, respectively, such that $d(o, p'; \alpha, \beta, \gamma) = d(o, q'; \alpha, \beta, \gamma) = n\beta + \gamma$ having a shortest path between o and p' containing a γ- and n β-movements, $(n \in \mathbb{N}, n \geq 1)$.*

    Then there is a lane containing both p and q and that lane does not contain any of the points o, o_x, o_y, o_z. Moreover p and q have the same parity.

By special equilateral triangles formed by pixels one can prove the forthcoming lemmas.

Lemma 3. *Let $p \in L_{i,-j}$ and $q \in \triangleright L_{i,-j}$ such that $d(o, p; \alpha, \beta, \gamma) = n\alpha = d(o, q; \alpha, \beta, \gamma)$ for a value $n \in \mathbb{N}, n \geq 2$ and there is a shortest path between o and p containing only α-movements; and let $p' \in L^*_{pq}$. Then $d(o, p'; \alpha, \beta, \gamma) = n\alpha$.*

Notice that L^*_{pq} may contain only the points p and q in case of $n \in \{2, 3\}$.

Lemma 4. *Let $p \in L_{i,-j}$ and $q \in \triangleright L_{i,-j}$ and $p' \in L^*_{pq}$ such that $d(o, p; \alpha, \beta, \gamma) = d(o, q; \alpha, \beta, \gamma) = n\beta$, and there is a shortest path between o and p containing n β-movements, $(n \in \mathbb{N}, n \geq 2)$. Then $d(o, p'; \alpha, \beta, \gamma) = d(o, p; \alpha, \beta, \gamma)$.*

3.1   Case $2\alpha \leq \beta$ and $3\alpha \leq \gamma$

In this case α-movements are used.

Theorem 1. *If $2\alpha \leq \beta$ and $3\alpha \leq \gamma$, then the convex hull of $D(o, r; \alpha, \beta, \gamma)$ is H-convex and*

(a) *a point if $r < \alpha$.*

(b) *a triangle if $\alpha \leq r < 2\alpha$.*

(c) *a hexagon if $r \geq 2\alpha$.*

Proof. The proof goes by the cases indicated in the theorem.

(a) When $r < \alpha$ the set $D(o, r; \alpha, \beta, \gamma)$ contains only the Origin o, therefore its convex hull is also just a point and therefore, trivially H-convex.

(b) When $\alpha \leq r < 2\alpha$ the set $D(o, r; \alpha, \beta, \gamma)$ contains 4 points: $(0, 0, 0), (1, 0, 0),$ $(0, 1, 0), (0, 0, 1)$. Its convex hull is a triangle (same size as the pixels of the grid), and it is straightforward to proof that it is H-convex. (See also Fig. 2(a).)

(c) Let us see the case $r \geq 2\alpha$: Under the condition of the theorem, the shortest weighted paths are built up by only α-movements. The points that have maximal Euclidean distance from o among the points having at most distance r are on half-lanes $L_{i,-j}$. Let $p_{i,-j}$ represent the point on $L_{i,-j}$ with this property. Let $\triangleright p_{i,-j}$ represent the point on $\triangleright L_{i,-j}$. Hence $d(o, p_{i,-j}; \alpha, \beta, \gamma) = n\alpha$ and by Lemma 1 $p_{i,-j}$ and $\triangleright p_{i,-j}$ are in the same lane. Moreover, by Lemma 3, all the points from $L^*_{p_{i,-j} \triangleright p_{i,-j}}$ have digital distance of $n\alpha$. It means that a point is a corner if and only if it is one of the points $p_{i,-j}$ ($i, j \in \{x, y, z\}$) and, therefore, the shape of $\bar{D}$ is a hexagon (see also Fig. 2(b and c)). Moreover, $\bar{D}$ is H-convex, because it is shown that the points on the sides of $\bar{D}$ have digital distance $n\alpha$ from o and it is clear that the inner points have digital distance less than $n\alpha$ and outer points have larger distance than $n\alpha$.   □

In the following subsections we compute some geometrical measures of the obtained disks (based on their convex hulls).

Subcase $\alpha \leq r < 2\alpha$. In this subcase $r < 2\alpha$ and $2\alpha < \beta$, therefore from the Origin one can move in exactly one step and exactly to those points which have a side in common and it cost α. The convex hull is a triangle with same area than a pixel has (see Fig. 2(a)). The length of the sides of the pixels is one unit. Consequently, the length of the sides for this convex hull is $l = 1$, the perimeter is $P = 3$, the area is $A = \frac{\sqrt{3}}{4}$ and the isoperimetric ratio $\kappa = \frac{36}{\sqrt{3}} \approx 20.78$.

Subcase $2n\alpha \leq r < (2n + 1)\alpha$. By considering a larger value of r, the convex hull be regular or almost regular hexagon. In this subcase, it forms a regular hexagon, where n is a natural number. Also n can be derived from the value of r: $n = \lfloor \frac{r}{2\alpha} \rfloor$ with $l = n$ (every 2 consecutive α-movements correspond to distance of 1 unit), $P = 6n$, $A = \frac{3\sqrt{3}}{2}n^2$, and $\kappa = \frac{36}{\sqrt{3}} \approx 13.86$ (see Fig. 2(b)).

Subcase $(2n + 1)\alpha \leq r < 2(n + 1)\alpha$. In the third subcase we have $n = \lfloor \frac{r-\alpha}{2\alpha} \rfloor$ and the convex hull is a non regular hexagon with sidelengths n and $n + 1$ (three sides with length n and the other three has length $n + 1$). Further, $P = 6n + 3$, $A = (6n^2 + 6n + 1)\frac{\sqrt{3}}{4}$, and the isoperimetric ratio is a function of n that is $\kappa(n) = \frac{36(2n+1)^2}{\sqrt{3}(6n^2+6n+1)}$ (see Fig. 2(c)). As one may observe the hexagon is almost regular, in the sense that the difference of the length of sides is minimal in the grid (1 unit). Also, with larger and larger value of n it is closer and closer to the regular hexagon.

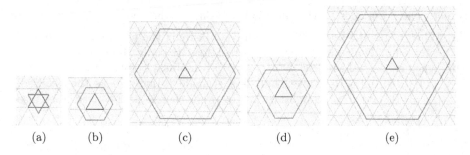

Fig. 2. (a) the convex hull of the digital disk in case $2\alpha \leq \beta$ and $3\alpha \leq \gamma$ with $\alpha \leq r < 2\alpha$ is shown by pink color (the Origin as pixel is marked by black color). The same disk is obtained in case $2\alpha > \beta$, $\alpha + \beta \leq \gamma$ with $\alpha \leq r < \beta$ and, also, in case $2\alpha > \beta$, $3\alpha > \gamma$, $\alpha + \gamma > 2\beta$, $\alpha + \beta > \gamma$ with $\alpha \leq r < \beta$. (b and c) convex hulls of digital disks in case $2\alpha \leq \beta, 3\alpha \leq \gamma$: (b) $2\alpha \leq r < 3\alpha$, (c) $8\alpha \leq r < 9\alpha$. (b and c) the same disks are obtained in case $2\alpha > \beta, \alpha + \beta \leq \gamma$: (b) $\beta \leq r < \alpha + \beta$, (c) $4\beta \leq r < \alpha + 4\beta$ and (b and c) in case $2\alpha > \beta$, $3\alpha > \gamma$, $\alpha + \gamma > 2\beta$, $\alpha + \beta > \gamma$: (b) $\beta \leq r < \gamma$, (c) $4\beta \leq r < 3\beta + \gamma$. (d and e) Examples of digital disks in case $2\alpha \leq \beta, 3\alpha \leq \gamma$: (d) $3\alpha \leq r < 4\alpha$, (e) $9\alpha \leq r < 10\alpha$. (d and e) Same disks are obtained in case $2\alpha > \beta, \alpha + \beta \leq \gamma$: (d) $\alpha + \beta \leq r < 2\beta$, (e) $\alpha + 4\beta \leq r < 5\beta$ and, also, (d and e) in case $2\alpha > \beta$, $3\alpha > \gamma$, $\alpha + \gamma > 2\beta$, $\alpha + \beta > \gamma$; (d) $\alpha + \beta \leq r < 2\beta$, (e) $\alpha + 4\beta \leq r < 5\beta$ (Color figure online).

3.2   Case $2\alpha > \beta$ and $\alpha + \beta \leq \gamma$

Actually, in this case we unite two of the cases of [16]: The formulae for distance are very similar in case $(2\alpha > \beta, 3\alpha \leq \gamma)$ and in case $(2\alpha > \beta, 3\alpha > \gamma, \alpha + \beta \leq \gamma)$. We have joined these cases with common condition: $(2\alpha > \beta$ and $\alpha + \beta \leq \gamma)$.

Lemma 5. *Let $p \in L_{i,-j}$ and $q \in \triangleright L_{i,-j}$ and $p' \in L_{pq}^*$ such that $d(o, p; \alpha, \beta, \gamma) = d(o, q; \alpha, \beta, \gamma) = n\beta + \alpha$, and there is a shortest path between o and p containing an α-movement and n β-movements ($n \in \mathbb{N}, n \geq 2$). Then $d(o, p'; \alpha, \beta, \gamma) = d(o, p; \alpha, \beta, \gamma)$.*

In these cases the obtained convex hulls form the same set by using identical intervals of the digital distance r. Here, again, there are three subcases, and the obtained convex hulls are the same as in the previous subsection.

In the case when $r < \beta$, actually we have exactly the same digital distances, digital discs and convex hulls as in the previous subsection (Theorem 1, case $r < 2\alpha$), and also for larger radii the disks obtained in this case are similar to the ones discussed earlier. Let us see how and why they are.

Theorem 2. *If $2\alpha > \beta$ and $\alpha + \beta \leq \gamma$, then the convex hull of $D(o, r; \alpha, \beta, \gamma)$ is H-convex and*

(a) a point if $r < \alpha$.
(b) a triangle if $\alpha \leq r < \beta$.
(c) a regular hexagon if $n\beta \leq r < \alpha + n\beta$ (for $n \in \mathbb{N}, n \geq 1$).
(d) a non regular hexagon if $\alpha + n\beta \leq r < (n+1)\beta$ (for $n \in \mathbb{N}, n \geq 1$).

The proof is analogous to the proof of Theorem 1 applying Lemmas 2, 4 and 5.

Subcase $\alpha \leq r < \beta$. The convex hull of this condition is an equilateral triangle with $l = 1$, $P = 3$, $A = \frac{\sqrt{3}}{4}$ and $\kappa = \frac{36}{\sqrt{3}}$ (see Fig. 2(a), also, but here, in this case, the condition $\alpha \leq r < \beta$ applies).

Subcase $n\beta \leq r < \alpha + n\beta$. The convex hull is a regular hexagon and $n = \lfloor \frac{d}{\beta} \rfloor$ with $l = n$, $P = 6n$, $A = \frac{3\sqrt{3}}{2}n^2$, and $\kappa = \frac{36}{\sqrt{3}} \approx 13.86$ (See Fig. 2(b)).

Subcase $\alpha + n\beta \leq r < (n+1)\beta$. The convex hull is an almost regular hexagon and $n = \lfloor \frac{r-\alpha}{\beta} \rfloor$ with three sides with length n and three sides with length $n+1$, $P = 6n + 3$, $A = (6n^2 + 6n + 1)\frac{\sqrt{3}}{4}$, and $\kappa(n) = \frac{36(2n+1)^2}{\sqrt{3}(6n^2+6n+1)}$ (see Fig. 2(c)).

3.3   Case $2\alpha > \beta$ and $\alpha + \gamma > 2\beta$ but $\alpha + \beta > \gamma$

In this case steps to strict 3-neighbor may also occur in some shortest paths, and therefore we have a larger variety of cases.

Lemma 6. *Let α, β and γ be given such that $2\alpha > \beta$, $\alpha + \gamma > 2\beta$ and $\alpha + \beta > \gamma$. Let $p \in L_{i,-j}$ and $q \in \triangleright L_{i,-j}$ and $\triangleright L_{i,-j} = L_{k,-j}$ (where $i, j, k \in \{x, y, z\}$ have distinct values) such that $d(o, p; \alpha, \beta, \gamma) = d(o, q; \alpha, \beta, \gamma) = n\beta$ with a shortest path between o and p containing only β-movements; and there are p_1, q_1 strict 3-neighbor points of p and q, respectively, such that $d(o, p_1; \alpha, \beta, \gamma) = d(o, q_1; \alpha, \beta, \gamma) = n\beta + \gamma$ ($n \in \mathbb{N}, n \geq 1$) and there is a shortest path between o and p_1 containing a γ and n β-movements. Further, let $p' \in L^*_{p_1 q_1}$. Then $d(o, p'; \alpha, \beta, \gamma) = d(o, p_1; \alpha, \beta, \gamma)$.*

The next theorem can be proven using Lemmas 2(c), 4 and 6.

Theorem 3. *If $2\alpha > \beta$, and $\alpha + \gamma > 2\beta$ but $\alpha + \beta > \gamma$, then the convex hull is H-convex and it is*

(a) *a point if $\alpha > r$.*
(b) *an equilateral triangle, if $\alpha \leq r \leq \beta$.*
(c) *a regular hexagon, if $n\beta \leq r < (n-1)\beta + \gamma$ (for $n \in \mathbb{N}, n \geq 1$).*
(d) *a non regular hexagon, if $n\beta + \alpha \leq r < (n+1)\beta$ (for $n \in \mathbb{N}, n \geq 1$).*
(e) *an enneagon, if $\gamma \leq r < \alpha + \beta$.*
(f) *a dodecagon, if $(n-1)\beta + \gamma \leq r < \alpha + n\beta$ (for $n \in \mathbb{N}, n \geq 2$).*

In the following subsections some geometric measures of the obtained disks are shown.

Subcase $\alpha \leq r < \beta$. Again, one can see the triangle with $l = 1$, $P = 3$, $A = \frac{\sqrt{3}}{4}$ and $\kappa = \frac{36}{\sqrt{3}}$ (see Fig. 2(a)).

Subcase $n\beta \leq r < (n-1)\beta + \gamma$. This convex hull, in this case, is a regular hexagon and $n = \lfloor \frac{r}{\beta} \rfloor$. Consequently, $l = n$, $P = 6n$, $A = \frac{3\sqrt{3}}{2}n^2$, and $\kappa = \frac{36}{\sqrt{3}} \approx$ 13.86 (see Fig. 2(b and c)).

Subcase $(n-1)\beta + \gamma \leq r < \alpha + n\beta$. In this special subcase $n = \lfloor \frac{r+\beta-\gamma}{\beta} \rfloor$ and the convex hull is dodecagon if $n > 1$. In case of $n = 1$ three pairs of the corners of the dodecagon become identical and thus three of the sides has 0 length, hence forming an enneagon. Generally, for these polygons (an enneagon and the dodecagons) the side lengths are $n - 1, n, \frac{\sqrt{3}}{3}$, $P = 6n + 2\sqrt{3} - 3$, $A = \frac{\sqrt{3}}{12}(18n^2 + 6n - 3)$ and $\kappa(n) = \frac{12(6n+2\sqrt{3}-3)^2}{\sqrt{3}(18n^2+6n-3)}$ (see Fig. 3).

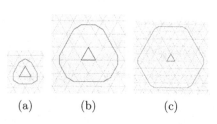

(a)   (b)   (c)   (a)   (b)

Fig. 3. Digital disks in case $2\alpha > \beta$ and $3\alpha > \gamma$, $\alpha + \gamma > 2\beta$ but $\alpha + \beta > \gamma$; (a) $\gamma \leq r < \alpha+\beta$, (b) $\beta+\gamma \leq r < \alpha+2\beta$, (c) $3\beta + \gamma \leq r < \alpha + 4\beta$.

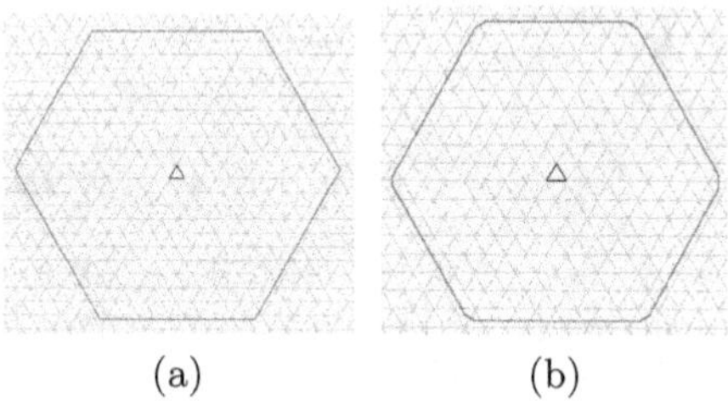

Fig. 4. (a) a non regular hexagon with $(\alpha, \beta, \gamma) = (1, 3, 4)$, $n = 19$, (b) dodecagon with $(\alpha, \beta, \gamma) = (3, 4, 5)$, $n = 8$

Subcase $\alpha + n\beta \leq r < (n+1)\beta$. In the last subcase we have non regular hexagons and $n = \lfloor \frac{d-\alpha}{\beta} \rfloor$ with sidelength $n, n+1$, with $P = 6n + 3$, $A = (6n^2 + 6n + 1)\frac{\sqrt{3}}{4}$, and $\kappa(n) = \frac{36(2n+1)^2}{\sqrt{3}(6n^2+6n+1)}$ (see Fig. 2(c)).

4   Approximation of Euclidean Circles

There are plenty of ways to measure approximation quality of a digital disk, see, e.g. [5]. In this section based on the isoperimetric ratio we show best approximations of the Euclidean circles/disks. The approximations in the limit $r \to \infty$ are also investigated. We also summarize the data about digital disks provided in the previous section in tables. These data are used to find optimal disks to provide the best approximation in each case and subcase. The isoperimetric ratio for any convex hull is greater than the isoperimetric ratio of the Euclidean circle which is 4π, because by using a fixed perimeter, the circle has the greatest area (and, also having a fixed area the circle has the lowest perimeter). Therefore, the smallest possible isoperimetric ratio gives the best approximation of the Euclidean disk.

Table 1. The side lengths l, the perimeter P and the area A at case $2\alpha \le \beta$ and $(3\alpha \le \gamma$ or $\alpha + \beta \le \gamma)$. The isoperimetric ratio κ is also shown.

Subcase	l	P	A	κ
1 $\alpha \le r < 2\alpha$	1	3	$\frac{\sqrt{3}}{4}$	$\frac{36}{\sqrt{3}} \approx 20.78$
2 $2n\alpha \le r < (2n+1)\alpha$	n	$6n$	$\frac{3\sqrt{3}}{2}n^2$	$\frac{36}{\sqrt{3}} \approx 13.86$
3 $(2n+1)\alpha \le r < 2(n+1)\alpha$	$n, n+1$	$6n+3$	$(6n^2+6n+1)\frac{\sqrt{3}}{4}$	$\frac{36(2n+1)^2}{\sqrt{3}(6n^2+6n+1)}$

- Case $2\alpha \le \beta$ and $3\alpha \le \gamma$ and Case $2\alpha \le \beta$ and $\alpha + \beta \le \gamma$: In this case Table 1 summarizes the data of the obtained convex hulls $\bar{D}$. As one can see, κ is constant at the first two subcases (triangle and regular hexagon). At the third case, at non regular hexagons, it depends on n, and κ is decreasing with n having $\lim_{n\to\infty} \kappa = \frac{24}{\sqrt{3}} \approx 13.86$ which is same as the isoperimetric ratio of the regular hexagon. These non regular hexagons look like a regular hexagon when n grows (see Fig. 4(a), for an example).
- Case $2\alpha > \beta$ and $3\alpha > \gamma$, but $\alpha + \beta > \gamma$: For this case three of the subcases are the same as the subcases of the previous case(s), see Table 2. However, there is a subcase that was not present in the other cases. In this subcase (line 3 of Table 2) $\kappa(n) = \frac{12(6n+2\sqrt{3}-3)^2}{\sqrt{3}(18n^2+6n-3)}$. This function has two critical value: at $n_1 = \frac{1}{6}(3-2\sqrt{3}) \approx -0.077$ and at $n_2 = \frac{-3-2\sqrt{3}}{12(-2+\sqrt{3})} \approx 2.01$. Since n is a non negative integer, its optimal value at $n_2 \approx 2$ is considered. Consequently, $\kappa(2) \approx 13.29$ is the minimum value of κ, i.e., the best approximation of an Euclidean disk. The isoperimetric ratio as the function of n is strictly monotonously increasing from n_2 to ∞, and in $\lim_{n\to\infty} k = \frac{24}{\sqrt{3}} \approx 13.86$: this dodecagon become similar to a regular hexagon in the limit $n \to \infty$. See, also, Fig. 4(b) for an example.

Comparing all these cases and subcases, the best κ is found at case $(2\alpha > \beta$ and $3\alpha > \gamma$, but $\alpha + \beta > \gamma)$ with the subcase of $(n-1)\beta + \gamma \le r < \alpha + n\beta$. Other cases/subcases result in triangles and hexagons.

Table 2. The side lengths l, the perimeter P, the area A and the isoperimetric ratio κ at case $2\alpha > \beta$ and $3\alpha > \gamma$, but $\alpha + \beta > \gamma$.

Subcase	l	P	A	κ
1 $\alpha \le r < \beta$	1	3	$\frac{\sqrt{3}}{4}$	$\frac{36}{\sqrt{3}}$
2 $n\beta \le r < (n-1)\beta + \gamma$	n	$6n$	$\frac{3\sqrt{3}}{2}n^2$	$\frac{36}{\sqrt{3}} \approx 13.86$
3 $(n-1)\beta + \gamma \le r < \alpha + n\beta$	$n-1, n, \frac{\sqrt{3}}{3}$	$6n+2\sqrt{3}-3$	$\frac{\sqrt{3}}{12}(18n^2+6n-3)$	$\frac{12(6n+2\sqrt{3}-3)^2}{\sqrt{3}(18n^2+6n-3)}$
4 $\alpha + n\beta \le r < (n+1)\beta$	$n, n+1$	$6n+3$	$(6n^2+6n+1)\frac{\sqrt{3}}{4}$	$\frac{36(2n+1)^2}{\sqrt{3}(6n^2+6n+1)}$

5   Conclusions

Approximation of the Euclidean distance and circles are frequent topics of papers in digital geometry connected to image processing. The literature about weighted/chamfer distances is also rich. This concept has appeared recently on the triangular grid, as well. In this paper, we have continued the work on this field by providing characterization of digital disks for some of the possible cases and also showing best approximations of the Euclidean circle/disk in these cases.

We believe that to find an appropriate digital distance for a given application can be more easily if theoretical results for the digital distances are already provided. One of the criteria of the usage of some distance is their metric properties. In [16] it is proven that all our weighted distances are metrics. Another criteria could be the approximation of the Euclidean distance, results on this line of research are shown here.

References

1. Celebi, M.E., Celiker, F., Kingravi, H.A.: On Euclidean norm approximations. Pattern Recogn. **44**(2), 278–283 (2011)
2. Das, P.P., Chakrabarti, P.P., Chatterji, B.N.: Generalised distances in digital geometry. Inform. Sci. **42**, 51–67 (1987)
3. Deutsch, E.S.: Thinning algorithms on rectangular, hexagonal and triangular arrays. Comm. ACM **15**, 827–837 (1972)
4. Eckhardt, U.: Digital lines and digital convexity. In: Bertrand, G., Imiya, A., Klette, R. (eds.) Digital and Image Geometry. LNCS, vol. 2243, pp. 209–228. Springer, Heidelberg (2002)
5. Farkas, J., Baják, Sz, Nagy, B.: Notes on approximating the Euclidean circle in square grids. Pure Math. Appl. - PU.M.A. **17**(3–4), 309–322 (2006)
6. Her, I.: Geometric transformations on the hexagonal grid. IEEE Trans. Image Proc. **4**, 1213–1221 (1995)
7. Klette, R., Rosenfeld, A.: Digital Geometry: Geometric Methods for Digital Picture Analysis. Morgan Kaufmann Publishers, Elsevier Science B.V., San Francisco (2004)
8. Luczak, E., Rosenfeld, A.: Distance on a hexagonal grid. IEEE Trans. Comput. **25**(5), 532–533 (1976)
9. Middleton, L., Sivaswamy, J.: Hexagonal Image Processing: A Practical Approach. Springer, London (2005)
10. Mukherjee, J.: Hyperspheres of weighted distances in arbitrary dimension. Pattern Recogn. Lett. **34**, 117–123 (2013)
11. Nagy, B.: Metrics based on neighbourhood sequences in triangular grids. Pure Math. Appl. - PU.M.A. **13**(1–2), 259–274 (2002)
12. Nagy, B.: Shortest path in triangular grids with neighbourhood sequences. J. Comput. Inf. Technol. **11**, 111–122 (2003)
13. Nagy, B.: Characterization of digital circles in triangular grid. Pattern Recogn. Lett. **25**, 1231–1242 (2004)
14. Nagy, B.: An algorithm to find the number of the digitizations of discs with a fixed radius. Electron. Notes Discrete Math. **20**, 607–622 (2005)

15. Nagy, B.: Distance with generalized neighbourhood sequences in nD and ∞D. Disc. Appl. Math. **156**, 2344–2351 (2008)

16. Nagy, B.: Weighted distances on a triangular grid. In: Barneva, R.P., Brimkov, V.E., Šlapal, J. (eds.) IWCIA 2014. LNCS, vol. 8466, pp. 37–50. Springer, Heidelberg (2014)

17. Nagy, B., Barczi, K.: Isoperimetrically optimal polygons in the triangular grid. In: Aggarwal, J.K., Barneva, R.P., Brimkov, V.E., Koroutchev, K.N., Korutcheva, E.R. (eds.) IWCIA 2011. LNCS, vol. 6636, pp. 194–207. Springer, Heidelberg (2011)

18. Nagy, B., Barczi, K.: Isoperimetrically optimal polygons in the triangular grid with Jordan-type neighbourhood on the boundary. Int. J. Comput. Math. **90**, 1629–1652 (2013)

19. Nagy, B., Strand, R.: Approximating Euclidean circles by neighbourhood sequences in a hexagonal grid. Theoret. Comput. Sci. **412**, 1364–1377 (2011)

20. Nagy, B., Strand, R., Normand, N.: A weight sequence distance function. In: Hendriks, C.L.L., Borgefors, G., Strand, R. (eds.) ISMM 2013. LNCS, vol. 7883, pp. 292–301. Springer, Heidelberg (2013)

21. Sintorn, I.-M., Borgefors, G.: Weighted distance transforms in rectangular grids. In: ICIAP, pp. 322–326 (2001)

22. Stojmenovic, I.: Honeycomb networks: topological properties and communication algorithms. IEEE Trans. Parallel Distrib. Syst. **8**(10), 1036–1042 (1997)

23. Svensson, S., Borgefors, G.: Distance transforms in 3D using four different weights. Pattern Recogn. Lett. **23**, 1407–1418 (2002)

24. Vainsencher, D., Bruckstein, A.M.: On isoperimetrically optimal polyforms. Theoret. Comput. Sci. **406**, 146–159 (2008)

Discrete Shape Representation, Recognition and Analysis

Representation of Imprecise Digital Objects

Isabelle Sivignon[1,2]([✉])

[1] University Grenoble Alpes, GIPSA-Lab, 38000 Grenoble, France
[2] CNRS, GIPSA-Lab, 38000 Grenoble, France
isabelle.sivignon@gipsa-lab.grenoble-inp.fr

Abstract. In this paper, we investigate a new framework to handle noisy digital objects. We consider digital closed simple 4-connected curves that are the result of an imperfect digital conversion (scan, picture, etc.), and call digital imprecise contours such curves for which an imprecision value is known at each point. This imprecision value stands for the radius of a ball around each point, such that the result of a perfect digitization lies in the union of all the balls. In the first part, we show how to define an imprecise digital object from such an imprecise digital contour. To do so, we define three classes of pixels: inside, outside and uncertain pixels. In the second part of the paper, we build on this definition for a volumetric analysis (as opposed to contour analysis) of imprecise digital objects. From so-called toleranced balls, a filtration of objects, called λ-objects is defined. We show how to define a set of sites to encode this filtration of objects.

Keywords: Digital geometry · Imprecision · Contour · Shape analysis

1 Introduction

Whatever the quality of the sensor of the acquisition device, digital images are always intrisically noisy because no device can reproduce exactly the result of a mathematically well-defined digitization framework that converts a mathematical object into a set of pixels. When image analysis focuses on the objects contained in images, the segmentation step used to extract the objects further increases this phenomenon. Many approaches have been proposed over the years in order to deal with the noise during digital objects analysis. Some approaches introduce a global "thickness" parameter that is used to twarth the noise effect [6]. This has two important drawbacks: first, digital objects with non uniform amount of noise are not handled correctly because of the global parameter - important details in smooth parts of the object may be lost; second, cancelling the noise effect is somehow a loss of information as an ill-defined denoised object is implicitly studied. Even if methods using adaptive thickness instead of a global constant parameter have been proposed [17], the result is always a unique

I. Sivignon—Partially founded by the French *Agence Nationale de la Recherche* (Grant agreement ANR-11-BS02-009).

© Springer International Publishing Switzerland 2016
N. Normand et al. (Eds.): DGCI 2016, LNCS 9647, pp. 401–414, 2016.
DOI: 10.1007/978-3-319-32360-2_31

structuring of the curve, that can be seen as a "denoised" curve (the noise is hidden in the structure). Instead of analysing one possible denoised object, why not analysing all the possible objects at once?

In the last decade, many researches have been conducted on imprecise or uncertain data and related geometric problems. The terms "imprecise" [13] and "uncertain" [8] or "fuzzy" [12,15] data can be found in the literature: uncertain or fuzzy data is endowed with a probabilistic information, while imprecise data only contains geometric information. In this work, we will focus on imprecise data: each point is then replaced by a *region* that models the imprecision.

Dealing with imprecise data has been an active field in computational geometry recently [13]. Results concern for instance upper and lower bounds on geometric quantities (smallest enclosing balls, area of convex hull), or pre-processing of geometric structures (Delaunay triangulation). In digital geometry, geometric predicates on digital segments (concurrency, parallelism) were introduced in the context of imprecise data in [16].

In this paper, we present a general framework that lays the foundations for a geometrical analysis of imprecise digital contours and imprecise digital objects by integrating the imprecision instead of discarding it. In Sect. 2, we show how to compute an imprecise digital object from an imprecise closed digital contour. Then, in a second part, we show how to define a family of objects from an imprecise digital object, and how to encode this family.

2   From an Imprecise Digital Contour to an Imprecise Digital Object

In this work, digital objects live in the 2D cellular grid space [11]. Thus, a digital object is a set of 2-cells - the pixels - together with an adjacency relation (4 or 8 adjacency). The contour of a digital object can be defined as an ordered sequence of 1-cells - the linels - such that the intersection of two consecutive linels is a 0-cell - a pointel. In the following, pointels will be refered to as digital points or points for short, whereas the term pixel will be used for 2-cells.

2.1   Imprecise Digital Contour

Given a curve $\mathscr{C}$ in $\mathbb{R}^2$ and a digitization process $\mathbb{D}$ on $\mathbb{Z}^2$ (e.g. Gauss digitization), we call perfect digitization of $\mathscr{C}$ the curve $\mathbb{D}(\mathscr{C})$. Let $\mathcal{C}$ be a simple closed 4-connected digital curve. In our framework we assume that the digitization process that led to $\mathcal{C}$ (image acquisition, image segmentation) from a curve $\mathscr{C}$ is not perfect, so that the perfect digitization $\mathcal{C}_0$ of $\mathscr{C}$ lies somewhere more or less close to $\mathcal{C}$.

In order to model the result of this "imperfect" digitization process, we define the input data as follows. We suppose that each point p_i of the digital curve $\mathcal{C}$ is endowed with a positive integer weight r_i. Therefore, the input data is an ordered set of weighted or *imprecise points* (p_i, r_i), $i \in [0, n[$, $p_i \in \mathbb{Z}^2$, $r_i \in \mathbb{Z}^+$. The weight value stands for the confidence in the input data at each point: the

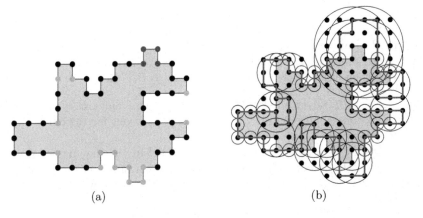

(a)   (b)

Fig. 1. (a) A synthetic imprecise digital contour, color indicates the imprecision value assigned to each point: point p_i is black when $r_i = 1$, green when $r_i = 2$ and blue when $r_i = 3$. (b) Corresponding balls: each ball is depicted by its set of digital points $B_i(p_i, r_i)$ and the Euclidean ball enclosing it; for the sake of lisibility, the Euclidean ball depicted is not of radius r_i, but of radius r_i' such that $r_i' = \min\{r|B_i(p_i, r_i) \subset b(p_i, r)\} + \varepsilon$ (Color figure online).

greater the weight, the more imprecise the contour around this point. Points p_i for which the position is exact (point p_i belongs to C_0) are assigned with a weight equal to one. More precisely, imprecise points are modeled as regions: the weight of the point p is the radius of an open Euclidean ball centered on p, such that the curve C_0 may actually go through the digital points in this ball. Other models could of course be considered. The digital curve C together with the weights is called an *imprecise digital contour* (see Fig. 1 for an illustration).

In this part, we see how to define an imprecise digital object from this imprecise contour. The goal here is to identify each pixel as *inside*, *outside*, or *uncertain*, depending on whether the object enclosed by C_0 may include the pixel or not. A precise definition and characterization of these pixels need some more work, which is developed in the following pages.

Experimental validation will be performed on the data computed using the *a posteriori* computation of noise level (called meaningful scale) proposed in [9,10]: the meaningful scale computed for each point of a digital contour stands for the imprecision at this point.

2.2   Family of Tours

We denote by $B_i(p_i, r_i)$ the set of digital points included in the open Euclidean ball centered in p_i and of radius r_i. Since the digital curves we consider are simple and closed, all indices on the points p_i are considered modulo n.

Observations. Since by hypothesis for all i p_i and p_{i+1} are 4-connected, and all r_i are integers, we have:

- if $r_i > r_{i+1}$ (or conversely), then $B_{i+1} \subset B_i$ (or conversely);
- if $r_i = r_{i+1} = r$, then if $r = 1$, $B_i \cap B_{i+1} = \emptyset$, otherwise $B_i \cap B_{i+1} \neq \emptyset$ and B_i and B_{i+1} are not included in one another.
- let $x \in B_i$ and $y \in B_{i+1}$ such that x and y are 4-connected. If $r_i \neq 1$ or $r_{i+1} \neq 1$, then either x or y (or both) belongs to $B_i \cap B_{i+1}$.

Let $\cup B = \cup_i B_i$. Given the knowledge of the curve $\mathcal{C}$ and $\cup B$, we now define the family of curves representing all the possible positions for the curve $\mathcal{C}_0$. We call these curves *tours*.

A tour γ of $\mathcal{C}$ is a closed simple curve included in $\cup B$ that passes through the balls B_i in the "right order", so that the original order on the points p_i is somehow preserved in the tour. The precise definition is given below.

Definition 1 (Tour). *A tour of $\mathcal{C}$ is a digital curve that:*

(i) is closed, simple, 4-connected and included in $\cup B$;
(ii) can be decomposed into parts $X_0\, X_1\, \ldots\, X_i\, \ldots\, X_n\, X_0$, where for all i, X_i is connected, $X_i \subset B_i$ and $X_i \neq \emptyset$.

This notion of tour is similar, but more restrictive, than the one introduced in [13] in the context of computational geometry. The curve depicted in red in Fig. 1(b) is an example of a tour.

2.3   Definition of an Imprecise Object: Pixel Labeling

Since tours are closed simple curves, Jordan theorem applies and the complementary of each tour is composed of two connected components: the closed one is the object associated to the tour, the other one will be called its complementary. Any pixel in $\mathcal{S}$ is in one of the following classes:

- *inside pixels*: pixels that are in the object associated with a tour for any tour;
- *outside pixels*: pixels that are in the complementary of the object associated with a tour for any tour;
- *uncertain pixels*: pixels for which there exist two tours such that the pixel is in the object for one of the tours, and in its complementary for the other one.

From the definition, tours are oriented: they must go from ball to ball following the order of the points of $\mathcal{C}$. In the following, we assume w.l.o.g. that the digital curve $\mathcal{C}$ is counter-clockwise oriented.

Definition 2 (Valid Arc). *A directed arc between two 4-connected digital points is valid if there exist a tour that uses this arc.*

Definition 3 (Mandatory Arc). *A directed arc xy is said to be mandatory if x belongs to a ball B_i of radius 1 and y belongs to a ball B_{i+1} of radius 1.*

Proposition 1 (Necessary Conditions). *A directed arc xy between two 4-connected points x and y is valid only if:*

(i) *either xy is a mandatory arc (Note that the reverse arc yx is not valid);*

(ii) *or x and y do not belong to any ball of radius one, and there exists an i such that either x and y belong to B_i or x belongs to B_i and y belongs to B_{i+1};*

(iii) *or x belongs to a ball B_i of radius 1, y does not belong to any ball of radius 1 but belongs to the ball B_{i+1};*

(iv) *or y belongs to a ball B_i of radius 1, x does not belong to any ball of radius 1 but belongs to the ball B_{i-1}.*

The arcs verifying one of these properties are called graph arcs.

Proof. Consider an arc a between two 4-connected points x and y in $\cup B$, such that, by contradiction, a does not fulfill any of the conditions above. We prove that this arc is not valid considering the differents cases below.

(a) Suppose that x and y belong to balls of radius 1. By (i), xy is not mandatory. Then x belongs to B_i of radius 1, and y belongs to B_j of radius 1 with $j \neq i + 1$. By Definition 1, any tour must go through $B_i = x = X_i$ and the edge of source x must have its target in $X_{i+1} \subset B_{i+1}$. Thus, no tour can go through xy, and xy is not valid.

(b) Suppose that x and y do not belong to a ball of radius 1. By (ii), there is no i such that x and y belong to B_i or x belongs to B_i and y to B_{i+1}. Thus, x and y neither belong to the same X_j nor to two consecutive X_j and X_{j+1} $\forall j$. From Definition 1, the edge is not valid.

(c) Suppose that x belongs to a ball B_i of radius 1, and y does not belong to any ball of radius 1 (or conversely). By (iii) (or (iv)), y does not belong either to the ball B_{i+1} (or conversely, x does not belong to B_{i-1}). Using a similar argument as in (a), xy cannot be valid.   □

Note that the reciprocal is not true: some graph arcs may not be valid arcs, which means that for some of them, there may not exist a valid tour taking this arc. However, we have the following property:

Property 1. Mandatory arcs are valid arcs.

Figure 2(a) is an illustration of graph arcs and mandatory arcs for an imprecise digital contour. We now introduce some properties that enable to classify the pixels using these graph arcs.

Proposition 2 (Initialization). *Let a be a mandatory arc, let p be the pixel to the left of a, and q the pixel to the right of a. Then p is inside and q is outside.*

Property 2. If the input curve C is a simple closed 4-connected curve, there is no pixel p which is both to the left and to the right of mandatory arcs.

Proposition 3 (Diffusion). *Let p' be a pixel 4-connected to an inside pixel p, such that the linel shared by p and p' does not support a graph arc. Then p' is inside. (similar property if q' is 4-connected to q which is outside).*

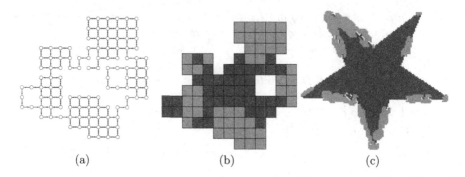

(a)   (b)   (c)

Fig. 2. (a) Graph arcs in blue, mandatory arcs in red for the imprecise digital contour depicted in Fig. 1. (b)–(c) Inside pixels in blue, uncertain pixels in orange, outside pixels in white. In (c), the input imprecise digital contour is from [9,10]: imprecision values are equal to the meaningful scale (Color figure online).

Proof. Suppose p' is not inside. Then there exist a valid tour γ such that p' is outside. Since p is inside, p is in the object associated to γ. Then, since γ is a valid tour, it is a simple closed curve, and thus a Jordan curve: any path between inside and outside must cross γ. As a result, γ must go through the linel shared by p' and p, which is not possible since this linel does not support a graph arc. □

As a corollary, any pixel bounded by four non mandatory arcs will never be labeled using the diffusion of Proposition 3.

These two propositions directly lead to an algorithm (see Algorithm 1) to label the pixels. Note that if there is no mandatory arc to use Proposition 2, then the initialization can still be done if inside and outside pixels are known as part of the input. Results of this labeling are presented in Fig. 2(b)(c).

We denote by $\mathcal{I}$ the set of inside pixels and $\mathcal{U}$ the set of uncertain pixels, and call **imprecise digital object** the pair of sets $(\mathcal{I}, \mathcal{U})$.

3   Volumetric Analysis

Any object bounded by a tour lies somewhere "in between" the digital object defined by $\mathcal{I}$ and the digital object defined by $\mathcal{I} \cup \mathcal{U}$. In the following, we present some tools to study this family of objects.

3.1   Preliminary Definitions

Toleranced balls were introduced in [3] in the context of computational geometry for molecular modelling. They are defined as follows.

Definition 4 (Toleranced Ball). *A toleranced ball $b(c, r^-, r^+)$ is a pair of concentric balls defined by a point c and two radii r^- and r^+ such that $0 \leq r^- \leq r^+$. We denote b^- (resp. b^+) the open ball of center c and radius r^- (resp. r^+).*

Algorithm   1.   PixelLabeling(Set of Pixels P, mandatory arcs M, graph arcs G)

---

foreach *arc $a \in M$* **do**
    label p to the left of a as `inside` and q to the right of a as `outside`;
end
foreach *unlabeled pixel $p \in P$* **do**
    if *the four linels of p are in G* **then**
        label p as `uncertain`
    end
end
Let L be the set of labeled pixels.
foreach *connected component C of $P \backslash L$* **do**
    Let N be the set of pixels p labeled as `inside` or `outside`, 4-connected to C
    and such that one of the linels between p and C is not in G.
    if $N = \emptyset$ **then**
        label all pixels in C as `uncertain`
    else
        if *there is $p \in N$ of label `inside` and $q \in N$ of label `outside`* **then**
            label all pixels in C as `uncertain`
        else all the pixels of N have the same label l
            label all pixels in C as l
        end
    end
end

In the original setting, the constraint on r^- and r^+ was $0 < r^- < r^+$. But as we will see in the next paragraph, we need to relax this constraint in our framework.

Given a collection $\mathcal{B}$ of toleranced balls, let $B^- = \cup_{b \in \mathcal{B}} b^-$ and $B^+ = \cup_{b \in \mathcal{B}} b^+$. We define a **filtration** of the object B^+ as a nested sequence of objects $B^- = S^0 \subseteq S^1 \subseteq \cdots \subseteq S^m = B^+$. The sequence of objects defining the filtration is parameterized using a function $\lambda_{\mathcal{B}} : \mathbb{R}^2 \to \mathbb{R}$ (later on, the function will be restricted to $\mathbb{Z}^2$). For a given $\lambda^i \in \mathbb{R}$, let us define the λ^i-**object** as the set of points p such that $\lambda_{\mathcal{B}}(p) < \lambda^i$. For a given increasing sequence of values $0 = \lambda^0 < \lambda^1 < \cdots < \lambda^m$, the λ^i-objects define a filtration of B^+ if and only if (i) λ^0-object $= B^-$, (ii) for all i, λ^i-object $\subseteq \lambda^{i+1}$-object, and (iii) λ^m-object $= B^+$.

The function $\lambda_{\mathcal{B}}$ can be seen as a function that governs the way toleranced balls grow from b^- to b^+. In the next section we show how to define (i) a collection $\mathcal{B}$ of toleranced balls from an imprecise digital object and (ii) the function $\lambda_{\mathcal{B}}$.

3.2   Toleranced Balls of an Imprecise Digital Object

Distance transformation is a classical tool in digital geometry and more generally image analysis for volumetric analysis. The distance transformation of a digital object $\mathcal{S}$ consists in computing for each pixel of $\mathcal{S}$ the distance to the closest pixel in $\bar{S}$. We denote $dt_S(p) = \min_{q \in \bar{S}}(d(p, q))$, where d is a distance on $\mathbb{Z}^2$. In [5], a very efficient algorithm is presented to compute the exact distance transformation according to the Euclidean distance.

In our setting, the digital object contour is imprecise, so that the distance between a pixel of the object and its complementary is also imprecise. In this context, we define two distance transformations: $dt_\mathcal{I}$ and $dt_{\mathcal{I} \cup \mathcal{U}}$. For each pixel p in $\mathcal{I}$ these distance transformations provide information on the smallest and the greatest possible distance between p and a pixel in $\bar{S}$ for any object S bounded by a tour. For any pixel p in $\mathcal{U}$, the distance transformation $dt_{\mathcal{I} \cup \mathcal{U}}(p)$ is the smallest distance between p and a pixel of $\bar{S}$ for any object S bounded by a tour and that contains p.

We use these two distance transformations to define a toleranced ball for each pixel $p \in \mathcal{I} \cup \mathcal{U}$ as follows. For any pixel p in $\mathcal{I}$ we define the toleranced ball $b(p, dt_\mathcal{I}(p), dt_{\mathcal{I} \cup \mathcal{U}}(p))$. Similarly, for any pixel p in $\mathcal{U}$ we define the toleranced ball $b(p, 0, dt_{\mathcal{I} \cup \mathcal{U}}(p))$. Taking a look back at Definition 4, we note that in this case, we may have $r^- = dt_\mathcal{I}(p) = dt_{\mathcal{I} \cup \mathcal{U}}(p) = r^+$, and for toleranced balls centered on uncertain pixels, we have $r^- = 0$.

The collection of all the toleranced balls thus defined for a given imprecise digital object is denoted by $\mathcal{B}$. By construction, for any $b \in \mathcal{B}$, the open ball $b^-(c, r^-) \subset I$ and the open ball $b^+(c, r^+) \subset \mathcal{I} \cup \mathcal{U}$. With the notations introduced in the previous section, we have $B^- = \mathcal{I}$ and $B^+ = \mathcal{I} \cup \mathcal{U}$ (see Fig. 4(a)).

3.3 Distances

In order to define the function $\lambda_\mathcal{B}$ which governs the ball growing process, the first thing is to specify the distance between a toleranced ball and a point. Several distances can be defined. In [3], the additively-multiplicatively distance is used: each ball grows linearly with respect to the difference between r^- and r^+. This definition assumes that $r^- \neq r^+$.

Definition 5 (Additively-Multiplicatively Distance). *Let $b(c, r^-, r^+)$ a toleranced ball, p a point, and d the Euclidean distance between points of $\mathbb{Z}^2$. Then $d_{am}(b, p) = \frac{d(p, c) - r^-}{r^+ - r^-}$.*

This signed distance is actually based on the additively weighted distance between a ball $b(c, r)$ and a point p defined as $d(c, p) - r$. The additively-multiplicatively distance is associated with the so-called compoundly-weighted Voronoi diagram [14], which falls in the class of non-linear Voronoi diagrams (bissectors are not linear).

In our framework, some of the toleranced balls may have equal radii, so that we have to take this case into account in the distance definition. When the two radii of the toleranced ball are equal, the ball does not grow (nor deflate), and then can never touch a given point p which is not on its boundary.

Let d_b be a signed distance between a ball b and a point p, such that $d_b(b, p) < 0$ if $p \in b$, $d_b(b, p) = 0$ si $p \in \partial b$ and $d_b(b, p) > 0$ if $p \notin b$. We propose a new distance between a toleranced ball and a point, that generalizes the additively-multiplicatively distance and introduces a threshold.

Definition 6. *Let* $b(c, r^-, r^+)$ *a toleranced ball,* p *a point and* d_b *a distance between a ball and a point. Then the function* $d_{tb} : \mathcal{B} \times \mathbb{Z}^2 \to \mathbb{R}$ *is defined as:*

$$d_{tb}(b,p) = \begin{cases} +\infty & \text{if } p \notin b^+ \\ -\infty & \text{if } p \in b^- \\ \frac{d_b(b^-,p)}{d_b(b^-,p)-d_b(b^+,p)} & \text{if } b^- \neq b^+ \\ 0 & \text{if } b^- = b^+ \text{ and } d_b(b^-,p)=0 \end{cases}$$

Note that his definition is quite general since many distances or even semi-metrics d_b can be used. For instance, we recall that the power of a point p with respect to a ball $b(c, r)$ is given by $d(c, p)^2 - r^2$. Injecting this signed semimetric in the generalized distance defined above, we can make the balls grow linearly with respect to the difference between the squares of the two radii.

However, we have to keep in mind that there are two types of toleranced balls: some toleranced balls which center is in $\mathcal{I}$ and others for which the center lies in $\mathcal{U}$. If we make all these balls grow simultaneously using the d_{tb} distance, the balls centered on uncertain points will start to grow at the same time as the balls centered on inside points. So that if the distance d_{tb} is used straightforwardly to define $\lambda_\mathcal{B}$, objects may grow in an unexpected way. In order to take into consideration these different types of balls, we introduce an index η which controls the moment a given toleranced ball begins to grow (see Fig. 4(b)).

Definition 7 (Index). *Given a imprecise digital object* $(\mathcal{I}, \mathcal{U})$ *and its collection of toleranced balls* $\mathcal{B}$, *we define the index* $\eta : \mathcal{I} \cup \mathcal{U} \to \mathbb{Z}^+$ *as:*

$$\eta(p) = \begin{cases} 0 & \text{if } \exists b \in \mathcal{B}, \ p \in b^-, \\ 1 & \text{if } \exists b \in \mathcal{B}, \ b^- \neq \emptyset \text{ and } p \in b^+ \setminus b^- \\ \min_{b(c,r^-,r^+)\in\mathcal{B}}\{\eta(c)+1 \mid p \in b^+\} & \text{otherwise.} \end{cases}$$

For some specific pixels in $\mathcal{U}$, the recursive definition above may not converge. Indeed, a very simple example is the following: consider a pixel $p \in \mathcal{U}$ such that its toleranced ball is $b(p, 0, 1)$ and p belongs to no other toleranced ball. In this case, the recursion loops. More generally, the only pixels p for which the recursion does not end are pixels in $\mathcal{U}$ such that for all b containing p, the center of b is a pixel for which the recursion does not end either. An example of such pixels is given in Fig. 3. In order to prevent such configurations, we would have to make some hypothesis on the sets $\mathcal{U}$ and $\mathcal{I}$. To overcome this problem without constraining the construction of these sets, we define a maximal value $\hat{\eta}$ for η, which is equal to the maximum of the well defined values $\eta(p)$ plus one.

We put together the index and the distance d_{tb} to define a new distance λ.

Definition 8. *Let* d_{tb} *be the distance defined in Definition 6 and* η *the index map defined in Definition 7. Then the function* $\lambda : \mathcal{B} \times \mathbb{Z}^2 \to \mathbb{R}$ *is defined as:*

$$\lambda(b,p) = \eta(c) + d_{tb}(b,p).$$

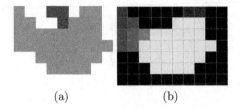

(a)   (b)

Fig. 3. (a) Uncertain pixels in orange, inside pixels in blue, outside ones in white; (b) For the pixels in white the recursive definition of η does not end: the centers of the toleranced balls containing these pixels are never covered during the growing process (Color figure online).

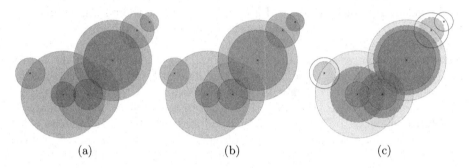

(a)   (b)   (c)

Fig. 4. (a) Set B^- in blue, B^+ in orange and blue; (b) Indices: 0 for the green regions, 1 for the yellow ones, 2 for the orange ones and 3 for the red ones; (c) λ-objects using the additively weighted distance: for $\lambda = 0$ in blue, $\lambda = 0.5$ in purple and blue, $\lambda = 1$ in pink, purple and blue and $\lambda = 1.75$ in yellow, pink, purple and blue (Color figure online).

3.4   Toleranced Balls Growing Process

We can now introduce the growth-governing function $\lambda_\mathcal{B}$.

$$\lambda_\mathcal{B} \colon \mathbb{Z}^2 \to \mathbb{R}$$
$$p \mapsto \min_{b \in \mathcal{B}} \lambda(b, p).$$

Proposition 4. *Given an imprecise object $(\mathcal{I}, \mathcal{U})$ and the collection $\mathcal{B}$ of its toleranced balls, the λ^i-objects defined using function $\lambda_\mathcal{B}$, with an increasing sequence of λ^i, with $\lambda^0 = 0$ and $\lambda^m = \hat{\eta} + 1$ defines a filtration of $\mathcal{I} \cup \mathcal{U}$.*

Proof. First, we prove that the λ^0-object is equal to $B^- = \mathcal{I}$. Let p be a pixel in $\mathcal{I}$, and consider the toleranced ball $b(p, r^-, r^+)$ centered on p. By definition, $b \in \mathcal{B}$, and we have $d_b(b^-, p) < 0$. Then, $\lambda_\mathcal{B}(p) < 0$ and p belongs to the λ^0-object. Conversely, let p be a pixel such that $\lambda_\mathcal{B}(p) < 0$. This implies that there exist a toleranced ball $b(c, r^-, r^+) \in \mathcal{B}$ such that $d_{tb}(b, p) < 0$ and $\eta(c) = 0$. Thus, p belongs to a ball $b^-(c, r^-)$ with $c \in \mathcal{I}$, which proves that p belongs to $\mathcal{I}$.

Proving that for all i and $\lambda^i < \lambda^{i+1}$, the λ^i-object is included in the λ^{i+1}-object is straightforward since a λ^i-object is defined as threshold on the $\lambda_\mathcal{B}$ function.

Last, we shall prove that the λ^m-object with $\lambda^m = \hat{\eta} + 1$ is equal to $B^+ = \mathcal{I} \cup \mathcal{U}$. Let $p \notin \mathcal{I} \cup \mathcal{U}$. Then for all $b \in \mathcal{B}$, $p \notin b^+$, so that $d_{tb}(b,p) = +\infty$ for all b, and thus $\lambda_\mathcal{B}(p) = +\infty$ and $p \notin \lambda^m$-object. Conversely, let $p \in \mathcal{I} \cup \mathcal{U}$. Then there exist $b \in \mathcal{B}$ such that $p \in b^+$. If $p \in b^-$, then $\lambda_\mathcal{B}(p) = -\infty$ and $p \in \lambda^m$-object. Otherwise, for any $b(c_0, r^-, r^+) \in \mathcal{B}$ such that $p \in \mathcal{B}$, we have $\eta(c_0) \leq \hat{\eta}$ and $d_{tb}(b,p) < 1$. As a result, $\lambda_\mathcal{B}(p) < \hat{\eta} + 1$ and $p \in \lambda^m$-object. $\qquad\square$

Let us study a little bit more in details which balls compose λ^i-objects. Given a toleranced ball $b(c, r^-, r^+)$, the ball b grown by a factor α is defined as $b_\alpha = \{p \mid d_{tb}(b,p) < \alpha\}$. If the distance between balls and pixels is the additively weighted distance, then $b_\alpha = b(c, r^- + \alpha(r^+ - r^-))$. If the power distance is used, then $b_\alpha = b(c, r^{-2} + \alpha(r^{+2} - r^{-2}))$. Then we have:

Proposition 5. *Given an imprecise object $(\mathcal{I}, \mathcal{U})$ and the collection $\mathcal{B}$ of its toleranced balls, the λ^i-object is equal to*

$$\left(\bigcup_{b \in \mathcal{B}, \eta(c) < \lfloor \lambda^i \rfloor} b^+ \right) \cup \left(\bigcup_{b \in \mathcal{B}, \eta(c) = \lfloor \lambda^i \rfloor} b_{\lambda^i - \lfloor \lambda^i \rfloor} \right)$$

Proof. Let p be a pixel in the λ^i-object. This is equivalent to say that there exist a toleranced ball $b(c, r^-, r^+) \in \mathcal{B}$ such that $\eta(c) + d_{tb}(b,p) < \lambda^i$. Only two cases are possible:

1. either $\eta(c) < \lfloor \lambda^i \rfloor$ and $dt_{tb}(b,p) < 1$, so that $p \in b^+$:
2. or $\eta(c) = \lfloor \lambda^i \rfloor$ and $dt_{tb}(b,p) < \lambda^i - \lfloor \lambda^i \rfloor$ and thus $p \in b_{\lambda^i - \lfloor \lambda^i \rfloor}$. $\qquad\square$

An illustration of a sequence of λ-objects for a given collection of toleranced balls is given in Fig. 4 (c), for $\lambda^0 = 0$, $\lambda^1 = 0.5$, $\lambda^2 = 1$ and $\lambda^3 = 1.75$. Figure 5(c) shows the λ-objects for the same sequence of λ^i for the imprecise digital *Star* object. Moreover, in Fig. 5(a) and (b) we can see the difference between two growth models using the additively weighted distance in (a) and the power distance in (b) (for λ values equal to $0.2, 0.5, 0.7$ and 0.9).

3.5   Compact Representation of the Filtration

The $\lambda_\mathcal{B}$ function defined in the previous section gives an information, for each pixel of the imprecise digital object, on the instant this pixel will be reached during the ball growing process. The collection $\mathcal{B}$ of toleranced balls used in this process counts exactly one toleranced ball per pixel of the imprecise digital object. However, during the growing process, some of these toleranced balls may never be the first to reach a pixel of the object. As a result, these toleranced balls carry redundant information about the imprecise digital object and could be discarded.

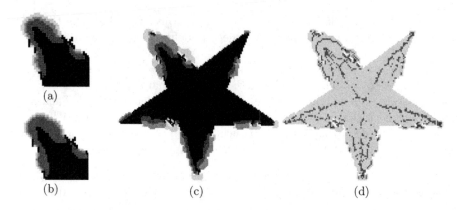

Fig. 5. (a)-(b) λ-objects obtained for $\lambda \in \{0.2, 0.5, 0.7, 0.9\}$ using the additively weighted distance in (a) and the power distance in (b); (c) λ-objects for $\lambda \in \{0, 0.5, 1, 1.75\}$ using the additively weighted distance; (d) Sites extracted from the λ-Voronoi Map based on the additively weighted distance. The colormap from blue to red and then yellow represents the level τ of each site (Color figure online).

Definition 9 (λ-Voronoi Map). *Given a collection $\mathcal{B}$ of toleranced balls, the λ-Voronoi region of a ball $b \in \mathcal{B}$ is defined as $R_b = \{p \mid \lambda(b, p) < \lambda(b', p) \; \forall b' \in \mathcal{B}\}$.*

The λ-Voronoi Map is defined on the set of pixels B^+ for a given collection of toleranced balls $\mathcal{B}$. The set of regions is a tessellation of B^+.

Definition 10 (Sites). *Given a collection $\mathcal{B}$ of toleranced balls, a site is a toleranced ball with a non empty λ-Voronoi region. The level τ of a site b_s is defined as $\tau(b_s) = min\{\lambda(b_s, p) \mid p \in R_{b_s}\}$.*

This approach is similar to the extraction of the medial axis from the squared Euclidean distance transformation [5]: the points of the medial axis are the ones with a non empty region in the power diagram defined with all the balls computed from the distance transformation. Similarly to the medial axis, it is straightforward to show that any λ-object can be reconstructed from the set of sites (see Fig. 5(d) for an illustration).

Proposition 6 (Reconstruction). *Given a collection $\mathcal{B}$ of toleranced balls, any λ^i-object can be reconstructed from the set of sites b_s such that $\tau(b_s) < \lambda^i$ using Proposition 5.*

4   Discussion

In this paper, we introduced a new framework to undertake contour and volumetric analysis of digital objects taking into account the imprecision of the input data. To conclude, we would like to open the discussion on several key points and potential perspectives.

Model. Section 3 was devoted to the definition of a growth model based on the d_{tb} distance function. Using this function, the growth speed of each toleranced ball depends somehow on the difference between the two radii. We could easily define a similar function that would lead to a uniform growth speed for all the balls. With such a setting, we may build on well-known results on the Voronoi Diagram stability through growth process. Indeed, if the rule to grow each ball is to increase the square radius r^2 to $r^2 + t$ at time t, then the Power Voronoi Diagram of these balls is constant at all times [7].

Finally, every growth model leads to different λ^i-objects and there is a need to define criteria (maybe depending on the application) to compare these models.

Algorithms. The results were obtained using brute-force algorithms, which leads to a worst-case complexity of $\mathcal{O}(n^2)$ for an image with n pixels for all the computations (graph, λ values, sites). When volumetric analysis is concerned, all comes down to the computation of the λ-Voronoi Diagram for a collection of toleranced balls. Indeed, the λ values for each point can be easily retrieved from this diagram. Unfortunately, the distance λ used to compute the λ-Voronoi diagram is not separable [4] so that very efficient separable algorithms cannot be applied straightforwardly.

Implementation. Algorithms were implemented using the open-source libraries DGtal [1] for the digital geometry part, and Lemon [2] for the graph part. In particular, distance transformations according to the exact Euclidean distance were computed using the state-of-the-art algorithm from [5] available in DGtal. However, the distance introduced in this work involve non integer computations, that may lead to slightly incorrect results (for instance incorrect reconstruction) due to unintentional rounding operations.

To Go further. The second part of this work about volumetric analysis could actually be applied to other imprecise digital objects inputs: the only requirement is for the imprecise digital object to be defined as a pair of sets of inside and uncertain pixels. One could imagine to extract this kind of information using a segmentation algorithm on gray level images. Similarly, this volumetric study may also be extended to 3D imprecise digital objects quite easily.

Finally, an ultimate objective when volumetric analysis is concerned is to compute the medial axis of the object. A first step towards the definition of the medial axis of an imprecise digital object could be to compute the medial axis for well chosen sample of λ^i-objects, and study the evolution of the positions and radii of the centers computed.

References

1. DGtal: Digital Geometry Tools and Algorithms Library. http://dgtal.org
2. Lemon Graph Library. https://lemon.cs.elte.hu/trac/lemon

3. Cazals, F., Dreyfus, T.: Multi-scale geometric modeling of ambiguous shapes with: toleranced balls and compoundly weighted alpha-shapes. Comput. Graph. Forum **29**(5), 1713–1722 (2010)
4. Coeurjolly, D.: 2D subquadratic separable distance transformation for path-based norms. In: Barcucci, E., Frosini, A., Rinaldi, S. (eds.) DGCI 2014. LNCS, vol. 8668, pp. 75–87. Springer, Heidelberg (2014)
5. Coeurjolly, D., Montanvert, A.: Optimal separable algorithms to compute the reverse euclidean distance transformation and discrete medial axis in arbitrary dimension. IEEE Trans. Pattern Anal. Mach. Intell. **29**(3), 437–448 (2007)
6. Debled-Rennesson, I., Feschet, F., Rouyer-Degli, J.: Optimal blurred segments decomposition of noisy shapes in linear time. Comput. Graph. **30**(1), 30–36 (2006)
7. Goodman, J.E., O'Rourke, J. (eds.): Handbook of Discrete and Computational Geometry. CRC Press Inc, Boca Raton (1997)
8. Jooyandeh, M., Mohades, A., Mirzakhah, M.: Uncertain voronoi diagram. Inf. Process. Lett. **109**(13), 709–712 (2009)
9. Kerautret, B., Lachaud, J.: Meaningful scales detection along digital contours for unsupervised local noise estimation. IEEE Trans. Pattern Anal. Mach. Intell. **34**(12), 2379–2392 (2012)
10. Kerautret, B., Lachaud, J.: Meaningful scales detection: an unsupervised noise detection algorithm for digital contours. IPOL J. **4**, 98–115 (2014)
11. Khalimsky, E., Kopperman, R., Meyer, P.R.: Computer graphics and connected topologies on finite ordered sets. Topology Appl. **36**(1), 1–17 (1990)
12. Lindblad, J., Sladoje, N.: Linear time distances between fuzzy sets with applications to pattern matching and classification. IEEE Trans. Image Process. **23**(1), 126–136 (2014)
13. Löffler, M.: Data imprecision in computational geometry. Ph.D. thesis, Utrecht University (2009)
14. Okabe, A., Boots, B., Sugihara, K.: Spatial Tessellations: Concepts and Applications of Voronoi Diagrams. Wiley, New York (2000)
15. Sladoje, N., Nyström, I., Saha, P.K.: Measurements of digitized objects with fuzzy borders in 2D and 3D. Image Vis. Comput. **23**(2), 123–132 (2005)
16. Veelaert, P.: Graph-theoretical properties of parallelism in the digital plane. Discrete Appl. Math. **125**(1), 135–160 (2003)
17. Zrour, R., Kenmochi, Y., Talbot, H., Buzer, L., Hamam, Y., Shimizu, I., Sugimoto, A.: Optimal consensus set for digital line and plane fitting. Int. J. Imaging Syst. Technol. **21**(1), 45–57 (2011)

Recognition of Digital Polyhedra
with a Fixed Number of Faces

Yan Gérard[(✉)]

LIMOS - UMR 6158 CNRS, Université d'Auvergne,
Clermont-Ferrand, France
yan.gerard@udamail.fr

Abstract. The main task of the paper is to investigate the question of the recognition of digital polyhedra with a fixed number of facets: given a finite lattice set $S \subset \mathbb{Z}^d$ and an integer n, does there exist a polyhedron P of $\mathbb{R}^d$ with n facets and $P \cap \mathbb{Z}^d = S$? The problem can be stated in terms of polyhedral separation of the set S and its complementary $S^c = \mathbb{Z}^d/S$. The difficulty is that the set S^c is not finite. It makes the classical algorithms intractable for this purpose. This problem is overcome by introducing a partial order *"is in the shadow of"*. Its minimal lattice elements are called the *jewels*. The main result of the paper is within the domain of the geometry of numbers: under some assumptions on the lattice set S (if $S \subset \mathbb{Z}^2$ is not degenerated or if the interior of the convex hull of $S \subset \mathbb{Z}^d$ contains an integer point), it has only a finite number of lattice jewels. In this case, we provide an algorithm of recognition of a digital polyhedron with n facets which always finishes.

Keywords: Pattern recognition · Geometry of numbers · Polyhedral separation · Digital polyhedron · Convex sets · Polytopes

1 Introduction

Since Bresenham's pioneer work in the 1960s years, many digital objects or primitives have been defined [1] while a large attention has been given to the question of their recognition. Surprisingly, only small interest has been paid on the digital counterparts of the old Euclidean two-dimensional figures such as triangles, quadrangles, squares, rectangles or more generally regular or irregular polyhedra with n faces in the field of digital geometry (Fig. 1).

There exist many mathematical results dealing with polytopes or polyhedra and lattices. From Pick's Theorem [4] to Ehrhart quasi-polynomials [8] via Barvinok polynomial time algorithm for counting the number of vertices in a polyhedron [3] with a fixed dimension, one of the mathematical stakes of the geometry of numbers is to count the integer points in a convex polytope. There are also many general results about lattice polytopes, namely polytopes with their vertices in a lattice [5,13].

A deep interest on lattices and polyhedra comes from computer science with Integer Programming and Operations Research. In Integer Linear Programming,

© Springer International Publishing Switzerland 2016
N. Normand et al. (Eds.): DGCI 2016, LNCS 9647, pp. 415–426, 2016.
DOI: 10.1007/978-3-319-32360-2_32

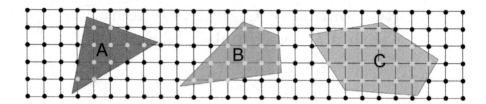

Fig. 1. Digital polyhedra are the intersection of the classical Euclidean polyhedra with the integer lattice. A is a digital triangle, B is a digital quadrangle, C a digital diamond.

a polyhedron is described by linear constraints and the question is to find integer solutions. The problem considered in this paper is the converse: a set $S \subset \mathbb{Z}^d$ of integer points is given and the request is to find a minimal number of real linear constraints which characterize it. A variant is to fix a number n of constraints and to determine if it is possible to characterize S with n linear constraints. This problem is a completely natural question of pattern recognition: given an integer n and a finite subset S of $\mathbb{Z}^d$, is S the Gauss digitization of a convex polyhedron of $\mathbb{R}^d$ with n faces? In other words, does there exist a real polyhedron P verifying $P \cap \mathbb{Z}^d = S$? Such a natural question can wait an easy solution with a quite simple algorithm. Unfortunately, even by adding some assumptions, we only provide a weak result: the problem is decidable.

We prove it in several steps. First, in Sect. 2, we state the problem of recognition of a digital polyhedron with n faces as a question of polyhedral separation. Secondly, in Sect. 3, we investigate the geometry of the problem by introducing a partial order and prove that under some assumptions, the number of minimal lattice elements is finite. At last, in Sect. 4, we present an algorithm and prove that it terminates (under the same assumptions as in Sect. 3).

2   Problem Statement

Let us first recall some basics of the geometry of polytopes and polyhedra: a polytope of $\mathbb{R}^d$ is the convex hull of a finite set of points of $\mathbb{R}^d$ while a polyhedron is usually defined as the intersection of finitely many closed half-spaces $v_i.x \geq h_i$ for i going from 1 to n. The main difference is that polyhedra can be unbounded while polytopes are always compact. Nevertheless Minkowski-Weyl theorem [14] states that bounded polyhedra (bounded means here for instance that they do not contain any half-lines) are polytopes and conversely.

Now, we introduce the lattice $\mathbb{Z}^d$. The objects investigated in the paper are the intersection between a real convex polyhedron P and $\mathbb{Z}^d$. This intersection is usually called the Gauss digitization of P and we prefer to introduce a specific notation taking into account the number of constraints:

Definition 1. *A set $S \subset \mathbb{Z}^d$ is a **digital n-polyhedron** if there exist n closed half-spaces H_i defined by linear inequalities $v_i.x \geq h_i$ with $v_i \in \mathbb{R}^d/_{\{(0)_{1 \leq i \leq d}\}}$*

and $h_i \in \mathbb{R}$, such that the polyhedron $P = \bigcap_{1 \leq i \leq n} H_i$ has an intersection with the lattice $\mathbb{Z}$ equal to S:

$$S = P \cap \mathbb{Z}^d = \{x \in \mathbb{Z}^d | \forall i \in \{1 \cdots n\}, v_i.x \geq h_i\}.$$

Definition 1 is restricted to polyhedra and not polytopes although the paper deals mainly with finite sets S. The reason is that it is more convenient for our purpose and the difference is slight. There exist some finite subsets of $\mathbb{Z}^d$ which are digital n-polyhedra and not n-polytopes (the intersection of a polytope with n faces and the lattice). These digital polyhedra remain, however, pathological. Their recession cone is for instance degenerated with an irrational direction (otherwise, it can be proved that an unbounded digital n-polyhedron contains an infinite number of integer points).

We notice now that if S is a digital n-polyhedron, it is also a m-polyhedron for all $m \geq n$ (just by adding useless linear constraints). Thus, given a finite subset S of $\mathbb{Z}^d$, a question is to determine the minimum n for which S is a digital n-polyhedron. There is of course a first condition on the convexity of S. If S is not digitally convex (the intersection of the real convex hull conv(S) of S with the lattice $\mathbb{Z}^d$ is exactly S), then S is not a digital n-polyhedron. Moreover, if S is convex, it is by definition the intersection of its convex hull with the lattice. This convex hull conv(S) is a polytope. If its affine dimension is d, then, it has some faces of dimension $d-1$. Their cardinality provides an upper bound for the minimal number of faces n of a polyhedron P verifying $P \cap \mathbb{Z}^d = S$.

The problem investigated in this paper is the following:

Problem 1 (DigitalPolyhedronRecognition(d, n, S)).
Input: A dimension d, an integer n and a finite subset S of $\mathbb{Z}^d$.
Output: Is S a digital n-polyhedron? If yes, find a satisfying polyhedron $P \subset \mathbb{R}^d$ with n faces and $S = P \cap \mathbb{Z}^d$.

We can also state Problem 1 as a specific instance of a more generic problem of separation of two sets by a polyhedron with n faces.

Problem 2 (PolyhedralSeparation(d, n, S, T)).
Input: A dimension d, an integer n and two subsets S and T of $\mathbb{R}^d$.
Output: Find a polyhedron P defined by n linear inequalities with S inside P and T outside.

In problem 2, the question is to find n vectors $v_i \in \mathbb{R}^d$ and constants $h_i \in \mathbb{R}$ verifying $\forall x \in S, \forall i \in \{1 \cdots n\}$, $v_i.x \leq h_i$ and $\forall x \in T, \exists i \in \{1 \cdots n\}$, $v_i.x > h_i$. This problem of polyhedral separability has been intensively investigated in the 1980s and 1990s years. The question of separating two sets by an hyperplane, namely *PolyhedralSeparation$(d, 1, S, T)$* can be solved with linear programming with a worst case linear-time complexity in fixed dimension [11]. In dimension $d = 2$, an algorithm with a worst case complexity in $O(k \, log(k))$ (k is the total number of points $|S| + |T|$) is given in [7] for finding the minimum feasible number of faces n. It solves *PolyhedralSeparation$(2, n, S, T)$* in polynomial time. In arbitrary dimension, the problem *PolyhedralSeparation(d, n, S, T)* becomes

NP-complete [12] even with $n = 2$: the 2-separability *PolyhedralSeparation* $(d, 2, S, T)$ is NP-complete. In fixed dimension $d = 3$, as far as we know, the complexity of the problem remains open.

There exist other results in the framework of nested polyhedra introduced by Victor Klee. In this class of problems, the two sets of points S and T given in the input are replaced by two polyhedra S' and T'. The solution has to contain S' and be contained by T'. The problem is solved efficiently in dimension $d = 2$ [2], and is NP-complete from the fixed dimension $d = 3$ [6].

For general polyhedral separation *PolyhedralSeparation*(d, n, S, T), a brute-force approach is to proceed in the following way:

- Solve *PolyhedralSeparation*$(d, 1, S, T')$ with linear programming for all subsets T' of T.
- Find n subsets T^i with $T = \bigcup_{1 \leq i \leq n} T^i$ and where for each index $i \in \{1 \cdots n\}$, the instance *PolyhedralSeparation*$(d, 1, S, T^i)$ is feasible.

Even if *PolyhedralSeparation*(d, n, S, T) is NP-complete, we need an algorithm for solving the problem with finite sets S and T because it is used as routine in Sect. 4. *PolyhedralSeparation*(d, n, S, T) provides also a generic way to express the problem *DigitalPolyhedronRecognition*(d, n, S):

Lemma 1. *Let S be a subset of $\mathbb{Z}^d$ and S^c its complementary in $\mathbb{Z}^d$. The problems DigitalPolyhedronRecognition(d, n, S) and PolyhedralSeparation(d, n, S, S^c) are equivalent.*

Lemma 1 is straightforward. It shows that our task is to tackle a specific instance of *PolyhedralSeparation*(d, n, S, T) with the integer points outside from S as outliers. Its main feature is that the set S^c is infinite. It makes this instance intractable for the usual algorithms of computational geometry. Hence, the goal of the paper is now to determine how we can deal with it.

A first idea is to reduce the set of outliers $T = S^c$ to a smaller subset, for instance, by using of the outer contour of S (we do not precise the notion of neighborhood yet). We consider the instance *Separation*$(d, n, S, \text{OuterContour}(S))$ where the set of outliers is the outer contour of S. As OuterContour(S) is a subset of the complementary S^c of S, a solution *Separation*(d, n, S, S^c) is also a solution of *Separation*$(d, n, S, \text{OuterContour}(S))$ but the converse is false. Although such counterexamples are specific (in dimension $d = 2$, the counterexamples S should have an acute angle $\theta < \frac{\pi}{4}$), they can be easily built (Fig. 2). Thus, the infinite set S^c can not be replaced by the outer contour of S without precautions. It requires at least a better understanding of the geometry of the problem and of the properties of the lattice sets in this framework.

3   The Shadows and the Jewels

Let us consider the general problem *PolyhedralSeparation*(d, n, S, T) of polyhedral separation of two subsets S and T by a polyhedron with n faces.

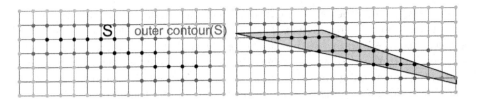

Fig. 2. On the left, a finite subset $S \subset \mathbb{Z}^2$ and its outer contour, in red. On the right, the polyhedron P is a solution of $Separation(2, 3, S, \text{OuterContour}(S))$ but it is not a valid solution of $Separation(2, 3, S, S^c)$ because there are integer points x in P outside from the outer contour of S (Color figure online).

3.1 A Partial Order

We first notice that, in an instance $PolyhedralSeparation(d, n, S, T)$, the set S can be replaced by its real convex hull conv(S) without changing the set of solutions. The convex hull of S and its vertices play an important role in the problem. On the side of T, there is no interest in considering its convex hull conv(T) but there exists another interesting structure -a partial order- that we introduce now. Let us start with the notion of shadow. Given a point $x \in \mathbb{R}^d$, we define its *shadow* relatively to S as the points y of $\mathbb{R}^d$ partially hidden from S by x, as if the convex hull of S was a light source and x a point generating a shadow (Fig. 3):

Definition 2. *Let S be a subset of $\mathbb{R}^d$ and x a point of $\mathbb{R}^d$. The **shadow** of x relatively to S is the set of points $y \in \mathbb{R}^d$ whose convex hull with S contains x:*

$$\text{shadow}_S(x) = \{y \in \mathbb{R}^d / x \in \text{conv}(S \cup \{y\})\}.$$

If x is in the convex hull of S, its shadow is the whole space $\mathbb{R}^d$. The definition is interesting only if the point x does not belong to the convex hull of S. We provide another characterization of the shadow of x:

Lemma 2. *Let S be a subset of $\mathbb{R}^d$ and x, y two points of $\mathbb{R}^d$ different from each other. The point y belongs to the shadow of x if and only if the half-line starting from y and passing though x crosses the convex hull of S.*

Proof. By definition, y is in the shadow of x if the convex hull of $\{y\} \cup S$ contains x. It follows from Caratheodory theorem that there exists a m-simplex Δ with its vertices in $\{y\} \cup S$ containing x. Its vertices are denoted x_k with k going from 0 to m. If we assume that x is not in the convex hull of S, then we can state $x_0 = y$ since y is necessarily one of the vertices of the simplex. The property $x \in \Delta$ provides the existence of $m + 1$ reals $\lambda_k \in [0, 1]$ with a sum $\sum_{k=0}^{d} \lambda_k = 1$ and $x = \sum_{k=0}^{m} \lambda_k x_k$. As x_0 is y, it follows $x - \lambda_0 y = \sum_{k=1}^{m} \lambda_k x_k$. After reorganization, we have

$$y + \frac{1}{\sum_{k=1}^{m} \lambda_k}(x - y) = \frac{1}{\sum_{k=1}^{m} \lambda_k} \sum_{k=1}^{m} \lambda_k x_k$$

which proves the lemma. □

We observe many other nice properties. First, the shadow of a point is convex (Lemma 2 allows to prove that if y and y' are in the shadow of x, then the segment $[y, y']$ is also included in it). Secondly, the set S being fixed, the binary relation "is in the shadow of" defined on $(\mathbb{R}^d / \text{conv}(S))^2$ is reflexive, transitive and antisymmetric (again with Lemma 2). It is a partial order and the set of its minimal elements is going to play an important role (cf the jewels).

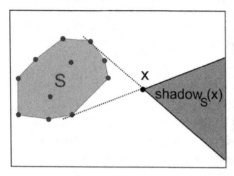 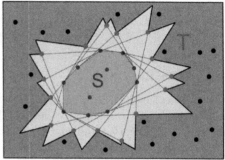

Fig. 3. On the left, the *shadow* of x is the set of points whose convex hull with S contains x. On the right, given two subsets S and T, the shadow of T (in grey) is the union of the shadows of the points of T. Some of the points of T are in the shadow of others. The minimal elements of T according to this partial order are called the *jewels* of T. They are colored in red (Color figure online).

The shadow of a set T is defined as the union of the shadows of its elements: $\text{shadow}_S(T) = \bigcup_{x \in T} \text{shadow}_x(T)$. It is fully determined by the minimal elements of T (minimal according to the relation "is in the shadow of"). We call these minimal elements the *jewels* of T:

Definition 3. *Let S and T be two subsets of $\mathbb{R}^d$. A point $x \in T$ is a **jewel** of T if there exist no point $x' \in T$ verifying $x \in \text{shadow}_S(x')$ (Fig. 3). We denote $\text{jewels}_S(T)$ their set.*

The jewels of T are the counterparts of the vertices of the convex hull of S on the side of T. By construction, we have the following equivalence:

Lemma 3. *Let S and T be two subsets of $\mathbb{R}^d$. We have the equivalence between PolyhedralSeparation(d, n, S, T) and PolyhedralSeparation$(d, n, S, \text{jewels}_S(T))$.*

Proof. Let the polyhedron P be a solution of *PolyhedralSeparation*(d, n, S, T). As the set $\text{jewels}_S(T)$ of the jewels is a subset of T, P is also a solution of *PolyhedralSeparation*$(d, n, S, \text{jewels}_S(T))$. To prove the converse, let us assume that P is a solution of *PolyhedralSeparation*$(d, n, S, \text{jewels}_S(T))$. If P is not a solution of *PolyhedralSeparation*(d, n, S, T), it means that P contains a point x of T. Two cases can occur: First case, x is a jewel, but it contradicts that P is a solution of *PolyhedralSeparation*$(d, n, S, \text{jewels}_S(T))$. Second case, x is not a jewel. Then, it is not minimal in T according to the partial order. Let y be a jewel of T such that y in the convex hull of $\{x\} \cup S$. It follows that y is in P. It contradicts that P is solution of $PolyhedralSeparation(d, n, S, jewels_S(T))$.   □

3.2   About the Number of Lattice Jewels

We are mainly interested in the problem $PolyhedralSeparation(d, n, S, T)$ with a lattice set $S \subset \mathbb{Z}^d$ and its complementary $T = S^c$ in $\mathbb{Z}^d$. The main difficulty is that the set $T = S^c$ is infinite which makes it intractable for usual algorithm. Lemma 3 leads to consider the jewels of the complementary of S in the lattice. As there is no ambiguity in the following, we provide some lighter expressions and notations:

Definition 4. *Let S be a subset of $\mathbb{R}^d$. The **lattice jewels** of S are the jewels of the complementary $S^c = \mathbb{Z}^d / (\mathbb{Z}^d \cap S)$ of S in the lattice $\mathbb{Z}^d$ relatively to S (Fig. 4). We denote $jewels_S$ the set $jewels_S(S^c)$.*

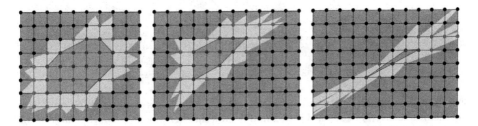

Fig. 4. Three lattice sets in dimension $d = 2$ in blue, their convex hulls in green, their lattice jewels in red and their crown in grey (Color figure online).

Thanks to Lemma 3, we can replace the set $T = S^c$ by its lattice jewels $jewels_S$ and solve $PolyhedralSeparation(d, n, S, jewels_S)$. There are some cases where this reduction is not sufficient to make the instance tractable: the set of the lattice jewels is still infinite (Fig. 5). We can, nevertheless, prove that it can not happen under some assumptions. This result is within the area of the geometry of numbers:

Theorem 1. *Let S be a finite subset of $\mathbb{Z}^d$.*

– *If the interior of the convex hull of S contains an integer point (i)*
– *or, if $d = 2$, and the affine dimension of S is 2 (ii),*

the set of the lattice jewels $jewels_S$ of S is finite.

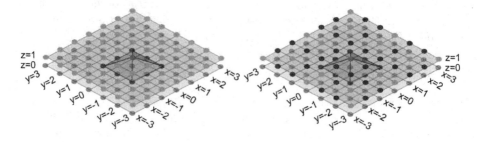

Fig. 5. An example of a finite lattice set with an infinite number of lattice jewels: on the left, the pyramid $S = \{(-1,-1,0),(-1,0,0),(-1,1,0),(0,-1,0),$ $(0,0,0),(0,1,0),(1,-1,0),(1,0,0),(1,1,0),(0,0,1)\}$. On the right, the complementary of the pyramid S^c has infinitely many jewels colored in brown: each point $(a,b,1) \in \mathbb{Z}^3$ with coprime a and b is a lattice jewel (Color figure online).

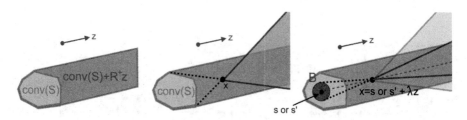

Fig. 6. Case (i): On the left, a set S of $\mathbb{Z}^d$, its convex hull $\mathrm{conv}(S)$ and the polyhedron $\mathrm{conv}(S) + \mathbb{R}^+ z$. In the middle, an integer point x in the interior of $(\mathrm{conv}(S) + \mathbb{R}^+ z)/\mathrm{conv}(S)$. On the right, a ball in the convex hull of S and an integer point on the half-line from its center in direction z. By construction, the shadow $\mathrm{shadow}_B(x)$ of x relatively to the ball B contradicts the existence of jewels tending in direction z.

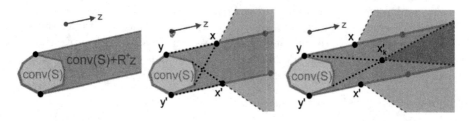

Fig. 7. Case (ii) with a rational z: On the left, the set $S \subset \mathbb{Z}^2$, its convex hull $\mathrm{conv}(S)$ and the polyhedron $\mathrm{conv}(S) + \mathbb{R}^+ z$. In the middle, two integer points x and x' on the boundary of $\mathrm{conv}(S) + \mathbb{R}^+ z$ and their shadows. On the right, the only way to have integer points going in the direction z is that they belong to the strip between $y + \mathbb{R}z$ and $y' + \mathbb{R}z$ but the shadow $\mathrm{shadow}_S(x_k)$ of x_k in the interior of the strip covers its extremity and leads to a contradiction.

Proof. We assume that the set of the lattice jewels of S is infinite and prove a contradiction with (i) or (ii). Under this assumption, we can consider a sequence of lattice jewels $x_k \in jewels_S$ with x_k different from $x_{k'}$ if $k \neq k'$ and for any $k \in \mathbb{N}, x_k \neq (0)_{1 \leq i \leq d}$. We consider the direction $y_k = \frac{x_k}{||x_k||_2}$ which belongs to the hypersphere $\mathbb{S}^d$ centered at the origin and of radius 1. As $\mathbb{S}^d$ is compact, the sequence y_k admits a convergent subsequence y'_k with a non null limit z: $lim_{k \to +\infty} y'_k = z$. It means that the corresponding subsequence of points $x'_k \in jewels_S$ is going in direction z (A).

In the case of (i), where the interior of the convex hull of S contains an integer point s, let us prove that there exists an integer point in the interior of $(\mathrm{conv}(S) + \mathbb{R}^+ z)/\mathrm{conv}(S)$ (B1). If the direction z is rational (the line $\mathbb{R}z$ contains integer points), then by translation, the half-line $s + \mathbb{R}^+ z$ contains infinitely many integer points. The ones which are not in the convex hull of S and on the good side towards it are in the interior of $(\mathrm{conv}(S) + \mathbb{R}^+ z)/\mathrm{conv}(S)$. Let's take such a point and call it x (B1). We have $x = s + \lambda z$ and we notice that by definition of s, there is a ball B centered at s with a strictly positive radius r included in the convex hull of S (B2). Now let us investigate the sub-case of an irrational direction: if z is not a rational direction, we use a classical result of simultaneous rational approximation. As s is in the interior of the convex hull of S, there exists a ball $B_r(s)$ centered at s of radius $r > 0$ included in the interior of the convex hull of S. Then, we consider the Minkowski sum $B_r(s) + \mathbb{R}^+ z$. By construction this set is in the interior of $\mathrm{conv}(S) + \mathbb{R}^+ z$. The simultaneous rational approximation, for instance with Dirichlet's approximation theorem, allows to prove that $B_r(s) + \mathbb{R}^+ z$ contains infinitely many integer points. Another proof of this result can be obtained with Minkowski's theorem by using a sequence of symmetric cylinders $B_{r_k}(s) + \mathbb{R}z$ with r_k tending to 0. Due to its infinite volume, $B_{r_k}(s) + \mathbb{R}z$ contains at least two symmetric integer points x_k and $2s - x_k$ different from s and by stating $r_{k+1} < d(x_k, s + \mathbb{R}z)$, we obtain two new ones... and so on. It proves that $B_r(s) + \mathbb{R}^+ z$ and then also $\mathrm{conv}(S) + \mathbb{R}^+ z$ contain infinitely many integer points. As they cannot be all in the convex hull of S, it proves (B1) in the sub-case of an irrational direction: we have an integer point x in the interior of $(\mathrm{conv}(S) + \mathbb{R}^+ z)/\mathrm{conv}(S)$. We can also notice that x can be written $s' + \lambda z$ with a ball B centered at s' contained by the convex hull of S (B2).

With (A), (B1) and (B2), the last step of the proof is to show that the shadow of x in the interior of $(\mathrm{conv}(S) + \mathbb{R}^+ z)/\mathrm{conv}(S)$ contradicts the existence of lattice jewels x'_k going to infinite in direction z (Fig. 6). The sequence of the integer points x'_k is necessarily going to infinite since there is only a finite number of lattice points in discrete balls of finite radius. It follows that the equality $lim_{k \to +\infty} \frac{x'_k - x}{||x'_k - x||} = lim_{k \to +\infty} \frac{x'_k}{||x'_k||} = z$. Since the convex hull of S contains the ball B centered at s (rational sub-case) or s' (irrational sub-case), the shadow $shadow_S(x)$ of x contains the shadow $shadow_B(x)$. As the points x'_k are lattice jewels of S, they don't belong to the shadow of any other integer point and thus, they are not in $shadow_B(x)$ which is a cone centered around the direction z and with x as vertex. It is in contradiction with the limit $lim_{k \to +\infty} \frac{x'_k - x}{||x'_k - x||} = z$.

In the case of (ii), the same arguments are used. If S is not degenerated, since we are in dimension $d = 2$, the polyhedron $\mathrm{conv}(S) + \mathbb{R}^+ z$ is bounded by two half-lines $y + \mathbb{R}^+ z$ and $y' + \mathbb{R}^+ z$. If the direction z is rational (Fig. 7), then there are integer points x and x' on each half-line. Their shadows $shadow_S(x)$ and $shadow_S(x')$ cover two angular sectors around the polyhedron $\mathrm{conv}(S) + \mathbb{R}^+ z$. As the jewels are not in the shadows of other integer points, the only possibility for the sequence of jewels x'_k to go to infinite in direction z is to be in the strip $\mathrm{conv}(S) + \mathbb{R}^+ z$. It follows that we have an integer point x'_k in the interior of $\mathrm{conv}(S) + \mathbb{R}^+ z$ and outside from the convex hull. As in case (i), it leads to a contradiction. If the direction z is irrational, the existence of an integer point in $(\mathrm{conv}(S) + \mathbb{R}^+ z)/\mathrm{conv}(S)$ is a consequence of the density of $\mathbb{Q}$ in $\mathbb{R}$. It leads again to a contradiction as in case (i).   □

In dimension $d = 2$, Theorem 1 determines whether or not a finite set S has an infinite number of vertices. If S is on a straight line, its complementary S^c has an infinite set of jewels. Otherwise, according to Theorem 1, $jewels_S$ is finite.

From dimension $d = 3$, the condition which guarantees that the set of jewels is finite is the existence of an integer point in the interior of the convex hull of S. There exists however many convex sets and in particular simplexes which do not satisfy it. These lattice sets and their convex hulls are the subject of a great interest in geometry of numbers and operational research [9,13]. One of these results on empty lattice simplexes -Khinchine's flatness theorem- has for instance been used by Lenstra in 1983 to prove that Integer Programming can be solved in polynomial time in fixed dimension [10]. The possibility that they could have an infinite number of lattice jewels makes them harder to investigate, at least in the framework of $PolyhedralSeparation(d, n, S, S^c)$.

4   Algorithm

We provide a pseudo-algorithm to solve $DigitalPolyhedronRecognition(d, n, S)$ for a finite subset $S \subset \mathbb{Z}^d$. Due to Lemmas 1 and 3, this instance is equivalent with the instance $PolyhedralSeparation(d, n, S, jewels_S)$. It means that, if the set of the lattice jewels was previously computed and is finite, it leads to solve the problem of polyhedral separation. It could be a strategy, but it remains the problem of the computation of the jewels which is an open question.

We suggest another strategy: we start with the outer contour of S as initial set of outliers T and use a routine $F_{\mathrm{separation}}(d, n, S, T)$ for solving the first instance $PolyhedralSeparation(d, n, S, T)$. If there is no solution, it ends the problem. Otherwise, the function $F_{\mathrm{separation}}(d, n, S, T)$ provides a solution P. If P contains no other integer solution than the points of S, it is a solution of $DigitalPolyhedron-Recognition(d, n, S)$. Otherwise, there is a non empty set X of integer points x which are not in S: $X = P \cap S^c$. In this case, we update T by adding one or more point of X. This addition is however not sufficient to ensure at the end that the algorithm finishes. We add in T all the integer points of the complementary of S which are in a ball B_X centered at the barycenter of S (the only thing which matters here is that the center of the ball is fixed, it could even be outside from S) and containing at least a point x of X. We proceed again to the resolution $PolyhedralSeparation(d, n, S, T)$ and repeat the process until finding a negative answer or a valid polyhedron P.

Algorithm 1. Incremental recognition : $DigitalPolyhedronRecognition(d, n, S)$

1: $T \leftarrow$ OuterContour(S)
2: polyhedron $\leftarrow F_{\text{separation}}(d, n, S, T)$
3: **if** (polyhedron $=$ null) **then**
4:     **return** *null*
5: $X \leftarrow$ polyhedron $\cap S^c$
6: **while** $card(X) > 0$ **do**
7:     $T \leftarrow T \cup B_X$
8:     polyhedron $\leftarrow F_{\text{separation}}(d, n, S, T)$
9:     **if** (polyhedron $=$ null) **then**
10:         **return** null
11:     $X \leftarrow$ polyhedron $\cap S^c$
12: **return** polyhedron

Theorem 2. *Let S be a finite subset of $\mathbb{Z}^d$.*

- *If the interior of the convex hull of S contains an integer point (i)*
- *or, if $d = 2$, and the affine. dimension of S is 2 (ii),*

then Algorithm 1 ends in a finite time.

Proof. Under the assumptions (i) or (ii), according to Theorem 1, the set S has only a finite set of lattice jewels. As the set T is containing growing balls B_X at each step of the algorithms, if it is running an infinite number of times, there is a step after which T is going to contain all the lattice jewels of S. According to Lemma 3, a polyhedron P obtained at this step cannot contain any other integer point than the ones of S.

4.1   Conclusion

The main task of the paper is to investigate the question of the recognition of digital polyhedra with a fixed number of facets. This natural question of pattern recognition leads to the world of the geometry of numbers which has been deeply investigated in the framework of Integer Programming. We have introduced the new notion of lattice jewel. They are the minimal lattice points under the partial order "is in the shadow of". Our main result is that under some assumptions, a lattice set S has only a finite number of lattice jewels. It is significant because it allows to reduce a problem of polyhedral separation with an infinite set of outliers to a finite set. Even if we don't compute explicitly the lattice jewels of S, it provides an algorithm of recognition of digital n-polyhedra with the guarantee that it finishes (under some assumptions on S).

This work should be continued in several directions:

- Determine the complexity of the problem in fixed dimensions $d > 2$ and provide some algorithms, perhaps based on the explicit computation of the jewels.
- Provide weaker assumptions in Theorem 1, for instance with the lattice width.
- Consider the problem of recognition of rectangles, squares, rotated polyominos or other figures in dimension 2 or 3.

This work is a first step in a very wide range of open questions that relate to both the geometry of numbers, computational geometry and digital geometry.

References

1. DGtal: Digital geometry tools and algorithms library. http://dgtal.org
2. Aggarwal, A., Booth, H., O'Rourke, J., Suri, S., Yap, C.: Finding minimal convex nested polygons. Inf. Comput. **83**(1), 98–110 (1989)
3. Barvinok, A.I.: Lattice points and lattice polytopes. In: Handbook of Discrete and Computational Geometry, 2nd edn., pp. 153–176. Chapman and Hall/CRC, Boca Raton (2004)
4. Beck, M., Robins, S.: Computing the Continuous Discretely. Undergraduate Texts in Mathematics. Springer, New York (2007)
5. Blichfeldt, H.F.: A new principle in the geometry of numbers, with some applications. Trans. Am. Math. Soc. **15**, 227–235 (1914)
6. Das, G., Joseph, D.: The complexity of minimum convex nested polyhedra. In: Proceedings of the 2nd Canadian Conference of Computational Geometry. pp. 296–301 (1990)
7. Edelsbrunner, H., Preparata, F.: Minimum polygonal separation. Inf. Comput. **77**(3), 218–232 (1988)
8. Ehrhart, E.: Sur les polyédres rationnels homothétiques á n dimensions. Comptes Rendus de l'Académie des Sciences **254**, 616–618 (1962)
9. Haase, C., Ziegler, G.: On the maximal width of empty lattice simplices. Eur. J. Comb. **21**, 111–119 (2000)
10. Lenstra, H.: Integer programming with a fixed number of variables. Math. Oper. Res. **8**, 538–548 (1983)
11. Megiddo, N.: Linear-time algorithms for linear programming in r^3 and related problems. SIAM J. Comput. **12**(4), 759–776 (1983)
12. Megiddo, N.: On the complexity of polyhedral separability. Discrete Comput. Geom. **3**(4), 325–337 (1988)
13. Scarf, H.E.: Integral polyhedra in three space. Math. Oper. Res. **10**, 403–438 (1985)
14. Ziegler, G.: Lectures on Polytopes. Graduate Texts in Mathematics. Springer, New York (1995)

Centerlines of Tubular Volumes Based on Orthogonal Plane Estimation

Florent Grélard[1,2][✉], Fabien Baldacci[1,2], Anne Vialard[1,2],
and Jean-Philippe Domenger[1,2]

[1] Univ. Bordeaux, LaBRI, UMR 5800, 33400 Talence, France
florent.grelard@labri.fr
[2] CNRS, LaBRI, UMR 5800, 33400 Talence, France

Abstract. In this article we present a new method to compute a center-line on tubular volumes. The curve-skeleton is central to many applications in discrete geometry, since it captures interesting geometrical and topological properties with respect to the initial volume. Although there are numerous algorithms for skeleton computation, they are not necessarily well-suited for tubular volume specificities, and can lead to unexpected results (faulty branches, not complete). Our method works on tubular-like volumes with junctions (tree structures) and varying diameter. It is based on the center of mass of cross-sections computed using a Voronoï covariance measure on the volume. Our method adapts its parameters to the shape of the tube. The results on both synthetic and real tubular structures with non-constant diameter illustrate the method's efficiency.

Keywords: Centerline · Voronoï covariance measure · Tubular structure analysis

1 Introduction

Simplification of 3D models is of major importance in shape analysis. By reducing dimensionality in particular, skeletonization algorithms applied to 3D objects yield a set of curves and/or surfaces. The particular problem of reducing a volume to a set of curves, known as centerline and curve-skeleton extraction, is key to many applications. For example, the centerline is used for the geometric analysis of 3D shapes [4,15], virtual endoscopy [3] and object matching [2]. Tubular volumes can be found in medical applications (vessels, airways, neurons and colons) but also in industrial applications [4,12]. The afore-mentioned tubes are characterized by elongation in one direction, varying diameter and varying aspect-ratio. They are topologically equivalent to a tree or a graph (see Fig. 1). In [5], the authors describe various desirable properties for a curve-skeleton, such as: same topology as the initial object, centeredness, invariance under isometric transformations, robustness to noise, thinness, junction-detection and smoothness.

N. Normand et al. (Eds.): DGCI 2016, LNCS 9647, pp. 427–438, 2016.
DOI: 10.1007/978-3-319-32360-2_33

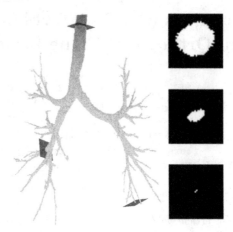

Fig. 1. Example of a tubular volume and three cross-sections, and their 2D equivalent on the right showing variations in diameter and ellipticity.

Satisfying all these wanted properties on any volume is not achievable, due to the data variability and complexity. However, given that our volumes are tubular and according to existing applications, only a subset of these properties are required. Indeed, for length and curvature estimation the curve-skeleton must be thin, for virtual endoscopy the points must be centered and the junctions detected, and for object matching the topology must be preserved. Moreover, since our volumes have varying diameter, the skeleton must be complete i.e. capture information at the lowest local levels. In this paper, an original approach for computing a centerline on any kind of tubes is proposed. It is based on the tracking of centers of mass which are extracted from orthogonal planes computed on the volume.

In Sect. 2 related works with different approaches towards curve-skeleton extraction are presented. Section 3 introduces orthogonal plane computation on a volume. In Sect. 4 the proposed method is explained. Finally, results obtained on both synthetic and real data are described and discussed in Sect. 5, with quantitative comparison with other methods.

2   Related Works

We focus here only on the skeletonization methods which process 3D voxel objects and compute curve-skeletons. According to the survey [5] we can classify the curve-skeleton algorithms for discrete objects in the following classes: (a) thinning algorithms and (b) distance field or general field based algorithms.

Among existing methods, we cite as examples [7,14] for the thinning class. The first one can handle any 3D voxel volume while the second is dedicated to three-dimensional tubular structures. The authors in [7] propose to compute the skeleton by filtering the medial axis with a new bisector function. The resulting filtered medial axis is used as constraint set for the computation of the euclidean skeleton by homotopic thinning. However, the skeleton computation depends on two parameters: r which allows to filter small maximal balls, and α, a threshold for the angle of the bisector function. This method, like other thinning methods, can create small faulty branches (see Fig. 2a), often removed by a pruning step, which are, in our case, difficult to distinguish from small branches that are representative of the volume.

As examples of the second class of curve-skeleton algorithms, we can cite the methods of [6,11] where the skeleton is extracted from a potential field computed on the volume. In [6], the authors compute a hierarchical skeleton based on a newtonian potential field. Each boundary point is considered as an electric charge, repelling interior points of the volume. From the potential vector field, critical points (points where the magnitude of the force vector vanishes) are extracted and connected thanks to paths, yielding a first hierarchy in the skeleton. Then, two levels of hierarchy are added through high divergence value and high-curvature points. These points are used as seeds to recompute new potential fields and extract details on the underlying shape. The main drawback of this kind of algorithm is that the resulting skeleton is not necessarily connected and can be incomplete (see Fig. 2b). Other methods can be found in the recent survey [17]. Among them, the method described in [16] computes a multiscale skeleton by defining an importance measure for each volume point. This method is not adapted to the previously described data: since the importance measure is based on an area computation, small branches in the volume do not appear in the skeleton.

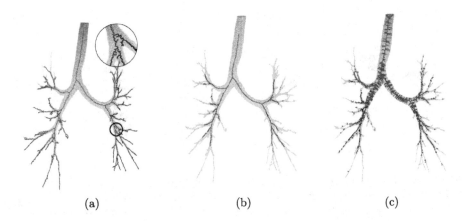

(a)   (b)   (c)

Fig. 2. (a) Skeleton computed using the method described in [7]: parameter values were chosen to obtain the most complete skeleton with the least amount of faulty branches, but there are some regardless, see closeup view. (b) (c) Skeleton computed using potential field [6]: two levels of hierarchy on a skeleton obtained with two different parameter values. In both cases, obtaining a skeleton containing all the branches in the volume as well as no faulty branches is not possible.

This paper addresses the shortcomings of the afore-mentioned methods, and proposes an accurate curve-skeleton extraction on tubular volumes which adapts its parameters to the shape of the tubular volume.

3 Orthogonal Plane Estimation

In [10], we described a way to compute orthogonal planes from a voxel set. Our method is based on a Voronoï covariance measure (VCM). The VCM was first introduced in [13] on point clouds to estimate surface normals, and extended in digital geometry in [8]. Moreover the VCM is proven to be multigrid convergent and resilient to Hausdorff noise. The principal direction of a Voronoï cell corresponds to the surface normal, defined in the computed covariance matrix by the eigenvector with the largest eigenvalue. The VCM computes at a point y in a digital object O a covariance matrix of row vectors between points in a domain of integration DI centered at y and the sites of their respective Voronoï cells, denoted by $p_O(x)$ (see Fig. 3a). The covariance matrix is given as:

$$\mathcal{V}_O(y) = \sum_{x \in DI_O(y)} (x - p_O(x))(x - p_O(x))^t$$

where DI is for example a ball of radius R.

In [10], we compute orthogonal planes on a volume by summing the contribution of all the normal cones $\mathcal{N}$ in the domain of integration (see Fig. 3b). The orthogonal plane is defined as the plane spanned by the two eigenvectors with the highest eigenvalues in the covariance matrix, i.e. those which define the normal cone.

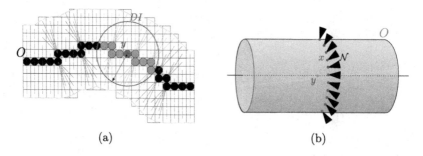

(a)   (b)

Fig. 3. (a) 2D digital object (in black, O) and computation of the VCM around y by integration of vectors (in orange) in the domain of integration, denoted by DI. (b) Computation of an orthogonal plane for a point (in green) inside the volume (cylinder), and relationship with flat normal cones (in dark gray, denoted by $\mathcal{N}$) at different points on the surface inside the domain of integration (in red). For sake of clarity, the normal cones are separated by blank areas, without these the union of all the normal cones corresponds to the expected orthogonal plane (Color figure online).

In this paper, a new approach based on orthogonal planes for curve-skeleton computation on tubular volumes is presented.

4   Proposed Method

Let $O \subseteq \mathbb{Z}^3$ be an object, $\mathcal{P}(p)$ an orthogonal plane at a point p, R the radius of the domain of integration of the VCM used for the orthogonal plane computation, as described in Sect. 3. The main idea of our algorithm is that the curve-skeleton C of O is the set of centers of mass of the orthogonal planes computed on the volume. C is obtained by tracking the centers of mass iteratively (as described in detail in the "Tracking" paragraph).

The parameter R must be "well-adjusted" to capture locally the shape of the object, while being robust to shape irregularities. In other words, the domain of integration should be adjusted such that it contains (a) surface points for which normal cones are aligned with the expected orthogonal plane and (b) the least amount of irrelevant surface points. In the following paragraph, using a ball as domain of integration, we describe how to obtain the minimal value for its radius R automatically.

The size of the domain of integration is initialized with the distance transform (DT) value. Obviously, this value does not correspond to the expected value for R in the case of a tube with elliptic cross-sections, or when the orthogonal plane is computed at a point p near the boundary of the object. The process is the following: an initial orthogonal plane $\mathcal{P}_0(p)$ is computed with $R = DT(p)$. If there is at least one point x in $O \cap \mathcal{P}(p)$, such that $d(p, x) > R$, then R is incremented. This process is repeated until no points in $O \cap \mathcal{P}(p)$ are found at $d > R$. It allows to converge towards both the expected orthogonal plane and the minimal value for R to include all the surface points in $O \cap \mathcal{P}(p)$.

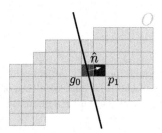

Fig. 4. Tracking performed by our algorithm. From a determined center of mass g_0, we propagate to the next point p_1 thanks to the plane normal $\hat{n}$.

Tracking. It is computationally expensive and redundant to go through all points in the volume. The intuitive idea to solve this problem is that the volume is processed layer by layer, orthogonal plane by orthogonal plane. The layer contains a unique curve-skeleton point so it does not need to be processed further. As a result, points belonging to each computed orthogonal plane are stored in a set M, which are not considered in the rest of the computation.

The starting point p_0 of the algorithm is the point in the volume which has the highest DT value, as it is centered. After the orthogonal plane computation at a point p_i we obtain a center of mass g_i. We then propagate to the next point p_{i+1} in the 26-neighborhood of g_i with the normalized plane normal:

$$p_{i+1} = \begin{cases} g_i + n_{\mathcal{P}(p_i)} & \text{if } p_{i+1} \notin M \\ \max_{q \in (O \backslash M)} DT(q) & \text{otherwise} \end{cases}$$

where $n_{\mathcal{P}(p_i)}$ is the plane normal vector. Since $g_i + n_{\mathcal{P}(p_i)} \approx g_{i+1}$ in regular (i.e. convex and flat) portions of the object, propagation is done either on the next center of mass or on its neighborhood (see Fig. 4). This ensures the algorithm prioritizes points which are well-centered and for which the orthogonal plane normal is close or equivalent to the plane normal after convergence. This process is done for the two normal directions ($p_0 + n_{\mathcal{P}(p_0)}$ and $p_0 - n_{\mathcal{P}(p_0)}$). If $p_{i+1} \in M$, the set of marked vertices, a new point p_{i+1} is assigned from the highest DT value.

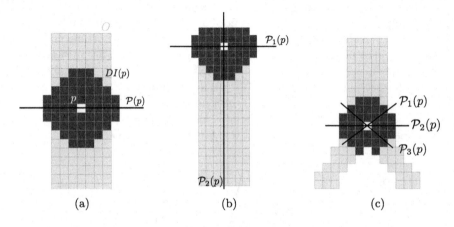

Fig. 5. Computation of orthogonal lines on 2D objects (in light gray). Different configurations arise: (a) regular case in the tube, and resulting orthogonal line (b) endpoint: contribution of two orthogonal lines $\mathcal{P}_1(p)$ and $\mathcal{P}_2(p)$ and (c) junction area: contribution of three orthogonal lines (one per branch).

For some points $p \in O$, the orthogonal plane is not defined because there are several incompatible contributions on the surface. This is illustrated in Fig. 5b for endpoints, and in Fig. 5c for points in junctions. The following paragraphs describe how to specifically detect and handle these two types of points.

Junctions. A junction consists in the intersection of three or more tubes. The reason why orthogonal plane computation performs so poorly in such areas is because the branches forming the junctions have different orthogonal planes, which are all relevant, but cannot be computed in a unique operation. Junctions correspond to high-curvature points on the surface of the volume. In our method, high-curvature points are processed differently from the rest of the points.

Our approach was inspired from the sharp-edge detection method described in [13]. The purpose is to detect such high-curvature points, and more particularly the concave parts of the object. In such parts, and under good sampling conditions, the eigenvalues of every vector in the covariance matrix of the Voronoï cells are roughly the same (i.e. small, see Fig. 6a). As a result, we considered the following method to detect concave points: from the eigenvalues, we compute the concave feature given as:

$$f_c(p) = \frac{\lambda_0}{\sum_i \lambda_i}$$

where λ_0 is the lowest eigenvalue. This captures the fact that the lowest eigenvalue is in the same order of magnitude as the rest of the eigenvalues only for concave points. We consider p a high-curvature point if $f_c(p) > t$ where t is a threshold. This threshold can be detected automatically using the Otsu thresholding method, since the classes of concave and non-concave points are clearly separated. In order to deal with the junctions parts specifically, when the domain of integration of the VCM contains a high curvature point, no further processing is done and the tracking resumes (see "Tracking" paragraph above).

At this stage, the curve-skeleton consists in various connected curves. Each curve needs to be connected to another in a second pass, such that the result is one connected component, i.e. the resulting centerline. For this, we reconstruct sub-volumes taking into account two tubes (out of the three or more in the junctions) and use the main tracking algorithm on these to compute missing parts of the skeleton. The computation follows two steps. Firstly, each extremity of an individual curve is connected to another by a line segment (see white line in Fig. 6b). For a junction consisting in the intersection of three tubes, three points have to be linked. There are three possible line segments linking each pair of points, one of which is not oriented along the main axis of a branch. The line segment which goes "near" a branching point is discarded as follows. Let L be the line segment linking two extremities e_1 and e_2, and B the set of high-curvature points. A line segment is discarded if $\exists (l, b) \in L \times B$ s.t. $d(l, b) < \min(DT(e_1), DT(e_2))$. As a second step, tubular sub-volumes are created by intersecting balls of radius $r = DT(p)$ with the volume, centered in $p \in L$. The main tracking algorithm is then used on this sub-volume, which yields a connected and geometrically-sound curve-skeleton.

Endpoints. The endpoints of the volume must also be treated specifically. This is illustrated in Fig. 5b where the Voronoï cells at the end part of the tube have an orthogonal orientation compared with the expected plane. One way to detect these points was presented in [1], where the authors compute a metric graph and detect the degree of a node in the graph. They compute the shell at p, defined by:

$$S(p) = (B_r(p) \setminus B_R(p)) \cap O$$

where $B_R(p)$ denotes a ball of radius R centered at p, and $r > R$. Here the parameter R is the same as the one described for the domain of integration

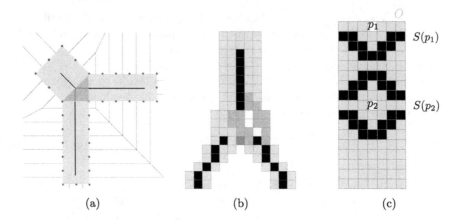

Fig. 6. (a) 2D Voronoï diagram showing concave areas (blue point) have a limited Voronoï cell and normal cone in all directions. The orange area is not considered in the first pass of the curve-skeleton computation. (b) Curves in the skeleton after the first pass (in dark gray), which are connected by a straight line (in white) avoiding a high-curvature point (in blue). From these points, a sub-volume is reconstructed (in orange) and used to compute the skeleton in junctions. (c) Shells (in dark grey) with one connected component corresponding to an endpoint (p_1) and with two components for a point in the volume (p_2) (Color figure online).

size. The number of connected components in $S(p)$ gives the degree $deg(p)$ of p (see Fig. 6c). When this number is equal to one, p is an endpoint, otherwise, when it is greater than one, p corresponds to a point inside the volume. Unlike high-curvature points, endpoints are not processed further because they do not provide relevant information for the curve-skeleton extraction. Theoretically, for junction points, the degree is equal to the number of branches. However this property cannot be used reliably for junction detection on varying-diameter tubes. In fact, given a junction, we want to set r and R such that $deg(p) = 3$. Nonetheless, for junctions where two branches have a very large diameter compared to a third branch limited in length, it is not possible to find solutions.

5   Results

In this section, our method's efficiency is evaluated by comparison to ground-truth curve-skeletons and to state-of-the-art methods namely euclidean skeleton thinning approach [7] and the potential field method [6] presented in Sect. 2. Although the limits of both these methods have been discussed, they satisfy interesting properties related to our applications and are tested within their field of application. Our method is tested on both synthetic and real data.

5.1   Synthetic Data

Various simple tubular volumes have been generated using a parametric curve: cylinder, torus, curved tube and curved tube with noise. Noise was added with

Table 1. Hausdorff distance (first column, in white) and F-measure (second column, in gray) for each method on different volumes.

Data	Euclidean		Potential field		Our method	
Cylinder	**0**	0.684	1.41	0.658	**0**	**0.809**
Torus	**0**	1	1.41	0.494	1	0.817
Curved tube	1.41	0.590	1	**0.740**	1.41	0.635
Noisy curved tube	1.41	0.503	1.41	0.596	1.41	**0.683**

a simplified version of Kanungo's method, which switches the voxel's value with probability α^d where d is the distance to the boundary. For each volume, the expected curve-skeleton corresponds to the parametric curve used to generate the volume. The expected skeleton is compared to the computed skeletons using the Hausdorff distance. The Hausdorff distance allows to measure the dissimilarity between two curves. Here, it is defined by the maximal distance between a point in the computed skeleton and its nearest point in the expected skeleton, i.e.:

$$d_H(T, C) = \max_{c \in c} \left\{ \min_{t \in T} d(t, c) \right\}$$

where T and C are the sets of points in the expected and computed skeleton respectively and d is the euclidean distance. The skeleton is centered if $d_H(T, C) \leq \sqrt{3}$, that is to say if the points in the computed skeleton are "connected" to those of the expected skeleton. Otherwise, the skeleton is thick or contains faulty branches. This distance does not allow to capture the completeness and sensitivity of our method. These two properties are estimated by the precision p and the recall r. The precision estimates whether our skeleton contains faulty points, and ranges from 0 (only faulty points) to 1 (no faulty points). The recall estimates whether the centerline is complete, and ranges from 0 (fully incomplete), to 1 (complete). The precision and recall allow to define the F-measure, which is a compromise between completeness and sensitivity. The results are presented in Table 1.

Various parameters for the state-of-the art methods are tested in order to obtain no faulty branches and the most complete skeleton. Only results with the best parameters are shown here.

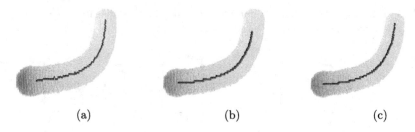

(a)   (b)   (c)

Fig. 7. Curve-skeletons computed on the curved tube using (a) thinning with euclidean skeleton (b) the potential field method and (c) our method.

For all volumes, our skeleton is centered $(d_H(T, C) \leq \sqrt{3})$, and that is the case of the other methods as well. Regarding the F-measure, our method produces the best results in average. The curve-skeleton is close to complete on the cylinder and the torus. Results on the curved tube are comparable to the euclidean skeleton. As for the noisy curved tube, the skeleton is more complete than both the potential field and the euclidean skeletons. Our skeleton is robust to deformations in the volume because our algorithm is designed to work on varying-diameter tubes. The resulting skeletons on these volumes are presented in Fig. 7.

5.2   Real Data

Our method was designed to work on varying-diameter tubular structures with junctions: results obtained from a human airway-tree acquired from CT-scan are shown. The airway-tree segmentation yields a volume topologically equivalent to a binary tree.

Figure 8a and b show the resulting skeletons on two different volumes. A general visual inspection shows the skeleton does not have the defects of the state-of-the-art methods illustrated in Sect. 2. Closeups show the resulting skeleton can capture local information (junctions, see Fig. 9a) and be exempt of faulty branches through the majority of the volume. Moreover, our skeleton is well centered and does not suffer from distortions. In addition to the fact the skeleton is obtained automatically, these properties are interesting in light of the various applications mentioned in Sect. 1. However, when the diameter is as small as one voxel, points might not be taken into account (see Fig. 9b) as they are marked and put in the set M wrongfully. For these particular small branches, the information is lost. Nonetheless, considering the level at which this problem appears, it does not seem like an issue: the branches are too small to provide relevant information. The skeleton was also computed on a tube acquired from laser scan (see Fig. 8c) with concave parts. By visual inspection, the skeleton corresponds precisely to what we would expect. More examples and images are available at [9].

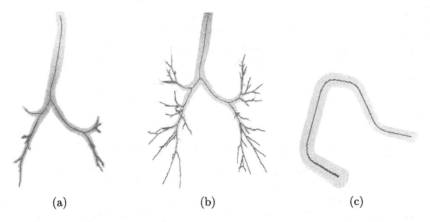

Fig. 8. (a) (b) Centerline extraction on two airway tree volumes. (c) Tube acquired from laser scan, courtesy of [12].

Fig. 9. Closeups of areas in the extracted centerline (a) example of a junction. (b) defect in the skeleton: the smaller branches on the left of the volume are not extracted.

6   Conclusion and Prospects

In this article, we have presented a new approach to estimate a centerline on a tubular volume. We have shown it possesses interesting properties in relation with various applications, is fully automatic, and has an interesting compromise between completeness and sensitivity. Moreover, it is centered and does not suffer from distortions. We have shown interesting results on tubular volumes, that is to say volumes which are elongated in one direction and restricted in the two other directions. This algorithm does not apply on non-tubular objects since orthogonal planes are not properly defined for these objects. This method will be developed further to ensure missing properties are satisfied in the future. For instance, connectivity is not ensured: three types of connectivity 6-, 18- and 26-connectivity are mixed in the skeleton as of yet. This property can be guaranteed easily by deleting or adding points in order to obtain the desired connectivity. We cannot ensure our method preserves topology for genus ≥1 objects. In particular, if there is a hole in a junction area, it will not be preserved in the resulting curve-skeleton. As another prospect, one could think about the relevance of choosing a ball as a domain of integration for the VCM. Only surface points yielding relevant information for orthogonal plane computation (i.e. points on the surface belonging to the plane) are interesting. The domain of integration could be defined as a structuring element depending on the local shape of the object.

References

1. Aanjaneya, M., Chazal, F., et al.: Metric graph reconstruction from noisy data. Int. J. Comput. Geom. Appl. **22**(04), 305–325 (2012)
2. Bai, X., Latecki, L.J.: Path similarity skeleton graph matching. IEEE Trans. PAMI **30**(7), 1282–1292 (2008)
3. Bauer, C., Bischof, H.: Extracting curve skeletons from gray value images for virtual endoscopy. In: Dohi, T., Sakuma, I., Liao, H. (eds.) MIAR 2008. LNCS, vol. 5128, pp. 393–402. Springer, Heidelberg (2008)
4. Coeurjolly, D., Svensson, S.: Estimation of curvature along curves with application to fibres in 3D images of paper. In: Bigun, J., Gustavsson, T. (eds.) SCIA 2003. LNCS, vol. 2749, pp. 247–254. Springer, Heidelberg (2003)
5. Cornea, N.D., Silver, D., Min, P.: Curve-skeleton properties, applications, and algorithms. IEEE Trans. Vis. Comput. Graph. **13**, 530–548 (2007)
6. Cornea, N.D., Silver, D., Yuan, X., Balasubramanian, R.: Computing hierarchical curve-skeletons of 3D objects. Vis. Comput. **21**(11), 945–955 (2005)
7. Couprie, M., Coeurjolly, D., Zrour, R.: Discrete bisector function and Euclidean skeleton in 2D and 3D. Image Vis. Comput. **25**(10), 1519–1698 (2007)
8. Cuel, L., Lachaud, J.-O., Thibert, B.: Voronoi-based geometry estimator for 3D digital surfaces. In: Barcucci, E., Frosini, A., Rinaldi, S. (eds.) DGCI 2014. LNCS, vol. 8668, pp. 134–149. Springer, Heidelberg (2014)
9. Grélard, F.: January 2016. http://www.labri.fr/perso/fgrelard/research.html# skeleton
10. Grélard, F., Baldacci, F., Vialard, A., Lachaud, J.-O.: Precise cross-section estimation on tubular organs. In: Azzopardi, G., Petkov, N., Effenberg, A.O. (eds.) CAIP 2015. LNCS, vol. 9257, pp. 277–288. Springer, Heidelberg (2015). doi:10. 1007/978-3-319-23117-4_24
11. Hassouna, M.S., Farag, A.A.: Variational curve skeletons using gradient vector flow. IEEE Trans. Pat. Anal. Mach. Intell. **31**(12), 2257–2274 (2009)
12. Kerautret, B., Krähenbühl, A., Debled-Rennesson, I., Lachaud, J.-O.: 3D geometric analysis of tubular objects based on surface normal accumulation. In: Murino, V., Puppo, E. (eds.) ICIAP 2015. LNCS, vol. 9279, pp. 319–331. Springer, Heidelberg (2015)
13. Mérigot, Q., Ovsjanikov, M., Guibas, L.: Voronoi-based curvature and feature estimation from point clouds. Vis. CG **17**(6), 743–756 (2011)
14. Palágyi, K., Tschirren, J., Hoffman, E., Sonka, M.: Quantitative analysis of pulmonary airway tree structures. Comput. Biol. Med. **36**(9), 974–996 (2006)
15. Postolski, M., Janaszewski, M., Kenmochi, Y., Lachaud, J.O.: Tangent estimation along 3D digital curves. In: 2012 21st International Conference on Pattern Recognition (ICPR), pp. 2079–2082. IEEE(2012)
16. Reniers, D., van Wijk, J.J., Telea, A.: Computing multiscale curve and surface skeletons of genus 0 shapes using a global importance measure. IEEE Trans. Vis. Comput. Graph. **14**(2), 355–368 (2008)
17. Sobiecki, A., Jalba, A., Telea, A.: Comparison of curve and surface skeletonization methods for voxel shapes. Pattern Recogn. Lett. **47**, 147–156 (2014)

Adaptive Tangential Cover for Noisy Digital Contours

Phuc Ngo[1,2](✉), Hayat Nasser[1,2], Isabelle Debled-Rennesson[1,2],
and Bertrand Kerautret[1,2]

[1] Université de Lorraine, LORIA, UMR 7503, 54506 Vandœuvre-lés-Nancy, France
{hoai-diem-phuc.ngo,hayat.nasser,isabelle.debled-rennesson,
bertrand.kerautret}@loria.fr
[2] CNRS, LORIA, UMR 7503, 54506 Vandœuvre-lés-Nancy, France

Abstract. The notion of tangential cover, based on maximal segments, is a well-known tool to study the geometrical characteristics of a discrete curve. However, it is not adapted to noisy digital contours. In this paper, we propose a new notion, named Adaptive Tangential Cover, to study noisy digital contours. It relies on the meaningful thickness, calculated at each point of the contour, which permits to locally estimate the noise level. The Adaptive Tangential Cover is then composed of maximal blurred segments with appropriate widths, deduced from the noise level estimation. We present a parameter-free algorithm for computing the Adaptive Tangential Cover. Moreover an application to dominant point detection is proposed. The experimental results demonstrate the efficiency of this new notion.

Keywords: Maximal blurred segment · Tangential cover · Noise level · Digital contour · Dominant point

1 Introduction

For more than ten years, the notion of maximal segment has been widely used in discrete geometry to analyze the contour of digital shapes. Based on the definition of discrete line [15], the sequence of all maximal segments along a digital contour C is called the *tangential cover* and a very interesting property is that it can be computed in $O(N)$ time complexity [4].

In [5], F. Feschet studies the structure of discrete curves with tangential cover and shows that the tangential cover has the property of being unique and canonical when computed on closed curves. Tangential cover and maximal segments induce numerous discrete geometric estimators (see [9] for a state of the art): length, tangent, curvature estimators, detection of convex or concave parts of a curve, minimum length polygon of a digital contour, detection of the noise level possibly damaging the shape [6,7], ...

However, the tangential cover is not adapted to noisy digital contours. To deal with this issue, several approaches [3,16,17] have been proposed to obtain

© Springer International Publishing Switzerland 2016
N. Normand et al. (Eds.): DGCI 2016, LNCS 9647, pp. 439–451, 2016.
DOI: 10.1007/978-3-319-32360-2_34

a better model of tangent cover, adapted to noise. One of them consists in using the notion of maximal blurred segments (MBS) which is an extension of maximal segments with a width parameter [3,11]. It was used in several geometric estimators: curvature estimator [11], dominant point detection [10,13], circularity detection, arc and segment decomposition [12,14], ... Nevertheless, the width parameter needs to be manually adjusted and the method is not adaptive to local amount of noise which can appear on real contours.

In this paper, we propose a new notion, named **Adaptive Tangential Cover (ATC)**, to study noisy digital contours. An ATC of a digital contour is composed of MBS with appropriated widths, deduced from the noise level detected in the contour. The meaningful thicknesses [8], local noise estimation at each point of the discrete contour, permits to determine the widths of MBS composing the ATC. Therefore the algorithm to compute ATC is parameter-free. We apply the ATC to dominant point detection [10] and present experimentations showing the interest of this new notion.

The paper is organized as follows: in Sect. 2, we recall definitions and results used in this paper about blurred segments and meaningful thickness. Then, in Sect. 3, we introduce the ATC definition associated to the meaningful thickness (ATC_{MT}) and illustrate its construction algorithm. In Sect. 4, an application and experimental results are presented.

2   Geometrical Tools for Discrete Curves Analysis

We recall in this section several notions of discrete geometry, very useful in the study of discrete curves. The main ideas of previous works are presented here and we refer the reader to the given references for more details.

2.1   Maximal Blurred Segments

As previously described, the discrete primitives as discrete lines [15], blurred segments [3] and maximal blurred segments [11] have been used in numerous works to determine geometrical characteristics of discrete curves.

Definition 1. *A **discrete line** $\mathcal{D}(a, b, \mu, \omega)$, with a main vector (b, a), a lower bound μ and an arithmetic thickness ω (with a, b, μ and ω being integer such that $gcd(a, b) = 1$) is the set of integer points (x, y) verifying $\mu \leq ax - by < \mu + \omega$. Such a line is denoted by $\mathcal{D}(a, b, \mu, \omega)$.*

Let us consider $\mathcal{S}_f$ as a sequence of integer points.

Definition 2. *A discrete line $\mathcal{D}(a, b, \mu, \omega)$ is said to be **bounding** for $\mathcal{S}_f$ if all points of $\mathcal{S}_f$ belong to $\mathcal{D}$.*

Definition 3. *A bounding discrete line $\mathcal{D}(a, b, \mu, \omega)$ of $\mathcal{S}_f$ is said to be **optimal** if the value $\frac{\omega-1}{max(|a|,|b|)}$ is minimal, i.e. if its vertical (or horizontal) distance is equal to the vertical (or horizontal) thickness of the convex hull of $\mathcal{S}_f$.*

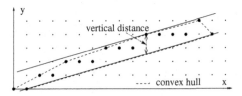

Fig. 1. $\mathcal{D}(2, 7, -8, 11)$, the optimal bounding line of the set of points (vertical distance $= \frac{10}{7} = 1.42$).

This definition is illustrated in Fig. 1 and leads to the definition of the blurred segments.

Definition 4. *A set S_f is a **blurred segment of width** ν if its optimal bounding line has a vertical or horizontal distance less than or equal to ν i.e. if $\frac{\omega-1}{max(|a|,|b|)} \leq \nu$.*

The notion of maximal blurred segment was introduced in [11]. Let C be a discrete curve and $C_{i,j}$ a sequence of points of C indexed from i to j. Let us suppose that the predicate "$C_{i,j}$ is a blurred segment of width ν" is denoted by $BS(i, j, \nu)$.

Definition 5. *$C_{i,j}$ is called a **maximal blurred segment of width** ν and noted $MBS(i, j, \nu)$ iff $BS(i, j, \nu)$, $\neg BS(i, j+1, \nu)$ and $\neg BS(i-1, j, \nu)$.*

The following important property was proved:

Property 1. *Let $MBS_\nu(C)$ be the set of width ν maximal blurred segments of the curve C. Then, $MBS_\nu(C) = \{MBS(B_0, E_0, \nu), MBS(B_1, E_1, \nu),\dots, MBS(B_{m-1}, E_{m-1}, \nu)\}$ and satisfies $B_0 < B_1 < \dots < B_{m-1}$. So we have: $E_0 < E_1 < \dots < E_{m-1}$.*

Deduced from the previous property, an incremental algorithm was proposed in [11] to determine the set of all maximal blurred segments of width ν of a discrete curve C. The main idea is to maintain a blurred segment when a point is added (or removed) to (from) it. The obtained structure for a given width ν can be considered as an extension of the tangential cover [4] and we name it *width ν tangential cover of C*. Examples of tangential covers for different widths are given in Fig. 5(c–f).

2.2   Meaningful Thickness

In [6,7], a notion, called *meaningful scale*, was designed to locally estimate what is the best scale to analyze a digital contour. This estimation is based on the study of the asymptotic properties of the discrete length L of maximal segments. In particular, it has been shown that the lengths of maximal segments covering a point P located on the boundary of a C^3 real object should be between $\Omega(1/h^{1/3})$ and $O(1/h^{1/2})$ if P is located on a strictly concave or convex part

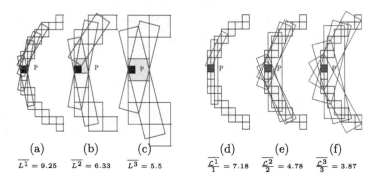

(a)   (b)   (c)   (d)   (e)   (f)

$\overline{L^1} = 9.25$ $\overline{L^2} = 6.33$ $\overline{L^3} = 5.5$ $\frac{\overline{\mathcal{L}^1}}{1} = 7.18$ $\frac{\overline{\mathcal{L}^2}}{2} = 4.78$ $\frac{\overline{\mathcal{L}^3}}{3} = 3.87$

Fig. 2. Images (a–c) illustrate the maximal segments (with their mean discrete length $\overline{L}$) used in the meaningful scale estimation computed by subsampling the initial contour (a). The equivalent blurred segments defined with different thicknesses illustrate the primitives used in the notion of meaningful thickness (d–f). The mean lengths $\overline{\mathcal{L}^k}$ of the blurred segments are given for each thickness/width k.

and near $O(1/h)$ elsewhere (where h represents the grid size). This theoretical property defined on finer and finer grid sizes was used by taking the opposite approach with the computation of the maximal segment lengths obtained with coarser and coarser grid sizes (from subsampling). Such a strategy is illustrated on figure Fig. 2(a–c) with a source point P and its tangential cover defined from subsampling grid size equals to 2 (image (b)) and 3 (image (c)). From the graph of the maximal segment mean lengths $\overline{L^i}$ obtained at different scales, the method consists in recognizing the maximal scale for which the lengths follow the previous theoretical behavior.

The previous method of meaningful scale detection [6,7] has been extended to the detection of the **meaningful thickness (MT)** [8]. This method mainly differs by the choice of the blurred segment primitive and by the scale definition which is given by the thickness/width parameter of the blurred segment. Such a strategy presents the first advantage to be easier to implement without the need to apply different subsamplings.

The length variation of the maximal blurred segments obtained at different thicknesses/widths follows the equivalent properties than for the maximal segment defined from sub-sampling. Figure 3 shows the comparison of the length variations obtained with the maximal segments (b) and the maximal blurred segments (c). In the two cases, the evolution of lengths presents equivalent slopes which are included in the same interval. More formally, if we denote by t_i the thickness of value i, a **multi-thickness profile** $\mathcal{P}_n(P)$ of a point P is defined as the graph $(\log(t_i), \log(\overline{\mathcal{L}}^{t_i}/t_i))_{i=1,...,n}$. The following property has been experimentally checked.

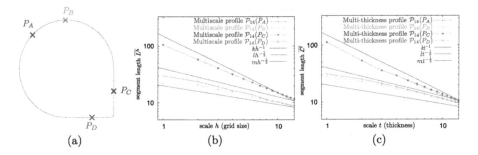

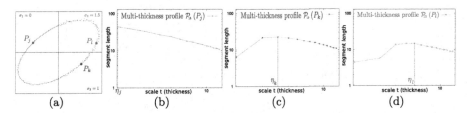

Fig. 3. Comparison between multiscale (b) and multi-thickness (c) profiles on different types of points defined on a shape (a) containing curved (P_A, P_B) and flat (P_C, P_D) parts (Color figure online).

Fig. 4. Multi-thickness profiles (b–d) obtained on different points: P_j with no noise (graph (b)), with low noise (P_k, graph (c)) and important noise (P_l, graph (d)). The meaningful thickness η_j, η_k and η_l are represented on each multi-thickness profile $\mathcal{P}_{15}$ (Color figure online).

Property 2 *(Multi-thickness). The plots of the lengths $\mathcal{L}_j^{t_i}/t^i$ in log-scale are approximately affine with negative slopes s located between $-\frac{1}{2}$ and $-\frac{1}{3}$ for a curved part and around -1 for a flat part.*

Such a profile is illustrated on Fig. 4(a), (b) where a multi-thickness profile is given on a point located on a contour part presenting no noise.

From the multi-thickness profile, a ***meaningful thickness*** is defined as a pair (i_1, i_2), $1 \leq i_1 < i_2 \leq n$, such that for all i, $i_1 \leq i < i_2$, $\frac{Y_{i+1}-Y_i}{X_{i+1}-X_i} \leq T_m$, and the property is not true for $i_1 - 1$ and i_2. As suggested in [8], the value of the parameter T_m is set to 0. In the following, we will denote by η_j the first meaningful thickness (i_1) of a point P_j. Figure 4 illustrates the meaningful thickness obtained for different types of point P_j, P_k and P_l which present respectively the following thicknesses: $\eta_j = 1$, $\eta_k = 3$ and $\eta_l = 5$. Another illustration of meaningful thickness result is proposed in Fig. 5(b).

This notion is used in the next section to define an adaptive tangential cover by taking into account the amount of noise on the curve.

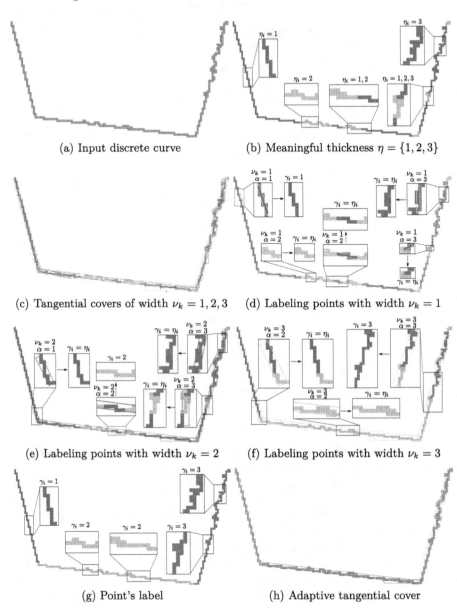

(a) Input discrete curve

(b) Meaningful thickness $\eta = \{1, 2, 3\}$

(c) Tangential covers of width $\nu_k = 1, 2, 3$

(d) Labeling points with width $\nu_k = 1$

(e) Labeling points with width $\nu_k = 2$

(f) Labeling points with width $\nu_k = 3$

(g) Point's label

(h) Adaptive tangential cover

Fig. 5. Illustration of Algorithm 1. (a) Input discrete curve C. (b) Noise level at each point C_i of C detected by meaningful thickness method; the red, green and violet points correspond to the meaningful thickness η_i of 1, 2 and 3 respectively. The label of each point C_i is initialized by its corresponding η_i. (c) Tangential covers of three different widths $\nu_k = 1$, 2 and 3 in yellow, blue and cyan. (d) (e) and (f) Labeling all points C_i of C in function of its meaningful thickness and the tangent covers of widths 1, 2 and 3 respectively; The label γ_i of each point C_i is updated to ν_k if the maximal meaningful thickness, namely α, of points that belong to the $MBS(B_i, E_i, \nu_k)$ passing by C_i is equal to ν_k, and stayed as γ_i otherwise. (g) Label γ_i associated to each point of the considering curve. (h) Adaptive tangential cover obtained from the tangential covers and the labels of points (Color figure online).

3   Adaptive Tangential Cover

The tangential covers applied for dominant point [10] and arc/circle detection [14] use mostly mono-width value, denoted by ν_1. Such a parameter ν_1 allows to take into account the amount of noise present in digital contours. This method has two drawbacks. Firstly, the value of ν_1 is manually adjusted in order to obtain a relevant approximating polygon of the contours w.r.t. the noise. Secondly, the noise appearing along the contour can be random. In other words, different noise levels can be presented along the contours. Figure 5(b) illustrates the different noise levels detected by the meaningful thickness method. Thus, using mono-width value for tangential covers is inadequate in case of noisy curves.

To overcome these issues, we present the definition of *adaptive tangential cover* which is a tangential cover with different width values. To this end, we first introduce the notion of *inclusion* of two MBS.

Definition 6. *Let C be a discrete curve and $MBS_i = MBS(B_i, E_i, .)$, $MBS_j = MBS(B_j, E_j, .)$ two distinct maximal blurred segments on C. MBS_j is said to be **included** in MBS_i if $B_i \leq B_j$ and $E_i \geq E_j$, and noted by $MBS_j \subseteq MBS_i$.*

Definition 7. *Let $MBS(C)$ be a set of maximal blurred segment of a discrete curve C. $MBS_i = MBS(B_i, E_i) \in MBS(C)$ is said **largest** if for all $MBS_j \in MBS(C)$ with $i \neq j$, $MBS_j \not\subseteq MBS_i$.*

Definition 8. *Let $C = (C_i)_{0 \leq i \leq n-1}$ be a discrete curve. Let $\eta = (\eta_i)_{0 \leq i \leq n-1}$ be the vector of meaningful thickness associated to each C_i of C. Let $MBS(C) = \{MBS_{\nu_k}(C)\}$ be the sets of MBS for the different values ν_k in η. An **adaptive tangential cover associated to meaningful thickness** (ATC_{MT}) of C is defined as the set of the largest MBS of $\{MBS_j = MBS(B_j, E_j, \nu_k) \in MBS(C) \mid \nu_k = \max\{\eta_t \mid t \in [\![B_j, E_j]\!]\}\}$.*

As stated in Sect. 2.2, the method of meaningful thickness allows to prevent/estimate locally the noise level at each point of the discrete contour. Such a framework is thus integrated in the construction of ATC_{MT} to provide the information of noise along the contour. More precisely, the ATC_{MT} contains the MBS with width values varying in function of the perturbations obtained by the meaningful thickness values. Since the noise levels are different along the contour curve, accordingly, the obtained ATC_{MT} has the MBS with bigger width values at noisy zones, and with smaller width values in zones with less or no noise (see Fig. 5(h)). Furthermore, this framework is parameter-free. The method for computing ATC_{MT} is described in Algorithm 1. This algorithm is divided into two steps: (1) labelling the point from the meaningful thickness values, and (2) building the ATC_{MT} of the curve from the labels previously obtained.

Algorithm 1. Calculation of adaptive tangential cover associated to meaningful thickness ATC_{MT}.

Input   : $C = (C_i)_{0 \leq i \leq n-1}$ discrete curve of n points,
$\eta = (\eta_i)_{0 \leq i \leq n-1}$ vector of meaningful thickness associated to each point C_i of C,
$\nu = \{\nu_k \mid \nu_k \in \eta\}$ ordered set of meaningful thickness value of η, and $MBS(C) = \{MBS_{\nu_k}(C)\}_{k=0}^{m-1}$ sets of maximal blurred segments of C for each width value $\nu_k \in \nu$

Output   : $ATC_{MT}(C)$ adaptive tangential cover associated to C

Variables: α maximum meaningful thickness of η in a given interval,
$\gamma = (\gamma)_{0 \leq i \leq n-1}$ vector of labels associated to each point C_i of C w.r.t. $MBS(C)$,

---

1 **begin**
  　/* Initialization   */
2 　　$ATC_{MT}(C) = \emptyset$;
3 　　$\gamma_i = \eta_i$ for $i \in [\![0, n-1]\!]$;
  　/* Step 1: Label each point of C with the maximum meaningful thickness w.r.t the MBS of width ν_k passing through the point */
4 　　**foreach** $\nu_k \in \nu$ **do**
5 　　　**foreach** $MBS(B_i, E_i, \nu_k) \in MBS_{\nu_k}(C)$ **do**
6 　　　　$\alpha = \max\{\eta_i \mid i \in [\![B_i, E_i]\!]\}$;
7 　　　　**if** $\alpha = \nu_k$ **then**
8 　　　　　$\gamma_i = \nu_k$ for $i \in [\![B_i, E_i]\!]$;

  　/* Step 2: Calculate ATC_{MT} by keeping the MBS that contain at least one point whose label is equal to the width of the MBS */
9 　　**foreach** $\nu_k \in \nu$ **do**
10 　　　**foreach** $MBS(B_i, E_i, \nu_k) \in MBS_{\nu_k}(C)$ **do**
11 　　　　**if** $\exists \gamma_i$, for $i \in [\![B_i, E_i]\!]$, such that $\gamma_i = \nu_k$ **then**
12 　　　　　$ATC_{MT}(C) = ATC_{MT}(C) \cup \{MBS(B_i, E_i, \nu_k)\}$;

13 **end**

More precisely, the algorithm is initialized with an empty ATC_{MT} and the labels associated to each point are the same as meaningful thickness values (Lines 2–3). In the first step (Lines 4–8), the tangential covers with widths corresponding to all different noise levels are examined in order to find the label of each point. At each level ν_k, the label of a point is updated to ν_k if the MBS passing through the point has the maximal meaningful thickness being equal to ν_k. It should be noted that the number of noise levels overall the contour is much smaller than the number of points on the contour. Thus, the number of considered tangential covers is often small. Then, in the second step (Lines 9–12), the ATC_{MT} is composed of the MBS with widths being the label associated to points constituting the MBS. An illustration of the algorithm is given in Fig. 5.

4   Application to Dominant Point Detection

Tangential covers, as stated previously, are involved in applications of dominant point detection [10]. The previous approaches use tangential covers composed of maximal blurred segments with a constant width along the curve. Such a width allows a flexible segmentation of discrete curves with respect to the noise. In general, this parameter needs to be manually adjusted to obtain a good result of detection algorithm. Therefore, such approaches are not adaptive to discrete contours with irregular noise.

In this section, we present a dominant point detection algorithm using ATC_{MT}. The reason is twofold: (1) the ATC_{MT} takes into account the amount

---

Algorithm 2. Dominant point detection.

Input   : $C = (C_i)_{0 \leq i \leq n-1}$ discrete curve of n points,
              $ATC_{MT}(C) = \{MBS(B_j, E_j)\}_{j=0}^{m-1}$ adaptive tangential cover of C
Output   : D set of dominant points
1 **begin**
2 $\quad$ $q = 0$; $p = 1$; $D = \emptyset$;
3 $\quad$ **while** $p < m$ **do**
4 $\quad\quad$ **while** $E_q > B_p$ **do** $p++$
      $\quad\quad D = D \cup \min\{Angle(C_{B_q}, C_i, C_{E_{p-1}}) \mid i \in [\![B_{p-1}, E_q]\!]\}$;
5 $\quad\quad$ $q = p - 1$;

6 **end**

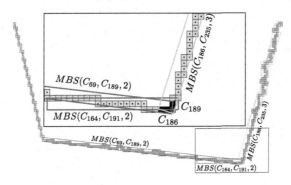

Fig. 6. Illustration of Algorithm 2 with the adaptive tangential cover obtained by Algorithm 1 in Fig. 5. Considering the maximal blurred segments $MBS(C_{69}, C_{189}, 2)$, $MBS(C_{164}, C_{191}, 2)$ and $MBS(C_{186}, C_{235}, 3)$, the common zone determined by these three segments are four points: $C_{186}, C_{187}, C_{188}$ and C_{189} (green and red points in the zoom). The left and right extremities of the common zone are C_{69} and C_{235} respectively. The angle between each point in the common zone and the two extremities are respectively 102.708°, 101.236°, 99.7334° and 100.817°. The dominant point is the point having the smallest angle measure, *i.e.*, C_{188} (red point in the zoom) (Color figure online).

of noise on the curve and thus allows a better model of curve segmentation, and (2) the algorithm for computing ATC_{MT} is parameter-free.

4.1 Dominant Point Detection Algorithm

The algorithm for dominant point detection proposed in this section is the same as the one presented by [10]. It should be mentioned that the modified part is to

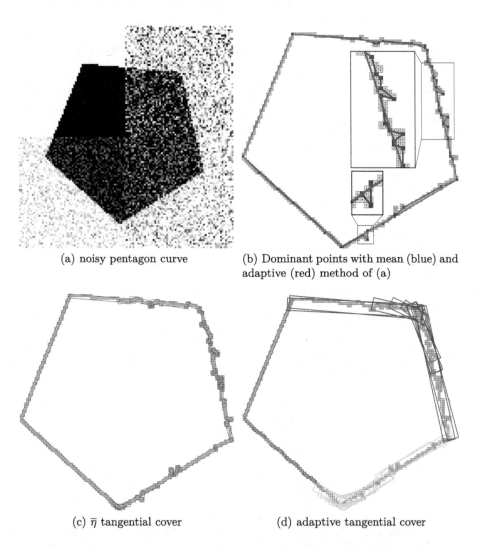

(a) noisy pentagon curve

(b) Dominant points with mean (blue) and adaptive (red) method of (a)

(c) $\bar{\eta}$ tangential cover

(d) adaptive tangential cover

Fig. 7. Dominant point detection of noisy pentagon curve. Blue and red points are dominant points detected by the mean tangential cover and ATC methods, respectively. Blue and red lines are the polygonal approximation from the dominant points detected (Color figure online).

use an ATC_{MT} instead of a classical tangential cover with mono-width for segmenting the digital curve. The algorithm consists in first finding the candidates of dominant points in the smallest common zone induced by successive maximal blurred segments. Then, the dominant point of each common zone is identified as the point having the smallest angle with the two extremities of the left and right of the maximal blurred segments composing the zone. The algorithm is described in Algorithm 2, and illustrated in Fig. 6.

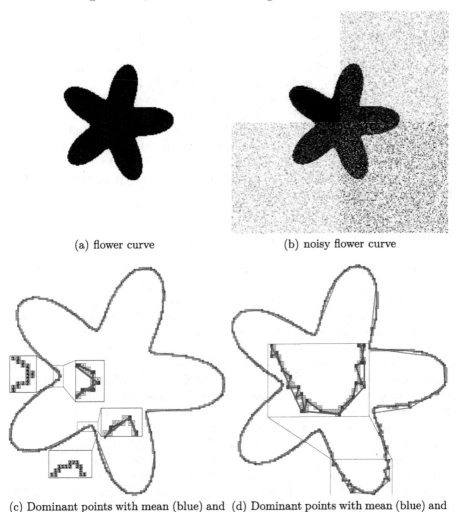

(a) flower curve

(b) noisy flower curve

(c) Dominant points with mean (blue) and adaptive (red) method of (a)

(d) Dominant points with mean (blue) and adaptive (red) method of (b)

Fig. 8. Dominant point detection of flower curves, with and without noise. Blue and red points are dominant points obtained with the mean tangential cover and the proposed ATC methods, respectively. Blue and red lines are the polygonal approximation from the dominant points detected (Color figure online).

4.2   Experimentations

In this section, we present experimental results of the dominant point detection algorithm using the proposed notion of ATC_{MT}. In order to compare the current parameter-free method, we consider in our experiments the **mean tangential cover** with MBS of width-$\overline{\eta}$ equals to the average of the obtained meaningful thicknesses at each point of the studied curve. In fact, this width-$\overline{\eta}$ parameter was proposed and used in [14].

The experiments are carried out on both data with and without noise. From Figs. 7 and 8, it can be seen that using the mean tangential cover is not always a relevant strategy, particularly in the high noisy zones of curves. This is due to the fact that the width-$\overline{\eta}$ parameter could not capture the local noise on curve, contrary to the ATC_{MT} method (see Figs. 7 and 8(b),(d)).

In Fig. 8(a), (c), the flower curve seems to be a discrete curve without noise. Though, the meaningful thickness method detects two noise levels, 1 and 2 (see the zooms in Fig. 8(c)). In the 2-meaningful thickness zones, the dominant point detection with ATC_{MT} method fits better the corners, whereas the mean method induces a decomposition very close to the studied curve and detects more dominant points. In other words, in the curved zones, the ATC_{MT} method simplifies the representation of the curve.

5   Conclusion and Perspectives

We present in this paper a new notion, the Adaptive Tangential Cover deduced from the meaningful thickness. The obtained decomposition in MBS of various widths transmits the noise levels and the geometrical structure of the given discrete curve. Moreover the method to compute the ATC is parameter free. An online demonstration based on the DGtal [1] and ImaGene [2] library, is available at the following website:

http://ipol-geometry.loria.fr/~kerautre/ipol_demo/ATC_IPOLDemo/

The ATC is used in a dominant point detection algorithm and permits to obtain a parameter-free method with very good results on the polygonal shapes with or without noise. For the shapes with convex and/or concave parts, the algorithm simplifies the shapes in a polygonal way.

In this article, we have considered the ATC definition based on the notion of the meaningful thickness. In the further work, we may consider the generalization of ATC using other width estimations. The proposed approach opens numerous perspectives, for example the use of ATC in geometric estimators or in the decomposition of a curve in arcs and segments.

Acknowledgement. The authors would like to thank the reviewers for their valuable comments.

References

1. DGtal: Digital Geometry tools and algorithms library. http://libdgtal.org
2. Imagene, Generic digital Image library. http://gforge.liris.cnrs.frs/projects/imagene
3. Debled-Rennesson, I., Feschet, F., Rouyer-Degli, J.: Optimal blurred segments decomposition of noisy shapes in linear time. Comput. Graph. **30**(1), 30–36 (2006)
4. Feschet, F., Tougne, L.: Optimal time computation of the tangent of a discrete curve: application to the curvature. In: Bertrand, G., Couprie, M., Perroton, L. (eds.) DGCI 1999. LNCS, vol. 1568, pp. 31–40. Springer, Heidelberg (1999)
5. Feschet, F.: Canonical representations of discrete curves. Pattern Anal. Appl. **8**(1–2), 84–94 (2005)
6. Kerautret, B., Lachaud, J.O.: Meaningful scales detection along digital contours for unsupervised local noise estimation. IEEE Trans. Pattern Anal. Mach. Intell. **34**(12), 2379–2392 (2012)
7. Kerautret, B., Lachaud, J.O.: Meaningful scales detection: an unsupervised noise detection algorithm for digital contours. Image Process. On Line **4**, 98–115 (2014)
8. Kerautret, B., Lachaud, J.O., Said, M.: Meaningful thickness detection on polygonal curve. In: Proceedings of the 1st International Conference on Pattern Recognition Applications and Methods, pp. 372–379. SciTePress (2012)
9. Lachaud, J.-O.: Digital shape analysis with maximal segments. In: Köthe, U., Montanvert, A., Soille, P. (eds.) WADGMM 2010. LNCS, vol. 7346, pp. 14–27. Springer, Heidelberg (2012)
10. Ngo, P., Nasser, H., Debled-Rennesson, I.: Efficient dominant point detection based on discrete curve structure. In: Barneva, R.P., et al. (eds.) IWCIA 2015. LNCS, vol. 9448, pp. 143–156. Springer, Heidelberg (2015)
11. Nguyen, T.P., Debled-Rennesson, I.: On the local properties of digital curves. Int. J. Shape Model. **14**(2), 105–125 (2008)
12. Nguyen, T.P., Debled-Rennesson, I.: Decomposition of a curve into arcs and line segments based on dominant point detection. In: Heyden, A., Kahl, F. (eds.) SCIA 2011. LNCS, vol. 6688, pp. 794–805. Springer, Heidelberg (2011)
13. Nguyen, T.P., Debled-Rennesson, I.: A discrete geometry approach for dominant point detection. Pattern Recogn. **44**(1), 32–44 (2011)
14. Nguyen, T.P., Kerautret, B., Debled-Rennesson, I., Lachaud, J.-O.: Unsupervised, fast and precise recognition of digital arcs in noisy images. In: Bolc, L., Tadeusiewicz, R., Chmielewski, L.J., Wojciechowski, K. (eds.) ICCVG 2010, Part I. LNCS, vol. 6374, pp. 59–68. Springer, Heidelberg (2010)
15. Reveillès, J.P.: Géométrie discrète, calculs en nombre entiers et entiersgorithmique, thèse d'état. Université Louis Pasteur, Strasbourg (1991)
16. Rodríguez, M., Largeteau-Skapin, G., Andres, E.: Adaptive pixel resizing for multiscale recognition and reconstruction. In: Wiederhold, P., Barneva, R.P. (eds.) IWCIA 2009. LNCS, vol. 5852, pp. 252–265. Springer, Heidelberg (2009)
17. Vacavant, A., Roussillon, T., Kerautret, B., Lachaud, J.: A combined multiscale/irregular algorithm for the vectorization of noisy digital contours. Comput. Vis. Image Underst. **117**(4), 438–450 (2013)

Author Index

Printed in the United States
By Bookmasters